Lecture Notes in Computer Science 1627

Edited by G. Goos, J. Hartmanis and J. van Leeuwen

Springer

Berlin
Heidelberg
New York
Barcelona
Hong Kong
London
Milan
Paris
Singapore
Tokyo

Takao Asano   Hiroshi Imai   D.T. Lee
Shin-ichi Nakano   Takeshi Tokuyama (Eds.)

Computing and Combinatorics

5th Annual International Conference, COCOON'99
Tokyo, Japan, July 26-28, 1999
Proceedings

Springer

Volume Editors

Takao Asano
Department of Information and System Engineering
Faculty of Science and Engineering, Chuo University
1-13-27, Kasuga, Bunkyo-ku, Tokyo, 112-8551 Japan
E-mail: asano@ise.chuo-u.ac.jp

Hiroshi Imai
Department of Information Science, University of Tokyo
7-3-1 Hongo, Bunkyo-ku, Tokyo, 113-0033 Japan
E-mail: imai@is.s.u-tokyo.ac.jp

D.T. Lee
Institute of Information Science, Academia Sinica
Nankang, Taipei, Taiwan
E-mail: dtlee-nu@iis.sinica.edu.tw

Shin-ichi Nakano
Department of Computer Science
Faculty of Engineering, Gunma University
1-5-1 Tenjin-cho, Kiryu, Gunma, 376-8515 Japan
E-mail: nakano@msc.cs.gunma-u.ac.jp

Takeshi Tokuyama
IBM Tokyo Research Laboratory
1623-14, Shimo-Tsuruma, Yamato Kanagawa, 242-0001 Japan
E-mail: ttoku@trl.ibm.co.jp

Cataloging-in-Publication data applied for

Die Deutsche Bibliothek - CIP-Einheitsaufnahme

Computing and combinatorics : 5th annual international conference
; proceedings / COCOON '99, Tokyo, Japan, July 26 - 28, 1999.
Takao Asano ... (ed.). - Berlin ; Heidelberg ; New York ; Barcelona ;
Hong Kong ; London ; Milan ; Paris ; Singapore ; Tokyo : Springer,
1999
   (Lecture notes in computer science ; Vol. 1627)
   ISBN 3-540-66200-6

CR Subject Classification (1998): F.2, G.2.1-2, I.3.5, C.2.3-4, E.1

ISSN 0302-9743
ISBN 3-540-66200-6 Springer-Verlag Berlin Heidelberg New York

Typesetting: Camera-ready by author
SPIN: 10703294   06/3142 – 5 4 3 2 1 0   Printed on acid-free paper

Preface

The abstracts and papers in this volume were presented at the Fifth Annual International Computing and Combinatorics Conference (COCOON '99), which was held in Tokyo, Japan from July 26 to 28, 1999. The topics cover most aspects of theoretical computer science and combinatorics pertaining to computing.

In response to the call for papers, 88 high-quality extended abstracts were submitted internationally, of which 46 were selected for presentation by the program committee. Every submitted paper was reviewed by at least three program committee members. Many of these papers represent reports on continuing research, and it is expected that most of them will appear in a more polished and complete form in scientific journals. In addition to the regular papers, this volume contains abstracts of two invited plenary talks by Prabhakar Raghavan and Seinosuke Toda. The conference also included a special talk by Kurt Mehlhorn on LEDA (Library of Efficient Data types and Algorithms).

The Hao Wang Award (inaugurated at COCOON '97) is given to honor the paper judged by the program committee to have the greatest scientific merit. The recipients of the Hao Wang Award 1999 were Hiroshi Nagamochi and Toshihide Ibaraki for their paper "An Approximation for Finding a Smallest 2-Edge-Connected Subgraph Containing a Specified Spanning Tree".

We wish to thank all who have made this meeting possible: the authors for submitting papers, the program committee members and external referees for their excellent work in reviewing the papers, the sponsors, the local organizers, ACM SIGACT and the University of Tokyo for handling electronic submissions, and Springer-Verlag for their support and assistance.

July 1999

Takao Asano
Hiroshi Imai
D. T. Lee
Shin-ichi Nakano
Takeshi Tokuyama

Sponsoring Institutions

COCOON '99 was sponsored by Chuo University and the Algorithm Engineering Project, Grant-in-Aid of MESSC Japan. It was organized in cooperation with the Special Interest Group on Algorithms of the Information Processing Society of Japan (SIGAL, IPSJ), and the Technical Group on Computation of the Institute of Electronics, Information, and Communication Engineers of Japan (TGCOMP, IEICE).

Conference Organization

Program Committee Co-chairs

D. T. Lee (Academia Sinica, Taiwan)
Takeshi Tokuyama (IBM Tokyo Research Laboratory, Japan)

Program Committee

Jean-Daniel Boissonnat (INRIA, France)
Zhi-Zhong Chen (Tokyo Denki University, Japan)
Xiaotie Deng (Hong Kong City University, Hong Kong)
David Eppstein (UC Irvine, USA)
Uriel Feige (Weizmann Institute, Israel)
Harold Gabow (University of Colorado, USA)
Ronald Graham (UC San Diego, USA)
Frank Hwang (Chiao Tung University, Taiwan)
Tao Jiang (McMaster University, Canada)
Howard Karloff (Georgia Institute of Technology, USA)
Samir Khuller (University of Maryland, USA)
Rao Kosaraju (Johns Hopkins University, USA)
Xuemin Lin (University of New South Wales, Australia)
Bruce Maggs (Carnegy Melon University, USA)
Kurt Mehlhorn (Max Planck Institut für Informatik, Germany)
Satoru Miyano (University of Tokyo, Japan)
Seffi Naor (Technion, Israel)
Günter Rote (Freie Universität, Germany)
Madhu Sudan (MIT, USA)
Roberto Tamassia (Brown University, USA)
Jeff Vitter (INRIA, France and Duke University, USA)
Guoliang Xue (University of Vermont, USA)

Organizing Committee

Takao Asano (Confernce Co-chair, Chuo University, Japan)
Hiroshi Imai (Conference Co-chair, Univeristy of Tokyo, Japan)
Shin-ichi Nakano (Publicity, Gunma University, Japan)

Referees

Miklós Ajtai
Hiroki Arimura
Woyciech Banaszczyk
Bob Beals
Arnon Boneh
Dan Boneh
Jin-Yi Cai
Mao-cheng Cai
Claude Carlet
Eranda Cela
Barun Chandra
Kevin Christian
Hossam El-Gindy
Ran El-Yaniv
Vladimir Estivill-Castro
Stefan Felsner
Amos Fiat
Bill Gasarch
Qian-Ping Gu
Sudipto Guha
Xin He
Kouichi Hirata
Daisuke Ikeda
Russell Impagliazzo
Michael Kaufmann
Sanjeev Khanna
Bettina Klinz
Pil Joong Lee
Daniel Lehmann

Weifa Liang
Guo-Hui Lin
Eric Martin
Hiroshi Matsuno
Balaji Raghavenchari
Franz Rendl
Leonard Schulman
Steve Seiden
Jose Sempere
Henry Shapiro
Bruce Shepherd
Yaoyun Shi
Shinichi Shimozono
Takayoshi Shoudai
Arcot Sowmya
Yuji Takada
Seinosuke Toda
Ron van der Meyden
Lusheng Wang
Todd Wareham
Lorenz Wernisch
Gerhard Woeginger
Rebecca Wright
Jinhui Xu
Neal Young
Yuzhong Zhang
An Zhu
Huafei Zhu

Table of Contents

Invited Talks

Hao Wang Award Paper

Data Structures

Computational Biology

Graph Drawing

Discrete Mathematics

Graph Algorithms 1

Automata and Language

Complexity Theory and Learning

Combinatorial Optimization 1

Graph Algorithms 2

Number Theory

Distributed Computing

Combinatorial Optimization 2

Network Routing Problems

Computational Geometry

Online Algorithms

Rewriting Systems

Parallel Computing

Combinatorial Optimization 3

The Web as a Graph: Measurements, Models, and Methods

Jon M. Kleinberg[1], Ravi Kumar[2], Prabhakar Raghavan[2],
Sridhar Rajagopalan[2], and Andrew S. Tomkins[2]

[1] Department of Computer Science, Cornell University, Ithaca, NY 14853.
[2] IBM Almaden Research Center K53/B1, 650 Harry Road, San Jose CA 95120.

Abstract. The pages and hyperlinks of the World-Wide Web may be
viewed as nodes and edges in a directed graph. This graph is a fascinating
object of study: it has several hundred million nodes today, over a billion
links, and appears to grow exponentially with time. There are many reasons — mathematical, sociological, and commercial — for studying the
evolution of this graph. In this paper we begin by describing two algorithms that operate on the Web graph, addressing problems from Web
search and automatic community discovery. We then report a number of
measurements and properties of this graph that manifested themselves
as we ran these algorithms on the Web. Finally, we observe that traditional random graph models do not explain these observations, and we
propose a new family of random graph models. These models point to
a rich new sub-field of the study of random graphs, and raise questions
about the analysis of graph algorithms on the Web.

1 Overview

Few events in the history of computing have wrought as profound an influence
on society as the advent and growth of the World-Wide Web. For the first time,
millions — soon to be billions — of individuals are creating, annotating and
exploiting hyperlinked content in a distributed fashion. A particular Web page
might be authored in any language, dialect, or style by an individual with any
background, culture, motivation, interest, and education; might range from a
few characters to a few hundred thousand; might contain truth, falsehood, lies,
propaganda, wisdom, or sheer nonsense; and might point to none, few, or several
other Web pages. The hyperlinks of the Web endow it with additional structure;
and the network of these links is rich in latent information content. Our focus
in this paper is on the directed graph induced by the hyperlinks between Web
pages; we refer to this as the *Web graph*.

For our purposes, nodes represent static html pages and hyperlinks represent
directed edges. Recent estimates [4] suggest that there are several hundred million nodes in the Web graph; this quantity is growing by a few percent a month.
The average node has roughly seven hyperlinks (directed edges) to other pages,
making for a total of several billion hyperlinks in all.

T. Asano et al. (Eds.): COCOON'99, LNCS 1627, pp. 1–17, 1999.

There are several reasons for studying the Web graph. The structure of this graph has already led to improved Web search [6,8,21,29], more accurate topic-classification algorithms [11] and has inspired algorithms for enumerating emergent cyber-communities [23]. The hyperlinks themselves represent a fecund source of sociological information. Beyond the intrinsic interest of the structure of the Web graph, measurements of the graph and of the behavior of users as they traverse the graph, are of growing commercial interest.

1.1   Guided tour of this paper

In Section 2 we review two algorithms that have been applied to the Web graph: Kleinberg's HITS method [21] and the enumeration of certain bipartite cliques [23]. We have chosen these algorithms here because they are both driven by the presence of certain structures in the Web graph. These structures (which we will detail below) appear to be a fundamental by-product of the manner in which Web content is created.

In Section 3 we summarize a number of measurements we have made on the entire Web graph, and on particular local subgraphs of interest. We show, for instance, that the in- and out-degrees of nodes follow Zipfian (inverse polynomial) distributions [12,17,26,31]. This and other measurements of the frequency of occurrence of certain structures suggest that traditional random graph models such as $\mathcal{G}_{n,p}$ [7] are likely to do a poor job of modeling the Web graph.

In Section 4 we lay down a framework for a class of random graph models, and give evidence that at least some of our observations about the Web (for instance, the degree distributions) can be established in these models. A notable aspect of these models is that they embody some version of a *copying* process: we add links to a node v by picking a random (other) node u in the graph, and copying some links from u to v (i.e., we add edges from v to some of the nodes that u points to). Such copying operations seem to be fundamental both to the process of content-creation on the Web and to the explanation of the statistics we have observed. One consequence is that the mathematical analysis of these graph models promises to be far harder than in traditional graph models in which the edges emanating from a node are drawn independently.

We conclude in Section 5 with a number of directions for further work.

1.2   Related work

Analysis of the structure of the Web graph has been used to enhance the quality of Web search [5,6,8,9,21,29]. The topics of pages pointed to by a Web page can be used to improve the accuracy of determining the (unknown) topic of this page in the setting of supervised classification [11].

Statistical analysis of the structure of the academic citation graph has been the subject of much work in the Sociometrics community. As we discuss below, Zipf distributions seem to characterize Web citation frequency. Interestingly, the same distributions have also been observed for citations in the academic literature. This fact, known as *Lotka's law*, was demonstrated by Lotka in 1926 [26].

Gilbert [17] presents a probabilistic model explaining Lotka's law, which is similar in spirit to our proposal, though different in details and application. The field of bibliometrics [14,16] has studied phenomena in citation; some of these insights have been applied to the Web as well [25].

A view of the Web as a semi-structured database has been advanced by many authors. In particular, LORE [1] and WebSQL [27] use graph-theoretic and relational views of the Web respectively. These views support structured query interfaces to the Web (Lorel [1] and WebSQL [27]) that are evocative of and similar to SQL. An advantage of this approach is that many interesting queries can be expressed as simple expressions in the very powerful SQL syntax. The corresponding disadvantage is that the generality comes with an associated computational cost which can be prohibitive until we develop query optimizers for Web graph computations that similar to those available for relational data. LORE and WebSQL are but two examples of projects in this space. Some other examples are W3QS [22], WebQuery [8], Weblog [24], and ParaSite/Squeal [29].

Traditional data mining research (see for instance Agrawal and Srikant [2]) focuses largely on algorithms for finding association rules and related statistical correlation measures in a given dataset. However, efficient methods such as *a priori* [2] or even more general methodologies such as *query flocks* [30], do not scale to the numbers of "items" (pages) in the Web graph. This number is already two to three orders of magnitude more than the number of items in a typical market basket analysis.

The work of Mendelzon and Wood [28] is an instance of structural methods in mining. They argue that the traditional SQL query interface to databases is inadequate in its power to specify several structural queries that are interesting in the context of the Web. They provide the example of path connectivity between nodes subject to some constraints on the sequence of edges on the path (expressible in their case as a regular expression). They describe G^+, a language with greater expressibility than SQL for graph queries.

2 Algorithms

We have observed the following recurrent phenomenon on the Web. For any particular topic, there tend to be a set of "authoritative" pages focused on the topic, and a set of "hub" pages, each containing links to useful, relevant pages on the topic. This observation motivated the development of two algorithms which we describe below. The first is a search algorithm that distils high-quality pages for a topic query (Section 2.1), and the second enumerates all topics that are represented on the Web in terms of suitably dense sets of such hub/authority pages (Section 2.2).

2.1 The HITS algorithm

Beginning with a search topic, specified by one or more query terms, Kleinberg's HITS algorithm [21] applies two main steps: a *sampling* step, which constructs a focused collection of several thousand Web pages likely to be rich in relevant authorities; and a *weight-propagation* step, which determines numerical estimates of hub and authority scores by an iterative procedure. The pages with the highest scores are returned as hubs and authorities for the search topic.

Any subset S of nodes induces a *subgraph* containing all edges that connect two nodes in S. The first step of the HITS algorithm constructs a subgraph expected to be rich in relevant, authoritative pages, in which it will search for hubs and authorities. To construct this subgraph, the algorithm first uses keyword queries to collect a *root set* of, say, 200 pages from a traditional index-based search engine. This set does not necessarily contain authoritative pages; however, since many of these pages are presumably relevant to the search topic, one can expect some to contain links to prominent authorities, and others to be linked to by prominent hubs. The root set is therefore expanded into a *base set* by including all pages that are linked to by pages in the root set, and all pages that link to a page in the root set (up to a designated size cut-off). This follows the intuition that the prominence of authoritative pages is typically due to the endorsements of many relevant pages that are not, in themselves, prominent. We restrict our attention to this base set for the remainder of the algorithm; this set typically contains roughly 1000–3000 pages, and that (hidden) among these are a large number of pages that one would subjectively view as authoritative for the search topic.

We begin with one modification to the subgraph induced by the base set. Links between two pages on the same Web site very often serve a purely navigational function, and typically do not represent conferral of authority. We therefore delete all such links from the subgraph induced by the base set, and apply the remainder of the algorithm to this modified subgraph.

Good hubs and authorities can be extracted from the base set by giving a concrete numerical definition to the intuitive notions of hub and authority from the beginning of this section. The algorithm associates a non-negative *authority weight* x_p and a non-negative *hub weight* y_p with each page $p \in V$. We will only be interested in the *relative* values of these weights, not their actual magnitudes; so in the manipulation of the weights, we apply a normalization to prevent the weights from overflowing. (The actual choice of normalization does not affect the results; we maintain the invariant that the squares of all weights sum to 1.) A page p with a large weight x_p (resp. y_p) will be viewed as a "better" authority (resp. hub). Since we do not impose any *a priori* estimates, all x- and y-values are set to a uniform constant initially. As will be seen later, however, the final results are essentially unaffected by this initialization.

The authority and hub weights are updated as follows. If a page is pointed to by many good hubs, we would like to increase its authority weight; thus for a page p, the value of x_p is updated to be to be the sum of y_q over all pages q that link to p:

$$x_p = \sum_{q \text{ such that } q \to p} y_q, \tag{1}$$

where the notation $q \to p$ indicates that q links to p. In a strictly dual fashion, if a page points to many good authorities, its hub weight is increased via

$$y_p = \sum_{q \text{ such that } p \to q} x_q. \tag{2}$$

There is a more compact way to write these updates, and it turns out to shed more light on the mathematical process. Let us number the pages $\{1, 2, \ldots, n\}$ and define their *adjacency matrix* A to be the $n \times n$ matrix whose (i, j)th entry is equal to 1 if page i links to page j, and is 0 otherwise. Let us also write the set of all x-values as a vector $x = (x_1, x_2, \ldots, x_n)$, and similarly define $y = (y_1, y_2, \ldots, y_n)$. Then the update rule for x can be written as $x \leftarrow A^T y$ and the update rule for y can be written as $y \leftarrow Ax$. Unwinding these one step further, we have

$$x \leftarrow A^T y \leftarrow A^T Ax = (A^T A)x \tag{3}$$

and

$$y \leftarrow Ax \leftarrow AA^T y = (AA^T)y. \tag{4}$$

Thus the vector x after multiple iterations is precisely the result of applying the *power iteration* technique to $A^T A$ — we multiply our initial iterate by larger and larger powers of $A^T A$ — and a standard result in linear algebra tells us that this sequence of iterates, when normalized, converges to the principal eigenvector of $A^T A$. Similarly, the sequence of values for the normalized vector y converges to the principal eigenvector of AA^T. (See the book by Golub and Van Loan [18] for background on eigenvectors and power iteration.)

In fact, power iteration will converge to the principal eigenvector for any "non-degenerate" choice of initial vector — in our case, for example, for any vector all of whose entries are positive. This says that the hub and authority weights computed are truly an *intrinsic* feature of the collection of linked pages, not an artifact of the choice of initial weights or the tuning of arbitrary parameters. Intuitively, the pages with large weights represent a very "dense" pattern of linkage, from pages of large hub weight to pages of large authority weight. This type of structure — a densely linked *community* of thematically related hubs and authorities — will be the motivation underlying Section 2.2 below.

Finally, the output of HITS algorithm for the given search topic is a short list consisting of the pages with the largest hub weights and the pages with the largest authority weights. Thus the algorithm has the following interesting feature: after using the query terms to collect the root set, the algorithm completely ignores textual content thereafter. In other words, HITS is a purely link-based computation once the root set has been assembled, with no further regard to the query terms. Nevertheless, HITS provides surprisingly good search results for a wide range of queries. For instance, when tested on the sample query "search engines", the top authorities returned by HITS were Yahoo!, Excite, Magellan, Lycos, and AltaVista — even though none of these pages (at the time of the experiment) contained the phrase "search engines."

In subsequent work [5,9,10], the HITS algorithm has been generalized by modifying the entries of A so that they are no longer boolean. These modifications take into account the content of the pages in the base set, the internet domains in which they reside, and so on. Nevertheless, most of these modifications retain the basic power iteration process and the interpretation of hub and authority scores as components of a principal eigenvector, as above.

2.2   Trawling the Web for cyber-communities

In this section we turn to a second algorithm developed for the Web graph. In contrast to HITS, which is a search algorithm designed to find high-quality pages about a fixed topic, the *trawling* algorithm described below seeks to enumerate *all* topics (under a certain definition), and therefore processes the entire Web graph.

We begin with a more concrete definition of the types of topic we wish to enumerate. Recall that a complete bipartite clique $K_{i,j}$ is a graph in which every one of i nodes has an edge directed to each of j nodes (in the following treatment it is simplest to think of the first i nodes as being distinct from the second j; in fact this is not essential to our algorithms or models). We further define a *bipartite core* $C_{i,j}$ to be a graph on $i + j$ nodes that contains at least one $K_{i,j}$ as a subgraph. The intuition motivating this notion is the following: on any sufficiently well represented topic on the Web, there will (for some appropriate values of i and j) be a bipartite core in the Web graph. Figure 1 illustrates an instance of a $C_{4,3}$ in which the four nodes on the left have hyperlinks to the home pages of three major commercial aircraft manufacturers. Such a subgraph

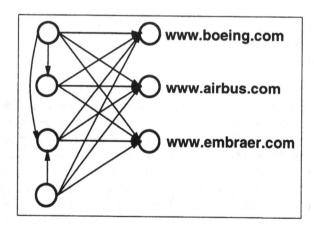

Fig. 1. A bipartite core.

of the Web graph would be suggestive of a "cyber-community" of aficionados of commercial aircraft manufacturers who create hub-like pages like the four on the left side of Figure 1. These pages co-cite the authoritative pages on the right.

Loosely speaking, such a community emerges in the Web graph when many (hub) pages link to many of the same (authority) pages. In most cases, the hub pages in such communities may not co-cite *all* the authoritative pages for that topic. Nevertheless, it is tempting to subscribe to the following weaker hypothesis: every such community will contain a bipartite core $C_{i,j}$ for non-trivial values of i and j. Turning this around, we could attempt to identify a large fraction of cyber-communities by enumerating all the bipartite cores in the Web for, say $i = j = 3$; we call this process *trawling*. Why these choices of i and j? Might it not be that for such small values of i and j, we discover a number of coincidental co-citations, which do not truly correspond to communities?

In fact, in our experiment [23] we enumerated $C_{i,j}$'s for values of i ranging from 3 to 9, for j ranging from 3 to 20. The results suggest that (1) the Web graph has several hundred thousand such cores, and (2) it appears that only a minuscule fraction of these are coincidences — the vast majority do in fact correspond to communities with a definite topic focus. Below we give a short description of this experiment, followed by some of the principal findings.

From an algorithmic perspective, the naive "search" algorithm for enumeration suffers from two fatal problems. First, the size of the search space is far too large — using the naive algorithm to enumerate all bipartite cores with two Web pages pointing to three pages would require examining approximately 10^{40} possibilities on a graph with 10^8 nodes. A theoretical question (open as far as we know): does the work on fixed-parameter intractability [13] imply that we cannot – in the worst case – improve on naive enumeration for bipartite cores? Such a result would argue that algorithms that are provably efficient on the Web graph must exploit some feature that distinguishes it from the "bad" inputs for fixed-parameter intractability. Second, and more practically, the algorithm requires random access to edges in the graph, which implies that a large fraction of the graph must effectively reside in main memory to avoid the overhead of seeking a disk on every edge access.

We call our algorithmic methodology the *elimination-generation* paradigm. An algorithm in the elimination/generation paradigm performs a number of sequential passes over the Web graph, stored as a binary relation. During each pass, the algorithm writes a modified version of the dataset to disk for the next pass. It also collects some metadata which resides in main memory and serves as state during the next pass. Passes over the data are interleaved with sort operations, which change the order in which the data is scanned, and constitute the bulk of the processing cost. We view the sort operations as alternately ordering directed edges by source and by destination, allowing us alternately to consider out-edges and in-edges at each node. During each pass over the data, we interleave *elimination* operations and *generation* operations, which we now detail.

Elimination. There are often easy necessary (though not sufficient) conditions that have to be satisfied in order for a node to participate in a subgraph of interest to us. Consider the example of $C_{4,4}$'s. Any node with in-degree 3 or smaller cannot participate on the right side of a $C_{4,4}$. Thus, edges that are directed into such nodes can be pruned from the graph. Likewise, nodes with out-degree 3 or smaller cannot participate on the left side of a $C_{4,4}$. We refer to these necessary conditions as *elimination filters*.

Generation. Generation is a counterpoint to elimination. Nodes that barely qualify for potential membership in an interesting subgraph can easily be verified to either belong in such a subgraph or not. Consider again the example of a $C_{4,4}$. Let u be a node of in-degree exactly 4. Then, u can belong to a $C_{4,4}$ if and only if the 4 nodes that point to it have a neighborhood intersection of size at least 4. It is possible to test this property relatively cheaply, even if we allow the in-degree to be slightly more than 4. We define a *generation filter* to be a procedure that identifies barely-qualifying nodes, and for all such nodes, either outputs a core or proves that such a core cannot exist. If the test embodied in the generation filter is successful, we have identified a core. Further, regardless of the outcome, the node can be pruned since all potential interesting cores containing it have already been enumerated.

Note that if edges appear in an arbitrary order, it is not clear that the elimination filter can be easily applied. If, however, the edges are sorted by source (resp. destination), it is clear that the outlink (resp. inlink) filter can be applied in a single scan. Details of how this can be implemented with few passes over the data (most of which is resident on disk, and must be streamed through main memory for processing) may be found in [23].

After an elimination/generation pass, the remaining nodes have fewer neighbors than before in the residual graph, which may present new opportunities during the next pass. We can continue to iterate until we do not make significant progress. Depending on the filters, one of two things could happen: (1) we repeatedly remove nodes from the graph until nothing is left, or (2) after several passes, the benefits of elimination/generation "tail off" as fewer and fewer nodes are eliminated at each phase. In our trawling experiments, the latter phenomenon dominates.

Why should such algorithms run fast? We make a number of observations about their behavior:

1. The in/out-degree of every node drops monotonically during each elimination/generation phase.
2. During each generation test, we either eliminate a node u from further consideration (by developing a proof that it can belong to no core), or we output a subgraph that contains u. Thus, the total work in generation is linear in the size of the Web graph plus the number of cores enumerated, assuming that each generation test runs in constant time.
3. In practice, elimination phases rapidly eliminate most nodes in the Web graph. A complete mathematical analysis of iterated elimination is beyond the scope of this paper, and requires a detailed understanding of the kinds of random graph models we propose in Section 4.

Kumar *et al.* [23] report trawling a copy of the Web graph derived from a Web crawl obtained from Alexa, inc. This experiment generated well over 100,000 bipartite cores $C_{3,3}$. Note that since these cores are the result of enumeration (rather than querying), they lack any form of context or topic with which one can tag them. Indeed, as noted in [23], the only certain way of determining whether a core is coincidental or real is manual inspection. The results of [23] suggest that over 90% of the cores enumerated in this experiment are not coincidental, but in fact bear definite themes.

We conclude this section by remarking that bipartite cores are not necessarily the only subgraph enumeration problems that are interesting in the setting of the Web graph. The subgraphs corresponding to Webrings look like bidirectional stars, in which there is a central page with links to and from a number of "spoke" pages. Cliques, and directed trees, are other interesting structures for enumeration. Devising general paradigms for such enumeration problems appears to be difficult, unless one understands and exploits the peculiarities of the Web graph. The next two sections address this issue.

3   Measurements

In the course of experiments with the algorithms of Section 2, we were able to study many of the local properties of the Web graph. In this section we survey these observations, and point out that traditional random graph models like $\mathcal{G}_{n,p}$ would do a poor job of explaining them.

3.1   Degree distributions

We begin with the in- and out-degrees of nodes in the Web graph. Figure 2 is a log-log plot (with the x-axis negated) of the in-degree distribution. The plot suggests that the probability that a node has degree i is proportional to $1/i^\alpha$, where α is approximately 2. Such a Zipfian distribution [31] cannot arise in a model such as $\mathcal{G}_{n,p}$, where (due to the superposition of Bernoulli trials) in-degrees exhibit either a Poisson or a binomial distribution. Consider next the out-degree distribution (Figure 3). Again the distribution looks faintly Zipfian,

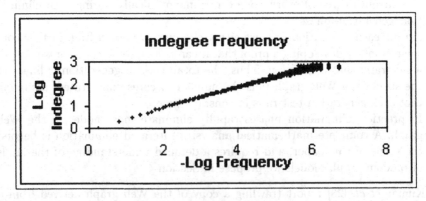

Fig. 2. In-degree distribution.

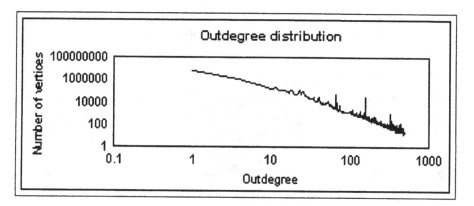

Fig. 3. Out-degree distribution.

although here the variations seem larger. The average out-degree we observed is about 7.2. A natural question now arises: if $\mathcal{G}_{n,p}$ will not result in such Zipfian distributions, what is a natural stochastic process that will? We provide a partial answer in Section 4.

3.2   Number of bipartite cores

We turn next to the distribution of cores $C_{i,j}$, based on the numbers discovered in the trawling experiment [23]. Figures 4 and 5 depict these distributions as functions of i and j; these quantities are from a crawl of the Web that is roughly

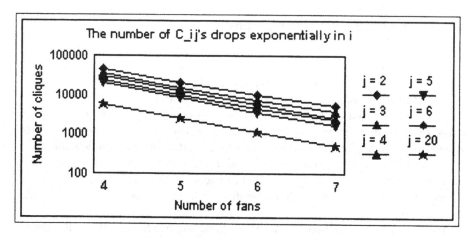

Fig. 4. Core distribution by left side.

two years old, obtained from Alexa, inc. The number of Web pages in this crawl is roughly 100 million. How does this compare with the numbers one might observe

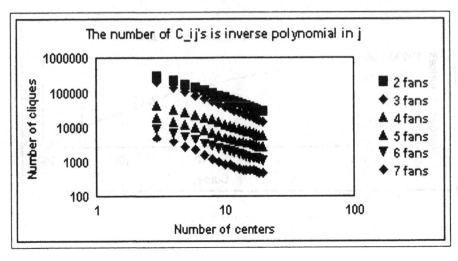

Fig. 5. Core distribution by right side.

in a graph generated using $\mathcal{G}_{n,p}$, say for $np = 7.2$ (our observed out-degree)? A simple calculation yields that the expected number of $C_{i,j}$'s is

$$\binom{n}{i}\binom{n}{j}\left(\frac{7.2}{n}\right)^{ij},$$

which is negligible for $ij > i + j$. Clearly, one cannot explain the multitude of $C_{i,j}$'s in the Web graph using $\mathcal{G}_{n,p}$; once again, we hope that the models we propose in Section 4 can explain these observations.

3.3 Connectivity of local subgraphs

We now consider some relating to the connectivity of local subgraphs of the Web graph. We begin by fixing our set of local subgraphs to be the base sets arising in the HITS algorithm of Section 2. To recapitulate, we begin with c nodes obtained by issuing a keyword query to a text-based search engine. We then expand this root set to the base set by adding any page linked to by a page in the root set, and any page linking to a page in the root set (up to a cut-off of d pages per element of the root set). In experiments, we have applied this with c, the size of the root set, equal to 200; and d, the cut-off, equal to 50. As above, the resulting base set typically consists of roughly 1000 to 3000 pages, and since we wish to focus on cross-site links we discard links between two pages within the same site.

We now ask how well-connected these local subgraphs are. We will view G primarily as a directed graph, but we also define the undirected version G_u obtained by ignoring the direction of all edges.

A collection of graphs constructed in this way for an ensemble of fifty sample query terms reveals some consistently occurring structural properties; we will discuss several of these here at a qualitative level.

A range of connectivity relations. A first observation is that the graphs G_u are not, in general, connected. This is intuitively very natural: the initial root set typically induces very few edges; and while the expansion to the base set serves to connect many of these nodes, others remain in small isolated components.

A graph G_u of this form, however, does typically contain a "giant component" — a connected component that contains a significant constant fraction of all the nodes. In all the cases considered, there was only a single giant component of G_u.

Turning to a stronger measure, we consider *biconnectivity*: we say that two nodes u and v are *biconnected* if there is no third node w so that w lies on all u-v paths. We will call the equivalence classes of nodes under this relation the *biconnected components*. The graph G_u typically has a "giant biconnected component," again unique. Thus, we can intuitively picture the structure of G_u as consisting of a central biconnected "nucleus," with small pieces connected to this nucleus by cut-nodes, and with other small pieces not connected to this nucleus at all.

The biconnected nucleus contains much of the interesting structure in G; for example, it generally contains all the top hubs and authorities computed by HITS. We will use H to denote the subgraph of G induced on this biconnected nucleus; H is thus a directed graph, and we use H_u to denote its undirected version.

We can try to further refine the structure of H by looking at *strongly connected components*. In this we make crucial use of the orientation of edges: u and v are related under strong connectivity if each can reach the other by a directed path. For this relation, however, we discover the following: the subgraphs H do not contain "giant strongly connected components." Indeed, for many of the graphs we considered, the largest strongly connected component had size less than 20.

Thus, while G is not connected when viewed in its entirety, it can be viewed as having a large subgraph that is biconnected as an undirected graph. G, however, does not generally contain any large strongly connected subgraphs.

Alternating connectivity. Biconnectivity yielded a giant component; strong connectivity pulverized the graph into tiny components. It is natural to ask whether there is a natural connectivity measure that takes some account of the orientation of edges and still results in large "components." We now describe one such measure, *alternating connectivity*, and observe a sense in which it is difficult to find the right definition of "component" under this measure.

If u and v are nodes, we say that a sequence P of edges in G is an *alternating path* from u to v if (1) P is a path in G_u with endpoints u and v, and (2) the orientations of the edges of P strictly alternate in G. Thus, P leads from u to v via a sequence of link traversals that alternate forward and backward directions. This definition corresponds closely to the HITS algorithm. Intuitively, the strict alternation of edge orientations parallels the way in which hub and authority weight flows between nodes; and indeed, the hub and authority weights effectively compute the relative growth rates of alternating paths from each node.

Furthermore, two steps in an alternating path connect two nodes that either cite the same page, or are cited by the same page — this notion of co-citation has been used as a similarity measure among documents [14] and among web pages (e.g. [11,21,25]).

Suppose we write $u \sim v$ if there is an alternating path between u and v. It is clear immediately from the definition that the relation $\sim$ is symmetric; however, it is easy to construct examples with nodes u, v, w so that $u \sim v$ and $v \sim w$, but $u \not\sim w$. As a result, $\sim$ is not transitive and hence not an equivalence relation.

However, we can show a sense in which $\sim$ is "nearly" an equivalence relation: we can prove that if a node u is related to three nodes v_1, v_2, v_3, then at least one pair among $\{v_1, v_2, v_3\}$ is also related. We can call such a relation *claw-free* — no node is related to three nodes that are themselves mutually unrelated.

In tests on the subgraphs H, we find that a large fraction of all pairs of nodes $u, v \in H$ are related by alternating connectivity. Among pairs that are related, we can define their *undirected distance* as the length of the shortest u-v path in H_u; and we can define their *alternating distance* as the length of the shortest alternating u-v path in H. We find that the average alternating distance between related pairs is generally at most a factor of two more than the average undirected distance between them; this indicates that the biconnected nuclei H are rich in short, alternating paths.

4   Model

In this section we lay the foundation for a class of plausible random graph models in which we can hope to establish many of our observations about the local structure of the Web graph. There are a number of reasons for developing such a model:

1. It allows us to model various structural properties of the Web graph — node degrees, neighborhood structures, etc. Of particular interest to us is the distribution of Web structures such as $C_{i,j}$'s which are signatures of Web communities or other Web phenomena of interest.
2. It allows us to predict the behavior of algorithms on the Web; this is is of particular interest when these algorithms are doomed to perform poorly on worst-case graphs.
3. It suggests structural properties of today's Web that we might be able to verify and then exploit.
4. It allows us to make predictions about the shape of the Web graph in the future.

Desiderata for a Web graph model. We begin by reviewing some criteria that are desirable in such a graph model; many of these are motivated by empirical observations on the structure of the Web graph. Next we present our model. Note that we do not seek to model the text or the sizes of the pages; we are only interested here in the interconnection patterns of links between pages. We would like our model to have the following features:

1. It should have a succinct and fairly natural description.
2. It should be rooted in a plausible macro-level process for the creation of content on the Web. We cannot hope to model the detailed behavior of the many users creating Web content. Instead, we only desire that the aggregate formation of Web structure be captured well by our graph model. Thus, while the model is described as a stochastic process for the creation of individual pages, we are really only concerned with the aggregate consequences of these individual actions.
3. It should not require some *a priori* static set of "topics" that are part of the model description — the evolution of interesting topics and communities should instead be an emergent feature of the model.[1] Such a model has several advantages:
   - It is extremely difficult to characterize the set of topics on the Web; thus it would be useful to draw statistical conclusions without such a characterization.
   - The set of topics reflected in Web content has proven to be fairly dynamic. Thus, the shifting landscape of actual topics will need to be addressed in any topic-aware model of time-dependent growth.
4. We would like the model to reflect many of the structural phenomena we have observed in the Web graph.

4.1   A class of random graph models

In our model, we seek to capture the following intuition:

- some page creators on today's Web may link to other sites without regard to the topics that are already represented on the Web, but
- *most* page creators will be drawn to Web pages covering existing topics of interest to them, and will link to pages within some of these existing topics.

We have already observed that the Web graph has many hub pages that contain resource lists focused on a topic. Here is the dominant phenomenon for link-creation in our model: a user encounters a resource list for a topic of interest to him, and includes many links from this list in his/her page.

We reiterate that this process is *not* meant to reflect individual user behavior on the Web; rather, it is a local procedure which in aggregate works well in describing page creation on the Web and which implicitly captures topic creation as follows: first, a few scattered pages begin to appear about the topic. Then, as users interested in the topic reach critical mass, they begin linking these scattered pages together, and other interested users are able to discover and link to the topic more easily. This creates a "locally dense" subgraph around the topic of interest. This intuitive view summarizes the process from a page-creator's standpoint; we now recast this formulation in terms of a random graph model that — again, on aggregate — captures the above intuition.

---

[1] In particular, we avoid models of the form "Assume each node is some combination of topics, and add an edge from one page to another with probability dependent on some function of their respective combinations."

Indeed, it is our thesis that random copying is a simple, plausible stochastic mechanism for creating Zipfian degree distributions. Below, we state at a high level our model for the evolution of the Web graph. We are unable to provide complete probabilistic analyses of the graphs generated by even the simplest concrete instantiations of such models. Heuristic calculations, however, yield distributions for degree sequences, $C_{i,j}$'s and other local structures that conform remarkably well with our observations.

Our model is characterized by four stochastic processes — *creation* processes C_v and C_e for node- and edge-creation, and *deletion* processes D_v and D_e for node- and edge-deletion. These processes are discrete-time processes. Each process is a function of the time step, and of the current graph.

A simple node-creation process would be the following: independently at each step, create a node with probability $\alpha_c(t)$. We could have a similar Bernoulli model with probability $\alpha_d(t)$ for node deletion: upon deleting a node, we also delete all its incident edges. Clearly we could tailor these probabilities to reflect the growth rates of the Web, the half-life of pages, etc.

The edge processes are rather more interesting. We begin with the edge-creation process C_e. At each step we sample a probability distribution to determine a node v to add edges out of, and a number of edges k that will be added. With probability β we add k edges from v to nodes chosen independently and uniformly at random. With probability $1 - \beta$, we *copy* k edges from a randomly chosen node to v. By this we mean that we choose a node u at random, and create edges from v to nodes w such that (u, w) is an edge. One might reasonably expect that much of the time, u will not have out-degree exactly k; if the out-degree of u exceeds k we pick a random subset of size k. If on the other hand the out-degree of u is less than k we first copy the edges out of u, then pick another random node u' to copy from, and so on until we have enough edges. Such a copying process is not unnatural, and consistent with the qualitative intuition at the beginning of this section.

A simple edge-deletion process D_e would again be a Bernoulli process in which at each step, with probability δ, we delete a randomly chosen node. The probability that a particular node v is deleted would ideally be non-increasing in its in-degree.

We illustrate these ideas with a very simple special case. Consider a model in which a node is created at every step. Nodes and edges are never deleted, so the graph keeps on growing. Consider the following edge process: for some $\beta \in (0, 1)$, at each step the newly-created node points to a node chosen uniformly at random. With probability $1 - \beta$, it copies a uniform random edge out of a random node. Simulations (and heuristic calculations) suggest that under this model, the probability that a node has in-degree i converges to $i^{-1/(1-\beta)}$. Similar calculations suggest that the numbers of cores $C_{i,j}$ are significantly larger than random graphs in which edges go to uniform, independent random destinations.

Clearly the processes creating these graphs, as well as the statistics and structures observed, differ significantly from those of traditional random graph models. This is both a feature and a challenge. On the one hand, the relationship between copying and Zipfian distributions is of intrinsic interest for a variety of reasons (given such distributions also arise in a number of settings outside of the Web — in term frequencies, in the genome, etc.). On the other hand, the process of copying also generates a myriad of dependencies between the random variables of interest, so that the study of such graphs calls for a whole new suite of analytical tools.

5   Conclusion

Our work raises a number of areas for further work:

1. How can we annotate and organize the communities discovered by the trawling process of Section 2.2?
2. What extensions and applications can be found for the connectivity measures discussed in Section 3.3?
3. What are the properties and evolution of random graphs generated by specific versions of our models in Section 4? This would be the analogue of the study of traditional random graph models such as $\mathcal{G}_{n,p}$.
4. How do we devise and analyze algorithms that are efficient on such graphs? Again, this study has an analogue with traditional random graph models.
5. What can we infer about the distributed sociological process of creating content on the Web?

We thank Byron Dom and Ron Fagin for their comments. The work of Jon Kleinberg was supported in part by an Alfred P. Sloan Research Fellowship and by NSF Faculty Early Career Development Award CCR-9701399.

References

1. S. Abiteboul, D. Quass, J. McHugh, J. Widom, and J. Weiner. The Lorel Query language for semistructured data. *Intl. J. on Digital Libraries*, 1(1):68-88, 1997.
2. R. Agrawal and R. Srikanth. Fast algorithms for mining association rules. *Proc. VLDB*, 1994.
3. G. O. Arocena, A. O. Mendelzon, G. A. Mihaila. Applications of a Web query language. *Proc. 6th WWW Conf.*, 1997.
4. K. Bharat and A. Broder. A technique for measuring the relative size and overlap of public Web search engines. *Proc. 7th WWW Conf.*, 1998.
5. K. Bharat and M. R. Henzinger. Improved algorithms for topic distillation in a hyperlinked environment. *Proc. ACM SIGIR*, 1998.
6. S. Brin and L. Page. The anatomy of a large-scale hypertextual Web search engine. *Proc. 7th WWW Conf.*, 1998.
7. B. Bollobás. *Random Graphs*, Academic Press, 1985.
8. J. Carrière and R. Kazman. WebQuery: Searching and visualizing the Web through connectivity. *Proc. 6th WWW Conf.*, 1997.
9. S. Chakrabarti, B. Dom, D. Gibson, J. Kleinberg, P. Raghavan and S. Rajagopalan. Automatic resource compilation by analyzing hyperlink structure and associated text. *Proc. 7th WWW Conf.*, 1998.

10. S. Chakrabarti, B. Dom, S. R. Kumar, P. Raghavan, S. Rajagopalan, and A. Tomkins. Experiments in topic distillation. *SIGIR workshop on hypertext IR*, 1998.
11. S. Chakrabarti and B. Dom and P. Indyk. Enhanced hypertext classification using hyperlinks. *Proc. ACM SIGMOD*, 1998.
12. H. T. Davis. *The Analysis of Economic Time Series*. Principia press, 1941.
13. R. Downey, M. Fellows. Parametrized Computational Feasibility. In *Feasible Mathematics II*, P. Clote and J. Remmel, eds., Birkhauser, 1994.
14. L. Egghe, R. Rousseau, *Introduction to Informetrics*, Elsevier, 1990.
15. D. Florescu, A. Levy and A. Mendelzon. Database techniques for the World Wide Web: A survey. *SIGMOD Record*, 27(3): 59-74, 1998.
16. E. Garfield. Citation analysis as a tool in journal evaluation. *Science*, 178:471–479, 1972.
17. N. Gilbert. A simulation of the structure of academic science. *Sociological Research Online*, 2(2), 1997.
18. G. Golub, C. F. Van Loan. *Matrix Computations*, Johns Hopkins University Press, 1989.
19. M. R. Henzinger, P. Raghavan, and S. Rajagopalan. Computing on data streams. *AMS-DIMACS series*, special issue on computing on very large datasets, 1998.
20. M. M. Kessler. Bibliographic coupling between scientific papers. *American Documentation*, 14:10–25, 1963.
21. J. Kleinberg. Authoritative sources in a hyperlinked environment, *J. of the ACM*, 1999, to appear. Also appears as IBM Research Report RJ 10076(91892) May 1997.
22. D. Konopnicki and O. Shmueli. Information gathering on the World Wide Web: the W3QL query language and the W3QS system. *Trans. on Database Systems*, 1998.
23. S. R. Kumar, P. Raghavan, S. Rajagopalan and A. Tomkins. Trawling emerging cyber-communities automatically. *Proc. 8th WWW Conf.*, 1999.
24. L. V. S. Lakshmanan, F. Sadri, and I. N. Subramanian. A declarative approach to querying and restructuring the World Wide Web. *Post-ICDE Workshop on RIDE*, 1996.
25. R. Larson. Bibliometrics of the World Wide Web: An exploratory analysis of the intellectual structure of cyberspace. *Ann. Meeting of the American Soc. Info. Sci.*, 1996.
26. A. J. Lotka. The frequency distribution of scientific productivity. *J. of the Washington Acad. of Sci.*, 16:317, 1926.
27. A. Mendelzon, G. Mihaila, and T. Milo. Querying the World Wide Web, *J. of Digital Libraries* 1(1):68–88. 1997.
28. A. Mendelzon and P. Wood. Finding regular simple paths in graph databases. *SIAM J. Comp.*, 24(6):1235-1258, 1995.
29. E. Spertus. ParaSite: Mining structural information on the Web. *Proc. 6th WWW Conf.*, 1997.
30. D. Tsur, J. Ullman, S. Abiteboul, C. Clifton, R. Motwani, S. Nestorov, and A. Rosenthal. Query Flocks: A generalization of association rule mining. *Proc. ACM SIGMOD*, 1998.
31. G. K. Zipf. Human behavior and the principle of least effort. *New York: Hafner*, 1949.

Some Observations on the Computational Complexity of Graph Accessibility Problem (Extended Abstract)

Jun Tarui[1] and Seinosuke Toda[2]

[1] Dept. DENSHI-JYOUHOU, University of Electro-Communications, 1-5-1
Chofugaoka, Chofu-shi, Tokyo 182, JAPAN
[2] Dept. Applied Mathematics, College of Humanities and Sciences, Nihon University,
3-25-40 Sakurajyousui, Setagaya-ku, Tokyo 156, JAPAN

Abstract. We investigate the space complexity of the (undirected) graph accessibility problem (UGAP for short). We first observe that for a given graph G, the problem can be solved deterministically in space $O(\text{sw}(G)^2 \log_2 n)$, where n denotes the number of nodes and $\text{sw}(G)$ denotes the separation-width of G that is an invariant of graphs introduced in this paper. We next observe that for the class of all graphs consisting of only two paths, the problem still remains to be hard for deterministic log-space under the NC^1-reducibility. This result tells us that the problem is essentially hard for deterministic log-space.

1 Introduction

It is widely know that the (undirected/directed) graph accessibility problem (UGAP / GAP for short; this is alternatively called the st-connectivity problem)) captures the nature of space-bounded (symetric/nondeterministic) computations. Because of this, its computational complexity has been investigated for a long period since 1970's. It was shown by Savitch[Sav70] that GAP can be solved deterministically in space $O(\log^2 n)$. However, no improvement on the amount of space has been accomplished so far. At this point, many researchers have conjectured that $O(\log^2 n)$ space would be required to solve the problem. On the other hand, there have been several remarkable results on UGAP. Alleliunas et. al. [AKL79] showed that UGAP can be solved probabilistically in logarithmic space and simultenously in polynomial average time. Furthermore Nisan [Nis92] showed that UGAP can be solved deterministically in space $O(\log^{3/2} n)$ and Armoni et. al. [ATW97] recently improved this space bound to $O(\log^{4/3} n)$.

In this paper, we investigate the space complexity of UGAP. The aim of our investigation is to relate some invariant on graphs to the space complexity of UGAP. We first show that for a given graph G, the problem can be solved deterministically in space $O(\text{sw}(G)^2 \log_2 n)$, where n denotes the number of nodes and $\text{sw}(G)$ denotes the separation-width of G that is an invariant of graphs introduced in this paper. As an immediate consequence, for the class of all graphs with separation-width bounded above by a given constant, the problem can be solved

T. Asano et al. (Eds.): COCOON'99, LNCS 1627, pp. 18–30, 1999.
© Springer-Verlag Berlin Heidelberg 1999

deterministically in logarithmic space. We also observe that the separation-width is smaller than or equal to the path-width. Thus, we also obtain the same result for the class of graphs with bounded path-width. As far as the authors know, there was no nontrivial class of graphs, except the class of cycle-free graphs, for which the problem is solvable deterministically in logarithmic space. Thus, our result observe a second nontrivial class of graphs with that property. We next show that for the class of all graphs consisting of only two paths, the problem still remains to be hard for deterministic log-space under the NC^1-reducibility. This result observes that the problem is essentially hard for deterministic log-space.

2   Preliminaries

For any graph $G = (V, E)$ and any subsets X, Y of V, we simply say that there exists a path between X and Y on G if there is a path on G that connects some node in X with some node in Y. We alternatively say that X is connected to Y on G or say that Y is **reachable** from X on G. In this paper, we will use the terminology "incident" with an unusual meaning, defined as follows. For any subsets X, Y of V, we say that X is **incident** with Y if either $X \cap Y \neq \emptyset$ or there is an edge connecting a node in X with a node in Y. We will further define $N_G(u)$ of a node u to be the set $\{v \in V : uv \in E\} \cup \{u\}$. Furthermore, we denote by $N_G(X)$, for a subset X of V, the set $\bigcup_{x \in X} N_G(x)$. For a subset X of V, we define $E_G[X] = \{uv \in E : u \in X \text{ or } v \in X\}$, and for a subset F of E, we define $V_G[E] = \{u \in V : \exists v \in V[uv \in F]\}$.

We below define some invariant of graphs that is relevant to the present paper.

Definition 1. *We define a **layout** of G to be a bijection from V to the set $\{1, 2, \cdots, |V|\}$. We also call each element of the set $\{1, \cdots, |V|\}$ a **location** when dealing with any layout of G. Let φ be a layout of G. We say that an edge $uv \in E$ **crosses** a location i (w.r.t. φ) if either $\varphi(u) \leq i < \varphi(v)$ or $\varphi(v) \leq i < \varphi(u)$. We denote by $E_\varphi(i)$ the set of all edges crossing a location i. Furthermore, we say that a path on G crosses a location if the path contains an edge that crosses the location.*

For a location i, we define $V_\varphi^-(i)$, $V_\varphi^+(i)$, and $V_\varphi^*(i)$ as follows:

$$V_\varphi^-(i) = \{u \in V_G[E_\varphi(i)] : \varphi(u) \leq i\},$$
$$V_\varphi^+(i) = \{v \in V_G[E_\varphi(i)] : i < \varphi(v)\},$$

$$V_\varphi^*(i) = \begin{cases} V_\varphi^-(i) & \text{if } |V_\varphi^-(i)| \leq |V_\varphi^+(i)| \\ V_\varphi^+(i) & \text{otherwise} \end{cases}$$

*Then we define the **separation-width** of G with respect to φ, denoted by $\mathrm{sw}_\varphi(G)$, by*

$$\mathrm{sw}_\varphi(G) = \max \{|V_\varphi^*(i)| : 1 \leq i \leq |V|\}.$$

*We further define the **separation-width** $\mathrm{sw}(G)$ of G by $\mathrm{sw}(G) = \min_\varphi \mathrm{sw}_\varphi(G)$, where the minimum is taken over all possible layouts of G.*

It was shown in [Kin92] that for each graph, its vertex separation numbers is equal to its path-width (though we omit the definitions of those invariant), and it is easy to verify that for each graph, its separation-width is smaller than or equal to its vertex separation number. Though we don't use these facts in this extended abstract, this may give the reader a brief intuition on the separation-width since the path-width is much common in graph theory community.

Hereafter, we only deal with the separation-width. Thus, we say that a layout φ of G is *optimal* if $\mathrm{sw}_\varphi(G) = \mathrm{sw}(G)$.

For a set A and a nonnegative integer k, we define $2^{A,k}$ to be the set of all subsets whose number of elements is at most k. For a set Σ, we denote by Σ^* the set of all finite sequences of elements in Σ. We further denote a sequence in Σ^* consisting of elements $a_1, a_2, \cdots, a_m$ by $\langle a_1, a_2, \cdots, a_m \rangle$ or by $\langle a_j \rangle_{j=1}^m$. We sometimes abbreviate the range of the indices from the latter notation when it is clear in the context.

3   An algorithm

In this section, we show that UGAP can be solved deterministically in space $O(\mathrm{sw}(G)^2 \log_2 n)$ for a given graph G, where n denotes the number of nodes in G.

Definition 2. *For all nonnegative integers k, we define two predicates denoted REACH$_k$ and SIDE$_k$ respectively, and we define a function denoted NEXT$_k$, as described in Figure 1. For any nodes x, y in a graph G, we abbreviate REACH$_k$ $(G, \{x\}, \{y\})$, NEXT$_k$ $(G, \{x\}, \{y\})$, and SIDE$_k$ $(G, \{x\})$ by REACH$_k$ (G, x, y), NEXT$_k$ (G, x, y), and SIDE$_k$ (G, x) respectively.*

Let k be a nonnegative integer, let $G = (V, E)$ be a graph, and let S, T be any subsets of V. Then, we define a predicate REACH$_k(G, S, T)$ inductively as follows.

(I) REACH$_0(G, S, T) \Leftrightarrow S \cap T \neq \emptyset$.
(II) For all $k \geq 1$:   REACH$_k(G, S, T)$
$$\Updownarrow$$
Either REACH$_{k-1}(G, S, T)$ holds, or there are a node $u \in V$ and a sequence of sets $X_0, X_1, \ldots, X_m \in 2^{V,k}$ satisfying the conditions (R1) through (R4) below, where we define $X_{-1} = X_{m+1} = \emptyset$.
(R1) $X_0 = \{u\}$.
(R2) For all $p(0 \leq p \leq m)$, NEXT$_k(G, X_{p-1}, X_p) = X_{p+1}$.
(R3) For some $p(0 \leq p \leq m)$,
        REACH$_{k-1}(G - E[X_{p-1} \cup X_p \cup X_{p+1}], S, N_G(X_p))$.
(R4) For some $q(0 \leq q \leq m)$,
        REACH$_{k-1}(G - E[X_{q-1} \cup X_q \cup X_{q+1}], N_G(X_q), T)$.

Fig. 1(a). The definition of REACH$_k$.

In order to define NEXT_k, we suppose an arbitrary linear ordering over $2^{V,k}$ such that for all graphs $G = (V, E)$ and two subsets $X, Y \in 2^{V,k}$, we can test deterministically in space $O(\log_2 |V|)$ whether X is less than Y under the linear ordering.

Now let $G = (V, E)$ be a graph and let X, Y be any subsets of V. Then, we define $\text{NEXT}_k(G, X, Y)$, where $k \geq 1$, to be the minimum set under the ordering among all sets $Z \in 2^{V,k}$ satisfying the conditions (N1) through (N3) described below.

(N1) For all $z \in Z$, $\neg\text{SIDE}_k(G - E[X \cup Y], z)$.
(N2) For all $w \in N_G(Y)$, $\text{SIDE}_k(G - E[X \cup Y \cup Z], w)$.
(N3) For all $z \in Z$, $\text{REACH}_{k-1}(G - E[X \cup Y \cup Z], N_G(Y), N_G(z))$.

If there is no set in $2^{V,k}$ satisfying the above conditions, then we conventionally define $\text{NEXT}_k(G, X, Y) = \perp$.

Fig. 1(b). The definition of NEXT_k.

For all graphs G and all nodes w of G:
$$\text{SIDE}_k(G, w) \quad \text{ik} \geq \text{1j}$$
$$\updownarrow$$

(S1) For all $v \in V$, $\text{REACH}_{k-1}(G, w, N_G(v)) \Rightarrow \text{REACH}_{k-1}(G, w, v)$, and
(S2) For all $u, v \in V$, $\text{REACH}_{k-1}(G, w, u)$ and $\text{REACH}_{k-1}(G, w, v) \Rightarrow \text{REACH}_{k-1}(G, u, v)$.

Fig. 1(c). The definition of SIDE_k.

It is not difficult to see that the predicates and the function defined above can be realized by (recursive) algorithms that operate in space $O(k^2 \log_2 n)$. Our purpose of this section is to prove that for all graphs G with $sw(G) \leq k$ and all subsets S, T of the node set of G, $\text{REACH}_k(G, S, T)$ holds if and only if there exists a path on G between S and T. That is, we will prove the following theorem throughout the whole of this section. Note that the statement (a) and (b) below are the core claims of the theorem and the other statements are technically used to prove the core claims.

Theorem 1. *Let k be any nonnegative integer, let $G = (V, E)$ be any graph, and let S, T be any subset of V. Then we have the following.*

(a) If $REACH_k(G, S, T)$ holds, then there is a path on G between S and T.

(b) Conversely, if $\mathrm{sw}(G) \le k$ and there is a path on G between S and T, then $REACH_k(G, S, T)$ holds.

(c) Let s and t be any nodes in G and let H be any subgraph of G. We also suppose that H contains both the component of G including s and the component of G including t. Then, $REACH_k(G, s, t)$ holds if and only if $REACH_k(H, s, t)$ holds.

(d) Let s and t be any nodes in G and let S and T be any subsets of V. We also suppose that S contains s and T contains t. Then, if $REACH_k(G, s, t)$ holds, then $REACH_k(G, S, T)$ holds.

We prove this theorem by using an induction on k. It is obvious that the theorem holds for $k = 0$. Now, let k be any positive integer and let us assume, as an induction hypothesis, that the statements (a) through (d) in the above theorem hold for all nonnegative integers less than k. Note that we will use this induction hypothesis until the end of the present section. Since the proof for the induction step is involved, we conventionally denote the induction hypothesis by **Th[$k-1$]**.

The most difficult part of the induction step is to prove the statement (b) in the theorem. Proving the other statements are easy and are omitted. Hereafter until almost end of this section, we prove Theorem 1(b). To begin it, we first observe some elementary properties of SIDE_k that will be used frequently.

Lemma 1. *Let $G = (V, E)$ be a graph, let w be a node in G, let C_w denote a component of G that contains w, and let V_w denote the node set of C_w. Then we have the following.*

(a) $SIDE_k(G, w)$ holds if and only if for all nodes $u, v \in V_w$, $REACH_{k-1}(G, u, v)$ holds.

(b) $SIDE_k(G, w)$ holds if and only if for all nodes $u \in V_w$, $SIDE_k(G, u)$.

(c) If $\mathrm{sw}(C_w) < k$, then $SIDE_k(G, w)$ holds.

Before going into the proof of Theorem 1(b), we here give a brief explanation on the role and/or the motivation of the predicate SIDE_k, based on the lemma above. We further give a brief explanation on the role of the function NEXT_k. These may be helpful for the reader to understand how our algorithm works well.

For a given node w in a graph G, $\mathrm{SIDE}_k(G, w)$ has a role to check whether, for all pairs of nodes in the component containing w, REACH_{k-1} can correctly verify the existence of a path between the two nodes. This role can be seen from (a) of the last lemma. Note also, from the contrapositive of Th[$k-1$](a), that for all pairs u, v of nodes in G, if there is no path between the two nodes, then $\mathrm{REACH}_{k-1}(G, u, v)$ never holds. Therefore, in a happy case that $\mathrm{SIDE}_k(G, w)$ holds for all nodes w in G, we may use REACH_{k-1} recursively to check the existence of a path between any pair of nodes in G. From (c) of the last lemma, we know that this case happens when $\text{sw}(G) < k$. However, in a bad case that $\text{sw}(G) \geq k$, REACH_{k-1} might not work well; that is, there might be some node w for which $\text{SIDE}_k(G, w)$ did not hold and hence there might be two nodes u, v for which $\text{REACH}_{k-1}(G, u, v)$ did not hold while there was a path between the two nodes. In this bad case, we know that there is a component of G such that for all nodes w in the component, $\text{SIDE}_k(G, w)$ does not hold. Hereafter, we call such a component **fat** and in contrast, we call a component **slim** if for some node w in the component, $\text{SIDE}_k(G, w)$ holds.

On a fat component, we must give a help to REACH_{k-1} in order to let it work well. Roughly speaking, our algorithm helps it as follows. For a fat component C of G, our algorithm finds a sequence $X_0, X_1, \cdots, X_m \subseteq V_C$, where V_C denotes the node set of C, such that by deleting all edges incident with $\bigcup_{j=0}^{m} X_j$ from C, we can make C slim; more precisely speaking, by deleting all of the edges, we can divide C into several components that are all slim. Then, after the deletion, our algorithm uses REACH_{k-1} to check the accessibility between all pairs of nodes in C.

What our algorithm does is slightly different from one mentioned in the above paragraph because of the amount of space available. Since it must work within a small amount of space, it computes the sequence of subsets as above sequentially and gradually one after another for $j = 0, 1, \ldots, m$ rather than finding all of them for once. Furthermore, it keeps only three subsets at a same time. This is a motivation behind NEXT_k.

Now let us return to proving the induction step of Theorem 1(b). The following definition is a convention for making our presentation concise.

Definition 3. *Let $G = (V, E)$ be a graph. For all subsets X, Y of V, we denote by* $\text{path}(G, X, Y)$ *a predicate that is true if and only if there is a path on G between X and Y.*

The following definition formalizes a notion of sequences like one mentioned above, which is suitable for our purpose.

Definition 4. *Let $G = (V, E)$ be a graph and let $\langle X_j \rangle_{j=0}^{m}$ be a sequence in $(2^V)^*$. Then, we say that the sequence is **k-fat tissue** if and only if it satisfies the following conditions:*

(0) *$|X_j| \leq k$ for all $j (0 \leq j \leq m)$ and there is an optimal layout φ of G such that $X_0 = \{\varphi^{-1}(1)\}$.*

(1) *For all $p (0 \leq p \leq m)$ and all $w \in N_G(X_p)$, $\text{SIDE}_k(G - E[\bigcup_{j=0}^{m} X_j], w)$ holds.*

(2) *For all $p (0 \leq p < m)$ and all $v \in X_{p+1}$, $\text{path}(G - E[\bigcup_{j=0}^{m} X_j], N_G(X_p), N_G(v))$ holds.*

(3) *For all $p (0 < p < m)$, $\text{path}(G - E[X_p], \bigcup_{j=0}^{p-1} X_j, \bigcup_{j=p+1}^{m} X_j)$ does not hold.*

*We call each X_j a **cell** of the k-fat tissue, and for any p, q with $0 \leq p, q \leq m$, we call $E[\bigcup_{j=p}^{q} X_j]$ a **k-fat** of the k-fat tissue.*

We below give a brief explanation about the intention behind each condition in the above definition, for the sake of an intuitive understanding on how our algorithm works well.

The condition in (0) above that each cell is of size at most k is intended to ensure that they are not so large compared with the parameter k. The other condition $X_0 = \{\varphi^{-1}(1)\}$ in (0) together with the condition (2) above guarantees that a single component of G contains all cells.

The condition (1) above guarantees that by deleting all edges incident with the k-fat tissue, we can divide the component containing the tissue into several slim components. We note here that each of the slim components obtained after the deletion must be incident with some cell, and hence we also note that the node w mentioned in (1) above are intended as a "representative" for one of the slim components.

The condition (3) together with the condition (2) is intended to ensure that each cell in the k-fat tissue can be computed sequentially and "locally" in the sense that each X_{p+1} can be found only by using X_{p-1} and X_p. To understand this intention more, let us consider a situation that we could find a part of the k-fat tissue, say $X_0, X_1, \cdots, X_p$ for some $p < m$, and that we wanted to find out a next cell X_{p+1}. If $\mathrm{SIDE}_k(G, w)$ holds for all nodes $w \in \bigcup_{j=0}^{p} X_j$, then we need nothing to do; that is, we may set X_{p+1} to the empty set. Otherwise, by the condition (2), we must seek X_{p+1} among the nodes reachable in $G - E[\bigcup_{j=0}^{p} X_j]$ from $N_G(X_p)$. However, we meet a difficulty here in that task. It is that we have no space to keep all X_j's found so far. On the other hand, we are allowed in that task to deal with $G - E[X_{p-1} \cup X_p]$ instead of $G - E[\bigcup_{j=0}^{p} X_j]$. This is because the condition (3) ensures that all nodes in G reachable from $\bigcup_{j=0}^{p-2} X_j$ is not reachable from any nodes in X_p without passing through X_{p-1}. This is the intention behind the condition (3). We also note that this consideration leads us to the definition of NEXT_k.

Hereafter for a while, we will discuss about the existence of a k-fat tissue for all components of G with separation-width at most k.

Definition 5. *Let $G = (V, E)$ be a graph and let $\langle X_j \rangle_{j=0}^{m}$ be a sequence in $(2^V)^*$. Then, we say that $\langle X_j \rangle$ is a **k-fat substance** of G if and only if it satisfies the conditions (0), (2), and (3) of being a k-fat tissue and satisfies the following condition (1') instead of (1):*

(1') For all $p (0 \leq p < m)$ and all nodes $w \in N_G(X_p)$, $\mathrm{SIDE}_k(G - E[\bigcup_{j=0}^{m} X_j], w)$ holds.

Only the difference from the condition (1) of being a k-fat tissue is that the last set X_m in the sequence may not satisfy the conditions on SIDE_k. Note that any k-fat tissue is also a k-fat substance.

We can show that a sequence is k-fat substance if and only if it is a prefix of a k-fat tissue. However, this is not trivial. We will observe this fact in Theorem 2.

As mentioned previously, any k-fat tissue is in a single component. More generally, we can observe the following lemma. We will use this fact everywhere without specifying explicitly (since this is almost obvious).

Lemma 2. *Let $G = (V, E)$ be a graph and let $\langle X_j \rangle_{j=0}^m$ be a k-fat substance of G. Then, there exists a single component of G that contains all nodes in $\bigcup_{j=0}^m X_j$.*

The next proposition tells us the existence of a k-fat substance for all connected graphs.

Proposition 1. *Let C be a connected graph and let φ be an optimal layout of C. Then, the sequence $\langle \{\varphi^{-1}(1)\} \rangle$ is a k-fat substance of C.*

The next lemma says that for all k-fat substances $\langle X_j \rangle_{j=0}^p$ of a connected graph C, if there exists a fat sub-component after deleting all edges incident with $\bigcup_{j=0}^p X_j$, then the fat sub-component must be incident only with X_p but not incident with other X_j's.

Lemma 3. *Let $C = (V, E)$ be a connected graph and let $\langle X_j \rangle_{j=0}^p$ be a k-fat substance of C. Then we have that for all nodes $w \in V$, if $SIDE_k(C - E[\bigcup_{j=0}^p X_j], w)$ does not hold, then there is no path on $C - E[X_p]$ between $\bigcup_{j=0}^{p-1} X_j$ and V_w, where V_w denotes the node set of the component of $C - E[\bigcup_{j=0}^p X_j]$ which contains w.*

Now we show that each k-fat substance can be extended to some longer k-fat substance and that each k-fat substance is a prefix of a k-fat tissue.

Theorem 2. *Let $C = (V, E)$ be a graph and suppose $sw(C) \leq k$. Furthermore, let $\alpha = \langle X_j \rangle_{j=0}^p \in (2^{V.k})^*$ be a k-fat substance of C. Then, we have the following:*

(I)   *There is a set $X_{p+1} \in 2^{V.k}$ satisfying the following conditions.*
    (i)   *For all nodes $z \in X_{p+1}$, $\neg SIDE_k(C - E[\bigcup_{j=0}^p X_j], z)$.*
    (ii)   *For all nodes $w \in N_C(X_p)$, $SIDE_k(C - E[\bigcup_{j=0}^p X_j \cup X_{p+1}], w)$.*
    (iii) *For all nodes $z \in X_{p+1}$, $path(C - E[\bigcup_{j=0}^p X_j \cup X_{p+1}], N_C(X_p), N_C(z))$.*
(II)   *For all sets $X_{p+1} \in 2^{V.k}$ satisfying the conditions (i) through (iii) above, $\beta = \langle X_0, \cdots, X_p, X_{p+1} \rangle$ is a k-fat substance of C.*
(III) *Suppose $X_{p+1} = \emptyset$ satisfies the conditions (i) through (iii) above. Then, only the empty set satisfies the conditions and α is a k-fat tissue of C.*

Proof. We give an outline of the proof of (I) and omit the proofs of (II) and (III).

In the case in which $SIDE_k(C - E[\bigcup_{j=0}^p X_j], w)$ holds for all nodes $w \in V$, we can easily verify that $X_{p+1} = \emptyset$ satisfies all of the conditions (i) through (iii) above. Thus, until the end of this proof, we consider the case that $SIDE_k(C - E[\bigcup_{j=0}^p X_j], w)$ does not hold for some node $w \in V$. Note that in this case, $sw(C) = k$. This can be seen from Lemma 1(c) (together with the monotonic property of the separation-width with respect to the subgraph relationship).

Let φ be an optimal layout of C and let $D = (V_D, E_D)$ denote a fat component of $C - E[\bigcup_{j=0}^p X_j]$. Note here that $sw(D) = k$ and for all nodes w in D, $SIDE_k(C - E[\bigcup_{j=0}^p X_j], w)$ does not hold. We will use this fact frequently. Below, we observe several facts required for the current proof.

We can see that D has no node common with $\bigcup_{j=0}^{p} X_j$, as follows. Assume, to a contrary, that D has a common node with $\bigcup_{j=0}^{p} X_j$. Then D must consist of a single node because before defining D, we have deleted all edges incident with $\bigcup_{j=0}^{p} X_j$. Thus we see $sw(D) = 0$. However, this is impossible because $sw(D) = k > 0$.

To show another fact, we define h_X by $h_X = \max\{\varphi(x) : x \in \bigcup_{j=0}^{p} X_j\}$, and we define h_D to be any location i satisfying $V_\varphi^*(i) \subseteq V_D$ and $|V_\varphi^*(i)| = k$. The existence of such a location is guaranteed by the fact $sw(D) = k$. We further note that all edges crossing the location h_D belong to E_D. Now we can show the following fact.

Fact A $h_X < h_D$.
Proof. Let X_q, for some $q(0 \le q \le p)$, contain a node whose location is h_X and let u_q denote the node. Then, it follows from the conditions (2) and (3) of $\langle X_j \rangle$ being a k-fat substance that there exists a path $P = u_0, P_1, u_1, P_2, \cdots, u_{q-1}, P_q, u_q$, where $u_0 = \varphi^{-1}(1)$ and for each $i \ge 1$, u_i belongs to X_i and each P_i is a path on $C - E[\bigcup_{j=0}^{p} X_j]$ between $N_C(u_{i-1})$ and $N_C(u_i)$. Since each P_i is a path on a component of $C - E[\bigcup_{j=0}^{p} X_j]$ that is incident (on C) with X_{i-1}, we see, from the condition (1') of $\langle X_j \rangle$ being a k-fat substance together with Lemma 1(b), that for all nodes v on P_i, $SIDE_k(C - E[\bigcup_{j=0}^{p} X_j], w)$ holds. By this fact together with the fact on D mentioned previously, we see that P has no common node with D. This implies that any path crossing h_D cannot cross all location i with $1 \le i \le h_X$ since P crosses all of those locations. Thus we have $h_X < h_D$.
(End of FACT A)

We next observe that there is at most one fat component in $C - E[\bigcup_{j=0}^{p} X_j]$.
Fact B There is at most one fat component in $C - E[\bigcup_{j=0}^{p} X_j]$.
Proof. Assume, to a contrary, that there are two fat components, say $D_1 = (V_1, E_1)$ and $D_2 = (V_2, E_2)$, in $C - E[\bigcup_{j=0}^{p} X_j]$. For $t = 1, 2$, let h_t denote a location i satisfying $V_\varphi^*(i) \subseteq V_D$ and $|V_\varphi^*(i)| = k$. The existence of such a location is guaranteed by the fact $sw(D_t) = k$. We further assume $h_1 < h_2$ without loss of generality. From the Fact A, we have $h_X < h_1 < h_2$. Now consider a path on C between $\varphi^{-1}(1)$ and a node in D_2 with location greater than or equal to h_2. Then the path must cross the location h_1. This means that the path must pass through inside of D_1. Furthermore, when walking along the path starting from $\varphi^{-1}(1)$ until the node in D_2, we never meet any nodes in $\bigcup_{j=0}^{p} X_j$ after the last point where we cross the location h_1. This implies that D_2 are connected with D_1 in $C - E[\bigcup_{j=0}^{p} X_j]$. However, this contradicts that D_1 and D_2 are different components in $C - E[\bigcup_{j=0}^{p} X_j]$. **(End of FACT B)**

Let us denote again by $D = (V_D, E_D)$ a unique fat component of $C - E[\bigcup_{j=0}^{p} X_j]$. Furthermore, we define g_D to be the minimum among all locations i satisfying $V_\varphi^*(i) \subseteq V_D$ and $|V_\varphi^*(i)| = k$. We then define X_{p+1} as follows:
$$X_{p+1} = \{z \in V_\varphi^*(g_D) : \text{path}(C - E[\bigcup_{j=0}^{p} X_j \cup V_\varphi^*(g_D)], N_C(X_p), N_C(z)\}.$$
We here note that $|X_{p+1}| \le k$ since $|V_\varphi^*(g_D)| \le k$. We below show that this X_{p+1} satisfies the conditions (i) through (iii).

Since all nodes in X_{p+1} are in V_D and D is a fat component of $C - E[\bigcup_{j=0}^{p} X_j]$, we have that for all nodes $z \in X_{p+1}$, $SIDE_k(C - E[\bigcup_{j=0}^{p} X_j], z)$ does not hold.

Thus we have the condition (i). Since $C - E[\bigcup_{j=0}^{p} X_j \cup V_\varphi^*(g_D)]$ is a subgraph of $C - E[\bigcup_{j=0}^{p} \cup X_{p+1}]$, we see from the definition of X_{p+1} that for all nodes $z \in X_{p+1}$, there is a path on $C - E[\bigcup_{j=0}^{p} \cup X_{p+1}]$ between $N_C(X_p)$ and $N_C(z)$. Thus we have the condition (iii).

To see the condition (ii) holds, let $g_X = \max \{\varphi(x) : x \in X_p\}$, let w be a node in $N_C(X_p)$, and let D_w be a component of $C - E[\bigcup_{j=0}^{p} \cup X_{p+1}]$ that contains the node w. If D_w does not contain any nodes in D, then we see that D_w is also a component of $C - E[\bigcup_{j=0}^{p} X_j]$. Hence, we see, from Fact B that D is the only fat component of $C - E[\bigcup_{j=0}^{p} X_j]$, that D_w is not a fat component and hence $\mathrm{SIDE}_k(C - E[\bigcup_{j=0}^{p} X_j \cup X_{p+1}], w)$ holds. We below assume that D_w contains a node in D. We note that in this case, D_w is a subgraph of D.

Now consider two cases. The first case is when D_w contains a node in $X_p \cup X_{p+1}$. Then by its definition, D_w consists of a single node and hence $\mathrm{sw}(D_w) = 0 < k$. This together with Lemma 1(c) implies that $\mathrm{SIDE}_k(C - E[\bigcup_{j=0}^{p} X_j \cup X_{p+1}], w)$ holds.

Now consider the second case that D_w does not contain a node in $X_p \cup X_{p+1}$. In this case, all edges connecting X_p and D_w cannot cross the location g_D. For, assuming that such an edge cross g_D, one end-point of the edge must belong to X_{p+1} and also to D_w but this is a contradiction. We also see that D_w never contains an edge crossing the location g_D, by its definition. Thus, we have that either for all nodes y in D_w, $\varphi(y) \leq g_D$, or for all nodes y in D_w, $\varphi(y) > g_D$. On the other hand, since $g_X < g_D$ by Fact A and there must be the edge connecting X_p with w, we see that $\varphi(w) < g_D$. Thus, we can observe that for all nodes y in D_w, $\varphi(y) \leq g_D$. This together with the minimality of g_D implies that $\mathrm{sw}(D_w) < k$ (recall D_w is a subgraph of D in our current case). Thus, we have that $\mathrm{SIDE}_k(C - E[\bigcup_{j=0}^{p} X_j \cup X_{p+1}], w)$ holds.

By the discussion above, we have the condition (ii) too.

Proposition 1 and Theorem 2 tell us how to compute a k-fat tissue of a given connected graph C, when we could find the top node of an optimal layout of C. On the other hand, we don't need find the top node. Instead, we may enumerate all nodes in C and may check whether we could extend each node to a k-fat tissue. REACH_k, NEXT_k, and SIDE_k is defined along this line. Particularly, NEXT_k realizes the function to extend a k-fat substance to a longer one. We want the reader to note the similarity between the conditions mentioned in Theorem 2(I) and those used in NEXT_k. However, there is still a gap between them. In Theorem 2(I), the conditions are global in a sense that they are dependent on the whole sets of a k-fat substance under consideration, while the conditions (N1) through (N3) used in NEXT_k are local in a sense that they are dependent only on a few sets in the k-fat substance. The theorem below fills in this gap.

Theorem 3. Let $G = (V, E)$ be a graph with $\mathrm{sw}(G) \leq k$ and let $\alpha = \langle X_j \rangle_{j=0}^{p} \in (2^{V,k})^*$ be a k-fat substance of G. Then, we have the following.

(I)   There is a set $X_{p+1} \in 2^{V,k}$ with $\mathrm{NEXT}_k(G, X_{p-1}, X_p) = X_{p+1}$; that is, $\mathrm{NEXT}_k(G, X_{p-1}, X_p) = \bot$ never holds.

(II) For all sets $X_{p+1} \in 2^{V,k}$ such that $NEXT_k(G, X_{p-1}, X_p) = X_{p+1}$, $\beta = \langle X_0, \cdots, X_p, X_{p+1} \rangle$ is also a k-fat substance of G.

(III) If $NEXT_k(G, X_{p-1}, X_p) = \emptyset$, then the set $X_{p+1} = \emptyset$ is the only set that satisfies $NEXT_k(G, X_{p-1}, X_p) = X_{p+1}$. Furthermore, in this case, α is a k-fat tissue.

Corollary 1. Let $G = (V, E)$ be a graph with separation-width at most k, let φ be an optimal layout of G, and let $\alpha = \langle X_j \rangle_{j=0}^m$ be a sequence in $(2^{V,k})^*$. Furthermore, α satisfies the following conditions.

(R1) $X_0 = \{\varphi^{-1}(1)\}$.

(R2) For all $p(0 \le p \le m)$, $NEXT_k(G, X_{p-1}, X_p) = X_{p+1}$.

Then the sequence α is a k-fat tissue of (a component of) G. Conversely, any k-fat tissue α of (a component of) G satisfies the above conditions.

Now, we are ready to prove the induction step for Theorem1(b).

Proof of Theorem1(b) Assume there exists a path on G between S and T. Then, let $C = (V_C, E_C)$ denote a component of G that contains some node in S and some node in T simultaneously. If $sw(C) < k$, then we have the claim immediately from the induction hypothesis. Thus, we below assume $sw(C) = k$.

Let $\alpha = \langle X_j \rangle_{j=0}^m$ be a k-fat tissue of C. From the assumption that there exists a path on C between S and T, there must be some component of $C - E[\bigcup_{j=0}^m X_j]$ that contains some node in S and that is incident with some X_p. Similarly, there must be some component of C that contains some node in T and that is incident with some X_q. Let D_p and D_q denotes the two components respectively.

Then, from the condition (1) of α being a k-fat tissue, $SIDE_k(C - E[\bigcup_{j=0}^m X_j], w)$ holds for all nodes in $V_p \cap N_G(X_p)$, where V_p denotes the node set of D_p. Combining this with Lemma 1(a), we have the following.

(a) For some $p(0 \le p \le m)$, $REACH_{k-1}(C - E[\bigcup_{j=0}^m X_j], S, N_G(X_p))$.

By a similar argument, we also have the following.

(b) For some $q(0 \le q \le m)$, $REACH_{k-1}(C - E[\bigcup_{j=0}^m X_j], N_G(X_q), T)$.

In below, we modify these conditions into other equivalent ones.

When D_p contains some node in $\bigcup_{j=0}^m X_j$, we easily see that D_p consists of a single node in X_p. Thus, in this case, we see that D_p is a component of $G - E[\bigcup_{j=p-1}^{p+1} X_j]$. On the other hand, when D_p does not contain any nodes in $\bigcup_{j=0}^m X_j$, we see, from the condition (3) of α being a k-fat tissue, that D_p is not incident in $C - E[\bigcup_{j=0}^m X_j]$ with $\bigcup_{j=0}^{p-2} X_j$ and with $\bigcup_{j=p+2}^m X_j$. Thus, in this case, we also have that D_p is a component of $G - E[\bigcup_{j=p-1}^{p+1} X_j]$. From this (with Th[$k-1$](c)), we have the following claim, which is the same as the condition (R3) in the definition of $REACH_k(G, S, T)$.

(R3) For some $p(0 \le p \le m)$, $REACH_{k-1}(G - E[\bigcup_{j=p-1}^{p+1} X_j], S, N_G(X_p))$.

By a similar argument, we also have the following.

(R4) For some $q(0 \leq q \leq m)$, $\mathrm{REACH}_{k-1}(G - E[\bigcup_{j=q-1}^{q+1} X_j], N_G(X_q), T)$.

Since α is a k-fat tissue of G, we have, from Collorary 1, that it satisfies the conditions (R1) and (R2) in the definition of $\mathrm{REACH}_k(G,S,T)$. Thus, $\mathrm{REACH}_k(G,S,T)$ holds. ∎

Corollary 2. *Let k be any positive integer. let G be any graph, and let S and T be any subsets of the node set of G. Suppose $\mathrm{sw}(G) \leq k$. Then, $REACH_k(G, S, T)$ if and only if there exists a path on G between S and T.*

When we compute $\mathrm{SIDE}_k(G, v)$ for a graph G, each node v of G, and each integer $k = 1, 2, \ldots$ one after another, we can see that for some integer $k \leq \mathrm{sw}(G) + 1$, $\mathrm{SIDE}_k(G, v)$ holds for all nodes v of G. This is guaranteed by Lemma 1(c). Furthermore, at that point, we can correctly decide the reachability in G by using REACH_k. This is guaranteed by the above corollary. This implies the following corollary.

Corollary 3. *UGAP is decidable deterministically in space $O(\mathrm{sw}(G)^2 \log_2 n)$, where G denotes a given graph and n denotes the number of its nodes.*

4   Hardness of UGAP

When a given graph is of separation-width at most one, we see that the graph consists of several paths. For such a graph, we might hope that UGAP would be solvable more efficiently than deterministic log-space, for example, via an NC^1 circuit. We further might hope that for all graphs with bounded separation-width, UGAP could be solved, say, in NC^1. However, our second result below refutes this hope unless deterministic log-space is contained in NC^1.

Let $\mathcal{P}$ be an arbitrary set of three-tuples $\langle G, s, t \rangle$ where G is any graph and s, t are nodes in G. Futhermore, suppose that $\mathcal{P}$ satisfies the following condition: for each graph G which consists only of two paths and for all nodes s, t in G, G has a path between s and t iff $\langle G, s, t \rangle$ is in $\mathcal{P}$. In other words, the set $\mathcal{P}$ gives us a correct answer on UGAP if a given graph consists only of two paths (but it may give us a wrong answer otherwise). This is a kind of a solution for a so-called *promise problem*. For a general notion of promise problems, refer to the paper [GS88].

Theorem 4. *Every set $\mathcal{P}$ as above is hard for deterministic log-space under NC^1 reducibility.*

The proof of this facts is based on the fact that UGAP on the cycle-free graphs is hard for deterministic log-space under NC^1 reducibility and on a well-known Eularian traversal method on such graphs. We omit the detail.

References

AKL79.   R.Alleliunas, R.Karp, R.Lipton, L.Lovasz and C.Rackoff: Random walks, universal sequences and the complexity of maze problems, FOCS, 1979.

ATW97.   R.Armoni, A.Ta-Shma, A.Wigderson and S.Shou: SL $\subseteq$ L$^{4/3}$, STOC, 1997.

GS88.   J.Grollman and A.L.Selman: Complexity measures for public-key cryptosystems, SIAM J. on Comput., bf 17,1988.

Nis92.   N.Nisan: RL $\subseteq$ SC, STOC, 1992.

NSW92.   N.Nisan, E.Szemerédi and A.Wigderson: Undirected connectivity in $O(\log^{1.5} n)$ space, FOCS, 1992.

Kin92.   N.G.Kinnersley: The vertex separation number of a graph equals its pathwidth, IPL, **42**, 1992.

Sav70.   W.J.Savitch: Relationships between nondeterministic and deterministic space complexities, JCSS, **4**, 1970.

An Approximation for Finding a Smallest 2-Edge-Connected Subgraph Containing a Specified Spanning Tree *

Hiroshi Nagamochi and Toshihide Ibaraki

Kyoto University, Kyoto, Japan 606-8501
{naga, ibaraki}@kuamp.kyoto-u.ac.jp

Abstract. Given a graph $G = (V, E)$ and a tree $T = (V, F)$ with $E \cap F = \emptyset$ such that $G + T = (V, F \cup E)$ is 2-edge-connected, we consider the problem of finding a smallest 2-edge-connected spanning subgraph $(V, F \cup E')$ of $G + T$ containing T. The problem, which is known to be NP-hard, admits a 2-approximation algorithm. However, obtaining a factor better than 2 for this problem has been one of the main open problems in the graph augmentation problem. In this paper, we present an $O(\sqrt{n}m)$ time $\frac{12}{7}$-approximation algorithm for this problem, where $n = |V|$ and $m = |E \cup F|$.

1 Introduction

Given a 2-edge-connected undirected multigraph $H = (V, E)$ with n vertices and m edges and a spanning subgraph $H_0 = (V, E_0)$, we consider the problem of finding a smallest 2-edge-connected spanning subgraph $H_1 = (V, E_1)$ that contains H_0. Note that the problem can be regarded as a graph augmentation problem of finding a smallest subset $E' \subseteq E - E_0$ of edges to augment H_0 to a 2-edge-connected graph $H_1 = (V, E_1 = E_0 \cup E')$. The problem is shown to be NP-hard [6] even if $E_0 = \emptyset$. In the case of $E_0 = \emptyset$, the problem, which is called *the minimum 2-edge-connected spanning subgraph problem* (2-ECSS), has been extensively studied and several approximation algorithms are known [2,3,10]. The currently best approximation ratio for 2-ECSS is $\frac{17}{12}$ due to Cheriyan *et al.* [2]. On the other hand, if H_0 is connected, H_0 can be assumed to be a spanning tree of H without loss of generality (since every 2-edge-connected component in H_0 can be contracted into a single vertex without losing the property of the problem). Let us call the problem with a tree H_0 *the minimum 2-edge-connected subgraph problem containing a spanning tree* (2-ECST). In the special case of H being a complete graph, 2-ECST is the problem of augmenting a tree H_0 to a 2-edge-connected graph by adding a minimum number of new edges, for which Eswaran and Tarjan [5] presented a linear time algorithm (which

* This research was partially supported by the Scientific Grant-in-Aid from Ministry of Education, Science, Sports and Culture of Japan, and the subsidy from the Inamori Foundation.

T. Asano et al. (Eds.): COCOON'99, LNCS 1627, pp. 31–40, 1999.

creates no multiple edges). If H is a general graph, we are permitted to add to H_0 only edges from $E - E_0$. For general 2-ECST, there is a 2-approximation algorithm [6,9], which relays on the minimum branching algorithm. However, as remarked by Khuller [8, p.263], one of the main open problems in the graph augmentation problem is to obtain a factor better than 2 for 2-ECST. In this paper, we present a $\frac{12}{7}$-approximation algorithm for 2-ECST. Our algorithm is based on the maximum matching algorithm and a certain decomposition of a tree. Its running time is $O(\sqrt{n}m)$, where $n = |V|$ and $m = |E|$.

As pointed out in [2,9], the following augmentation problem can be reduced to 2-ECST: given a k-edge-connected graph $H_0 = (V, E_0)$ for an odd integer k and an edge set E with $E \cap E_0 = \emptyset$, find a smallest set $E' \subseteq E$ to augment H_0 to a $(k + 1)$-edge-connected graph. It is known that all k-edge-cuts in H_0 can be represented by a tree $T(H_0)$ if k is odd [4] (see [7] for efficient algorithms for constructing such trees). Thus, the problem can be viewed as the 2-ECST in which the tree $T(H_0)$ (which represents all k-edge-cuts in H_0) is augmented to a 2-edge-connected graph by adding a minimum number of edges from E (see [2,9] for the detail). Therefore, by applying our result, we can obtain a $\frac{12}{7}$-approximation algorithm for this problem.

2   Preliminaries

A singleton set $\{x\}$ may be simply written as x. For an undirected graph $H = (V, E)$ and an edge set E', we denote by $H + E'$ (resp., $H - E'$) the graph obtained from H by adding (resp., removing) edges in E'. For a subset $X \subseteq V$, let $\overline{X}$ denote $V - X$, and $H[X]$ denote the subgraph induced from H by X.

Let $G = (V, E)$ be an undirected graph, and $T = (V, F)$ a tree on the same vertex set V, where $E \cap F = \emptyset$ is assumed, but there possibly exits a pair of edges $e \in E$ and $e' \in F$ such that e and e' have the same end vertices. We denote by $E(u)$ the set of edges in E which are incident to a vertex $u \in V$. For two vertices $u, v \in V$, let $P_T(u, v)$ denote the path connecting u and v in T. We say that an edge $e = (u, v) \in E$ covers an edge $e' \in F$ if $P_T(u, v)$ contains e', and that an edge set $E' \subseteq E$ covers an edge set $F' \subseteq F$ if each edge in F' is covered by an edge in E'. Clearly, $T + E'$ is 2-edge-connected for a subset $E' \subseteq E$ if and only if E' covers F.

We choose an arbitrary vertex $r \in V$ as a root of T, which defines a parent-child relation among vertices in V on T. For a vertex $u \in V$, let $Ch(u)$ denote the set of children of u, and $D(u)$ denote the set of all descendents of u (including u). The subgraph $T[D(u)]$ induced from T by $D(u)$ is called the subtree at u (which is connected). A vertex u is called a *leaf vertex* if u has no child, and is called a *fringe vertex* if all the children of u are leaf vertices. For a vertex $u \in V$, let $LEAF(u)$ (resp., $FRINGE(v)$) denote the set of all leaf vertices (resp., fringe vertices) in the subtree $T[D(u)]$. In a rooted tree T, we denote an edge $e \in F$ with end vertices u and v by an ordered pair (u, v) so that $u \in Ch(v)$ holds. An edge $e = (u, v) \in F$ in T is called an *leaf edge* (resp., *fringe edge*) of u if u is a leaf vertex (resp., a fringe vertex). The subtree $T[D(u)]$ at a vertex

u is called a *leaf tree* if u is a fringe vertex. We call a leaf tree with exactly two leaf vertices *prime*. For an edge $e = (u, v) \in E$, we denote by $lca(e)$ the least common ancestor of end vertices u and v in the rooted tree T.

3   Lower bounds

In this section, we introduce some lower bounds on the number of edges required to cover all leaf (and fringe) edges in a subtree in T.

Lemma 1. *Let* $G = (V, E)$ *and* $T = (V, F)$ *be a graph and a tree with* $E \cap F = \emptyset$, *respectively, and* v *be a non-leaf vertex* v *in* T. *Then we need at least* $|LEAF(v)| - |M^*|$ *edges in* E *to cover all leaf edges in the subtree* $T[D(v)]$, *where* $M^* \subseteq E$ *is a maximum matching in the induced graph* $G[LEAF(v)]$.

Proof: Let $E' \subseteq E$ be an arbitrary subset that covers all leaf edges in $T[D(v)]$, and let E_{leaf} be the set of all edges $e = (u, u') \in E$ with $u, u' \in LEAF(v)$. We choose a maximal matching $M \subseteq E'$ in the graph $(LEAF(v), E' \cap E_{leaf})$. For each unmatched vertex $w \in LEAF(v)$, there must be an edge $e_w \in E(w) \cap E^*$ to cover the leaf edge of w. Note that $e_w \neq e_{w'}$ holds for distinct unmatched vertices w, w' by the maximality of M. Therefore, we obtain $|E'| \geq |M \cup \{e_w \mid$ unmatched vertex $w \in LEAF(v)\}| = |M| + (|LEAF(v) - 2|M|) = |LEAF(v)| - |M| \geq |LEAF(v)| - |M^*|$. $\square$

Lemma 2. *Let* $G = (V, E)$ *and* $T = (V, F)$ *be a graph and a tree with* $E \cap F = \emptyset$, *and* v *be a vertex in* T *which is neither a leaf nor fringe vertex. For the subtree* $T[D(v)]$, *let* E_{prime} *be the set of edges* $(w, w') \in E$ *such that* $\{w, w'\} = Ch(u)$ *for some prime leaf tree* $T[D(u)]$. *Then we need at least*

$$\beta_v = \frac{2}{3}|LEAF(v)| - \frac{1}{3}|M^*|$$

edges in E *to cover all leaf and fringe edges in* $T[D(v)]$, *where* $M^* \subseteq E$ *is a maximum matching in the graph* $G[LEAF(v)] - E_{prime}$.

Proof: Let $E' \subseteq E$ be an arbitrary subset that covers all leaf edges and fringe edges in $T[D(v)]$, and let E_{leaf} be the set of all edges $e = (u, u') \in E$ with $u, u' \in LEAF(v)$. We choose a maximal matching $M \subseteq E'$ in the graph $(LEAF(v), E' \cap (E_{leaf} - E_{prime}))$, and let W be the set of unmatched vertices in $LEAF(v)$. Thus, $|W| = |LEAF(v)| - 2|M|$. For each unmatched vertex $w \in W$, E' must contain an edge $e_w = (w, z) \in E' \cap E(w)$ to cover the leaf edge of w; we fix an edge e_w even if $|E' \cap E(w)| \geq 2$. By the maximality of M, $e_w = e_{w'}$ occurs for some distinct $w, w' \in W$ if and only if w and w' are the children of a prime leaf tree $T[D(u)]$ (i.e., $(w, w') \in E_{prime} \cap E'$). We call such a prime leaf tree $T[D(u)]$ a *dangerous tree*, and denote by L_d the set of leaf vertices of all dangerous trees (hence $|L_d|/2$ is the number of all dangerous trees). Thus, we have $E' - M \supseteq E_{prime} \cap E'$, where $|E_{prime} \cap E'| = |L_d|/2$. Also the number of edges e_w with $w \in W - L_d$ is $|\{e_w \mid w \in W - L_d\}| = |W| - |L_d|$. Now consider the fringe edge $(u, u') \in F$ of a dangerous tree $T[D(u)]$. Let $\{w, w'\} = Ch(u)$. Note that $(w, w') \in E'$ does not cover (u, u'). To cover the fringe edge (u, u'), $E' - \{(w', w'')\}$ must contain an edge (say, $e^{(u)}$) between $D(u) = \{u, w', w''\}$ and $\overline{D}(u)$. Clearly $e^{(u)} \notin M \cup E_{prime}$.

Case-a. $|W| - |L_d| \geq \frac{1}{2}|L_d|$: Hence $\frac{1}{2}|L_d| \leq \frac{1}{3}|W|$. Since all edges e_w with $w \in W - L_d$ are distinct, we obtain $E' \supseteq M \cup (E_{prime} \cap E') \cup \{e_w \mid w \in W - L_d\}$. Thus, $|E'| \geq |M| + \frac{1}{2}|L_d| + (|W| - |L_d|) \geq |M| + \frac{2}{3}|W|$.

Case-b. $|W| - |L_d| < \frac{1}{2}|L_d|$: Hence $\frac{1}{2}|L_d| > \frac{1}{3}|W|$. There must be at least $\frac{1}{2}|L_d| - (|W| - |L_d|) = \frac{3}{2}|L_d| - |W|$ dangerous trees $T[D(u)]$ such that no edge e_w with $w \in W - L_d$ is incident to $T[D(u)]$. Hence we need at least $\frac{1}{2}(\frac{3}{2}|L_d| - |W|) = \frac{3}{4}|L_d| - \frac{1}{2}|W|$ edges to cover the fringe edges of those remaining dangerous trees (where we divided $\frac{3}{2}|L_d| - |W|$ by 2 since an edge in E can cover at most two such fringe edges). Therefore, we obtain $|E'| \geq |M| + \frac{1}{2}|L_d| + (|W| - |L_d|) + (\frac{3}{4}|L_d| - \frac{1}{2}|W|) \geq |M| + \frac{2}{3}|W|$.

In both cases, we have $|E'| \geq |M| + \frac{2}{3}|W|$. By $|W| = |LEAF(v)| - 2|M|$ and $|M| \leq |M^*|$, it holds $|E'| \geq |M| + \frac{2}{3}|W| = \frac{2}{3}|LEAF(v)| - \frac{1}{3}|M| \geq \frac{2}{3}|LEAF(v)| - \frac{1}{3}|M^*|$. $\square$

4   Approximation Algorithm

4.1   Outline

Based on the lower bounds in the previous section, we design an approximation algorithm for 2-ECST. The main idea of the algorithm is as follows. We first find a subtree $T[D(v)]$ satisfying the following conditions (i) and (ii), where E_{lf} denotes the set of edges in E which are incident to a leaf or fringe vertex in $T[D(v)]$:

(i) every edge in E_{lf} has its end vertices both in $T[D(v)]$, and
(ii) E_{lf} covers all edges in $T[D(v)]$.

Next we try to choose a small subset $E'_{lf} \subseteq E_{lf}$ which covers all edges in such a subtree $T[D(v)]$ (this is possible by (ii)). Notice that no edge in $E - E_{lf}$ can cover a leaf or fringe edge in $T[D(v)]$; we have to cover all leaf and fringe edges in $T[D(v)]$ by using edges in E_{lf}. Hence by Lemma 2 we need at least β_v edges from E_{lf} just to cover all leaf and fringe edges in $T[D(v)]$. Thus we obtain a performance ratio $|E'_{lf}|/\beta_v$ of the size of our solution E'_{lf} to the size of a smallest subset $E^*_{lf} \subseteq E_{lf}$ that covers all leaf and fringe edges in $T[D(v)]$.

It should be noted that, by (i), E_{lf} can cover only edges in $T[D(v)]$, and hence any edge not in $T[D(v)]$ must be covered by an edge in $E - E_{lf}$. Now all edges in $T[D(v)]$ are covered by the current $E'_{lf} \subseteq E_{lf}$. To consider the remaining problem, we contract all vertices in $D(v)$ into a single vertex, which becomes a new leaf vertex, and consider to cover the edges in the resulting tree. Then we repeatedly apply this argument until the tree becomes a single vertex. (More precisely, to boost up the performance ratio in the above, we treat some other reducible cases in section 4.2.)

To find a subtree $T[D(v)]$ satisfying the above (i) and (ii), we introduce some more definitions. A subtree $T[D(u)]$ is called l-*closed* in G if G has no edge between $LEAF(u)$ and $\overline{D(u)}$. Furthermore, the subtree $T[D(u)]$ is called lf-*closed* if G has no edge between $LEAF(u) \cup FRINGE(u)$ and $\overline{D(u)}$. Clearly, $T = T[D(r)]$ is l-closed and lf-closed. A subtree $T[D(v)]$ is called *minimally lf-closed* if $T[D(v)]$ is lf-closed and there is no proper subtree $T[D(u)]$ of $T[D(v)]$ which is lf-closed. An lf-closed subtree $T[D(v)]$ satisfies the above condition (i), and it also satisfies the condition (ii) if $T[D(v)]$ is minimally lf-closed (since if there is an edge $(u'.v') \in F$ in $T[D(v)]$ which is not covered by E_{lf} of $T[D(v)]$, then $T[D(u')]$ would be lf-closed).

4.2   Some reducible leaf trees

Let $u \in V$ be a leaf vertex, and v be its parent in T. Vertex u is called *isolated* if u is not adjacent (via edges in $E(u)$) to any other child of v (hence u is isolated if $|Ch(v)| = 1$). Vertex u is called *trivial* if $E(u) = \{(u,v)\}$; we must use the unique edge in $E(u)$ to cover the leaf edge (u,v). Let $E(u) = \{e_1 = (u,v_1), e_2 = (u,v_2),\ldots,e_p = (u,v_p)\}$, where $p = |E(u)| \geq 2$. An edge $e_i = (u,v_i)$ with $v_i = v$ is called *redundant* if $E(u)$ contains some $e_j = (u,v_j)$ with $v_j \neq v$. If all edges in $E(u)$ are multiple edges of $(u.v)$, then we choose an arbitrary edge (say e_1) in $E(u)$ and call the other edges e_i, $i = 2,\ldots,p$ *redundant*. (Even if G is originally simple, our algorithm will repeat contracting some vertices into a single vertex, which may produce multiple edges in G.) It is not difficult to see that there is an optimal subset $E^* \subseteq E$ that covers F without using any redundant edge.

We consider the following three cases, in each of which we can reduce the problem size by contracting a vertex subset.

Case-1. There is an l-closed leaf tree $T[D(v)]$: In this case, no edge in E can cover both an edge e_1 in $T[D(v)]$ and an edge e_2 not in $T[D(v)]$. Then we can find a smallest set $E_v^* \subseteq E$ that covers all leaf edges in $T[D(v)]$ as follows. First compute a maximum matching M^* in the induced graph $G[Ch(v)]$, and let W be the set of unmatched vertices in $Ch(v)$. Next we choose an arbitrary edge $e_w \in E(w)$ for each unmatched vertex $w \in W$ (where $E(w) \neq \emptyset$ by the 2-edge-connectivity of $T + E$). Obviously $E_v^* = M^* \cup \{e_w \mid$ unmatched vertex $w \in Ch(v)\}$ covers all leaf edges in $T[D(v)]$, and $|E_v^*| = |M^*| + |Ch(v)| - 2|M^*| = |Ch(v)| - |M^*|$. By Lemma 1, $|E_v^*|$ is the minimum among all subsets of E that cover all leaf edges in $T[D(v)]$. We retain E_v^* as part of solution to cover the current T. We then contract all vertices in $Ch(v) \cup \{v\}$ into a single vertex v' both in T and G, and delete any resulting self-loops. The vertex v' becomes a new leaf vertex in the resulting T.

Case-2. There is a leaf tree $T[D(v)]$ such that $T[D(v)]$ is not l-closed and there is an isolated leaf vertex $u \in Ch(v)$ (this includes the case of $|Ch(v)| = 1$): Let I_v denote the set of all isolated vertices in $Ch(v)$. For each trivial leaf vertex $u \in I_v$ (if any), we contract u and v into a vertex both in T and G (retaining the edge in $E(u)$ as part of solution to cover the original T). For each non-trivial leaf vertex $u \in I_v$ (if any), we remove all redundant edges in $E(u)$ from G. Now if there remains an isolated vertex $u \in I_v$, then any edge in E covering the leaf edge (u, v) also covers the fringe edge (v, v') of v, because $E(u)$ contains no redundant edge. For this reason, we contract v and v' into a single vertex both in T and G, and delete any resulting self-loops.

Case-3. There is a leaf tree $T[D(v)]$ such that $T[D(v)]$ is not l-closed, $Ch(v)$ contains no isolated vertex and $|Ch(v)| = 3$: We first remove all redundant edges incident to $u \in Ch(v)$. If there is a vertex $u \in Ch(v)$ with $|E(u)| = 1$, then choose such a vertex u. Now the edge $e \in E(u)$ connects u and another child $u' \in Ch(v)$ (since u is not isolated). To cover the leaf edge (u, v), the edge $e = (u, u')$ must be used. Therefore, we retain the edge (u, u') as part of solution, and contract $\{u, u', v\}$ into a single vertex v both in T and G, deleting any resulting self-loops.

On the other hand, if $|E(u)| \geq 2$ holds for all $u \in Ch(v)$, then we claim that v and its parent v' can be contracted without loss of generality. Let $Ch(v) = \{u_1, u_2, u_3\}$. Consider an arbitrary subset $E' \subseteq E$ that covers all edges in T. Suppose that E' contains no edge between $Ch(v)$ and $\overline{D(v)}$. That is, all leaf edges in $T[D(v)]$ are covered by (at least) two edges $e_1 = (u_i, u_j), e_2 = (u_j, u_h) \in E'$. Since $T[D(v)]$ is not l-closed, E contains an edge e_0 between a vertex $u \in Ch(v)$ and $w \in \overline{D(v)}$. If there is such an edge $e_0 = (u_i, w)$ (resp., $e_0 = (u_h, w)$), then we easily see that $\tilde{E} = (E' - e_1) \cup \{e_0\}$ (resp., $\tilde{E} = (E' - e_2) \cup \{e_0\}$) covers all edges in T. If all such edges e_0 are incident to u_j, then by $|E(u_i)| \geq 2$, E contains an edge $e_3 = (u_i, u_h)$. In this case, $\tilde{E} = (E' - \{e_1, e_2\}) \cup \{e_0, e_3\}$ covers all edges in T. In any case, we can assume that at least one edge between $Ch(v)$ and $\overline{D(v)}$ is used in E'. For this reason, we contract v and v' into a single vertex both in T and G, deleting any resulting self-loops.

4.3   Covering minimally lf-closed subtrees

Let $T[D(v)]$ be a minimally lf-closed subtree in T (where we can assume that v is not a fringe vertex in T by Case-1 in the previous subsection). Such a $T[D(u)]$ always exits, since $T = T[D(r)]$ is lf-closed. Assume that none of Cases-1,2 and 3 holds for all leaf trees in $T[D(v)]$. Thus each fringe vertex u in $T[D(v)]$ satisfies that

(a) the leaf tree $T[D(u)]$ is not l-closed,
(b) $|Ch(u)| \neq 1, 3$, and
(c) the induced graph $G[Ch(u)]$ contains at least one edge in E.

Let E_{lf} be the set of edges in E which are incident to a leaf or fringe vertex in $T[D(v)]$ (then by the lf-closedness, the both end vertices of any edge in E_{lf} are contained in $T[D(v)]$). In this section, we consider how to choose edges from E_{lf} to cover *all* edges in $T[D(v)]$. Before describing a procedure for choosing such edges, some notations are introduced. Let F_{leaf} denote the set of leaf edges in $T[D(v)]$. Let E_{leaf} be the set of all edges $e = (u, u') \in E$ with $u, u' \in LEAF(v)$, and E_{prime} be the set of edges $(w, w') \in E$ such that $\{w, w'\} = Ch(u)$ for some prime leaf tree $T[D(u)]$. The following procedure COVER computes a subset $E_1 \subseteq E_{lf}$ that covers all edges in $T[D(v)]$, which is minimally lf-closed.

Procedure COVER

1. Find a maximum matching $M^* \subseteq E$ in the graph $G[LEAF(v)] - E_{prime} = (LEAF(v), E_{leaf} - E_{prime})$. Let W be the set of unmatched vertices in $LEAF(v)$. We call a prime leaf tree $T[D(u)]$ a *dangerous tree* if its both leaf vertices are unmatched (i.e., $Ch(u) \subseteq W$), and denote by L_d the set of leaf vertices of all dangerous trees (hence $|L_d|/2$ is the number of all dangerous trees).

2. To cover the leaf edges of unmatched leaf vertices, we choose edges from E as follows.

   For each vertex $w \in W - L_d$, we choose an arbitrary edge $e_w = (w, z) \in E(w)$ $(E(w) \neq \emptyset$ by the 2-edge-connectivity of $T + E)$. For each dangerous tree $T[D(u)]$, we choose the edge $e_u = (w, w') \in E_{prime}$ with $\{w, w'\} = Ch(u)$, and denote by M' the set of these edges e_u for all dangerous trees $T[D(u)]$ (hence $|M'| = |L_d|/2$). The edge set $\{e_w \mid w \in W - L_d\} \cup M'$ covers all leaf edges of unmatched leaf vertices in $T[D(v)]$. ($M^* \cup \{e_w \mid w \in W - L_d\} \cup M'$ covers all leaf edges in $T[D(v)]$.)

3. To cover the all non-leaf edges in $T[D(v)]$, we choose edges from E_{lf} as follows.

   Construct the graph $\mathcal{G} = (LEAF(v) \cup FRINGE(v), F_{leaf} \cup M^* \cup M')$ which consists of leaf edges and edges in $M^* \cup M'$. Denote by $C_1, \ldots, C_h$ the connected components of $\mathcal{G}$ (there may be a C_i which contains v if $Ch(v) \cap LEAF(v) \neq \emptyset$). Note that there are exactly $\frac{1}{2}|L_d|$ components C_j such that C_j contains exactly two leaf vertices (i.e., such C_j corresponds to a dangerous tree).

   For each connected component C_i not containing v in $\mathcal{G}$, consider the edges $e_{C_i} \in E$ which connect a vertex in C_i and a vertex not in C_i, and choose such edge e_{C_i} so that $lca(e_{C_i})$ is closest to the root r. (Let e_{C_i} be empty if C_i does not have such edge.)

4. Let $E_1 = M^* \cup \{e_w \mid w \in W - L_d\} \cup M' \cup \{e_{C_i} \mid i = 1, \ldots, h\}$ $(\subseteq E_{lf})$. □

Lemma 3. *Let $G = (V, E)$ and $T = (V, F)$ be a graph and a tree with $E \cap F = \emptyset$ such that $T + E = (V, F \cup E)$ is 2-edge-connected. Let $T[D(v)]$ be a minimally lf-closed subtree satisfying the above $(a) - (c)$. Then the subset $E_1 \subseteq E_{lf}$ obtained by the procedure COVER covers all edges in $T[D(v)]$. Moreover $|E_1|$ is at most $\frac{12}{7}$ times of the smallest number of edges in E_{lf} to cover all leaf and fringe edges in $T[D(v)]$.*

Proof: It is easy to see that E_1 covers all leaf edges in $T[D(v)]$. Assume that $T[D(v)]$ contains a non-leaf edge $e^* = (u^*, v^*) \in F$ which is not covered by E_1. By the 2-edge-connectivity of $T + E$, there must be at least one edge $\tilde{e} = (\tilde{u}, \tilde{v}) \in E$ between $D(u^*)$ and $\overline{D(u^*)}$. By the minimal lf-closedness of $T[D(v)]$, there exists such an edge $\tilde{e} = (\tilde{u}, \tilde{v}) \in E$ for which $\tilde{u}$ is a leaf vertex or fringe vertex in $T[D(u^*)]$. Let C_j be the connected component containing the $\tilde{u}$ among $C_1, C_2, \ldots, C_h$ constructed in Step 3 of COVER. For such C_j (which clearly does not contain v), the procedure COVER have chosen an edge e_{C_j} among all edges in E incident to C_j so that $lca(e_{C_j})$ is closest to the root r. However, $lca(\tilde{e})$ is closer to r than v (and hence than $lca(e_{C_j})$), a contradiction. Therefore, E_1 covers all edges in $T[D(v)]$.

Next we estimate the size of $E_1 = M^* \cup \{e_w \mid w \in W - L_d\} \cup M' \cup \{e_{C_i} \mid i = 1, \ldots, h\}$. For a maximum matching M^* and the set W of unmatched vertices in $LEAF(v)$, we have $|W| = |LEAF(v)| - 2|M^*|$. Clearly, $|M'| = |L_d|/2$ and $|\{e_w \mid w \in W - L_d\}| = |W| - |L_d|$ by construction. Therefore we have $|E_1| \leq |M^*| + (|W| - |L_d|) + \frac{1}{2}|L_d| + h = |M^*| + |W| - \frac{1}{2}|L_d| + h = |M^*| + (|LEAF(v)| - 2|M^*|) - \frac{1}{2}|L_d| + h = |LEAF(v)| - \frac{1}{2}|L_d| + h - |M^*|$.

Now we derive an upper bound on h, which was defined in Step 3 of COVER. Recall that there are exactly $\frac{1}{2}|L_d|$ components C_i corresponding to dangerous trees. Let us call such C_i corresponding to a dangerous tree *prime*. Any non-prime component C_j contains a leaf tree having at least four leaf vertices (by (b)) or a pair of leaf trees joined by a matching edge in M^*. In either case, C_j has at least *four* leaf vertices. Hence there are at most $\frac{1}{4}(|LEAF(v)| - |L_d|)$ non-prime connected components, and we have

$$h \leq \frac{1}{2}|L_d| + \frac{1}{4}(|LEAF(v)| - |L_d|) = \frac{1}{4}|LEAF(v)| + \frac{1}{4}|L_d|. \qquad (4.1)$$

Moreover any non-prime C_j contains at least one matching edge in M^* (note that any leaf tree $T[D(u)]$ with at least four leaf vertices has an edge in $G[Ch(u)]$ by (c)). This means that the number of non-prime components is at most $|M^*|$, i.e.,

$$h - \frac{1}{2}|L_d| \leq |M^*|. \qquad (4.2)$$

To derive the approximation ratio, note that, by Lemma 2, at least $\beta_v = \frac{2}{3}|LEAF(v)| - \frac{1}{3}|M^*|$ edges in E are necessary to cover all leaf edges and fringe edges in $T[D(v)]$. Therefore

$$\frac{|E_1|}{\beta_v} = \frac{|LEAF(v)| - \frac{1}{2}|L_d| + h - |M^*|}{\frac{1}{3}(2|LEAF(v)| - |M^*|)} \leq \frac{|LEAF(v)| - \frac{1}{2}|L_d| + h - (h - \frac{1}{2}|L_d|)}{\frac{1}{3}(2|LEAF(v)| - (h - \frac{1}{2}|L_d|))}$$

$$\text{(by (4.2) and since } |LEAF(v)| - \frac{1}{2}|L_d| + h < 2|LEAF(v)| \text{ holds)}$$

$$\leq \frac{|LEAF(v)|}{\frac{1}{3}(2|LEAF(v)| - (\frac{1}{4}|LEAF(v)| + \frac{1}{4}|L_d|) + \frac{1}{2}|L_d|)} \quad \text{(by (4.1))}$$

$$= \frac{|LEAF(v)|}{\frac{7}{12}|LEAF(v)| + \frac{1}{12}|L_d|} \leq \frac{12}{7}.$$

□

4.4   Entire description

We are now ready to describe the entire algorithm. For a graph $H = (V, E')$ and a subset $X \subseteq V$, we denote by H/X the graph obtained from H by contracting X into a single vertex, and delete all the resulting self-loops.

APPROX

Input: A graph $G = (V, E)$ and a tree $T = (V, F)$ with $E \cap F = \emptyset$
such that $T + F = (V, F \cup E)$ is 2-edge-connected.

Output: A subset $E' \subseteq E$ that covers F.

   /* $\{F_v\}$ are computed only for the performance analysis described below. */
    $E' := \emptyset$;
    while T contains more than one vertex **do**
        while there is a leaf tree $T[D(v)]$ satisfying one of Cases-1,2 and 3 **do**
          Choose such a leaf tree $T[D(v)]$;
          if Case-1 holds for the $T[D(v)]$ **then**
            Compute a maximum matching M^* in $G[Ch(v)]$,
            and choose an edge $e_w \in E(w)$ for each unmatched vertex $w \in Ch(v)$;
            $F_v := \{\text{all leaf edges in } T[D(v)]\}$;
            $E_v := M^* \cup \{e_w \mid \text{unmatched vertex } w \in Ch(v)\}$; $E' := E' \cup E_v$;
            For $X = Ch(v) \cup \{v\}$, let $T := T/X$ and $G := G/X$
          else if Case-2 holds for the $T[D(v)]$ **then**
            Let I_v be the set of all isolated vertices in $Ch(v)$, and
            I_v^* be the set of trivial leaf vertices $u \in I_v$;
            if $I_v^* \neq \emptyset$ **then**
              $F_v := \{\text{the leaf edges } (u, v) \text{ for all } u \in I_v^*\}$;
              $E_v := \cup_{u \in I_v^*} E(u)$; $E' := E' \cup E_v$;
              For $X = I_v^* \cup \{v\}$, let $T := T/X$ and $G := G/X$;
            if $I_v - I_v^* \neq \emptyset$ **then**
              Remove redundant edges in $E(u)$ for all $u \in I_v - I_v^*$;
              For the fringe edge (v, v') of v, let $T := T/\{v, v'\}$ and $G := G/\{v, v'\}$
          else if Case-3 holds for the $T[D(v)]$ **then**
            Remove redundant edges edges in $E(u)$ for all $u \in Ch(v)$;
            if there is a vertex $u \in Ch(v)$ with $|E(u)| = 1$ **then**
              Choose such a vertex u and a vertex $u' \in Ch(v) - u$ adjacent to u;
              $F_v := \{(u, v)\}$; $E_v := \{(u, u')\}$; $E' := E' \cup E_v$;
              For $X = \{u, u', v\}$, let $T := T/X$ and $G := G/X$
            else /* $|E(u)| \geq 2$ holds for all $u \in Ch(v)$. */
              For the fringe edge (v, v') of v, let $T := T/\{v, v'\}$ and $G := G/\{v, v'\}$;
         end /* while */
        /* Conditions (a)-(c) hold for all leaf trees. */
        Choose a minimally lf-closed subtree $T[D(v)]$;
        Compute an edge set $E_1 \subseteq E$ by procedure COVER;
        $F_v := \{\text{all leaf and fringe edges in } T[D(v)]\}$; $E_v := E_1$; $E' := E' \cup E_v$;
        For $X = D(v)$, let $T := T/X$ and $G := G/X$;
      end /* while */
      Output E'. □

    Let $v_1, v_2, \ldots, v_p$ be the sequence of all vertices v such that its subtree $T[D(v)]$ has been chosen by APPROX as a leaf tree satisfying one of Cases-1,2 and 3 or a minimally lf-closed subtrees. For each v_i, F_v, is a subset of $F - \bigcup_{1 \leq j < i} F_{v_j}$. Let E^* be a minimum subset of E that covers F, and let $E_{v_i}^*$ be a minimum subset of E that covers F_{v_i}. By the choice of $F_{v_1}, F_{v_2}, \ldots, F_{v_p}$, we observe that no edge in E can cover two edges $e \in F_{v_i}$ and $e' \in F_{v_j}$ for distinct i, j. From this, it holds $|E^*| \geq |E^*_{v_1}| + |E^*_{v_2}| + \cdots + |E^*_{v_p}|$. For each v_i, E_{v_i} is a subset of $E - \bigcup_{1 \leq j < i} E_{v_j}$ and covers F_{v_i}. The output E' is then given by $E' = E_{v_1} \cup E_{v_2} \cup \cdots \cup E_{v_p}$. If $T[D(v_i)]$ has been chosen as a leaf tree, it holds $|E_{v_i}| = |E^*_{v_i}|$, because this is clear for Cases-2 and 3, and is valid for Case-1 by Lemma 1. On the other hand, if $T[D(v_i)]$ has been chosen as a minimally lf-closed tree, we have $|E_{v_i}| \leq \frac{12}{7}|E^*_{v_i}|$ by Lemma 2. Therefore, in total, we obtain the desired bound $|E'| \leq \frac{12}{7}|E^*|$.

It is not difficult to see that APPROX can be implemented to run in $O(\sqrt{n}m)$ time, where $n = |V|$ and $m = |E| + |F|$, based on the $O(\sqrt{n}m)$ time maximum matching algorithm [11] (the detail is omitted).

Theorem 1. *Given a graph $G = (V, E)$ and a tree $T = (V, F)$ with $E \cap F = \emptyset$ such that $T + E = (V, F \cup E)$ is 2-edge-connected, the problem of finding a smallest 2-edge-connected spanning subgraph $H = (V, F \cup E')$ containing T is $\frac{12}{7}$-approximable in $O(\sqrt{n}m)$ time, where $n = |V|$ and $m = |E \cup F|$.* □

References

1. J. Cheriyan, T. Jordán and R. Ravi: "On 2-coverings and 2-packings in laminar families," 7th Annual European Symposium on Algorithms, July 16-18, 1999, Prague, Czech Republic (1999) (to appear).

2. J. Cheriyan, A. Sebö and Z. Szigeti: "An improved approximation algorithm for minimum size 2-edge connected spanning subgraphs," *Lecture Notes in Computer Science*, 1412, Springer-Verlag, IPCO'98 (1998) 126-136.

3. J. Cheriyan and R. Thurimella: "Approximating minimum-size k-connected spanning subgraphs via matching," *Proc. 37th IEEE Symp. on Found. Comp. Sci.* (1996) 292-301.

4. E. A. Dinits, A. V. Karzanov and M. V. Lomonosov: "On the structure of a family of minimal weighted cuts in a graph," *Studies in Discrete Optimization* (in Russian) A.A. Fridman (Ed.) Nauka, Moscow (1976) 290-306.

5. K. P. Eswaran and R. E. Tarjan: "Augmentation problems," *SIAM J. Computing* 5 (1976) 653-665.

6. G. N. Frederickson and J. JáJá: "On the relationship between the biconnectivity augmentation problems," *SIAM J. Computing* 10 (1981) 270-283.

7. H. N. Gabow: "Applications of a poset representation to edge connectivity and graph rigidity," *Proc. 32nd IEEE Symp. on Found. Comp. Sci.* (1991) 812-821.

8. S. Khuller: "Approximation algorithms for finding highly connected subgraphs," in *Approximation Algorithms*, D. Hochbaum (Eds.), PWS publishing company (1997) 236-265.

9. S. Khuller and R. Thurimella: "Approximation algorithms for graph augmentation," *Proc. 19th International Colloquium on Automata, Languages and Programming Conference* (1992) 330-341.

10. S. Khuller and U. Vishkin: "Biconnectivity approximations and graph carvings," *J. ACM*, 41 (1994) 214-235.

11. S. Micali and V. V. Vazirani: "An $O(\sqrt{|V|}|E|)$ algorithm for finding maximum matching in general graph," *Proc. 21st IEEE Symp. on Found. Comp. Sci.* (1980) 17-27.

Theory of 2-3 Heaps

Tadao Takaoka

Department of Computer Science, University of Canterbury
Christchurch, New Zealand
tad@cosc.canterbury.ac.nz

Abstract. As an alternative to the Fibonacci heap, we design a new data structure called a 2-3 heap, which supports m decrease-key and insert operations, and n delete-min operations in $O(m + n \log n)$ time. The merit of the 2-3 heap is that it is conceptually simpler and easier to implement. The new data structure will have a wide application in graph algorithms.

1 Introduction

Since Fredman and Tarjan [6] published Fibonacci heaps in 1987, there has not been an alternative that can support n delete-min operations, and m decrease-key and insert operations in $O(m + n \log n)$ time. Logarithm here is with base 2, unless otherwise specified. Two representative application areas for these operations will be the single source shortest path problem and the minimum cost spanning tree problem. Direct use of these operations in Dijkstra's [5] and Prim's [7] algorithms with a Fibonacci heap will solve these two problems in $O(m + n \log n)$ time. A Fibonacci heap is a generalization of a binomial queue invented by Vuillemin [8]. When a key of a node v is decreased, the subtree rooted at v is removed and linked to another tree at the root level. If we perform this operation many times, the shape of a tree may become shallow, that is, the number of children from a node may become too many due to linkings without adjustment. If this happens at the root level, we will face a difficulty, when the node with the minimum is deleted and we need to find the next minimum. To prevent this situation, they allow loss of at most one child from any node. If one more loss is required, it will cause what is called cascading cut. This tolerance bound will prevent the degree of any node in the heap from getting more than $1.44 \log n$. The constant is the golden ratio derived from the Fibonacci sequence. Since this property will keep the degree in some bound, let us call this "horizontal balancing".

In the area of binary search trees, there are two well-known balanced tree schemes: the AVL tree [1] and the 2-3 tree [2]. When we insert or delete items into or from a binary search tree, we may lose the balance of the tree. To prevent this situation, we restore the balance by modifying the shape of the tree. As we control the path lengths, we can view this adjustment as vertical balancing. The AVL tree can maintain the tree height to be $1.44 \log n$ whereas the 2-3 tree will keep this to be $\log n$. As an alternative to the Fibonacci heap, we propose a

T. Asano et al. (Eds.): COCOON'99, LNCS 1627, pp. 41–50, 1999.

new data structure called a 2-3 heap, the idea of which is borrowed from the
2-3 tree and conceptually simpler than the Fibonacci heap. The degree of any
node in the 2-3 heap is bounded by $\log n$, better than the Fibonacci heap by a
constant factor. While the Fibonacci heap is based on binary linking, we base
our 2-3 heap on ternary linking; we link three roots of three trees in increasing
order according to the key values. We call this path of three nodes a trunk.
We allow a trunk to shrink by one. If there is requirement of further shrink,
we make adjustment by moving one or two subtrees from nearby positions. This
adjustment may propagate, prompting the need for amortized analysis. The word
"potential" used in [6] is misleading. It counts the number of nodes that lost one
child. This number reflects some deficiency of the tree, not potential. Thus we
use the word "deficit" for amortized analysis. We mark the trunk if it already
lost one node and its corresponding tree. We define the deficit of the 2-3 heap to
be the number of marked trunks. Amortized time for one decease-key or insert
is shown to be $O(1)$, and that for delete-min to be $O(\log n)$.

The concept of r-ary linking is similar to the product of graphs. When we
make the product of $G \times H$ of graphs G and H, we substitute H for every
vertex in G and connect corresponding vertices of H if an edge exists in G. See
Bondy and Murty [4], for example, for the definition. In the product of trees,
only corresponding roots are connected. The 2-3 heap is constructed by ternary
linking of trees repeatedly, that is, repeating the process of making the product
of a linear tree and a tree of lower dimension. This general description of r-ary
trees is given in Section 2. The precise definition of 2-3 heaps is given in Section
3. The description of operations on 2-3 heaps is given in Section 4. We also
give amortized analysis of those operations. In Section 5, we consider several
problems in implementation, and also some practical considerations for further
speed up. Section 6 concludes this report. Note that our computational model
is comparison-based. If we can use special properties of key values, there are
efficient data structures, such as Radix-heaps [3].

2   Polynomial of trees

We define algebraic operations on trees. We deal with rooted trees in the following. A tree consists of nodes and branches, each branch connecting two nodes.
The root of tree T is denoted by $root(T)$. A linear tree of size r is a liner list of r
nodes such that its first element is regarded as the root and a branch exists from
a node to the next. The linear tree of size r is expressed by bold face $\mathbf{r}$. Thus a
single node is denoted by $\mathbf{1}$, which is an identity in our tree algebra. The emplty
tree is nenoted by $\mathbf{0}$, which serves as the zero element. A product of two trees
S and T, $P = ST$, is defined in such a way that every node of S is replaced by
T and every branch in S connecting two nodes u and v now connects the roots
of the trees substituted for u and v in S. Note that $\mathbf{2} * \mathbf{2} \neq \mathbf{4}$, for example, and
also that $ST \neq TS$ in general. The symbol " $*$ " is used to avoid ambiguity.

The number of children of node v is called the degree of v and denoted by
$deg(v)$. The degree of tree T, $deg(T)$, is defined by $deg(root(T))$. A sum of two

trees S and T, denoted by $S + T$, is just the collection of two trees S and T if $deg(T) \neq deg(S)$. A polynomial of trees is defined next. Since the operation of product is associative, we use the notation of $\mathbf{r}^i$ for the products of i $\mathbf{r}$'s. An r-ary polynomial of trees of degree $k - 1$, P, is defined by

$$P = \mathbf{a}_{k-1}\mathbf{r}^{k-1} + ... + \mathbf{a}_1\mathbf{r} + \mathbf{a}_0 \ (1)$$

where $\mathbf{a}_i$ is a linear tree of size a_i and called a coefficient in the polynomial. Let $|P|$ be the number of nodes in P and $|\mathbf{a}_i| = a_i$. Then we have $|P| = a_{k-1}r^{k-1} + ... + a_1r + a_0$. We choose a_i to be $0 \leq a_i \leq r - 1$, so that n nodes can be expressed by the above polynomial of trees uniquely, as the k digit radix-r expression of n is unique with $k = \lceil \log_r(n + 1) \rceil$. The term $\mathbf{a}_i\mathbf{r}_i$ is called the i-th term. We call $\mathbf{r}^i$ the complete tree of degree i. Let the operation "$\bullet$" be defined by the tree $L = S \bullet T$ for trees S and T. The tree L is made by linking S and T in such a way that $root(T)$ is connected as a child of $root(S)$. Then the product $\mathbf{r}^i = \mathbf{r}\mathbf{r}^{i-1}$ is expressed by

$$\mathbf{r}^i = \mathbf{r}^{i-1} \bullet ... \bullet \mathbf{r}^{i-1} \ (r \text{ times}) \ (2)$$

The whole operation in (2) is to link r trees, called an i-th r-ary linking. The path of length $r - 1$ created by the r-ary linking is called the i-th trunk of the tree $\mathbf{r}^i$, which defines the i-th dimension of the tree in a geometrical sense. The j-th $\mathbf{r}^{i-1}$ in (2) is called the j-th subtree on the trunk. The path created by linking a_i trees of $\mathbf{r}_i$ in form (1) is called the main trunk of the tree corresponding to this term. A polynomial of trees is regarded as a collection of trees of distinct degrees, which are formed by repeating the linking process. We next define a polynomial queue. An r-nomial queue is an r-ary polynomial of trees with a label $label(v)$ attached to each node v such that if u is a parent of v, $label(u) \leq label(v)$. A binomial queue is a 2-nomial queue.

Example 1. A polynomial queue with an underlying polynomial of trees $P = 2 * 3^2 + 2 * 3 + 2$ is given below.

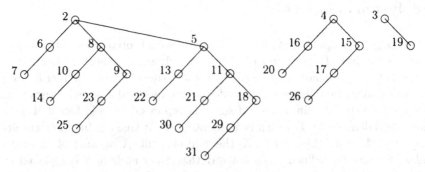

Fig. 1. Polynomial of trees with $r = 3$

Each term $\mathbf{a}_i\mathbf{r}_i$ in form (1) is a tree of degree $i + 1$ if $a_i > 1$. One additional degree is caused by the coefficient. The merging of two linear trees $\mathbf{r}$ and $\mathbf{s}$ is to merge the two lists by their labels. The result is denoted by the sum $\mathbf{r} + \mathbf{s}$. The merging of two terms $\mathbf{a}_i\mathbf{r}_i$ and $\mathbf{a}'_i\mathbf{r}_i$ is to merge the main trunks of the two trees by their labels. When the roots are merged, the trees underneath are moved accordingly. If $a_i + a'_i < r$, we have the merged tree with coefficient $\mathbf{a}_i + \mathbf{a}'_i$. Otherwise we have a carry tree $\mathbf{r}^{i+1}$ and the remaining tree with the main trunk of length $a_i + a'_i - r$. The sum of two polynomial queues P and Q is made by merging two polynomial queues in a very similar way to the addition of two radix-r numbers. We start from the 0-th term. Two i-th terms from both queues are merged, causing a possible carry to the $(i + 1)$-th terms. Then we proceed to the $(i + 1)$-th terms with the possible carry.

An insertion of a key into a polynomial queue is to merge a single node with the label of the key into the 0-th term, taking $O(r \log_r n)$ time for possible propagation of carries to higher terms. Thus n insertions will form a polynomial queue P in $O(nr \log_r n)$ time. The value of k in form (1) is $O(\log_r n)$ when we have n nodes in the polynomial of trees. We can take n successive minima from the queue by deleting the minimum in some tree T, adding the resulting polynomial queue Q to $P - T$, and repeating this process. This will sort the n numbers in $O(nr \log_r n)$ time after the queue is made. Thus the total time for sorting is $O(nr \log_r n)$. In the sorting process, we do not change key values. If the labels are updated frequently, however, this structure of polynomial queue is not flexible, prompting the need for a more flexible structure in the next section.

3   2-3 heaps

We linked r trees in form (2). We relax this condition in such a way that the number of trees linked is from l to r. Specifically an (l, r)-tree $T(i)$ of degree i is formed recursively by

$$T(0) = \text{a single node}$$
$$T(i) = T_1(i - 1) \bullet ... \bullet T_m(i - 1) \ (m \text{ is between } l \text{ and } r) \ (3)$$

Note that m varies for different linkings. The subscripts of $T(i)$ are given to indicate different trees of degree $i - 1$ are used. Note that the shape of an (l, r)-tree $T(i)$ is not unique for n such that $|T(i)| = n$. We say $root(T_1(i - 1))$ is the head node of this trunk. The dimension of non-head nodes is $i - 1$ and that of the head node is i or higher, depending on further linkings. We omit a subscript for $T(i)$; $T(i)$ sits for an (l, r)-tree of degree i. In this context, we sometimes refer to $T(i)$ as a type of tree. Then an extended polynomial of trees, P, is defined by

$$P = \mathbf{a}_{k-1}T(k - 1) + ... + \mathbf{a}_1T(1) + \mathbf{a}_0 \ (4)$$

The main trunk and the i-th trunk in each term of (4) are similarly defined to those for (1). So is the j-th subtree in the trunk. Let us assign labels with nodes in the tree in a similar way to the last section. That is, when the m-ary linking is made, the roots are connected in non-decreasing order of labels. Then the resulting polynomial is called an (l, r)-heap. When $l = 2$ and $r = 3$, we call it a 2-3 heap. We refer to the trees in (4) and their roots as those at the top level, as we often deal with subtrees at lower levels and need to distinguish them from the top level trees. The sum $P + Q$ of the two 2-3 heaps P and Q is defined similarly to that for polynomial queues. Note that those sum operations involve computational process based on merging.

Example 2. $P = 2T(2) + 2T(1) + 2T(0)$. A trunk is marked thick if it has only two nodes.

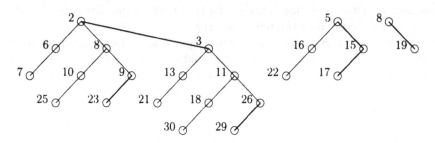

Fig. 2. A 2-3 tree

Lemma 1. *The number of nodes in 2-3 heap P in form (3), $|P|$, satisfies that* $2^k - 1 \leq |P| \leq 3^k - 1$.

¿From this we see $k = O(\log n)$. We informally describe delete-min, insertion, and decrease-key operations for a 2-3 heap. Precise definitions will be given in the next section. Let the dimension of node v be i, that is, the trunk of the highest dimension on which node v stands is the i-th. Suppose v is not at the top level. Then v is not the head node of the i-th trunk, which connects 2 or 3 trees of type $T(i-1)$. Let $tree(v)$ be one of these trees rooted at v. We define the work space for v to be a collection of nodes on the i-th trunk of v, the $(i+1)$-th trunk of the head node of v, and the other i-th trunks whose head nodes are on the $(i+1)$-th trunk. The work space has 4 to 9 nodes. Let the head node of the work space be the node at the first position of the $(i+1)$-th trunk. In this paper the words "key" and "label" are used with the same meaning.

Example 3. Let us denote the node with label x by $node(x)$ in Example 2. The work space for $node(18)$ is $\{node(3), node(13), node(21), node(11), node(18), node(30), node(26), node(29)\}$. The head node of this work space is $node(3)$. The work space for $node(9)$ is in higher dimensions and given by $\{node(2), node(8), node(9), node(3), node(11), node(26)\}$. The head node is $node(2)$.

Delete-min: Find the minimum by scanning the roots of the trees. Let $T(i)$ have the minimum at the root and Q be the polynomial resulting from $T(i)$ by removing the root. Then merge $P - T(i)$ and Q, i.e., $(P - T(i)) + Q$.
Marking of a trunk: A trunk is marked or unmarked if it has two or three nodes respectively.

Insertion: Perform $T + v$, where we insert v with its key into the 2-3 tree T. Here v is regarded as a 2-3 tree with only a term of type $T(0)$.

Removal of a tree: We do not remove $tree(v)$ for a node v if it is a root at the top level. Suppose we remove $tree(v)$ of type $T(i-1)$ for a node v. Consider the two cases where the size of the work space is greater than 4, and it is equal to 4. We start with the first case. Let the i-th trunk of node v be (u, v, w) or (u, w, v). Then remove $tree(v)$ and shrink the trunk. If the trunk is (u, v), remove $tree(v)$. And we would lose this trunk. To prevent this loss, we move one or two trees of type $T(i-1)$ from the work space.

Let the size of the work space be 4. Remove $tree(v)$ and rearrange the remaining three nodes in such a way that two will come under the head node of the work space to form the i-th trunk. Then we recover the i-th trunk to be of length 2, but lose the $(i+1)$-th trunk. Our work proceeds to a higher dimension, if it exists. Otherwise stop. Note that during the above modifications of the heap, we need to mark and unmark trunks, and need to compare key values of nodes to find out the correct positions for placing trees.

Decrease-key: Suppose the key of node v has been decreased. If v is not at the top level, remove $tree(v)$, and insert it to the j-th term at the top level with the new key, where $j = deg(tree(v))$. If v is at the top level, do nothing after you decrease the key.

Example 4. In Example 2, we name the node with label x by $node(x)$. Suppose we decrease labels 6 to 4, and 29 to 14. Then we have the following $T = 1T(3) + 1T(0)$.

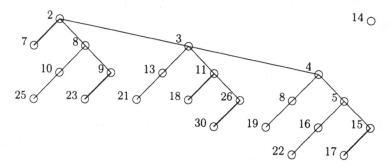

Fig. 3. The 2-3 tree after two decrease-key operations

We first remove $node(6)$, causing the move of $node(7)$ to the child position of $node(2)$ and the marking of the trunk between $node(2)$ and $node(7)$. The new node $node(4)$ is inserted into $2T(0)$, resulting in $T(1)$ with $node(8)$ and $node(19)$ and carrying over to $2T(1)$ to cause insertion. Then the newly formed $T(2)$ will be carried to $2T(2)$, resulting in $T(3)$. For the second decrease-key, we have a new node $node(14)$. Since the trunk is already marked, $node(30)$ is moved to the child position of $node(26)$, and the trunks between $node(11)$ and $node(18)$, and $node(26)$ and $node(30)$ are marked.

4   Analysis of 2-3 trees

For removal of a tree, let us classify the situation using parts of the tree structure. In the following figures, only work space is shown. Each node can be regarded as a tree of the same degree for general considerations. The left-hand side and the right-hand side are before and after conversion. The trunk going left-down is a trunk of i-th dimension (i-th trunk for short), and that going right-down is a trunk of $(i + 1)$-th dimension. We have two or three trunks of i-th dimension in the following figures. By removing a node and shrinking a trunk, we create a vacant position, which may or may not be adjusted. We classify the situation into several cases depending on the size w of the work space. In the following figures the i-th trunks are arranged in non-decreasing order of lengths for simplicity. Other cases are similar.

Case 1. $w = 9$. The removal of any of the six black nodes will bring the same new shape by adjusting the heap within the work space. We increase the number of marked trunks by one and spend no comparison. The case for $w = 8$ is similar, except for the case where v on a marked trunk is removed. In this case, we can rearrange the heap within the work space with one comparison and increase the marked trunks by one. See Fig, 4.

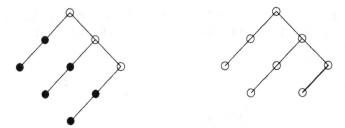

Fig. 4. The case of $w = 9$

Case 2. $w = 7$. We increase marked trunks by one and spend no comparison. See Fig. 5.

Fig. 5. The case of $w = 7$

Case 3. $w = 7$ We decrease marked trunks by one and spend at most one comparison. See Fig. 6.

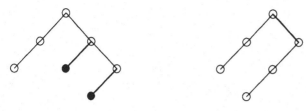

Fig. 6. Another case of $w = 7$

Case 4. $w = 6$. We spend at most one comparison and decrease marked trunks by one. See Fig. 7.

Fig. 7. The case of $w = 6$

Case 5. $w = 6$. We increase the number of marked trunks by one and spend no comparison. See Fig. 8.

Fig. 8. Another case of $w = 6$

Case 6. $w = 5$. We increase the number of marked trunks by one and spend at most one comparison. See Fig. 9.

Case 7. $w = 4$. The trunks of degree i and degree $i + 1$ are both marked. We make an unmarked i-th trunk for the head node and the situation becomes the loss of a node at the $(i + 1)$-th trunk. Make-up will be made in the work space defined by the $(i + 1)$-th trunk and the $(i + 2)$-th trunk. The make-up process stops in cases 1-6, or if no higher trunk exists. The $(i + 2)$-th trunk is drawn by a dotted line, and the by-gone $(i + 1)$-th trunk is drawn by a broken line at the right-hand side. We lose three marked trunks and spend at most one comparison. See Fig. 10.

Next we do analysis of top level insertions. Suppose we insert a tree of type $T(i)$ into the term $\mathbf{a}_i T(i)$ at the top level. We have three cases.

Fig. 9. Another case of $w = 5$

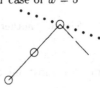

Fig. 10. The case of $w = 4$

Case A. $\mathbf{a}_i = \mathbf{0}$. We simply put the tree in the correct position.

Case B. $\mathbf{a}_i = \mathbf{1}$. We can form a new $2T(i)$ with one comparison and mark the new main trunk.

Case C. $\mathbf{a}_i = \mathbf{2}$. We make $T(i+1)$ with two comparisons and decrease marked trunks by one. Then proceed to the insertion at $a_{i+1}T(i+1)$.

We analyze the computing time by the number of comparisons between key values, based on amortized analysis. Other times are proportional to it. Define the deficit of a 2-3 heap by the number of marked trunks times 2. We regard the deficit value as the debt of comparisons. Suppose we perform n delete-min operations and m decrease-key and insert operations. The actual time and amortized time for the i-th operation are denoted by t_i and a_i, where $a_i = t_i + F_i - F_{i-1}$, and F_i is the deficit after the i-th operation. Let the total number of operation be N. Since $\Sigma a_i = \Sigma t_i + F_N - F_0$, and the deficits are assumed to be zero at the beginning and end, the total actual time is equal to the total amortized time.

Amortized time for delete-min: It takes $O(\log n)$ time to find the minimum and break apart the subtrees under the root with the minimum. The merging process for those subtrees at the top level takes $O(\log n)$ time. This process increases the deficit by $O(\log n)$. Thus one delete-min operation's amortized time is $O(\log n)$. The worst-case time is also $O(\log n)$.

Amortized time for decrease-key: It takes $O(1)$ time for decreasing the value of the key. After that, we perform various operations described above, all of which are bounded by 3, meaning that one decrease-key or insert operation's amortized time is $O(1)$. Note that the amortized times for case 7 and case C above are not greater than 0.

Amortized time for insertion: We insert a new node to the 0-th term of the top level. Thus the amortized time is $O(1)$.

If we perform n delete-min operations, and m decrease-key and insertion operations, the total time is now bounded by $O(m + n \log n)$. Thus we can solve the single source shortest path problem and the minimum cost spanning tree problem in $O(m + n \log n)$ time in a similar setting to [6].

5   Practical considerations

For the data structure of a 2-3 heap, we need to implement it by pointers. For the top level trees, we prepare an array of pointers of size $d = \lceil \log(n + 1) \rceil$, each of which points to the tree of the corresponding degree. Let a node v's highest trunk be the i-th. The data structure for node v consists of integer variables *key* and $dim = i$, a pointer to the head node of the i-th trunk, and an array of size d, whose elements are pairs $(second, third)$. The *second* of the j-th element of the array points to the root of the second node on the j-th trunk of v. The *third* is to the third node. It also has a Boolean array of size d. The j-th element shows whether the j-th trunk is marked or not. With this data structure, we can explore the work space for v. If we prepare the fixed size d for all the nodes, we would need $O(n \log n)$ space. By implementing arrays by pointers, we can implement our algorithm with $O(n)$ space, as is done in the Fibonacci heap, althogh this version will be less time-efficient.

6   Concluding remarks

We developed an alternative to the Fibonacci heap. If we measure the complexity of n delete-min's, and m decrease-key's and insert's by the number of comparisons, and showed it to be $O(m + n \log n)$. We will need to analyze the constant factors in this complexity and those of the Fibonacci heap to determine which of the 2-3 heap and Fibonacci heap is more efficient. The method for modifying the heap when a node on a marked trunk is removed is not unique. The method in Section 4 is designed to make as many unmarked trunks as possible to keep the packing ratio of the 2-3 heap better. To say this is the best method, we will need further study. The author expresses thanks to the referees who read the manuscript carefully and suggested many improvements.

References

1. Adel'son-Vel'skii, G.M, and Y.M. Landis, An algorithm for the organization of information, Soviet Math. Dokl. 3 (1962) 1259-1262.
2. Aho, A.V., J.E. Hopcroft, and J.D. Ullman, The Design and Analysis of Computer Algorithms, Addison-Wesley (1974).
3. Ahuja, K., K. Melhorn, J.B. Orlin, and R.E. Tarjan, Faster algorithms for the shortest path problem, Jour. ACM, 37 (1990) 213-223.
4. Bondy, J.A. and U.S.R. Murty, Graph Theory with Applications, Macmillan Press (1976).
5. Dijkstra, E.W., A note on two problems in connexion with graphs, Numer. Math. 1 (1959) 269-271.
6. Fredman, M.L. and R,E, Tarjan, Fibonacci heaps and their uses in inproved network optimization algorithms, Jour. ACM 34 (1987) 596-615
7. Prim, R.C., Shortest connection networks and some generalizations, Bell Sys. Tech. Jour. 36 (1957) 1389-1401.
8. Vuillemin, J., A data structure for manipulating priority queues, Comm. ACM 21 (1978) 309-314.

An External Memory Data Structure for Shortest Path Queries (Extended Abstract)[*]

David Hutchinson[1,**], Anil Maheshwari[1,**], and Norbert Zeh[1,2,***]

[1] School of Computer Science, Carleton University, Ottawa, Canada
[2] Fakultät für Math. und Inf., Friedrich-Schiller-Universität Jena, Germany
{hutchins,maheshwa,nzeh}@scs.carleton.ca

Abstract. We present results related to satisfying shortest path queries on a planar graph stored in external memory. In particular, we show how to store rooted trees in external memory so that bottom-up paths can be traversed I/O-efficiently, and we present I/O-efficient algorithms for triangulating planar graphs and computing small separators of such graphs. Using these techniques, we can construct a data structure that allows for answering shortest path queries on a planar graph I/O-efficiently.

1 Introduction

Motivation. Answering shortest path queries in graphs is an important and well studied problem. Applications include communication systems, transportation problems, scheduling, computation of network flows, and geographic information systems (GIS). Typically, an underlying geometric structure is represented by an equivalent combinatorial structure, which is often a weighted, planar graph.

Model of Computation. In many applications data sets are too large to fit into the main memory of existing machines. In such cases, conventional internal memory algorithms can be inefficient, accessing their data in a random fashion, and causing many data transfers between internal and external memory. This I/O-bottleneck is becoming more significant as parallel computing gains popularity and CPU speeds increase, since disk speeds are not keeping pace. Several computational models for estimating the I/O-efficiency of algorithms have been developed [10,11,3]. We adopt the *parallel disk model* PDM [10] as our model of computation for this paper due to its simplicity, and the fact that we consider only a single processor.

In the PDM, an *external memory*, consisting of D disks, is attached to a machine with memory size M data items. Each of the disks is divided into blocks of B consecutive data items. Up to D blocks, at most one per disk, can be transferred between internal and external memory in a single I/O operation. The complexity of an algorithm is the number of I/O operations it performs.

---

[*] For details see [12].
[**] Research partially supported by NSERC.
[***] Research partially supported by Studienstiftung des deutschen Volkes.

T. Asano et al. (Eds.): COCOON'99, LNCS 1627, pp. 51–60, 1999.
© Springer-Verlag Berlin Heidelberg 1999

Previous Results. Frederickson [5] proposed an $O\left(N\sqrt{\log N}\right)$ time algorithm to compute shortest paths in planar graphs using separators. This technique was extended by Djidjev [4] who developed an $O(S)$-space data structure ($N \leq S \leq N^2$) that answers distance queries on planar graphs in $O(N^2/S)$ time in internal memory. The corresponding shortest path can be reported in time proportional to the length of the reported path.

Lipton and Tarjan [6] presented a linear-time algorithm for finding a $\frac{2}{3}$-separator of size $O(\sqrt{N})$ for any planar graph.

In the PDM, sorting an array of size N takes $sort(N) = \Theta\left(\frac{N}{DB}\log_{\frac{M}{B}}\frac{N}{B}\right)$ I/Os [10,9]. Scanning an array of size N takes $scan(N) = \Theta\left(\frac{N}{DB}\right)$ I/Os. For a comprehensive survey of external memory algorithms, refer to [9]. The only external memory shortest path algorithm known to us is the single source shortest path algorithm by Crauser *et al.* [2], which takes $O\left(\frac{|V|}{D} + \frac{|E|}{DB}\log_{\frac{M}{B}}\frac{|E|}{B}\right)$ I/Os with high probability on a random graph with random weights. We do not know of previous work on computing separators in external memory. One can use the PRAM simulation results of Chiang *et al.* [1] together with known PRAM separator algorithms. Unfortunately, the PRAM simulation introduces $O(sort(N))$ I/Os for every PRAM step, and so the resulting I/O complexity is not attractive for our purposes.

Our Results. The main results of this paper are listed below. Details can be found in [12].

1. In Sect. 3, we present a blocking to store a rooted tree T of size N in at most $\left(2 + \frac{2}{1-\tau}\right)\frac{N}{B} + D$ blocks so that a path of length K *towards the root* can be traversed in at most $\left\lceil\frac{K}{\tau DB}\right\rceil + 1$ I/Os, for $0 < \tau < 1$. For fixed τ, the tree uses optimal $O(|T|/B)$ space and traversing a path takes optimal $O(K/DB)$ I/Os. Using the best previous result by Nodine *et al.* [8], the tree would use the same amount of space within a constant factor, but traversing a path would take $O(K/\log_d(DB))$ I/Os, where d is the maximal degree of the vertices in the tree.
2. In Sect. 4, we present an external memory algorithm to compute a separator of size $O(\sqrt{N})$ for an embedded planar graph in $O(sort(N))$ I/Os, provided that a breadth-first search tree (BFS-tree) of the graph is given. Our algorithm is based on the planar separator technique in [6]. The main challenge in designing an external memory algorithm for this problem is to determine a good separator corresponding to a fundamental cycle.
3. In Sect. 5, we describe an external memory algorithm which triangulates an embedded planar graph in $O(sort(N))$ I/O operations.
4. Results 1-3, above, are the main techniques that we use to construct an external memory data structure for answering shortest path queries online. Our data structure uses $O\left(N^{3/2}/B\right)$ blocks of external storage and answers online distance and shortest path queries in $O(\sqrt{N}/DB)$ and $O((\sqrt{N} + K)/DB)$ I/Os, respectively, where K is the number of vertices on the path.

The separator and triangulation algorithms may be of independent interest, since graph separators are used in the design of efficient divide-and-conquer graph algorithms and many graph algorithms assume triangulated input graphs.

2   Preliminaries

A *graph* $G = (V, E)$ is a pair of sets V and E, where V is called the *vertex set* and E is called the *edge set* of G. Each edge in E is an unordered pair $\{v, w\}$ of vertices v and w in V. A graph G is called *planar* if it can be drawn in the plane so that no two edges intersect, except possibly at their endpoints. Such a drawing defines, for each vertex v of G, an order of the edges incident to v clockwise around v. We call G *embedded* if we are given this order for every vertex of G. By Euler's formula, $|E| \leq 3|V| - 6$ for planar graphs. A *path* from a vertex v to a vertex w in G is a list $p = \langle v = v_0, \ldots, v_k = w \rangle$ of vertices, where $\{v_i, v_{i+1}\} \in E$ for $0 \leq i < k$. A graph G is *connected* if there is a path between any two vertices in G. A *subgraph* $G' = (V'. E')$ of G is a graph with $V' \subseteq V$ and $E' \subseteq E$. *Connected components* of G are the maximal connected subgraphs of G. Let $c : E \to \mathbb{R}^+$ be a mapping that assigns non-negative costs to the edges of G. The *cost* of a path $p = \langle v_0, \ldots, v_k \rangle$ is defined as $|p| = \sum_{i=0}^{k-1} c(\{v_i, v_{i+1}\})$. A *shortest path* $\pi(v, w)$ is a path from v to w of minimal cost. Let $w : V \to \mathbb{R}^+$ be a mapping that assigns non-negative weights to the vertices of G such that $\sum_{v \in V} w(v) \leq 1$. The *weight* of a subgraph H of G is the sum of the weights of the vertices in H. An ϵ-*separator*, $0 < \epsilon < 1$, of G is a subset C of V whose removal partitions G into two subgraphs, A and B, each of weight at most ϵ, so that there is no edge in G that connects any vertex in A to any vertex in B. We will describe results on paths in a tree which originate at an arbitrary node of the tree and proceed to the root. We will refer to such paths as *bottom-up* paths.

3   Blocking Rooted Trees

In this section we describe a blocking of a rooted tree T so that we can traverse a bottom-up path from a given vertex v of T in an I/O-efficient manner. We assume that all accesses to T are read-only. Thus, we can store each vertex an arbitrary number of times and use redundancy to reduce the number of blocks that have to be read. However, this increases the space requirements. The following theorem gives a trade-off between the space requirements of the data structure and the I/O-efficiency of the tree traversal. The proof follows from Lemmas 1 and 2. The proof of Lemma 2 and an algorithm to construct such a blocking for a given tree in $O(sort(N))$ I/Os are given in [12].

Theorem 1. *Given a rooted tree T of size N and a constant τ, $0 < \tau < 1$, we can store T in at most $\left(2 + \frac{2}{1-\tau}\right) \frac{N}{B} + D$ blocks on D parallel disks so that traversing any bottom-up path of length K in T takes at most $\left\lceil \frac{K}{\tau DB} \right\rceil + 1$ I/Os.*

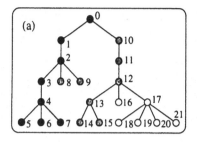 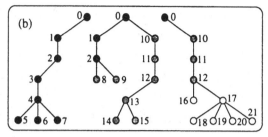

Fig. 1. (a) A rooted tree T_i with its vertices labelled with their preorder numbers. Assuming that $t = 8$, V_0, V_1, and V_2 are the sets of black, grey, and white vertices, respectively. (b) The subtrees $T_i(V_1)$, $T_i(V_2)$, and $T_i(V_3)$ from left to right.

Intuitively, our approach is as follows. We cut T into layers of height τDB. This divides every bottom-up path of length K in T into subpaths of length τDB; each subpath stays in a particular layer. We ensure that each such subpath is stored in a single block and can thus be traversed at the cost of a single I/O operation. This gives us the desired I/O-bound because any path of length K is divided into at most $\lceil \frac{K}{\tau DB} \rceil + 1$ subpaths.

More precisely, let $h(T)$ represent the height of T, and let $h' = \tau DB$ be the height of the layers to be created (we assume that h' is an integer). Let the level of a vertex v be the number of edges in the path from v to the root r of T. Cut T into layers $L_0, \ldots, L_{\lceil h(T)/h' \rceil - 1}$, where layer L_i is the subgraph of T induced by the vertices on levels ih' through $(i+1)h' - 1$. Each layer is a forest of rooted trees whose heights are at most h'. Suppose that there are r such trees, taken over all layers. Let $T_1, \ldots, T_r$ denote these trees for all layers of T.

Lemma 1. *Given a rooted tree T_i of height at most τDB, we can divide T_i into subtrees $T_{i,0}, \ldots, T_{i,s}$ with the following properties: (1) $|T_{i,j}| \leq DB$, for all $0 \leq j \leq s$, (2) $\sum_{j=0}^{s} |T_{i,j}| \leq \left(1 + \frac{1}{1-\tau}\right) |T_i|$, and (3) for every leaf l of T_i, there is a subtree $T_{i,j}$ containing the whole path from l to the root of T_i.*

Proof sketch. If $|T_i| \leq DB$, we "divide" T_i into one subtree $T_{i,0} = T_i$. Then Properties 1–3 trivially hold. So assume that $|T_i| > DB$. Given a preorder numbering of the vertices of T_i, let v_k be the vertex with preorder number k. Let $h' = \tau DB$, $t = DB - h'$, and $s = \lceil |T_i|/t \rceil - 1$. We define vertex sets $V_0, \ldots, V_s$, where $V_j = \{v_{jt}, \ldots, v_{(j+1)t-1}\}$ for $0 \leq j \leq s$ (see Fig. 1(a)). The subtree $T_{i,j} = T_i(V_j)$ is the subtree of T_i consisting of all vertices in V_j and their ancestors in T_i (see Fig. 1(b)). We claim that these trees $T_{i,j}$ have Properties 1–3.

Property 3 is ensured by including the ancestors in T_i of all vertices in V_j in $T_i(V_j)$. Property 2 follows, if we can prove Property 1 because then $\sum_{j=0}^{s} |T_{i,j}| \leq \sum_{j=0}^{s} DB = (s+1)DB \leq \left(\frac{1}{1-\tau} + 1\right) |T_i|$.

It can be shown that every vertex in $T_i(V_j)$ that is not in V_j is an ancestor of v_{jt}. As the height of T_i is at most h', there can be at most h' such ancestors of v_{jt}. Moreover, $|V_j| \leq DB - h'$. Thus, $|T_i(V_j)| \leq DB$. $\square$

Lemma 2. *If a rooted tree T of size N is partitioned into subtrees $T_{i,j}$ such that the properties 1–3 in Lemma 1 hold, then T can be stored using $\left(2 + \frac{2}{1-\tau}\right)\frac{N}{B} + D$ blocks of external memory, and any bottom-up path of length K in T can be traversed in at most $\left\lceil\frac{K}{\tau DB}\right\rceil + 1$ I/Os.*

4   Separating Embedded Planar Graphs

We now present an external memory algorithm for separating embedded planar graphs. It is based on Lipton and Tarjan's [6] linear-time separator algorithm. The input to our algorithm is an embedded planar graph G and a spanning forest F of G. Every tree in F is a rooted BFS-tree of the respective connected component. The graph G is represented by its vertex set V and its edge set E. To represent the embedding, let the edges incident to a vertex v be numbered in counterclockwise order around v. This defines two numbers $n_v(e)$ and $n_w(e)$, stored with every edge $e = \{v, w\}$. The spanning forest F is given implicitly by marking every edge of G as tree or non-tree edge and storing, with each vertex v in V, the name of its parent $p(v)$ in F. We prove the following theorem.

Theorem 2. *Given an embedded planar graph G with N vertices and a BFS-tree[1] T of G, a $\frac{2}{3}$-separator of G of size at most $2\sqrt{2}\sqrt{N}$ can be computed in $O(sort(N))$ I/Os.*

Proof sketch. W.l.o.g., we assume that the given graph is connected. If it is not, we can compute its connected components in $O(sort(N))$ I/Os [1] and compute a separator of the component with weight greater than $\frac{2}{3}$, if any. Moreover we assume that G is triangulated. If it is not, it can be triangulated in $O(sort(N))$ I/Os using the algorithm in Sect. 5.

The separator consists of two parts. First we compute two levels l_0 and l_2 in G's BFS-tree T whose removal divides G into three parts G_1, G_2, G_3 with $|G_1|, |G_3| \leq \frac{2}{3}$ and such that $L(l_0) + L(l_2) \leq 2\sqrt{2}\sqrt{N} - 2(l_2 - l_0 - 1)$, where $L(l)$ is the number of vertices on level l. Lipton and Tarjan [6] proved that such levels l_0 and l_2 exist. Computing levels l_0 and l_2 takes $O(sort(N))$ I/Os using a generalization of the list ranking algorithm in [1], sorting, and scanning.

To separate G_2 into components of weights at most $\frac{2}{3}$ each, we shrink levels 0 through l_0 to a single vertex, remove levels l_2 and below, and retriangulate the resulting graph. Call the resulting graph G'. We construct G', a spanning tree T' of G', and an embedding of G' in $O(sort(N))$ I/Os using sorting, scanning, and the triangulation algorithm in Sect. 5. The separator of G_2 is a simple cycle separator of G', which contributes at most $2(l_2 - l_0 - 1)$ vertices to the separator because the height of T' is $l_2 - l_0$. Thus, the total size of the separator is at most $2\sqrt{2}\sqrt{N}$. Lemma 3 states that a simple cycle separator of G' can be computed in $O(sort(N))$ I/Os. □

---

[1] The currently best known BFS-algorithm [7] takes $O\left(|V| + \frac{|E|}{|V|} sort(|V|)\right)$ I/Os.

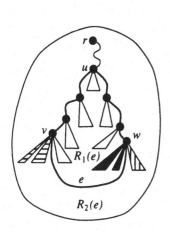

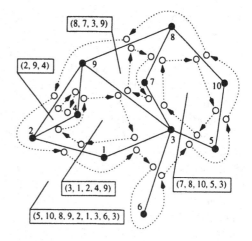

Fig. 2. A non-tree edge e and its fundamental cycle $c(e)$ shown in bold. $R_1(e)$ is the set of vertices embedded inside the cycle and $R_2(e)$ is the set of vertices embedded outside the cycle.

Fig. 3. A given graph G (black vertices and solid lines). White vertices and dotted arrows represent the graph $\hat{G}$ for G. Every face f of $\hat{G}$ is labelled with its corresponding vertex list F_f.

Finding a Small Simple Cycle Separator. Every non-tree edge $e = \{v, w\}$ in G' defines a fundamental cycle $c(e)$ consisting of e itself and the two paths in the tree T' from the vertices v and w to the lowest common ancestor (LCA) u of v and w (see Fig. 2). Any fundamental cycle $c(e)$ separates G' into two subgraphs $R_1(e)$ and $R_2(e)$, one induced by the vertices embedded inside $c(e)$ and the other induced by those embedded outside. Lipton and Tarjan showed that there is a non-tree edge e in G' such that $R_1(e)$ and $R_2(e)$ have weights at most $\frac{2}{3}$ each.

Lemma 3. *Given a triangulated graph G' with N vertices and a BFS-tree T' of G'. A $\frac{2}{3}$-simple cycle separator of size at most $2h(T') - 1$ for G' can be computed in $O(\text{sort}(N))$ I/Os.*

Proof sketch. For every vertex v in T', we compute a 4-tuple $A(v) = (n(v), \nu(v), \delta(v), W(v))$ of the following labels: (1) its preorder number $n(v)$, (2) its "weighted preorder number" $\nu(v) = \sum_{n(u) \leq n(v)} w(u)$, (3) the total weight $\delta(v)$ of v and all its ancestors in T', and (4) the weight $W(v)$ of all vertices in the subtree of T' rooted at v. It is essential that the preorder numbers $n(v)$ and $\nu(v)$ respect the embedding. That is, if $p(v)$, w_1, and w_2 appear in counterclockwise order around vertex v, then w_1 has a smaller preorder number than w_2. Less formally, this means that the subtrees in T' are labelled in left-to-right order. In [12] it is shown how to compute these labels in $O(\text{sort}(N))$ I/Os using known external memory techniques such as sorting, scanning, and time-forward processing [1].

We store the tuples $A(v)$, $A(w)$, and $A(u)$ with every edge $e = \{v, w\}$, where u is the LCA of v and w. Computing all LCAs takes $O(\text{sort}(N))$ I/Os [1]. To copy the appropriate vertex labels to all edges, we have to sort and scan the vertex and edge sets of G'.

For every vertex v of T', let $e_0, \ldots, e_k$ be the set of edges incident to it in counterclockwise order, and let $e_0 = \{v, p(v)\}$. We define $t(e_i) = 0$, if e_i is a non-tree edge, and $t(e_i) = W(w_i)$, if $e_i = \{v, w_i\}$ is a tree edge and $v = p(w_i)$. We compute labels $t_v(e_i) = t_v(e_{i-1}) + t(e_i)$, for $0 < i \leq k$, and $t_v(e_0) = 0$, sorting and scanning the edge set of G'.

Given a non-tree edge e, the weights of $R_1(e)$ and $R_2(e)$ can now be computed from the labels stored locally with e: Consider Fig. 2. A non-tree edge $e = \{v, w\}$ is shown and u is the LCA of v and w in T'. Four classes of subtrees are indicated by different patterns in Fig. 2. The vertices in the vertically hatched subtree are not important to the algorithm, but are included for completeness. The set of vertices $R_1(e)$, embedded inside the cycle, are the vertices in the white and black subtrees. The vertices in the horizontally hatched and white subtrees and on the tree path from u to w are exactly the vertices with preorder numbers between $n(v)$ and $n(w)$. Thus, their total weight is $\nu(w) - \nu(v)$. The total weight of the vertices in the horizontally hatched trees is $t_v(e)$; the total weight of the vertices on the path from u to w is $\delta(w) - \delta(u)$; the total weight of the black trees is $t_w(e)$. Thus, the total weight of vertices in $R_1(e)$ is $\nu(w) - \nu(v) - t_v(e) + t_w(e) - \delta(w) + \delta(u)$. The total weight of the vertices on $c(e)$ is $\delta(v) + \delta(w) - 2\delta(u) + w(u)$. Hence, the total weight of vertices in $R_2(e)$ is $w(G') - w(R_1(e)) - \delta(v) - \delta(w) + 2\delta(u) - w(u)$.

Thus, we can scan the edge set of G' and stop at the first non-tree edge with $w(R_1(e)), w(R_2(e)) \leq \frac{2}{3}$. The fundamental cycle $c(e)$ can be reported in $O(sort(N))$ I/Os by sorting the vertex set of G' by levels and preorder numbers in T' and then reporting on every level lower than that of u, those vertices with greatest preorder number less than $n(v)$ and $n(w)$, respectively.   □

5   Triangulating Embedded Planar Graphs

In this section we present an $O(sort(N))$-algorithm to triangulate a connected embedded planar graph $G = (V, E)$. We assume the same representation of G and its embedding as in the previous section. Our algorithm consists of two phases. First we identify the faces of G. We represent each face f by a list of vertices on its boundary, sorted clockwise around the face. In the second phase, we use this information to triangulate the faces of G. The following theorem follows from Lemmas 4 and 6 below.

Theorem 3. *An embedded planar graph can be triangulated in $O(sort(N))$ I/Os.*

Identifying Faces. We compute a list F which is the concatenation of vertex lists F_f, one for each face of G. For a given face f, the list F_f is the clockwise sequence of vertices around face f. The list F_f may contain more than one copy of the same vertex, depending on how often this vertex is visited in a clockwise traversal of the face boundary.

TRIANGULATEFACES(G, F):

1: Make all faces of G simple:

For each face f, (a) mark the first appearance of each vertex v in F_f, (b) append a marked copy of the first vertex in F_f to the end of F_f, and (c) scan F_f backward and remove each unmarked vertex v from f and F_f by adding a chord between its predecessor and successor in the current list.

2: Triangulate the simple faces:

Let $F_{\hat{f}} = \langle v_0, \ldots, v_k \rangle$. Add "temporary chords" $\{v_0, v_i\}$, $2 \le i \le k - 1$, to $\hat{f}$.

3: Mark conflicting chords:

Sort E lexicographically, ensuring that edge $\{v, w\}$ is stored before all "temporary chords" $\{v, w\}$. Scan E and mark all occurrences of each edge, except the first, as "conflicting". Restore the original order of all edges and "temporary chords".

4: Retriangulate conflicting faces:

For each face $\hat{f}$, let $D_{\hat{f}} = \langle \{v_0, v_2\}, \ldots, \{v_0, v_{k-1}\} \rangle$ be the list of "temporary chords". Scan $D_{\hat{f}}$ until we find the first conflicting chord $\{v_0, v_i\}$. Replace $\{v_0, v_i\}, \ldots, \{v_0, v_{k-1}\}$ by chords $\{v_{i-1}, v_{i+1}\}, \ldots, \{v_{i-1}, v_k\}$.

Algorithm 1: Triangulating the faces of G.

Lemma 4. *The list F can be constructed in $O(\mathrm{sort}(N))$ I/Os.*

Proof sketch. We first compute a graph $\hat{G}$ which is comprised of disjoint directed cycles. Each cycle represents a clockwise traversal of the boundary of a face f of G. Given a cycle in $\hat{G}$ that represents a face f of G, every vertex in this cycle represents an edge on the boundary of f. We construct F_f as the list of first endpoints of these edges in clockwise order around f (see Fig. 3).

Two vertices in $\hat{G}$ that are consecutive on a cycle of $\hat{G}$ represent two edges that are consecutive on the boundary of a face of G in clockwise order. Thus, these two edges are consecutive around a vertex of G in counterclockwise order. The graph $\hat{G}$ contains vertices $v_{(v,w)}$ and $v_{(w,v)}$ for every edge $\{v, w\}$ of G and edges $(v_{(u,v)}, v_{(v,w)})$, where $\{u, v\}$ and $\{w, v\}$ are consecutive counterclockwise around v. These vertex and edge sets can be computed sorting and scanning the vertex and edge sets of G.

We identify the cycles of $\hat{G}$ as its connected components using the algorithm in [1], sort and scan the edge set of $\hat{G}$ to remove an arbitrary edge from every such cycle. This transforms every cycle into a list, which can be ranked and then sorted by rank. As a result, the vertices of $\hat{G}$ are sorted clockwise around the faces of G. We scan these sorted lists to construct F. As we only use scanning, sorting, and the list ranking technique in [1], the complexity of this algorithm is $O(\mathrm{sort}(N))$.   □

Triangulating Faces. We triangulate each face f in four steps (see Algorithm 1). First, we reduce f to a simple face $\hat{f}$. (A face is simple if each vertex on its boundary is visited only once in a clockwise traversal of the boundary.) This reduces the list F_f to $F_{\hat{f}}$. In the second step, we triangulate $\hat{f}$. We ensure that there are no multiple edges in $\hat{f}$, but we might add *conflicting* edges (edges with

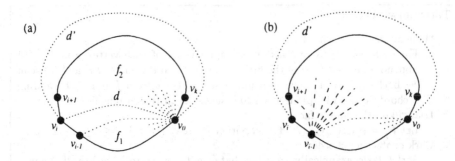

Fig. 4. (a) A simple face $\hat{f}$. Chord d conflicts with d' and divides $\hat{f}$ into two parts f_1 and f_2. One of them, f_1, is conflict-free. Vertex v_{i-1} is the third vertex of the triangle in f_1 that has d on its boundary. (b) The conflict-free triangulation of $\hat{f}$.

the same endpoints) to adjacent faces. (See Fig. 4 for an example.) In the third step, we detect all such conflicting edges. In the fourth step, we retriangulate all faces $\hat{f}$ so that conflicts are resolved and a final triangulation is obtained.

In [12] we show that, for each face f of G, the face $\hat{f}$, computed in Step 1 of Algorithm 1 is simple. The parts of f that are not in $\hat{f}$ are triangulated. Moreover, Step 1 does not introduce parallel edges. Step 2 triangulates all simple faces $\hat{f}$. However, we may add the same chord $\{v, w\}$ to several faces $\hat{f}_1, \ldots, \hat{f}_k$. It can also happen that $\{v, w\}$ is already an edge of G. If $\{v, w\} \in G$, we have to remove the chords $\{v, w\}$ from all k faces where we have added such a chord. Otherwise, we have to remove $k - 1$ of them. In Step 3, we mark the respective chords as conflicting. We have to show that the output of Step 4 is a conflict-free triangulation of G.

Lemma 5. *Step 4 makes all faces $\hat{f}$ conflict-free, i.e. the graph obtained after Step 4 is simple.*

Proof sketch. Let $d = \{v_0, v_i\}$ (see Fig. 4). Then d cuts $\hat{f}$ into two halves, f_1 and f_2. All chords $\{v_0, v_j\}$, $j < i$ are in f_1; all chords $\{v_0, v_j\}$, $j > i$ are in f_2. That is, f_1 does not contain conflicting chords. Vertex v_{i-1} is the third vertex of the triangle in f_1 that has d on its boundary. Step 4 removes d and all chords in f_2 and retriangulates f_2 with chords incident to v_{i-1}.

Let d' be the edge that is in conflict with d. Then d and d' form a closed curve, and v_{i-1} is outside this curve. All boundary vertices of f_2 excluding the endpoints of d are inside this curve. As no edge, except for the new chords in $\hat{f}$, can intersect this curve, the new chords in $\hat{f}$ are non-conflicting. The "old" chords in $\hat{f}$ were in f_1 and thus, by the choice of d and f_1, non-conflicting. Hence, $\hat{f}$ does not contain conflicting chords. □

We mark the first appearances of vertices in each list F_f in $O(sort(N))$ I/Os as follows: sort F_f by vertex numbers, scan F_f to mark the first appearance of every vertex, and restore the original order of F_f. The rest of Algorithm 1 takes $O(sort(N))$ I/Os.

Lemma 6. *Given the list F as defined in the previous section, Algorithm 1 triangulates the graph G in $O(sort(N))$ I/Os.*

In order to use this algorithm as part of our separator algorithm, we also have to embed the chords in the faces. Let $v_1, e_1, v_2, e_2, \ldots, v_k, e_k$ be the list of vertices and edges visited in a clockwise traversal of the boundary of a face f (i.e., $F_f = \langle v_1, \ldots, v_k \rangle$). We define labels $n_1(v_i) = n_{v_i}(e_{(i-1)\bmod k})$ and $n_2(v_i) = n_{v_i}(e_i)$, and store them with v_i in F_f. When we add a chord d incident to vertex v_i, we give it a label $n_{v_i}(d)$ which is a rational value between $n_1(v_i)$ and $n_2(v_i)$. (To avoid problems related to arithmetic precision, we assign the new label as an offset of $\frac{1}{N}$ from either $n_1(v_i)$ or $n_2(v_i)$.) This embeds d between $e_{(i-1)\bmod k}$ and e_i. After that, the labels $n_1(v_i)$ and $n_2(v_i)$ are updated to ensure that subsequent chords are embedded between $e_{(i-1)\bmod k}$ or e_i and d, depending on the current configuration. We can maintain labels $n_1(v)$ and $n_2(v)$ for all vertices in the lists F_f without increasing the number of I/O-operations by more than a constant factor.

Acknowledgements. We would like to thank Lyudmil Aleksandrov, Jörg-Rüdiger Sack, Hans-Dietrich Hecker, and Jana Dietel for helpful discussions.

References

1. Y.-J. Chiang, M. T. Goodrich, E. F. Grove, R. Tamassia, D. E. Vengroff, J. S. Vitter. External-memory graph algorithms. *Proc. 6th SODA*, Jan. 1995.
2. A. Crauser, K. Mehlhorn, U. Meyer. Kürzeste-Wege-Berechnung bei sehr großen Datenmengen. *Aachener Beitr. zur Inf.* (21). Verl. d. Augustinus Buchh. 1997.
3. F. Dehne, W. Dittrich, D. Hutchinson. Efficient external memory algorithms by simulating coarse-grained parallel algorithms. *Proc. 9th SPAA*, pp. 106–115, 1997.
4. H. N. Djidjev. Efficient algorithms for shortest path queries in planar digraphs. *Proc. of the 22nd Workshop on Graph-Theoretic Concepts in Comp. Sci.*, Lecture Notes in Comp. Sci., pp. 151–165. Springer Verlag, 1996.
5. G. N. Frederickson. Fast algorithms for shortest paths in planar graphs, with applications. *SIAM J. Comp.*, 16(6):1004–1022, Dec. 1987.
6. R. J. Lipton, R. E. Tarjan. A separator theorem for planar graphs. *SIAM J. Appl. Math.*, 36(2):177–189, 1979.
7. K. Munagala, A. Ranade. I/O-complexity of graph algorithms. *Proc. 10th SODA*, Jan. 1999.
8. M. Nodine, M. Goodrich, J. Vitter. Blocking for external graph searching. *Algorithmica*, 16(2):181–214, Aug. 1996.
9. J. Vitter. External memory algorithms. *Proc. 17th ACM Symp. on Principles of Database Systems*, June 1998.
10. J. Vitter, E. Shriver. Algorithms for parallel memory I: Two-level memories. *Algorithmica*, 12(2–3):110–147, 1994.
11. J. Vitter, E. Shriver. Algorithms for parallel memory II: Hierarchical multilevel memories. *Algorithmica*, 12(2–3):148–169, 1994.
12. N. Zeh. *An External-Memory Data Structure for Shortest Path Queries*. Diplommarbeit, Fak. f. Math. und Inf., Friedrich-Schiller-Univ. Jena, Nov. 1998.

Approximating the Nearest Neighbor Interchange Distance for Evolutionary Trees with Non-uniform Degrees

Wing-Kai Hon and Tak-Wah Lam[*]

Department of Computer Science and Information Systems, University of Hong Kong, Hong Kong, China, {wkhon,twlam}@csis.hku.hk

Abstract. The nearest neighbor interchange (nni) distance is a classical metric for measuring the distance (dissimilarity) between two evolutionary trees. The problem of computing the nni distance has been studied over two decades (see e.g., [16,3,7,12,8,4]). The long-standing conjecture that the problem is NP-complete was proved only recently, whereas approximation algorithms for the problem have appeared in the literature for a while. Existing approximation algorithms actually perform reasonably well (precisely, the approximation ratios are $\log n$ for unweighted trees and $4 \log n$ for weighted trees); yet they are designed for degree-3 trees only. In this paper we present new approximation algorithms that can handle trees with non-uniform degrees. The running time is $O(n^2)$ and the approximation ratios are respectively $(\frac{2d}{\log d} + 2) \log n$ and $(\frac{2d}{\log d} + 12) \log n$ for unweighted and weighted trees, where $d \geq 4$ is the maximum degree of the input trees.

1 Introduction

An *evolutionary tree* is a tree whose leaves are labeled with distinct symbols representing species. It is useful for modeling the evolutionary relationship of a set of species. Different theories of evolution often result in different evolutionary trees. A basic problem in computational biology is to compare the evolutionary trees arising from different theories. In particular, the metrics of maximum agreement subtree (*mast*) [6] and the nearest neighbor interchange (*nni*) distance [13,15] are widely used for measuring the similarity and dissimilarity of two evolutionary trees. Computing such metrics is however not trivial.

The past few years have produced a number of exciting results on computing such metrics. There are now very efficient algorithms for computing *mast* for unbounded-degree trees [5,10], as well as bounded-degree trees [1,9,10]. The long-standing conjecture (see e.g., [11]) that computing the *nni* distance is NP-hard was resolved in the affirmative recently [4]. In this paper we present new algorithms for approximating the *nni* distance of two evolutionary trees.

Let T_1 and T_2 be two evolutionary trees labeled with the same of set of species. The *nni* distance between T_1 and T_2 is defined to be the minimum

---

[*] Research supported in part by Hong Kong RGC Grant HKU-7027/98E.

T. Asano et al. (Eds.): COCOON'99, LNCS 1627, pp. 61–70, 1999.
© Springer-Verlag Berlin Heidelberg 1999

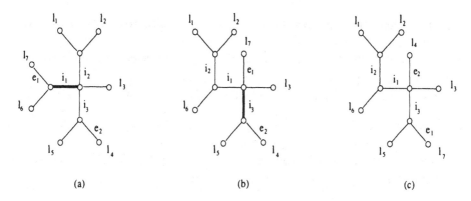

Fig. 1. (b) is transformed from (a) using an *nni* operation over the edge i_1, and (c) from (b) over the edge i_3. The symbols i_j, e_j, l_j denote an internal edge, external edge, and leaf label, respectively.

number of *nni* operations required to transform T_1 to T_2 (more precisely, T_1 is transformed to a tree isomorphic to T_2, while preserving the leaf labels). An *nni* operation refers to the transformation of a tree by swapping two subtrees attached to different ends of an internal edge. See Figure 1 for an example.

In the literature the study of the *nni* distance focuses on degree-3 trees [16,2,3,7,12,8,4]. Culik II and Wood [2] are the first to give a polynomial-time algorithm for approximating the *nni* distance. Then Li et al. [8] improved the approximation ratio from $4 \log n$ to $\log n$. An evolutionary tree may have weights on its edges. An edge weight could represent the evolutionary distance between two species. An *nni* operation is charged a cost equal to the weight of the internal edge over which the subtrees are swapped. The *nni* distance between two weighted evolutionary trees T_1 and T_2 is defined as the minimum cost of *nni* operations to transform T_1 into T_2 (more precisely, T_1 is transformed into a tree that is isomorphic to T_2, while preserving the edge weights and leaf labels). Generalizing the algorithm for unweighted trees [8], DasGupta et al. [4] recently devised an approximation algorithm for weighted degree-3 trees with an approximation ratio of $4 \log n$.

The studies of evolutionary trees often involve trees with degrees exceeding three and being not uniform over all nodes (see e.g., [14]). In this paper we initiates the study of the *nni* distance for such evolutionary trees. We formulate the necessary and sufficient conditions for transforming a tree to another tree using *nni* operations, and present new approximation algorithms. For the unweighted case, the new approximation algorithm attains an approximation ratio of $\left(\frac{2d}{\log d} + 2\right) \log n$, where d is the maximum degree (for the sake of analysis, we assume $d \geq 4$; the case when $d = 3$ becomes the problem addressed before). Technically speaking, our unweighted algorithm adapts the techniques for dealing weighted degree-3 trees [4] to handle unweighted non-uniform degree trees. When the trees are weighted and of non-uniform degree, existing techniques are however no longer sufficient as they cannot preserve degrees, edge weights, and leaf labels efficiently. We solve this problem by exploiting some interesting tree structures so that we can focus on one parameter at a time. The new approximation algorithm has an approximation ratio of $(\frac{2d}{\log d} + 12) \log n$. Like the previous algorithms, these approximation algorithms run in $O(n^2)$ time.

Note that an *nni* operation does not operate on an external edge (i.e. an edge attached with a leaf) and an external edge is never charged. To simplify our discussion, we assume that an external edge does not carry any weight. Unless otherwise specified, we assume throughout the paper that the input trees T_1 and T_2 have leaves labeled with the same set of species, and the degrees of internal nodes are at least three and may not be uniform.

2   Preliminaries

For any unrooted tree T, let $L(T)$ denote the set of leaf labels of T, let $D(T)$ denote the multi-set comprising the degrees of the internal nodes of T, and let $W(T)$ be the multi-set comprising the weights of the internal edges (i.e., edges not attached to leaves) of T.

2.1   Feasibility conditions

Suppose the input trees T_1 and T_2 are unweighted. If the internal nodes of both trees are of degree 3, it is well known that T_1 can always be transformed to T_2 using *nni* operations [15]. In general, if the nodes are not of uniform degree, T_1 can be transformed to T_2 if and only if $D(T_1) = D(T_2)$. The *only if* part is easy to observe since an *nni* operation does not change the degree of any internal node. The *if* part follows from a special case of Lemma 1, which considers internal edges to have weights.

In the weighted context, DasGupta et al. [4] proved that for any degree-3 trees T_1 and T_2, if $W(T_1) = W(T_2)$, then T_1 can be transformed to T_2. For trees with non-uniform degrees, we generalize the feasibility condition to $D(T_1) = D(T_2)$ and $W(T_1) = W(T_2)$ (see Lemma 1). Again, the condition is not only sufficient but also necessary. It is worth mentioning that the condition does not require two edges in T_1 and T_2 of the same weight to have end points with the same degree.

Lemma 1. *For any weighted trees T_1 and T_2, T_1 can be transformed to T_2 if $D(T_1) = D(T_2)$ and $W(T_1) = W(T_2)$.*

Proof. We use induction on the number of internal edges. The base case where there is one internal edge is obviously true (T_1 can be transformed into T_2 by swapping the leaves over the only internal edge). For the inductive case, we are going to show that T_1 can be transformed into a tree T' that has a subtree in common with T_2. Contracting this subtree into a new leaf would reduce one internal edge in both T' and T_2. Then, by the induction hypothesis, we can further transform T' to T_2, thus completing the proof.

To transform T_1 to T', we first identify an internal node v in T_2 which is attached to exactly one internal edge e. Let d_v be the degree of v and denote $l_1, l_2, \ldots, l_{d_v-1}$ as the labels of the leaves attached to v. Since T_1 and T_2 have the same sets of degrees and edge weights, T_1 also contain a node v' with degree equal to d_v and an edge e' with the same weight as e. Intuitively, we want to attach e' to v' if this is not already true. Denote x as the end point of e' that is closer to v', and let S be the subtree attached to x and including e'. To move e' to v', we repeatedly swap S with some subtrees attached to the path from x to v'. Similarly, we can attach the leaves in T_1 labeled with $l_1 \cdots l_{d_v-1}$ to v' one by one. At this point, T_1 is transformed to a tree that has a subtree (namely, e', v' and the attached leaves) in common with T_2. Contracting this subtree into a new leaf would reduce exactly one internal edge. This completes the reduction.

2.2   Lower bounds

Let T_1 and T_2 be any weighted trees satisfying the feasibility conditions. To derive a lower bound on the *nni* distance between T_1 and T_2, we extend the results on degree-3 trees by Li et al. [8] and DasGupta et al. [4].

Definition 1. *Consider any internal edge e of T_1 and any internal edge e' of T_2. We say that (e, e') is a* good *pair if the following conditions hold.*

- *The weight of e is the same as that of e'.*
- *The partition of the leaf labels, edge weights, and internal node degrees induced by e in T_1 is the same as the partition induced by e' in T_2.*

An edge e of T_1 is said to be *bad* if e does not form a good pair with any edge of T_2. Intuitively, in the course of transforming T_1, if no *nni* operation is operated on a bad edge, T_1 can never be transformed into T_2. Thus, we have the following lemma.

Lemma 2. *Let W_b denote the sum of the weights of the bad edges of T_1 (which is also equal to that of T_2). Then the cost of transforming T_1 into T_2 using* nni *operations is at least W_b.*

In particular, if T_1 and T_2 do not contain any good pair of edges, then their *nni* distance is at least the total weight of all the internal edges. When T_1 and T_2 are unweighted trees, the lower bound becomes the number of bad edges in either tree.

2.3   Structured trees

An *nni* operation is reversible. If a tree T_1 is transformed to another tree T_2 using k *nni* operations, T_2 can also be transformed to T_1 by k *nni* operations. Therefore, in order to transform T_1 to T_2 in $k_1 + k_2$ *nni* operations, it suffices to show that, for some tree T' (usually with a very simple structure), both T_1 and T_2 can be transformed to T' using k_1 and k_2 operations respectively. Below we define the structures of those intermediate trees T' which our algorithms make use of.

- A linear tree [8,4] is a tree in which all the internal edges form a single path.
- A (unweighted) compact tree is a tree whose diameter is minimized over all trees with the same multi-set of degrees. More precisely, a compact tree organizes its nodes in such a way that a breadth-first search starting from a designated node (the center) would first visit all internal nodes in descending order of the degrees, then followed by the leaves.

Lemma 3. *Define the* height *of a compact tree F to be the maximum number of internal edges on the path from a leaf to the center. If the average degree of the internal nodes is $\bar{d}$, then the height is at most $\log n / \log(\bar{d} - 1)$.*

The following lemmas show some important properties regarding the transformation of linear trees and compact trees.

Lemma 4. *[8,4] Any tree T can be transformed to some linear tree using m nni operations, where m is the number of internal edges of T.*

Proof. Let p be a path between two leaves in T. If p contains all internal edges, T is already linear. Otherwise, there exists an internal edge e attached to p. Using one *nni* operation, we can swap two subtrees over e (one overlaps with p and the other not) to include e into p. (For example, Figures 1(a) and 1(b) illustrate that an internal edge i_1 is included into the path between the leaves l_2 and l_4.) Thus, at most m *nni* operations are needed.

Lemma 5. *Let L_1 and L_2 be two linear trees each with m internal edges. Assume that $D(L_1) = D(L_2)$. Then $m \log(m + 1)$ nni operations are sufficient to transform L_1 into a tree that is structurally isomorphic to L_2 (but it may not preserve the leaf labels).*

Suppose the internal nodes of L_2 are labeled with integers in $[1, m + 1]$ from left to right. For each internal node i in L_2, we can assign the label i to a unique internal node in L_1 with the same degree. To transform L_1 to L_2, we adapt the merge-sort technique given in [8] (which was designed for arranging leaf labels of a degree-3 linear tree) to "sort" the internal nodes of L_1 according to the labels of L_2. The main idea is that given a linear tree with the left half of internal nodes sorted in descending order and the second half sorted in ascending order, we can "merge" them into a sorted linear tree by repeatedly swapping (a leaf and a subtree) over the internal edge at the middle. Details will be given in the full paper.

Lemma 6. *Let C_1 and C_2 be two compact trees each with n leaves. Assume that $L(C_1) = L(C_2)$ and $D(C_1) = D(C_2)$. Then C_1 can be transformed into C_2 using $2nh$ nni operations, where h is the height of C_1.*

By definition of compact trees, C_1 and C_2 are structurally isomorphic but the labels of the leaves may not match. We need to swap the leaves of C_1 so that their labels confront to that of C_2. Here we adapt the strategy used in [4] to handle general trees. We start with an arbitrary leaf l and move it to its correct position. Then we move the leaf l' previously occupying the correct position of l and so on. At first glance, such swapping seems to be simple. Yet after moving even one leaf, the structure of C_1 might change and, in particular, the height of the tree might increase. Nevertheless, with a careful analysis, we can show that the structure of C_1 can be restored at the end, and the number of nni operations used for moving each leaf is at most $2h$. Thus, the total number of nni operations is $2nh$.

3   Approximation algorithms for unweighted trees

This section serves as a warm-up, illustrating how to approximate the nni distance between two unweighted evolutionary trees in $O(n^2)$ time. Let T_1 and T_2 be the input trees and let n and m be the numbers of leaves and internal edges, respectively. We first devise an algorithm for transforming T_1 into T_2 using at most $\frac{2d}{\log d}m \log n + 2m \log n + 4m$ nni operations, where d is the maximum degree.

Lemma 7. *Let T_1 and T_2 be any two unweighted trees each with n leaves and m internal edges. Then T_1 can be transformed to T_2 using $\frac{2d}{\log d}m \log n + 2m \log n + 4m$ nni operations, where d is the maximum degree of T_1 and T_2.*

In the course of transforming T_1 into T_2, we need to cope with both the structure and leaf labels. With the sorting technique on linear trees, we can easily transform T_1 and T_2 to two trees C_1 and C_2 that are structurally isomorphic. To efficiently transform C_1 so that the leaf labels eventually confront to that of C_2, we choose C_1 and C_2 to be compact trees and make use of the result of Lemma 6. Details are as follows.

0. Construct a compact tree C with $D(C) = D(L)$. Transform C into some linear tree L_c. Denote ρ as the sequence of the nni operations. By Lemma 4, $|\rho| \le m$.
1. We first transform T_1 into some linear tree L_1, then use the sorting technique to transform L_1 into another linear tree L_1' which is structurally isomorphic to L_c. We further use the reverse of ρ to transform L_1' into a compact tree C_1. Similarly, we transform T_2 into a compact tree C_2.
2. Transform C_1 into C_2.

Let d and $\bar{d}$ denote the maximum degree and average degree of C_1. By Lemma 4 and Lemma 5, Step 1 needs at most $2(m + m \log(m + 1) + m)$ operations. Let h denote the height of C_1 and C_2. By Lemmas 3 and 6, Step 2 uses $2nh \le \frac{2n \log n}{\log(d-1)}$ operations. The total number of nni operations is at most

$$\frac{2n}{\log(\bar{d} - 1)} \log n + 2m \log n + 4m$$

$$= \frac{2((m+1)\bar{d} - 2m)}{\log(\bar{d} - 1)} \log n + 2m \log n + 4m \qquad \text{(because } n + 2m = (m+1)\bar{d}\text{)}$$

$$\leq \frac{2(m\bar{d} - m)}{\log(\bar{d} - 1)} \log n + 2m \log n + 4m \qquad \text{(assume that } h > 1, \text{ then } m \geq \bar{d})$$

$$\leq \frac{2m(\bar{d} - 1)}{\log(\bar{d} - 1)} \log n + 2m \log n + 4m \qquad \leq \frac{2d}{\log d} m \log n + 2m \log n + 4m.$$

To derive the desired bound on the approximation ratio, we first consider the special case where T_1 and T_2 contain bad edges only. By Lemma 2, the *nni* distance between T_1 and T_2 is at least m. Thus, Lemma 7 implies that the *nni* distance can be approximated within a factor of $(\frac{2d}{\log d} + 2) \log n + 4$. To analyze the general case when good edges are present in T_1 and T_2, we follow the approach in [8,4] to consider the individual components formed by the bad edges. Details are as follows.

First of all, we identify all the good pair of edges. This can be done in $O(n^2)$ time. (More precisely, we first find all the pairs of edges in T_1 and T_2 such that they partition the leaf labels in the same way. There are at most n such pairs and they can be found in $O(n^2)$ time [3]. Then, for each pair, we check whether they induce the same partition on node degrees.) Removing the good edges from T_1 (and T_2) would partition the tree into a number of connected components, each containing bad edges only. Each component in T_1 has a one-to-one correspondence in T_2. Such a pair of components has the same degree set and leaf labels. To transform T_1 to T_2, we apply Lemma 7 to transform each component of T_1 to the corresponding component of T_2. The total number of *nni* operations is bounded by $(\frac{2d}{\log d} + 2)m' \log n + 4m'$, where m' is the total number of internal edges in the components of T_1, or equivalently, the number of bad edges. Together with the lower bound result in Lemma 2, we can show that our algorithm approximates the actual *nni* distance between T_1 and T_2 within a factor of $(\frac{2d}{\log d} + 2) \log n + 4$.

Regarding the time complexity, we observe the bottleneck of the approximation algorithm above is identifying the components, which takes $O(n^2)$ time.

4 Approximation algorithm for weighted trees

This section shows how to approximate the *nni* distance between two weighted evolutionary trees T_1 and T_2. Let n, m and W be the number of leaves, number of internal edges and total weight of internal edges respectively in either tree. Below we devise an algorithm that can transform T_1 into T_2 with total cost at most $(\frac{2d}{\log d} + 12)W \log n + 8W$, where d is the maximum degree (see Lemma 8). Similar to the unweighted case, we can show that Lemma 8 leads to an algorithm for approximating the *nni* distance between two weighted trees within a factor of $(\frac{2d}{\log d} + 12) \log n + 8$, where d is the maximum degree of either tree.

Lemma 8. *Let T_1 and T_2 be any two weighted trees. Then T_1 can be transformed to T_2 with a total cost of at most $(\frac{2d}{\log d} + 12)W \log n + 8W$, where d is the maximum degree of either tree.*

To prove Lemma 8, we attempt to generalize the unweighted transformation described in the previous section. First of all, let us define a weighted version of compact tree.

Definition 2. *An increasing-weight compact tree is a compact tree with weights on its edges, satisfying an extra condition that the breadth-first search which visits all the internal nodes in descending order of their degrees also visits all the internal edges in* ascending *order of their weights.*

To derive a transformation from T_1 into T_2, we can first transform T_1 and T_2 into two increasing-weight compact trees C_1 and C_2, and then transform C_1 to C_2. By definition of increasing-weight compact trees, C_1 and C_2 are structurally isomorphic and the weights of their corresponding internal edges are the same. Thus, we can transform C_1 to C_2 by repeatedly swapping the leaves as before. In the full paper we will extend Lemma 6 to show that the *nni* operations involved in swapping of leaves incurs a total cost of $2nh\frac{W}{m}$, where h is the height of C_1.

The above approach seems to be straightforward; the only non-trivial step is transforming T_1 to an increasing-weight compact tree C_1. The difficulty arises from the fact that we have to cope with node degrees and edge weights at the same time. Applying the sorting technique twice (say, one on node degrees and the other on edge weights) cannot give us an increasing-weight compact tree at the end.

Let us first explain a simple but not efficient approach of transforming T_1 to an increasing-weight compact tree. Let C be any increasing-weight compact tree such that $D(C) = D(T_1)$ and $W(C) = W(T_1)$. The leaves of C are labeled arbitrarily. Denote ρ as the sequence of *nni* operations for transforming C into some linear tree, say, L_c. Note that the cost of ρ is at most W. To transform T_1 to an increasing-weight compact tree, we follow the following steps:

1. Transform T_1 into a linear tree L_1. Then use the sorting technique to transform L_1 into another linear tree L_1' such that the ordering of the internal edge weights of L_1' is the same as that of L_c. Note that the ordering of the degrees of the internal nodes of L_1' may not be the same as that of L_c.
2. We fix the node degrees of L_1' in a brute force manner. Suppose the first node of L_c has degree x. We locate a node with degree x in L_1' and repeatedly swap it with the previous node until it reaches the top. Note that the ordering of the edge weights remains the same. Next, we proceed to work on the second node of L_c and so on. Denote the resulting linear tree as L_1''.
3. Note that L_1'' differs from L_c only in the labels of the leaves. Applying the reverse of ρ to L_1'' can produce an increasing-weight compact tree.

In the above procedure, Step 2 is the most demanding step, using $O(m^2)$ *nni* operations and incurring a cost of $O(nW)$. This is too much if we aim at a logarithmic approximation ratio. The inefficiency of Step 2 is due to the linear structure of the trees L_1' and L_c. To improve the transformation to an increasing-weight compact tree, we introduce another intermediate structure, called an almost-binary tree.

Definition 3. *An* almost-binary *tree is a weighted tree such that the internal nodes themselves (i.e., excluding the leaves) form a degree-3 increasing-weight compact tree. That means, we can use a breadth-first search starting from a designated node (the center) to enumerate all the internal edges in ascending order of their weights. On the other hand, the 'actual' degree of an internal node is possibly greater than three.*

Intuitively, two almost-binary trees with the same set of edge weights organize their edge weights in the same manner. But the actual degrees of the internal nodes may not match.

With the notion of almost-binary trees, we can improve the procedure of transforming T_1 to an increasing-weight compact tree as follows.

1. Let B be any almost-binary tree with $W(B) = W(T_1)$ and $D(B) = D(T_1)$. Denote σ as the sequence of *nni* operations which can transform B into some linear tree, say, L_b.
2. Transform T_1 into some linear tree L_1, and then to another linear tree L'_1 whose ordering of edge weights is the same as that of L_b. Use the reverse of σ to transform L'_1 into an almost-binary tree B_1.
3. Transform L_c (i.e., the linearization of a compact tree C with $D(C) = D(T_1)$ and $W(C) = W(T_1)$) to another linear tree L'_c whose ordering of edge weights confronts to that of L_b. Transform L'_c to an almost-binary tree B_c using the reverse of σ.
4. Note that B_1 and B_c differ in the ordering of the node degrees. Transform B_1 to another almost-binary B'_1 using a brute force manner so that the B'_1 preserves the node degrees of B_c.
5. To transform B'_1 to an increasing-weight compact tree, we apply the reverse of transformation from C to B_c.

Before we go on, let us give the details of Step 4 above.

Lemma 9. *Given two almost-binary tree B_1 and B_2 with $W(B_1) = W(B_2)$ and $D(B_1) = D(B_2)$, then B_1 can be transformed into another almost-binary tree B'_1 preserving the node degrees of B_2 using cost at most $4W \log m$, where W and m are total weights and number of internal edges of either tree.*

Proof. To transform B_1 into B'_1, we need to move the internal nodes, while keeping the internal edges intact. We first fix the internal nodes on the bottom layer (i.e., farthest from the center), then those on the 2nd bottom layer and so on. To fix a particular layer, we identify the nodes in B_1 that belongs to that layer of B_2, and move them to their correct positions one by one. In the course of moving an internal node, all *nni* operations operates only on edges on the path between the node and its correct position. It is worth-mentioning that we use the trick of operating each internal edge twice every time to ensure the tree is immediately restored with respect to its binary structure and edge weights. The cost of all such movement is bounded by twice of the total distance (in terms of edge weights) between the leaves and their correct position. In an almost-binary tree there are at most $2 \log m$ edges between two internal nodes. We can further show that the total distance (in terms of edge weights) between the leaves and their correct position is at most $2W \log m$. Thus, the total cost of the nni operations is at most $4W \log m$.

By Lemma 9 and the weighted extension of Lemma 4 and Lemma 5, we can show that the total cost to transform T_1 into C_1 and T_2 into C_2 is at most $2(W + W \log m + W + 4W \log m + W + W \log m + W) \leq 12W \log n + 8W$. ¿From the previous section, we have $2nh \leq \frac{2d}{\log d} m \log n$. The cost to transform C_1 into C_2 is at most $\frac{2nhW}{m} \leq \frac{2d}{\log d} W \log n$. This completes the proof of Lemma 8.

References

1. R. Cole and R. Hariharan. An $O(n \log n)$ algorithm for the maximum agreement subtree problem for binary trees. In *Proceedings of the 7th Annual ACM-SIAM Symposium on Discrete Algorithms*, pages 323-332, 1996.
2. K. Culik II and D. Wood. A note on some tree similarity measures. *Information Processing Letters*, 15(1):39-42, 1982.
3. W.H.E. Day. Properties of the nearest neighbor interchange metric for trees of small size. *Journal of Theoretical Biology*, 101:275-288, 1983.
4. B. DasGupta, X. He, T. Jiang, M. Li, J. Tromp, and L. Zhang. On distances between phylogenetic trees. In *Proceedings of the 8th Annual ACM-SIAM Symposium on Discrete Algorithms*, pages 427-436, 1997.
5. M. Farach and M. Thorup. Optimal evolutionary tree comparison by sparse dynamic programming. In *Proceedings of the 35th Annual IEEE Symposium on Foundations of Computer Science* pages 770-779, 1994.
6. C. Finden and A. Gordon. Obtaining common pruned trees. *Journal of Classification*, 2:255-276, 1985.
7. J.P. Jarvis, J.K. Luedeman, and D.R. Shier. Comments on computing the similarity of binary trees. *Journal of Theoretical Biology*, 100:427-433, 1983.
8. M. Li, J. Tromp, and L.X. Zhang. Some notes on the nearest neighbour interchange distance. *Journal of Theoretical Biology*, 182, 1996.
9. M. Y. Kao. Tree contractions and evolutionary trees. *SIAM Journal on Computing*, 27(6):1592-1616, December 1998.
10. M. Y. Kao, T. W. Lam, T. M. Przytycka, W. K. Sung, and H. F. Ting. General techniques for comparing unrooted evolutionary trees. In *Proceedings of the 29th Annual ACM Symposium on Theory of Computing*, pages 54-65, 1997.
11. S. Khuller. Open Problems:10, *SIGACT News*, 24(4):46, 1994.
12. M. Křivánek. Computing the nearest neighbor interchange metric for unlabeled binary trees in NP-Complete. *Journal of Classification*, 3:55-60, 1986.
13. G.W. Moore, M. Goodman, and J. Barnabas. An interactive approach from the standpoint of the additive hypothesis to the dendrogram problem posed by molecular data sets. *Journal of Theoretical Biology*, 38:423-457, 1973.
14. C. Nielsen, *Animal Evolution*, Oxford University Press, 1995.
15. D.F. Robinson. Comparison of labeled trees with valency tree. *Journal of Combinatorial Theory*, 11:105-119, 1971.
16. M.S. Waterman. On the Similarity of Dendrograms. Journal of Theoretical Biology, 73:789-800, 1978.

Signed Genome Rearrangement by Reversals and Transpositions: Models and Approximations*

Guo-Hui Lin and Guoliang Xue

Department of Computer Science
University of Vermont
Burlington, VT 05405, USA
{ghlin, xue}@cs.uvm.edu

Abstract. An important problem in computational molecular biology is the genome rearrangement using reversals and transpositions. Analysis of genome evolving by inversions and transpositions leads to a combinatorial optimization problem of sorting by reversals and transpositions, i.e., sorting of a permutation using reversals and transpositions of arbitrary fragments. The reversal operation works on a single segment of the genome by reversing the selected segment. Two kinds of transpositions have been studied in the literature. The first kind of transposition operation deletes a segment of the genome and insert it into another position in the genome. The second kind of transposition operation deletes a segment of the genome and insert its inverse into another position in the genome. Both transposition operations can be viewed as operations working on two consecutive segments. A third transposition operation working on two consecutive segments is introduced which, together with reversal and the first two kinds of transposition operations, forms the complete set of operations on two consecutive segments. In this paper, we study the sorting of a signed permutation by reversals and transpositions. By allowing only the first kind of transpositions, or the first two kinds of transpositions, or all three kinds of transpositions, we have three problem models. After establishing a common lower bound on the numbers of operations needed, we present a unified 2-approximation algorithm for all these problems. Finally, we present a better 1.75-approximation for the third problem.

1 Introduction

The edit distance between two DNA sequences was traditionally taken as the measure of evolutionary distance between the two species which was measured as the minimum number of insertions, deletions and substitutions required to change one sequence to another. Things become different recently. In the late 1980's J. Palmer and his colleagues discovered a remarkable and novel pattern

* This research was supported in part by the Army Research Office grant DAAH04-96-10233.

T. Asano et al. (Eds.): COCOON'99, LNCS 1627, pp. 71-80, 1999.
© Springer-Verlag Berlin Heidelberg 1999

of evolutionary change in plant organelles. They compared the mitochondrial genomes of *Brassica oleracea* (cabbage) and *Brassica campestris* (turnip), which are very closely related (many genes are 99% – 99.9% identical). To their surprise, these molecules, which are almost identical in gene sequence, differ dramatically in gene order. This discovery and many other studies in the last decade convincingly proved that genome rearrangements is a common mode of molecular evolution in mitochondrial, chloroplast, viral and bacterial DNA [1]. That is, at the chromosome level, genetic sequences mutate by many global operations such as the reversal of a substring (*inversion*), the deletion and subsequent reinsertion of a substring far from its original site (*transposition*), duplication of a substring and the exchange of prefixes or suffixes of two sequences in the same organism (*translocation*). Here we are interested in inversions and transpositions.

The study of genome rearrangements by reversals and transpositions involves solving a combinatorial puzzle to find a shortest series of *reversals* and *transpositions* to transform one genome into another. This (minimum) number of operations required is then taken as the evolutionary distance between these two genomes. For genomes consisting of a small number of "conserved blocks", the most parsimonious scenarios may be found in a brute-force style. However, for genomes consisting of a large number of blocks, finding the optimal solution becomes difficult. Previously, much research focused on genome rearrangements by reversals only, which has been shown to be NP-hard [4]. In the design of approximation algorithms, Kececioglu and Sankoff [7] gave a 2-approximation. This performance ratio of 2 was further improved to 1.75 by Bafna and Pevzner [2]. The problem of sorting by transpositions only is also of interest. Bafna and Pevzner [3] designed a 1.5-approximation for this problem, although whose computational complexity is still unknown.

Any operation applied on a genome should not break the conserved blocks. From this point of view, we may represent a genome by a permutation π on set $\{1, 2, \cdots, n\}$, where n represents the number of conserved blocks in the genome and each element π_j in π represents a conserved block. In this way, it is easy to define reversals and transpositions performed on permutations which correspond to the reversals and transpositions performed on genomes; and thus a series of reversals and transpositions transforming one permutation π into another σ maps a series of reversals and transpositions transforming one genome into another, correspondingly. We will study the latter instead of the former for the sake of simplicity of exposition.

Let $\pi = (\pi_1, \pi_2, \cdots, \pi_n)$ be a permutation of $\{1, 2, \cdots, n\}$. For $1 \leq i < j \leq n + 1$, a *reversal* $r(i, j)$ is the permutation

$$\begin{pmatrix} 1, \cdots, i-1, \boxed{\text{i, i+1, } \cdots\text{, j-1}}, j, \cdots, n \\ 1, \cdots, i-1, \boxed{\text{j-1, } \cdots\text{, i+1, i}}, j, \cdots, n \end{pmatrix}.$$

Therefore, $\pi \cdot r(i, j) = (\pi_1, \cdots, \pi_{i-1}, \pi_{j-1}, \cdots, \pi_{i+1}, \pi_i, \pi_j, \cdots, \pi_n)$, i.e., $\pi \cdot r(i, j)$ has the effect of reversing the order of $\pi_i, \pi_{i+1}, \cdots, \pi_{j-1}$. For $1 \leq i < j \leq n + 1$

and $1 \leq k \leq n+1$ with $k \notin [i,j]$, a *transposition* $t(i,j,k)$ is the permutation

$$\begin{pmatrix} 1, \cdots, i-1, \boxed{\text{i, i+1, } \cdots \text{, j-1}}, \boxed{\text{j, } \cdots \text{, k-1}}, k, \cdots, n \\ 1, \cdots, i-1, \boxed{\text{j, } \cdots \text{, k-1}}, \boxed{\text{i, i+1, } \cdots \text{, j-1}}, k, \cdots, n \end{pmatrix}.$$

That is, $\pi \cdot t(i,j,k) = (\pi_1, \cdots, \pi_{i-1}, \pi_j, \cdots, \pi_{k-1}, \pi_i, \cdots, \pi_{j-1}, \cdots, \pi_k, \cdots, \pi_n)$. In other words, $\pi \cdot t(i,j,k)$ has the effect of moving $\pi_i, \pi_{i+1}, \cdots, \pi_{j-1}$ to a new location of π before π_k.

Given permutations π and σ, the *rearrangement distance problem* is to find a series of reversals and transpositions $\rho_1, \rho_2, \cdots, \rho_d$ such that $\pi \cdot \rho_1 \cdot \rho_2 \cdots \cdot \rho_d = \sigma$ and d is minimum. We call d the *rearrangement distance* between π and σ. Since the rearrangement distance between π and σ equals to the rearrangement distance between $\sigma^{-1}\pi$ and the *identity* permutation $\mathcal{I} = (1, 2, \cdots, n)$, sorting π is the problem of finding the rearrangement distance, denoted by $d(\pi)$, between π and $\mathcal{I}$.

Now we can define an *α-approximation algorithm*. An α-approximation algorithm for sorting a permutation is an algorithm which, given any π, finds a series of operations $\rho_1, \cdots, \rho_d$ such that $\rho_1, \cdots, \rho_d$ sort π into $\mathcal{I}$ and d satisfies $d(\pi) \leq d \leq \alpha d(\pi)$.

A signed permutation is a permutation π on $\{1, 2, \cdots, n\}$ with $+$ or $-$ sign associated with every element π_j of π. For example, $(-1, -4, +3, +2)$ is a signed permutation of $\{1, 2, 3, 4\}$. The identity signed permutation is $(+1, +2, \cdots, +n)$. It is noted that signed permutations are more relevant to genomes rearrangements, since genes are oriented in DNA sequences. For sorting signed genomes by reversals only, Bafna and Pevzner [2] designed a 1.5-approximation, and later Hannenhalli and Pevzner [6] presented a polynomial time algorithm which finds the minimum number of reversals for a signed permutation. This is the first polynomial algorithm for a realistic model of genome rearrangements problems.

For a signed permutation, a reversal on a block not only inverses the order of the elements in the block, but also changes the signs. It is clear that by transpositions only, sorting any signed permutation into the identity is impossible. Nonetheless, transpositions do help save some number of operations, if they are used together with reversals. See for example, the (signed) permutation $(-1, -4, +3, +2)$ in Figure 1 requires (exactly) four reversals, according to the algorithm by Hannenhalli and Pevzner [6]; while one reversal and two transpositions, if allowed, suffice to sort it into identity $\mathcal{I} = (+1, +2, +3, +4)$.

In a recent paper [5], Gu, Peng and Sudborough proposed a new kind of transpositions $rt(i,j,k)$, defined as the permutation (for $1 \leq i < j \leq n+1$ and $1 \leq k \leq n+1$ with $k \notin [i,j]$)

$$\begin{pmatrix} 1, \cdots, i-1, \boxed{\text{i, i+1, } \cdots \text{, j-1}}, \boxed{\text{j, } \cdots \text{, k-1}}, k, \cdots, n \\ 1, \cdots, i-1, \boxed{\text{j, } \cdots \text{, k-1}}, \boxed{\text{j-1, } \cdots \text{, i+1, i}}, k, \cdots, n \end{pmatrix}.$$

Thus, $\pi \cdot rt(i,j,k) = (\pi_1, \cdots, \pi_{i-1}, \pi_j, \cdots, \pi_{k-1}, \pi_{j-1}, \cdots, \pi_i, \cdots, \pi_k, \cdots, \pi_n)$, i.e., $\pi \cdot rt(i,j,k)$ has the effect of reversing $\pi_i, \pi_{i+1}, \cdots, \pi_{j-1}$ and moving it to a new

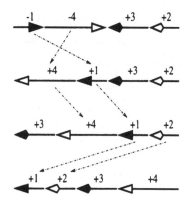

Fig. 1. Sorting $(-1, -4, +3, +2)$ into identity.

location of π before π_k. They then designed a 2-approximation algorithm for sorting signed permutations using these three kinds of operations, which runs in time $O(n^2)$ (n is the number of elements in the permutation).

In this paper, we propose a new kind of transpositions $rr(i, j, k)$, based on realistic applications, and consider three versions of sorting signed permutations: (1) by reversals and transpositions $t(i, j, k)$; (2) by reversals and transpositions $t(i, j, k)$ and $rt(i, j, k)$; (3) by reversals and transpositions $t(i, j, k)$, $rt(i, j, k)$ and $rr(i, j, k)$. After establishing a common lower bound on the numbers of operations needed, we present a unified approximation algorithm with performance guarantee of 2 for these three versions. In the next section, we present some definitions and conventions needed throughout this paper. The common lower bound and the unified 2-approximation algorithm are presented in Section 3. Section 4 describes a better approximation algorithm for the third version, whose performance ratio is 1.75. We conclude the paper in Section 5.

2   Preliminaries

For an unsigned permutation $\pi = (\pi_1, \pi_2, \cdots, \pi_n)$, (π_i, π_{i+1}) is called a breakpoint if $|\pi_i - \pi_{i+1}| \neq 1$. As there is no breakpoint in the identity permutation $\mathcal{I}$, the sorting problem is to reduce the number of breakpoints in π, using as few operations as possible.

Note that operation $r(i, j)$ acts on a segment of elements, while transpositions $t(i, j, k)$ and $rt(i, j, k)$ act on two consecutive segments of elements. If we take two consecutive segments and put them back with possible inversions, there are four possible operations. $t(i, j, k)$ and $rt(i, j, k)$ are two of the four operations. One of the other two reduces to reversal. The only operation which has not been studied in the literature is defined in the following, which we call the third kind of transpositions $rr(i, j, k)$. A transposition $rr(i, j, k)$ is the permutation (for

$1 \leq i < j \leq n + 1$ and $1 \leq k \leq n + 1$ with $k \notin [i, j]$)

$$\begin{pmatrix} 1, \cdots, i-1, \boxed{i, \ i+1, \ \cdots, \ j\text{-}1}, \boxed{j, \ j+1, \ \cdots, \ k\text{-}1}, k, \cdots, n \\ 1, \cdots, i-1, \boxed{j\text{-}1, \ \cdots, \ i+1, \ i}, \boxed{k\text{-}1, \ \cdots, \ j+1, \ j}, k, \cdots, n \end{pmatrix}.$$

Thus, $\pi \cdot rr(i, j, k) = (\pi_1, \cdots, \pi_{i-1}, \pi_{j-1}, \cdots, \pi_i, \pi_{k-1}, \cdots, \pi_j, \cdots, \pi_k, \cdots, \pi_n)$, i.e., $\pi \cdot rr(i, j, k)$ has the effect of reversing the segments $(\pi_i, \pi_{i+1}, \cdots, \pi_{j-1})$ and $(\pi_j, \pi_{j+1}, \cdots, \pi_{k-1})$ at the same time and then putting them back to the permutation.

In the case where π is a signed permutation of n elements, define its unsigned *image permutation* π^* of $2n$ elements as follows: Replace $+i$ with $(2i - 1, 2i)$ and replace $-i$ with $(2i, 2i - 1)$. In this way, the signed identity $\mathcal{I}$ has its image permutation $\mathcal{I}^* = (1, 2, 3, 4, \cdots, 2n - 1, 2n)$. In what follows, we assume that any operation on the image permutation π^* of a signed permutation π never breaks any pair of $(2i - 1, 2i)$ or $(2i, 2i - 1)$ which correspond to $+i$ or $-i$ in the corresponding signed permutation. It follows that any sequence of operations transforming π into $\mathcal{I}$ one-to-one corresponds to a sequence of operations transforming π^* into $\mathcal{I}^*$. In other words, the signed permutation π and its image (unsigned) permutation π^* are equivalent for our purpose. In the rest of this paper, by permutation we mean the image permutation for a signed permutation, unless otherwise specified. It is clear that an operation $t(i, j, k)$ $(r(i, j), rt(i, j, k), rr(i, j, k))$ performed on the signed permutation π corresponds to an operation $t^*(2i - 1, 2j - 1, 2k - 1)$ performed on the image permutation π^*. Therefore, in all later mentioned operations, the parameters i, j, k involved are all odd integers.

Note that every reversal and every one of the three kinds of transpositions acts on at most three breakpoints, and thus it can reduce at most three breakpoints. Therefore, let $b(\pi)$ denote the number of breakpoints in π, then $b(\pi)/3$ is a trivial lower bound for all three aforementioned sorting problems. We will derive a better lower bound in the following. Let us first define the *breakpoint graph* of a permutation.

Let $\pi = (\pi_1, \pi_2, \cdots, \pi_{2n})$ be an arbitrary permutation. Extend π by adding $\pi_0 = 0$ and $\pi_{2n+1} = 2n + 1$. Let $i \sim j$ if $|i - j| = 1$. A pair of consecutive elements (π_i, π_{i+1}) is an *adjacency* if $\pi_i \sim \pi_{i+1}$ (otherwise, a breakpoint). The *breakpoint graph* of π is an edge-colored graph $G(\pi)$:

- There are $2n + 2$ nodes $0, 1, 2, \cdots, 2n - 1, 2n, 2n + 1$ in $G(\pi)$;
- There is a *gray edge* between i and j if $i \sim j$ and i and j are not consecutive in π;
- There is a *black edge* between i and j if (i, j) is a breakpoint.

The graph $G(\pi)$ of $\pi = (-1, -4, +3, +2)$ is given in Figure 2.

A sequence of distinct nodes $v_1, v_2, \cdots, v_m$ is called a *segment* in $G(\pi)$ if $(v_i, v_{i+1}) \in E(G)$, for $1 \leq i \leq m - 1$. A sequence of nodes $v_1, v_2, \cdots, v_m = v_1$ is called a *cycle* in $G(\pi)$ if $(v_i, v_{i+1}) \in E(G)$, for $1 \leq i \leq m - 1$. A segment/cycle is *alternating* if the colors of edges in the segment/cycle alternate. Define the length of an alternating cycle the number of black edges (breakpoints) in the cycle.

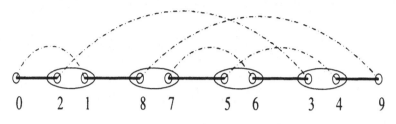

Fig. 2. $G(\pi)$: $\pi = (-1, -4, +3, +2)$.

Lemma 1. *For the breakpoint graph $G(\pi)$ of a permutation π,*

1. *the gray degree and black degree of each node are the same and equal to either 0 or 1;*
2. *each connected component of $G(\pi)$ is an alternating cycle;*
3. *each alternating cycle has at least two gray (black) edges;*
4. *there is no edge in $G(\mathcal{I})$, the breakpoint graph of the identity permutation $\mathcal{I}$.*

In what follows, by cycle we mean an alternating cycle. Call a cycle a k-cycle if its length is k. A k-cycle is an odd cycle if k is odd, otherwise it is called an even cycle. Let $o(\pi)$ be the number of odd cycles in $G(\pi)$.

3   A Unified 2-Approximation Algorithm

3.1   A lower bound

Let ρ be an operation, and $\pi' = \pi \cdot \rho$. Let $b(\pi')$ be the number of breakpoints in π'. Note that ρ acts on at most three breakpoints, $|b(\pi) - b(\pi')| \leq 3$. This implies $d(\pi) \geq b(\pi)/3$. In order to get a better lower bound of $d(\pi)$, we notice that only transpositions act on three breakpoints. Moreover, a transposition $\rho = t(i, j, k)$ $(rt(i, j, k), rr(i, j, k))$ reduces the number of breakpoints by three if and only if (π_{i-1}, π_i), (π_{j-1}, π_j) and (π_{k-1}, π_k) are breakpoints belonging to a 3-cycle, and this cycle has a configuration isomorphic to that shown in Figure 3(a) (3(b), 3(c), respectively). From this, to remove a cycle of length k, we need at least $\frac{k}{2}$ operations for k even and at least $\frac{k-1}{2}$ operations for k odd. This suggests that $d(\pi)$ is at least $(b(\pi) - o(\pi))/2$, where $o(\pi)$ is the number of odd cycles in $G(\pi)$. We show this in the following.

Lemma 2. *Any sequence of operations transforming permutation π into $\mathcal{I}$ can be regulated in such a way that each of the operations acts on breakpoints only.*

PROOF. Suppose that $\rho_1, \rho_2, \cdots, \rho_k$ is a sequence of operations transforming permutation π into identity $\mathcal{I}$. We may first regulate ρ_k to acts on breakpoints only. And then regulate ρ_{k-1} (and changing ρ_k accordingly) to acts on breakpoints only. Repeatedly, applying this process in the reversed order of operations, we

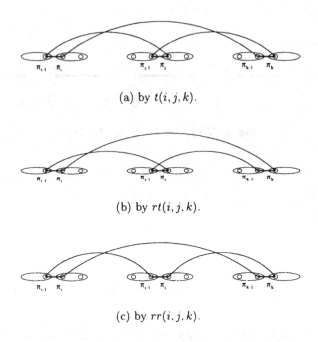

(a) by $t(i, j, k)$.

(b) by $rt(i, j, k)$.

(c) by $rr(i, j, k)$.

Fig. 3. Cycles being removed three breakpoints by a transposition.

will finally get another sequence of operations, which act on breakpoints only, transforming permutation π into identity $\mathcal{I}$.   □

From the above lemma, we need only to consider those operations acting on breakpoints only.

Lemma 3. *For a permutation π and an operation ρ with $\pi' = \pi \cdot \rho$, $\Delta(\rho) = (b(\pi) - o(\pi)) - (b(\pi') - o(\pi')) = -2, 0$ or 2, which is the* decrement *of operation ρ.*

PROOF. Note first that any operation reduces the parameter $b(\pi)$ by at most three, and whenever it reduces three the corresponding three breakpoints must belong to a same cycle and this cycle has only three black edges. Therefore, this operation reduces the parameter $(b(\pi) - o(\pi))$ by at most two, i.e., $\Delta(\rho) = (b(\pi) - o(\pi)) - (b(\pi') - o(\pi')) \geq -2$. Now the parity consideration implies $\Delta(\rho) = -2, 0$ or 2.   □

Theorem 1. *For any permutation π, $d(\pi) \geq (b(\pi) - o(\pi))/2$.*

PROOF. The theorem is clearly followed from Lemma 3 since $b(\pi) - o(\pi) = 0$ if and only if $\pi = \mathcal{I}$, the identity permutation.   □

3.2   2-approximation algorithm

Let $c(\pi)$ denote the number of cycles in the breakpoint graph $G(\pi)$ of a permutation π. Note that any reversal reduces the parameter $(b(\pi) - c(\pi))$ by at most one [2]. A gray edge g is *oriented* if a reversal acting on the two black edges incident to g reduces $(b(\pi) - c(\pi))$ by one, and *unoriented* otherwise. A cycle containing an oriented gray edge is *oriented* and otherwise *unoriented*.

Lemma 4. [6] *Let (π_i, π_j) be a gray edge incident to black edges (π_k, π_i) and (π_j, π_l). Then (π_i, π_j) is oriented if and only if $i - k = j - l$.*

Lemma 5. [5] *Let C be an unoriented cycle and $g = (\pi_{i_1}, \pi_{i_2})$ $(i_1 < i_2)$ be a gray edge of C. Then there is a black edge (π_{j_1}, π_{j_1+1}) of a different cycle satisfying $i_1 < j_1 < j_1 + 1 < i_2$.*

Lemma 5 implies that there is a black edge of another cycle which lies between the two black edges incident to g. For two black edges $b_1 = (\pi_{i_1}, \pi_{i_1+1})$ and $b_2 = (\pi_{j_1}, \pi_{j_1+1})$, we say $b_1 < b_2$ if $i_1 < j_1$. Given two cycles C and C', let $b_1, b_2, \cdots$ and $b'_1, b'_2, \cdots$ be the black edges in C and C', respectively, where $b_1 < b_2 < \cdots$ and $b'_1 < b'_2 < \cdots$. C and C' are *interleaving* if $b_1 < b'_1 < b_2 < b'_2$ or $b'_1 < b_1 < b'_2 < b_2$. From Lemma 5, we have

Lemma 6. [5] *Suppose in $G(\pi)$ there is no oriented cycle and $\pi \neq \mathcal{I}$, then there exist two cycles C and C' that are interleaving.*

Lemma 7. *Let C and C' be two interleaving unoriented cycles. Then there are two first kind of transpositions $t(i, j, k)$ which reduce the length of each of C and C' by one. Moreover, these two operations do not affect the length of other cycles; and in particular, if the length of C (C') is two, it is eliminated by these two operations.*

PROOF. Let $b_1 < b_2 < \cdots$ be the black edges in cycle C and $b'_1 < b'_2 < \cdots$ the black edges in cycle C'. By the definition of interleaving, we have, without loss of generality, $b_1 < b'_1 < b_2 < b'_2$, see Figure 4(a).

Applying $t(b, d, f)$ on the initial configuration, we get the configuration as shown in Figure 4(b). Then applying $t(d, f, h)$ on it will result in the configuration as shown in Figure 4(c).   □

Note that any first kind of transposition doesn't transform an unoriented gray edge into an oriented one. Lemmas 6 and 7 imply an algorithm which sorts a permutation π into $\mathcal{I}$ by using at most $b(\pi) - c(\pi)$ reversals $r(i, j)$ and first kind of transpositions $t(i, j, k)$. The algorithm is given in Figure 5. Since in this algorithm the other two kinds of transpositions $rt(i, j, k)$ and $rr(i, j, k)$ have not been used, we have the following theorem.

Theorem 2. *Algorithm SORT is an $O(n^2)$ time 2-approximation algorithm for all the three versions of sorting signed permutations.*

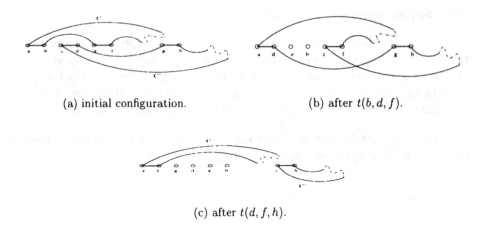

(a) initial configuration.   (b) after $t(b, d, f)$.

(c) after $t(d, f, h)$.

Fig. 4. Two first kind of transpositions acting on two interleaving unoriented cycles.

Algorithm SORT(π)
Input: a signed permutation π.
Output: a series of operations transforming π into $\mathcal{I}$.
Begin
1. Construct $G(\pi)$;
2. **While** (there is a black edge in $G(\pi)$) **Do** {
3.   **While** (there is an oriented gray edge g) **Do**
4.   apply a reversal on the two black edges incident to g;
5.   **If** (there is two interleaving unoriented cycles) **Then**
6.   apply Lemma 7 to reduce their lengths by one each;
7. }
End

Fig. 5. Sorting a signed permutation.

PROOF. As described in Figure 5, whenever there is an oriented gray edge, perform a reversal $r(i, j)$ on the two black edges incident to this gray edge. This is an operation reducing the parameter $b(\pi) - c(\pi)$ by one. In the case that there is no oriented gray edges, all the cycles are unoriented. Apply Lemmas 6 and 7 to get two first kind of transpositions $t(i, j, k)$ which reduce the parameter $b(\pi) - c(\pi)$ totally by two. This implies that each transposition reduces the parameter $b(\pi) - c(\pi)$ by one on average. As noted above, any first kind of transposition doesn't transform an unoriented gray edge into an oriented one. Therefore, after $(b(\pi) - c(\pi))$ of operations, we get a new permutation π' for which $b(\pi') - c(\pi') = 0$. This indicates that there is no black edge (as well as cycle) in the breakpoint graph $G(\pi')$. Therefore, $\pi' = \mathcal{I}$.

It takes linear time to find an oriented gray edge, and finding the interleaving unoriented cycles also takes linear time. It follows that Algorithm SORT takes $O(n^2)$ time.   □

By exploiting the special structures of the problem, we have also obtained the following result.

Theorem 3. *There exists a 1.75-approximation algorithm for sorting of signed permutations using reversals and all three kinds of transpositions.*

Because of the 10-page space limitation, we present the better approximation algorithm and its analysis in the appendix.

4   Concluding remarks

In this paper, we studied the sorting of a signed permutation by reversals and transpositions, a problem which adequately model genome rearrangements. Three variants of transpositions, and thus three corresponding rearrangements problems are considered. We establish a common lower bound on the numbers of operations needed, and give a unified 2-approximation algorithm for the three problems. By exploiting more structure properties of the possible configurations of short cycles, we presented a better 1.75-approximation for the third problem. Better approximations for all the three problems, as well as their computational complexities, are left open.

References

1. V. Bafna and P.A. Pevzner, Sorting by reversals: genome rearrangements in plant organelles and evolutionary history of X chromosome, *Molecular Biology Evolution*, 12(1995), 239–246.
2. V. Bafna and P.A. Pevzner, Genome rearrangements and sorting by reversals, *SIAM Journal on Computing*, 25(2)(1996), 272–289.
3. V. Bafna and P.A. Pevzner, Sorting by transpositions, *SIAM Journal on Discrete Mathematics*, 11(2)(1998), 224–240.
4. A. Caprara, Sorting by reversals is difficult, in *Proceedings of the First Annual International Conference on Computational Molecular Biology*, 1997, 75–83.
5. Q.-P. Gu, S. Peng and H. Sudborough, A 2-approximation algorithm for genome rearrangements by reversals and transpositions, *Theoretical Computer Science*, 210(1999), 327–339.
6. S. Hannenhalli and P. Pevzner, Transforming cabbage into turnip: polynomial algorithm for sorting signed permutations by reversals, to appear in *Journal of the ACM*. A preliminary version appeared in the *Proceedings of 27th Annual ACM Symposium on the Theory of Computing*, 1995.
7. S. Kececioglu and D. Sankoff, Exact and approximation algorithms for the inversion distance between two permutations, in *Proceedings of the 4th Annual Symposium on Combinatorial Pattern Matching*, LNCS 684, 1993, 87–105. Extended version appeared in *Algorithmica*, 13(1995), 180–210.

An Approximation Algorithm for the Two-Layered Graph Drawing Problem

Atsuko Yamaguchi and Akihiro Sugimoto

Advanced Research Laboratory, Hitachi, Ltd.
Hatoyama, Saitama 350-0395, Japan
atsuko@beri.co.jp, sugimoto@i.kyoto-u.ac.jp

Abstract. We present a polynomial-time approximation algorithm for the minimum edge crossings problem for two-layered graphs. We show the relationship between the approximation ratio of our algorithm and the maximum degree of the vertices in the lower layer of the input graph. When the maximum degree is not greater than four, the approximation ratio is two and this ratio monotonically increases to three as the maximum degree becomes larger. We also present our experiments, showing that our algorithm constructs better solutions than the barycenter method and the median method for dense graphs as well as sparse graphs.

1 Introduction

Graph drawing has been widely studied in many research fields including the graphical user interface [2]. Some aesthetic criteria in drawing graphs, such as minimizing the edge crossings or distributing the vertices uniformly, have been proposed and evaluating goodness of drawings based on these criteria has been reported [1,6,9].

A hierarchical graph is a graph $H = (L_1, \ldots, L_m, E)$, where L_i is a set of vertices called a layer and E is a set of directed edges from the vertices in L_j to those in L_k $(j < k)$. For given hierarchical graph $H = (L_1, \ldots, L_m, E)$, a hierarchical drawing[1] of H is to draw all the vertices in layer L_i on the ith horizontal line in the plane without overlapping. We therefore have to only determine the horizontal positions, i.e., the x coordinates, of the vertices that attain given aesthetic criteria. In this paper, we employ, as a aesthetic criterion, minimizing edge crossings and optimize it in drawing hierarchical graphs.

In drawing hierarchical graphs, the following approach is widely taken [10].

Step 1: For every edge spanning three or more layers, add a dummy vertex to every position where the edge and a layer cross so that all edges span only two layers.

Step 2: Iterate the following procedure from the first (top) layer to the last (bottom) layer in order.

---

[1] A hierarchical drawing is often used to draw a directed acyclic graph. That is, general methods for drawing directed acyclic graphs usually begin by transforming the input graph into a hierarchical graph, then applying a hierarchical graph drawing algorithm.

T. Asano et al. (Eds.): COCOON'99, LNCS 1627, pp. 81–91, 1999.
© Springer-Verlag Berlin Heidelberg 1999

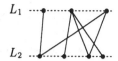

Fig. 1. Two-layered graph.

(1) Select two adjacent layers.
(2) Fix the ordering of the vertices in the higher layer[2], and reduce edge crossings by permuting the ordering of the vertices in the lower layer.
Step 3: Eliminate the dummy vertices from the graph, and then adjust the positions of the vertices to reduce the bends.

In this paper, we focus on Step 2 and present a polynomial-time approximation algorithm for drawing a two-layered graph with as few edge crossings as possible.

2   Minimum edge crossings problem for two-layered graphs

A two-layered graph is defined as a directed graph $G = (L_1, L_2, E)$, where L_1 and L_2 are disjoint sets of vertices and E is a subset of $L_1 \times L_2$ (see Fig. 1). We call L_1 the higher layer and L_2 the lower layer.

Let $G = (L_1, L_2, E)$ be a two-layered graph, x_1 be an ordering of the vertices in L_1, and x_2 be an ordering of the vertices in L_2. For given G, x_1 and x_2, we denote by $cross(G, x_1, x_2)$ edge crossings of G.

Definition 1. (The minimum edge crossings problem for two-layered graphs)
Instance: Two-layered graph $G = (L_1, L_2, E)$ and (fixed) ordering x_1 of the vertices in L_1.
Problem: Find an ordering x_2 of the vertices in L_2 such that $cross(G, x_1, x_2)$ is as small as possible.   □

For given a graph G and an ordering x_1 of the vertices in L_1, the minimum edge crossings for G and x_1 is denoted by

$$opt(G, x_1) := \min\{cross(G, x_1, x_2) \mid x_2 \text{ is an ordering of the vertices in } L_2\}.$$

The minimum edge crossings problem for two-layered graphs is shown to be NP-hard by Eades et al. [4][3] and therefore some heuristics [3,5,10] have been proposed for this problem.

The barycenter method [10] determines the x coordinate of each $u \in L_2$ to be the barycenter of the x coordinates of the adjacent vertices of u. Eades et al. [5]

---

[2] Though an alternative approach exists where the ordering of the vertices in the higher layer is not fixed either, the approach described here is used more often.
[3] Even if the maximum degree of the vertices in L_2 is not greater than 2, the problem is NP-hard.

showed that the approximation ratio of the barycenter method is $O(\sqrt{n})$, where $n = |L_1|$. They also gave an example that the barycenter method constructs a solution with $O(\sqrt{n})$ times as many edge crossings as the minimum edge crossings. This method may thus construct a worst solution in certain cases.

The median method [5], on the other hand, determines the x coordinate of each $u \in L_2$ to be the median of the x coordinates of the adjacent vertices of u. Eades $et\ al.$ [5] showed that the approximation ratio of the median method is $\dfrac{3 - c^2 + O(1/n)}{1 + c^2 - O(1/n^2)}$, where $n = |L_1| = |L_2|$ and $cn^2 (0 < c < 1)$ is the number of edges. This value approaches one as the number of vertices increases and the input graph becomes denser. This method therefore constructs good solutions that are very close to the optimum for dense graphs with large numbers of vertices. For sparse graphs, however, the approximation ratio has not been studied in detail.

Mäkinen [8] and Jünger $et\ al.$ [7] experimentally compared the barycenter method with the median method. They concluded that the averages of the edge crossings in the solutions constructed by the two methods are almost the same and that the barycenter method constructs slightly better solutions.

3   Approximation algorithm and error bound

We denote by (t, u) the edge whose initial vertex is t and terminal vertex is u. For two vertices $u, v \in L_2$, we define c_{uv} as follows:

$$c_{uv} := |\{< (t, u), (w, v) >|\ (t, u), (w, v) \in E,\ t, w \in L_1,\ x_1(t) > x_1(w)\}|,$$

which is the number of edge crossings when u is allocated to the left-hand side of v, i.e., $x_2(u) < x_2(v)$.

3.1   Algorithm

We propose the following polynomial-time approximation algorithm. In this algorithm, we use the greedy strategy of aligning the vertices so that edge crossings become as small as possible based on the allocation of any pair of vertices in L_2.

Input: Two-layered graph $G = (L_1, L_2, E)$ and (fixed) ordering x_1 of the vertices in L_1.
Output: An ordering x_2 of the vertices in L_2.

Step 0. $L := L_2,\quad i := 1$.
Step 1. Compute c_{uv} for all pairs of the vertices in L_2 $(u, v \in L_2)$.
Step 2. Find a vertex $u^* \in L$ that minimizes $\sum_{v \in L} c_{uv} / \sum_{v \in L} \min\{c_{vu}, c_{uv}\}$.

$$x_2(u^*) := i,\ L := L - \{u^*\}.$$

Step 3. If $L \neq \emptyset$, $i := i + 1$ and go to Step 2; otherwise stop.

The algorithm runs in polynomial-time. In Step 1, $O(|E|^2)$ time is needed because we compare the edges whose terminal vertex is u with the edges whose terminal vertex is v for all pairs of vertices $u, v \in L$. In Step 2, if we compute $\sum_{v \in L_2} c_{uv}$ and $\sum_{v \in L_2} \min\{c_{vu}, c_{uv}\}$ in advance, we can then compute $\sum_{v \in L_2} c_{uv} / \sum_{v \in L_2} \min\{c_{vu}, c_{uv}\}$ in constant time. In each iteration, we can renew $\sum_{v \in L} c_{uv}$ and $\sum_{v \in L} \min\{c_{vu}, c_{uv}\}$ in $O(|L|) \leq O(|L_2|)$ time by referring to $c_{u^* v}$ and c_{vu^*} $(u^* \in L \subseteq L_2)$. Step 2 therefore requires $O(|L_2|) \times |L_2| = O(|L_2|^2)$ time. Hence, our algorithm runs in $\max\{O(|E|^2), O(|L_2|^2)\}$ time.

Remark 1. The barycenter method and the median method both run in $O(|E|)$ time. Our algorithm requires more time than these methods. □

3.2 Error bound

For given positive integer d, $c(d)$ is defined as the minimum value that satisfies the following inequality: for any positive integers $d_1, d_2 \leq d$,

$$(\lceil \frac{c(d)-1}{c(d)+1} d_1 \rceil - 1) \times (\lceil \frac{c(d)-1}{c(d)+1} d_2 \rceil) + d_1 \times (\lfloor \frac{2}{c(d)+1} d_2 \rfloor) \leq \frac{c(d)}{c(d)+1} d_1 d_2$$
$$\text{and } c(d) \geq 2.$$

In this section, we show that the approximation ratio of our algorithm, i.e., the maximum ratio of the edge crossings in the solution constructed by our algorithm to the minimum edge crossings is $c(d)$ when the maximum degree of the vertices in L_2 is d.

It follows from the definition of c_{uv} that

$$cross(G, x_1, x_2) = \sum_{x_2(u) < x_2(v)} c_{uv}, \quad opt(G, x_1) \geq \sum_{u, v \in L_2} \min(c_{uv}, c_{vu}).$$

We see that for some ordering x_2 of the vertices in L_2, there exists some constant B, for which if $c_{uv} \leq B \cdot c_{vu}$ holds for any u, v such that $x_2(u) < x_2(v)$, then $B \cdot opt(G, x_1) \geq cross(G, x_1, x_2)$. To show that the approximation ratio of our algorithm is $c(d)$, it is sufficient to prove that $c_{uv} \leq c(d) \cdot c_{vu}$ for any u, v such that $x_2(u) < x_2(v)$. Since vertex u in L is already allocated in the leftmost position in L, where u minimizes $\sum_{v \in L} c_{uv} / \sum_{v \in L} \min\{c_{vu}, c_{uv}\}$ in each iteration of Step 2, it suffices to show the statement that for any $L \subseteq L_2$, there exists $u \in L$ that satisfies $c_{uv} \leq c(d) \cdot c_{vu}$ for any $v \in L - \{u\}$.

We have the following lemma to prove the above statement. For vertex u ($\in L_2$), let d_u be the degree of u, and $adj(u)_1, adj(u)_2, \ldots, adj(u)_{d_u}$ be the adjacent vertices aligned from the left side.

Lemma 1. For given positive integer d, let $u_1, \ldots, u_m \in L$ be a sequence of vertices such that $d_{u_i} \leq d$ $(i = 1, 2, \ldots, m)$ and $c_{u_j u_{j+1}} > c(d) \cdot c_{u_{j+1} u_j}$ $(j = 1, 2, \ldots, m-1)$. For vertex u, we define $l(u) = \frac{c(d)-1}{c(d)+1} d_u$ and $r(u) = \frac{2}{c(d)+1} d_u$ (Note that $l(u) + r(u) = d_u$). Then, we have the following properties (see Fig. 2).

(1) The x coordinate of $adj(u_1)_{\lceil l(u_1) \rceil}$ is not less than that of $adj(u_m)_{\lceil l(u_m) \rceil}$.
(2) The x coordinate of $adj(u_1)_{\lfloor r(u_1) \rfloor + 1}$ is not less than that of $adj(u_m)_{\lfloor r(u_m) \rfloor + 1}$. □

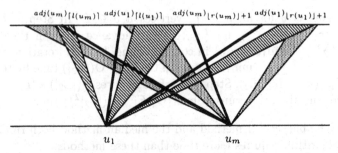

Fig. 2. Allocation of the adjacent vertices of $u_1, u_m \in L_2$.

We can easily show this lemma by noting that $c_{u_i u_{i+1}} \leq c(d) \cdot c_{u_{i+1} u_i}$, when vertices u_i, u_{i+1} satisfy neither (1) nor (2).

Theorem 1. If the maximum degree of the vertices in $L (\subseteq L_2)$ is not greater than d, then, for any $L \subseteq L_2$, there exists $u \in L$ satisfying $c_{uv} \leq c(d) \cdot c_{vu}$ for any $v \in L - \{u\}$.

Proof. For a sequence of vertices $u_0, \ldots, u_m$, let $adj(u_i)_1, adj(u_i)_2, \ldots, adj(u_i)_{d_{u_i}}$ be the adjacent vertices of u_i aligned from the left side. Let h_j^i denote the number of adjacent vertices of u_i whose x coordinates are between the x coordinate of $adj(u_{i+1})_j$ and that of $adj(u_{i+1})_{j+1}$. If $adj(u_{i+1})_j$ is one of the adjacent vertices of u_i, we define $h_j^{i'} = 1$, otherwise $h_j^{i'} = 0$.

For some $L \subseteq L_2$ and for any $u \in L$, we assume that there exists $v \in L - \{u\}$ satisfying $c_{uv} > c(d) \cdot c_{vu}$. Then, we consider a directed graph $G' = (L, E')$, where $E' = \{(u, v) \mid u, v \in L, c_{uv} > c(d) \cdot c_{vu}\}$. We see that cycle $C(u_0, \ldots, u_m)$ exists in $G' = (L, E')$ from the assumption. We show below that the existence of this cycle leads to a contradiction. See [11] for full details of the derivation.

We express $c_{u_i u_{i+1}}$ and $c_{u_{i+1} u_i}$, in terms of h_j^i and $h_j^{i'}$.

$$c_{u_i u_{i+1}} = \sum_{j=1}^{d_{u_i}} h_j^i \times j + \sum_{j=2}^{d_{u_i}} h_j^{i'} \times (j-1). \tag{1}$$

$$c_{u_{i+1} u_i} = \sum_{j=0}^{d_{u_i}-1} h_j^i \times (d_{u_i} - j) + \sum_{j=1}^{d_{u_i}-1} h_j^{i'} \times (d_{u_i} - j). \tag{2}$$

Since (u_0, u_1) exists in G', $c_{u_0 u_1} > c(d) \cdot c_{u_1 u_0}$. Substituting (1) and (2) into this inequality, we have

$$\sum_{j=\lceil \frac{c(d)}{c(d)+1} d_{u_1} \rceil}^{d_{u_1}} \{(c(d) + 1)j - c(d) \cdot d_{u_1}\} \times h_j^1$$

$$+ \sum_{j=\lceil \frac{c(d) \cdot d_{u_1}+1}{c(d)+1} \rceil}^{d_{u_1}} \{(c(d) + 1)j - c(d) \cdot d_{u_1} - 1\} \times h_j^{1'}$$

$$> \sum_{j=0}^{\lfloor \frac{c(d)}{c(d)+1} d_{u_1} \rfloor} \{c(d) \cdot d_{u_1} - (c(d) + 1) j\} \times h_j^1$$

$$+ \sum_{j=0}^{\lfloor \frac{c(d) \cdot d_{u_1} + 1}{c(d)+1} \rfloor} \{c(d) \cdot d_{u_1} - (c(d) + 1) j + 1\} \times h_j^{1'}. \tag{3}$$

Let h_j^* be the number of adjacent vertices of u_0 whose x coordinates are between the x coordinate of $adj(u_m)_j$ and that of $adj(u_m)_{j+1}$. If $adj(u_m)_j$ is one of the adjacent vertices of u_0, we define $h_j^{*'} = 1$, otherwise $h_j^{*'} = 0$. We apply Lemma 1 to path $(u_1, \ldots, u_m)$. Then, we have

$$h_0^1 + h_1^{1'} + \cdots + h_{u_{\lceil l(u_1) \rceil}}^1{}' \geq h_0^* + h_1^{*'} + \cdots + h_{u_{\lceil l(u_m) \rceil}}^*{}' \tag{4}$$

since Lemma 1 (1). We similarly obtain

$$h_{\lfloor r(u_m) \rfloor + 1}^* + \cdots + h_{u_{d u_m}}^*{}' + h_{u_{d u_m}}^*$$

$$\geq h_{u_{\lfloor r(u_1) \rfloor + 1}}^1 + \cdots + h_{u_{d u_1}}^1{}' + h_{u_{d u_1}}^1$$

$$> \frac{1}{d_{u_1}} \left[\sum_{j=0}^{\lfloor \frac{c(d)}{c(d)+1} d_{u_1} \rfloor} \{c(d) \cdot d_{u_1} - (c(d) + 1) j\} \times h_j^1 \right.$$

$$\left. + \sum_{j=0}^{\lfloor \frac{c(d) \cdot d_{u_1} + 1}{c(d)+1} \rfloor} \{c(d) \cdot d_{u_1} - (c(d) + 1) j + 1\} \times h_j^{1'} \right] \tag{5}$$

since Lemma 1 (2). Note that we use (3) for the inequality between the middle and the right-hand sides of (5). Computing the right-hand side of (5), we have

$$\text{RHS of (5)} \geq h_0^1 + h_1^{1'} + \cdots + h_{u_{\lceil l(u_1) \rceil}}^1{}',$$

where RHS means the right-hand side. We combine this with (4), which results in

$$h_{\lfloor r(u_m) \rfloor + 1}^* + \cdots + h_{u_{d u_m}}^*{}' + h_{u_{d u_m}}^* > h_0^* + h_1^{*'} + \cdots + h_{u_{\lceil l(u_m) \rceil}}^*{}'. \tag{6}$$

We now turn to compute $c(d) \cdot c_{u_0 u_m} - c_{u_m u_0}$.

$$c(d) \cdot c_{u_0 u_m} - c_{u_m u_0} \geq$$

$$c(d) \left\{ \sum_{j=\lceil l(u_m) \rceil + 1}^{\lfloor r(u_m) \rfloor + 1} (h_{j-1}^* + h_j^{*'}) \times \lceil l(u_m) \rceil + \sum_{j=\lfloor r(u_m) \rfloor + 2}^{d_{u_m}} (h_{j-1}^* + h_j^{*'}) \times \lceil r(u_m) \rceil \right\}$$

$$- \left\{ \sum_{j=0}^{\lceil l(u_m) \rceil} (h_{j-1}^* + h_j^{*'}) \times d_{u_m} + \sum_{j=\lceil l(u_m) \rceil + 1}^{\lfloor r(u_m) \rfloor + 1} (h_{j-1}^* + h_j^{*'}) \times \lfloor r(u_m) \rfloor \right.$$

$$+ \sum_{j=\lfloor r(u_m)\rfloor+2}^{d_{u_m}} (h^*_{j-1} + h^{*\prime}_j) \times \lceil l(u_m)\rceil \Bigg\}. \tag{7}$$

Note that we use the property shown in Fig. 3 for the inequality. We next apply $\lceil a\rceil - \lfloor b\rfloor > a - b$ (a, b are real numbers) to the first and the forth terms of the right-hand side of (7). We also apply $\lfloor a\rfloor + 1 - \lceil b\rceil > a - b - 1$ to the second and the fifth terms. It then follows from $c(d) \geq 2$ and $d_{u_m} \geq 1$ that

$$\text{RHS of (7)} \geq \sum_{j=\lfloor r(u_m)\rfloor+2}^{d_{u_m}} (h^*_{j-1} + h^{*\prime}_j) \times d_{u_m} - \sum_{j=0}^{\lceil l(u_m)\rceil} (h^*_{j-1} + h^{*\prime}_j) \times d_{u_m}.$$

Applying (6) to the above inequality leads to $c(d) \cdot c_{u_0 u_m} - c_{u_m u_0} \geq 0$, which contradicts our assumption. (From the assumption, $c_{u_m u_0} > c(d) \cdot c_{u_0 u_m}$ due to the existence of (u_m, u_0) in G'.) $\square$

Theorem 2. The approximation ratio of our algorithm is $c(d)$ if the maximum degree of the vertices in L_2 is d. $\square$

As we see, $c(d)$ is two if the maximum degree d of the vertices in L_2 is small. As d becomes larger, $c(d)$ monotonically increases to three. For example, we have $c(2) = c(3) = c(4) = 2$, $c(5) = c(6) = 7/3+\varepsilon$ (ε is an arbitrary positive number).

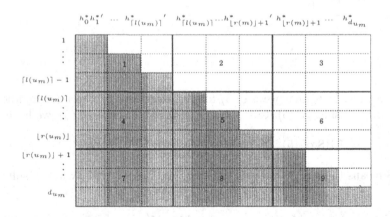

Fig. 3. Property for comparing greatness between $c(d) \cdot c_{u_0 u_m}$ and $c_{u_m u_0}$. (We generate a rectangle whose vertical-side length is d_{u_m} and whose horizontal-side length is d_{u_0}. We divide the vertical length uniformly into d_{u_m} and the horizontal length into d_{u_m}, each length of which is $h^*_{i-1} + h^{*\prime}_i$ ($i = 1, 2, \ldots, d_{u_m}$). Hence, the white region corresponds to the edge crossings when u_0 is allocated to the left-hand side of u_m while the black region corresponds to the sum of the adjacent vertices that u_0 and u_m share and the edge crossings when u_0 is allocated to the right-hand side of u_m. The thick lines are the $\lceil l(u_m)\rceil$th and the $\lfloor r(u_m)\rfloor$th lines of the adjacent edges of u_m aligned from the left side. Then, we see that "the area of the white region" $\geq$ "the sum of areas 2, 3 and 6" and that "the area of the black region" $\leq$ "the sum of areas 1, 4, 5, 7, 8 and 9".)

Corollary 1. The approximation ratio of our algorithm is two when the maximum degree of the vertices in L_2 is not greater than four, while it is $(7/3 + \varepsilon)$ (ε is an arbitrary positive number) when the maximum degree of the vertices in L_2 is not greater than six.   □

If the maximum degree d of the vertices in L_2 is even, we can construct an example in which the approximation ratio of the median method achieves $(3d - 2)/d$. When $d = 2$, the ratio is 2, while it is 2.5 when $d = 4$. The ratio is 8/3 for $d = 6$. Similar discussion can be done for the case of odd d. The approximation ratio of our algorithm is therefore strictly better than that of the median method when the maximum degree of the vertices in the lower layer is small.

4   Experimental results

We denote by n_1 the number of vertices in L_1 and by n_2 the number of vertices in L_2. In addition, We denote by p ($0 \leq p \leq 1$) the probability of generating edges between two vertices of interest. We set $5 \leq n_2 \leq 10$ and $n_1 = 2n_2$. For each n_2, we set p to be 0.2, 0.5 and 0.8, respectively. We generated two-layered graphs based on these three parameters n_1, n_2, p. To be more concrete, we generated n_1 vertices in L_1 and n_2 vertices in L_2, and then generated an edge (t, u) with probability p for each pair of vertices t, u ($t \in L_1, u \in L_2$) to generate a two-layered graph. For the generated graph, we fixed the ordering of the vertices in L_1. We then applied our algorithm (GM), the barycenter method (BM) and the median method (MM) to the graph, respectively, to compute edge crossings. Next, we computed the difference between the edge crossings obtained by each algorithm and the minimum edge crossings. We iterated the generation of a two-layered graph and the computation of the differences 50 times. We then computed the average and the standard deviation of the differences over the 50 times (Fig. 4).

In particular, we evaluated in detail the cases where the maximum degree of the vertices in L_2 is four and six, respectively (see Corollary 1). In a similar way, we iterated 50 times the generation of a two-layered graph whose maximum degree of the vertices in the lower layer is four and six, respectively. In this case, for each n_2 ($n_2 = 5, 6, \ldots, 10$; $n_1 = 2n_2$), we generated a two-layered graph by randomly generating an integer d_u between zero and four (six) for each vertex u ($\in L_2$), then generating d_u edges (t_i, u) ($i = 1, 2, \ldots, d_u$) where t_i is randomly selected from L_1. In addition to the averages of the differences, we also payed attention to the worst case among the 50 times. These results are shown in Figs. 5 and 6.

Figure 4 shows that our algorithm almost always constructs the optimum solutions for graphs with small numbers of vertices. This observation does not depend on p. The barycenter method and the median method, on the other hand, constructed the solutions with more edge crossings than the minimum edge crossings. This indicates that our algorithm is better than the other methods regardless of the number of vertices or edge density.

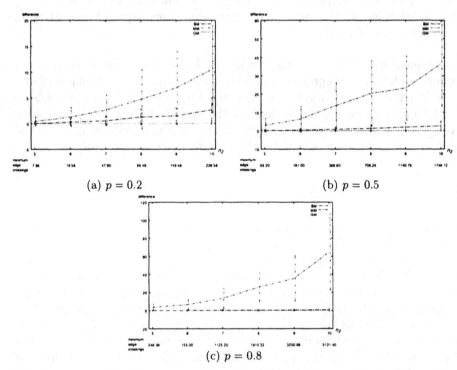

(a) $p = 0.2$ (b) $p = 0.5$

(c) $p = 0.8$

Fig. 4. Average of the differences in the edge crossings obtained by the three algorithms from the minimum crossings. (BM, MM and GM represent the barycenter method, the median method, and our algorithm, respectively.)

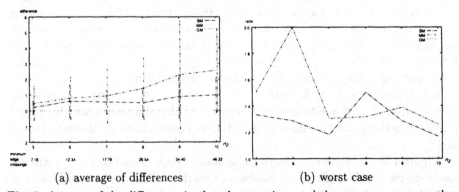

(a) average of differences   (b) worst case

Fig. 5. Average of the differences in the edge crossings and the worst case among the 50 times (the maximum degree of the vertices in L_2 is four). (The ratio in the worst case for GM is always 1.0.)

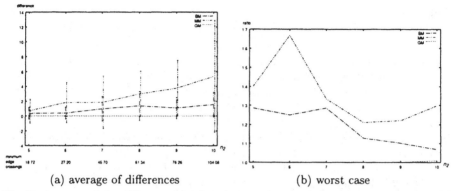

(a) average of differences   (b) worst case

Fig. 6. Average of the differences in the edge crossings and the worst case among the 50 times (the maximum degree of the vertices in L_2 is six).

Figures 5 and 6 show that our algorithm always constructs the optimum solutions for graphs whose maximum degree of the vertices in the lower layer is four or six. This is observed for any number of vertices of input graphs. (We have only one exceptional case where our algorithm failed to construct the optimum solution: see the case of $n_2 = 10$ in Fig. 6.) Comparison of the solutions in the worst case with the optimum solutions indicates that our algorithm always succeeds in constructing the optimum solutions, while the barycenter method and the median method do not.

5   Summary

We presented a polynomial-time approximation algorithm for the minimum edge crossings problem for two-layered graphs and showed that the number of edge crossings in solutions obtained by our algorithm is at most $c(d)$ times as large as the number of the minimum edge crossings, where d is the maximum degree of the vertices in the lower layer and $c(d)$ is a monotonic increasing function ranging between two and three (see Section 3.2 for its definition). Since the approximation ratio of the median method is three and an example exists in which the ratio reaches three, the error bound of the solutions constructed by our algorithm is strictly better than that of the median method.

We experimentally compared our algorithm with the barycenter method and the median method. The experimental results showed that our algorithm obtains better solutions than the other two methods. In particular, for the graphs with small numbers of vertices or with the small maximum degrees of the vertices in the lower layer, our algorithm almost always constructs the optimum solution regardless of their edge density.

Clarifying the approximation ratio independent of the maximum degree of the vertices in the lower layer is left for future research. Discussing the lower bound of the errors in solutions of the minimum edge crossings problem for two-layered graphs is also included in future investigations.

References

1. C. Batini, L. Furlani, and E. Nardelli: What Is a Good Diagram?: A Pragmatic Approach, *Proc. of the 4th International Conference on Entity-Relationship Approach*, pp. 312–319, 1985.
2. G. Di Battista, P. Eades, R. Tamassia, and I. G. Tollis: Algorithms for Drawing Graphs: An Annotated Bibliography, *Computational Geometry*, 4 (1994), pp. 235–282.
3. T. Catarci: The Assignment Heuristic for Crossing Reduction, *IEEE Transactions on Systems, Man, and Cybernetics*, 25 (1995), 3, pp. 515–521.
4. P. Eades, B. D. McKay, and N. Wormald: On an Edge Crossing Problem, *Proc. of the 9th Australian Computer Science Conference*, pp. 327–334, 1986.
5. P. Eades and N. Wormald: Edge Crossings in Drawings of Bipartite Graphs, *Algorithmica*, 11 (1994), 4, pp. 379–403.
6. C. Esposito: Graph Graphics: Theory and Practice, *Computers and Mathematics with Applications*, 15 (1988), 4, pp. 247–253.
7. M. Jünger and P. Mutzel: 2-Layer Straightline Crossing Minimization: Performance of Exact and Heuristic Algorithms, *Journal of Graph Algorithms and Applications*, 1 (1997), 1, pp. 1–25.
8. E. Mäkinen: Experiments on Drawing 2-level Hierarchical Graphs, *International Journal of Computer Mathematics*, 36 (1990), pp. 175–181.
9. H. C. Purchase, R. F. Cohen, and M. James: Validating Graph Drawing Aesthetics, *Proc. of Symposium on Graph Drawing, GD'95 (Lecture Notes in Computer Science, Vol. 1027)*, pp. 435–446, Springer, 1996.
10. K. Sugiyama, S. Tagawa, and M. Toda: Methods for Visual Understanding of Hierarchical Systems, *IEEE Transactions on Systems, Man, and Cybernetics*, 11 (1981), 2, pp. 109–125.
11. A. Yamaguchi and A. Sugimoto: *An Approximation Algorithm for the Two-layered Graph Drawing Problem*, ARL Research Report, 98-001, 1999.

Area Minimization for Grid Visibility Representation of Hierarchically Planar Graphs

Xuemin Lin[1] and Peter Eades[2]

[1] School of Computer Science & Engineering
The University of New South Wales
Sydney, NSW 2052, Australia. lxue@cse.unsw.edu.au.
[2] Department of Computer Science & Software Engineering
The University of Newcastle
Callaghan, NSW 2308, Australia.

Abstract. Hierarchical graphs are an important class of graphs for modelling many real applications in software and information visualization. In this paper, we shall investigate the computational complexity of constructing minimum area grid visibility representations of hierarchically planar graphs. Firstly, we provide a quadratic algorithm that minimizes the drawing area with respect to a fixed planar embedding. This implies that the area minimization problem is polynomial time solvable restricted to the class of graphs whose planar embeddings are unique. Secondly, we show that the area minimization problem is generally NP-hard.

Keywords: Graph Drawing, Hierarchically Planar Graph, Visibility Representation, Drawing Area.

1 Introduction

Automatic graph drawing plays an important role in many modern computer-based applications, such as CASE tools, software and information visualization, VLSI design, visual data mining, and internet navigation. Directed acyclic graphs are an important class [2] of graphs to be investigated in this area. The *upward* drawing convention for drawing directed acyclic graphs has received a great deal of attention since last decade; and a number of results for drawing *upward planar graphs* have been published [2,4,6,10,15,16].

Consider [7,8] that directed acyclic graphs are not powerful enough to model every real-life application. "Hierarchical" graphs are then introduced, where layering information is added to a directed acyclic graph. Consequently, the "hierarchical" drawing convention is proposed to display the specified layering information.

Due to the additional layering constraint, most problems in hierarchical drawing are inherently different to those in upward drawing. For example, testing for "upward planarity" of directed acyclic graphs is NP-Complete [10], while it can be done in linear time [3,11] for "hierarchically planarity". Therefore, issues, such as "planar", "straight-line", "convex", and "symmetric" representations,

T. Asano et al. (Eds.): COCOON'99, LNCS 1627, pp. 92–102, 1999.
© Springer-Verlag Berlin Heidelberg 1999

have been revisited [7,8,11,12,13] with respect to hierarchically planar graphs. In this paper, we shall investigate the problem of minimizing the drawing area for hierarchically planar graphs, where drawings are restricted to the 2-dimensional space.

Drawing a hierarchically planar graph involves two phases: 1) computing a "planar embedding", and 2) finding a good drawing "respecting" the embedding. A linear time algorithm [11] was proposed for phase 1. In [8], a simple and *force-directed* algorithm was developed that integrates the two phases and can deliver a convex and symmetric drawing. However, the results in [8] are applicable only to a special class of graphs - *well connected* graphs [8]. In [7], an efficient polynomial algorithm was provided for drawing hierarchically planar graph by the *straight-line* drawing standard (that is, using points to represent vertices and straight-line segments to represent arcs). Considering that these three algorithms [7,8,11] may produce drawings with exponential areas for a given *resolution requirement*, in our earlier work [13] we proved that exponential area is generally necessary for straight-line drawings. To resolve the exponential drawing area problem, a relaxation of the drawing standard such as allowing line segments to represent vertices was made, as with the upward planar graphs [2,6,16]. Particularly, in [4,13] it is shown that the drawing area can be always made within a quadratic area if the "grid visibility representation" is employed for hierarchically planar graphs. Moreover, an efficient grid visibility representation algorithm was presented that can achieve the minimum drawing area for a hierarchically planar graph with only one "source", only one "sink", and a fixed "planar embedding".

This paper presents a more general investigation than that in [13]. Firstly, we present an efficient (quadratic time) grid visibility representation algorithm that guarantees the minimum drawing area for a hierarchically planar graph with arbitrary number of sources and sinks, and a fixed planar embedding. This implies that for a hierarchical graph with the unique planar embedding, the area minimization problem for grid visibility representation is polynomial time solvable. This result is more general than that in [13]. The second contribution of the paper is to prove that the problem of area minimization is NP-hard for the grid visibility representation if a planar embedding is not fixed.

The rest of the paper is organized as follows. Section 2 gives the basic terminology and background, as well as the precise definition of our problem. Section 3 presents the first contribution and Section 4 presents the second contribution. This is followed by the conclusions and remarks.

2   Preliminaries

The basic graph theoretic definitions can be found in [1].

A *hierarchical graph* $H = (V, A, \lambda, k)$ consists of a simple directed acyclic graph (V, A), a positive integer k, and for each vertex u, an integer $\lambda(u) \in \{1, 2, ..., k\}$ with the property that if $u \to v \in A$, then $\lambda(u) > \lambda(v)$. For $1 \leq i \leq k$

the set $\{u : \lambda(u) = i\}$ of vertices is the *ith layer* of H and is denoted by L_i. An arc $a = u \rightarrow v$ in H is *long* if it spans more than two layers, that is, $\lambda(u) - \lambda(v) \geq 2$.

For each vertex u in H, we use A_u to denote the set of arcs incident to u, A_u^+ to denote the set of arcs outgoing from u, and A_u^- to denote the set of arcs incoming to u. A *sink* u of a hierarchical graph H is a vertex that does not have outgoing arcs; that is, $A_u^+ = \emptyset$. A *source* of H is a vertex that does not have incoming arcs; that is, $A_u^- = \emptyset$.

A hierarchical graph is *proper* if it has no long arcs. Clearly, adding $\lambda(u) - \lambda(v) - 1$ dummy vertices to each long arc $u \rightarrow v$ in an improper hierarchical graph H results in a proper hierarchical graph, denoted by H_p; H_p is called the *proper image* of H. Note that $H_p = H$ if H is proper.

To display the specified hierarchical information in a hierarchical graph, the *hierarchical drawing convention* is proposed, where a vertex in each layer L_i is separately allocated on the horizontal line $y = i$ and arcs are represented as curves monotonic in y direction; see Figures 1 (a)-(c). In this paper, we will discuss only hierarchical drawing convention.

A drawing is *planar* if no pair of no-incident arcs intersect. A hierarchical graph is *hierarchically planar* if it has a planar drawing admitting the hierarchical drawing convention.

An *embedding* E_H of a proper hierarchical graph H gives an ordered vertex set $\mathcal{L}_i$ for each layer L_i in H. For a pair of vertices $u, v \in \mathcal{L}_i$, u is on the *left side* of v if $u < v$. An *embedding* of an improper hierarchical graph H means an embedding of the proper image H_p of H, and is also denoted by E_H. Note that for an improper hierarchical graph H, $\mathcal{L}_i$ may contain more vertices than L_i due to additional dummy vertices.

A hierarchical drawing α of H *respects* E_H if for each pair of vertices u, v in a $\mathcal{L}_i$, the x-coordinate value $\alpha(u)$ is smaller than that of $\alpha(v)$ if and only if $u < v$. An embedding E_H is *planar* if a straight-line drawing of H_p respecting E_H is planar.

Besides straight line drawing, various drawing standards are available for drawing hierarchically planar graphs; see Figures 1(a) - 1(c) for example. In this paper we are focused only on a "visibility representation" of a hierarchically planar graph. In a *visibility representation* β, each vertex u is represented as a horizontal line segment $\beta(u)$ on $y = \lambda(u)$ and each arc $u \rightarrow v$ is represented as a vertical line segment connecting $\beta(u)$ and $\beta(v)$, such that:

- $\beta(u)$ and $\beta(v)$ are disjoint if $u \neq v$, and
- a vertical line segment and a horizontal line segment do not intersect if the corresponding arc and vertex are not incident.

See Figure 1(b), for example. Note that in a visibility representation, a line segment used to represent a vertex may be degenerate to a point.

A visibility representation is a *grid drawing* if horizontal line segments and vertical line segments use only grid points as their ends. The *drawing area* of a grid visibility representation β is the area of the minimum isothetic rectangle that contains β. The *width* and the *height* of β are the width and height respectively

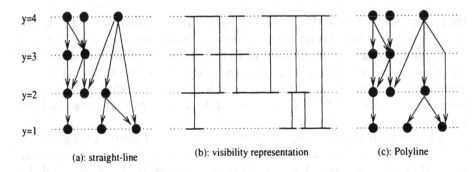

(a): straight-line   (b): visibility representation   (c): Polyline

Fig. 1. Various Representations

of this rectangle. In [13], we showed that a hierarchical graph is hierarchically planar if and only if it admits a grid visibility representation. In this paper, we shall study the following optimization problem.

Minimum Area of Grid Visibility Drawing (MAGVD)

INSTANCE: A hierarchical planar graph H is given.

QUESTION: Find a grid visibility representation of H such that the drawing area is minimized.

Without loss of generality, we assume that in H, there is no *isolated* vertex - a vertex without any incident arcs. Note that all grid visibility representations of a given hierarchically planar graph have the fixed height according to the drawing convention. Consequently, MAGVD is reduced to the width drawing minimization problem. In section 4, we will prove that MAGVD is NP-hard. Firstly, however, we show that it is polynomially solvable if the planar embedding is given as part of the input.

3   Area Minimization for a Fixed Planar Embedding

Di Battista and Tamassia [4] proposed an efficient grid visibility representation algorithm, VISIBILITY_DRAW, for drawing upward planar graphs. In fact the algorithm can be immediately applied to hierarchically planar graphs [13] with a fixed planar embedding. Below is the version for hierarchical graphs.

Algorithm VISIBILITY_DRAW

INPUT: a hierarchically planar graph H and its planar embedding E_H.
OUTPUT: a grid visibility representation of H respecting E_H.

Step 1: Labelling. Give each arc a an integer $l(a)$.

Step 2: Drawing. This step follows immediately Step 1 and draws H based on the output of Step 1. It consists of the following two phases: drawing vertices and drawing arcs of H.

Drawing vertices. For each vertex $u \in H$, let A_u represent the set of arcs in H which are incident to u. Assume $u \in L_i$. Represent u by the horizontal line segment from $(\min_{a \in A_u}\{l(a)\}, i)$ to $(\max_{a \in A_u}\{l(a)\}, i)$.

Drawing arcs. Represent an arc $a = u \rightarrow v$ with $u \in L_i$ and $v \in L_j$ by the vertical line segment from $(l(a), i)$ to $(l(a), j)$. $\square$

Suppose that the largest x-coordinate value assigned to a grid visibility representation β of H is N, and the smallest one is 1. Then the width of β is $N - 1$. Clearly, the key in applying the Algorithm VISIBILITY_DRAW to minimizing drawing width is to optimize Step 1. Note that neither the original labelling technique (*dual graph* technique) in [4] nor the labelling technique in [13] can guarantee the minimality of the drawing width for hierarchically planar graphs with arbitrary number of sources and sinks, and with a fixed planar embedding.

In this section, we provide a new algorithm OPTIMAL_LABELLING to Step 1, which guarantees the minimum drawing area for a hierarchically planar graph with a fixed planar embedding. The basic idea is simple - labelling each arc with the minimal possible integer.

To describe OPTIMAL_LABELING, the following notation is needed. For two different arcs $a_1 = u_1 \rightarrow v_1, a_2 = u_2 \rightarrow v_2 \in H$, a_1 is on the *left* side of a_2 with respect to E_H if and only if in E_H there are a vertex u on a_1 and a vertex v on a_2 such that u and v are in the same layer and u is on the left side of v. Note that E_H is a planar embedding of H_p, and u and v may be dummy vertices on the long arcs. The four possible cases are depicted in Figures 2 (a) - (d), where dotted lines indicate possible extensions to long arcs.

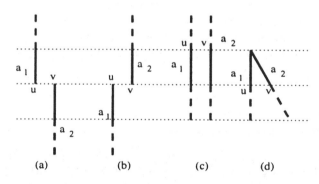

Fig. 2. 4 possible cases where a_1 is on the left side of a_2

An arc a in H is *left-most* with respect to E_H if there is no arc in H that is on the left side of a.

The algorithm OPTIMAL_LABELLING iteratively finds the left-most arcs (with respect to E_H) in H to label. In each iteration i:

S1: OPTIMAL_LABELLING scans the hierarchical graph H from the top layer to the bottom layer to label the left-most arcs in the current H with the integer i. Go to S2.

S2: OPTIMAL_LABELLING deletes all arcs labelled in this iteration; and deletes the isolated vertices resulted after arcs deletion in H. Go to $(i + 1)$th iteration.

The algorithm terminates if all arcs in H are labelled.

For instance, Figure 3(b) shows the result after applying the algorithm OPTIMAL_LABELLING to the graph with respect to the planar embedding depicted in Figure 3(a). Meanwhile, Figure 3(c) illustrates the result after applying Step 2 in VISIBILITY_DRAW to the output (Figure 3(b)) of OPTIMAL_LABELLING.

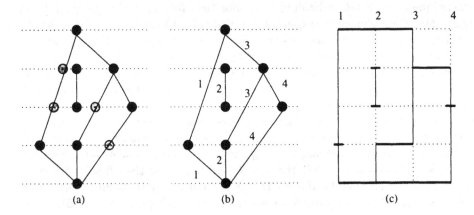

(a)   (b)   (c)

Fig. 3. OPTIMAL_LABELLING

It can be immediately verified that the drawing, given by a combination of OPTIMAL_LABELLING and Step 2 in VISIBILITY_DRAW, respects the given planar embedding E_H; that is,

Lemma 1. *The combination of OPTIMAL_LABELLING and Step 2 in VIS-IBILITY_DRAW gives a grid visibility representation of H respecting a given planar embedding E_H.*

Applying similar arguments as used in [4], we can immediately conclude that the grid visibility representation given by OPTIMAL_LABELLING occupies drawing area $O(n^2)$. Further, we can show:

Theorem 2. *Respecting a given planar embedding E_H of a hierarchically planar graph H, the grid visibility representation of H, produced by the combination of OPTIMAL_LABELLING and Step 2 in VISIBILITY_DRAW, has the minimum drawing width.*

Proof. **Sketch:** Basically, every arc has been allocated on the "left-most possible" vertical line by OPTIMAL_LABELLING. This can be immediately verified based on a mathematic induction, following the ordering of arc labelling. The full proof can be found in the full paper [14]. □

Suppose that vertices in each layer L_i in H are stored from left to right according to their ordering given by E_H, as well as the vertices in $\mathcal{L}_i$ do. Assume that for each vertex u, arcs in A_u^+ are also stored from left to right according their ordering. To execute OPTIMAL_LABELLING efficiently, S1 and S2 can be integrated together in each iteration. In each iteration i, start with the left most vertex u in the top layer of the remaining H, and search down along the leftmost arc $a = u \to v$ in A_u^+ to see if a is the leftmost arc in the current H:

case 1: If a is also the leftmost arc in the current H, then label a with i and delete a from H. Consequently, delete any resultant isolated vertices from H. Continue the iteration from the layer one level below the layer of v if A_v^+ is empty; otherwise continue the iteration from the layer of v.

case 2: If a is not the left most arc in the current H, in the remaining H there must be a vertex w such that the leftmost arc $b = u_1 \to v_1$ of A_w^+ is on the left side of a and u_1 is the leftmost vertex in the layer of u_1. Choose such a vertex u_1 that its layer number is maximized. Then continue the iteration i from the layer of u_1.

Clearly, the computation involved in the above two cases is proportional to a scan of the first vertices in the layers spanned by a. Consequently, for a $H = (V, A, \lambda, k)$ each iteration takes $O(k)$ time. Note that the number of iteration must be less than the number of arcs, because each iteration labels at least one arc. Therefore, the algorithm OPTIMAL_LABELLING runs in time $O(k|A|)$. As H is planar, $|A| = O(|V|)$; and thus the algorithm runs in $O(k|V|)$.

4   The Complexity of MAGVD

In this section, we prove the NP-hardness of MAGVD by showing the NP-completeness of the corresponding decision problem. As mentioned earlier, in MAGVD we need only to consider the drawing width minimization problem.

Decision Problem for MAGVD (DPMAGVD)

INSTANCE: A hierarchical planar graph H, and an integer K.

QUESTION: Find a grid visibility representation of H such that its width is not greater than K.

It is well known [9] that the 3-PARTITION problem is NP-complete. In our proof, we will transform 3-PARTITION to a special case of DPMAGVD.

3-PARTITION

INSTANCE: A finite set S of $3n$ elements, an integer B, and an integer weight $s(e)$ for each element $e \in S$ are given such that each $s(e)$ satisfies $\frac{B}{4} < s(e) < \frac{B}{2}$ and $\sum_{e \in S} s(a) = nB$

QUESTION: Can S be partitioned into n disjoint sets $S_1, S_2, \ldots, S_n$ such that for $1 \leq i \leq n$, $\sum_{e \in S_i} s(e) = B$?

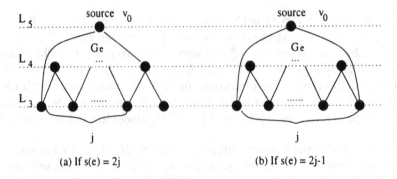

Fig. 4. Constructing G_e

Now, we transform an instance I_{3P} of 3-PARTITION to an instance $D_{I_{3P}} = (H_{I_{3P}}, K_{I_{3P}})$ of DPMAGVD. The hierarchically planar graph $H_{I_{3P}}$ has five layers and the top layer has only one source v_0. $H_{I_{3P}}$ is constructed as follows:

Each element $e \in S$ corresponds to a three-layered graph G_e that hangs over the source v_0 of $H_{I_{3P}}$, such that G_e takes one of the two possible graphs as depicted in Figures 4(a) and 4(b) depending on the odevity of $s(e)$. Further, in $H_{I_{3P}}$ we also duplicate n times a graph G_B. Here, G_B also takes one of the two shapes, depicted in Figures 5(a) and 5(b), subject to the odevity of B.

We assign $K_{I_{3P}}$ as $n(B + 2)$.

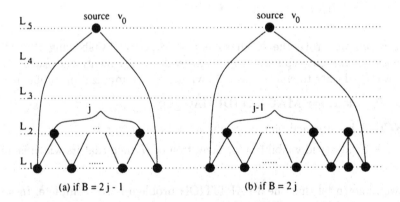

Fig. 5. Constructing G_B

Note that a G_e has a unique planar embedding up to a complete reversal, as well as G_B does. Consequently an application of OPTIMAL_LABELLING to a G_e (or G_B) can guarantee the minimality of the drawing width of G_e (or G_B). More specifically, the following Lemma can be immediately verified by the structures of G_e and G_B.

Lemma 3. *The minimum drawing width of a grid visibility representation of G_e is $s(e)$, and the minimum drawing width of G_B is $B + 2$.*

Corollary 4. *Let wid_β denote the width of a grid visibility representation β of $H_{I_{3P}}$. Then $wid_\beta \geq n(B + 2)$.*

Proof. The Corollary immediately follows Lemma 3 and the fact that $H_{I_{3P}}$ consists of n G_Bs. □

The following fact is the key to the proof of NP-Completeness of DPMAGVD.

Theorem 5. *3-PARTITION has a solution to I_{3P} if and only if DPMAGVD has a solution to $D_{I_{3P}} = (H_{I_{3P}}, K_{I_{3P}})$.*

Proof. Note that a pair of G_{e_1} and G_{e_2} cannot have any intervening apart from v_0 in a grid visibility representation α of $H_{I_{3P}}$. Further, in α a G_e must be either drawn inside a G_B or outside all G_Bs.

The above facts together with Lemma 3 and Corollary 4 imply that DP-MAGVD has a solution if and only if the set Π of subgraphs G_e of $H_{I_{3P}}$ can be divided into n disjoint sets $\Pi_1, \Pi_2, \Pi_3, \dots, \Pi_n$, and there exists a grid visibility representation β of $H_{I_{3P}}$, such that

- each Π_i consists of 3 different graphs $G_{e_{i_1}}$, $G_{e_{i_2}}$, and $G_{e_{i_3}}$;
- $s(e_{i_1}) + s(e_{i_2}) + s(e_{i_3}) = B$;
- in β, each Π_i is drawn inside the drawing of a G_B; and
- in β, the drawing of each G_B contains only one Π_i.

The theorem immediately follows. □

Furthermore, the following Lemma is immediate based on the construction of $D_{I_{3P}}$ from I_{3P}.

Lemma 6. *The transformation between each I_{3P} to $D_{I_{3P}}$ takes polynomial time with respect to $n + B$.*

Note that Lemma 6 does not necessary imply that there is a polynomial time transformation, with respect to the input size n of 3-PARTITION, between I_{3P} and $D_{I_{3P}}$, because B may be arbitrarily larger. However, it has been shown [9] that 3-PARTITION is strongly NP-complete; that is, it is NP-complete even if B is bounded by a polynomial of n. This, together with Theorem 5 and Lemma 6, imply that DPMAGVD is NP-Complete. Consequently [9]:

Theorem 7. *MAGVD is NP-hard.*

5   Conclusions

In this paper, we studied the drawing area minimization problem for hierarchically planar graph, restricted to the grid visibility representation. An efficient algorithm has been presented for producing a grid visibility representation with the minimal drawing area if a planar embedding is given and fixed. This implies that for a class of hierarchically planar graphs whose planar embeddings are unique, such as the well connected graphs [8], MAGVD is polynomial time solvable. However, we showed that in general, MAGVD is NP-hard. Note that a slight modification of the proof of NP-hardness can lead to a stronger result: MAGVD is NP-hard even restricted to hierarchically planar graphs with only one source [14]. For a possible future study, we are interested in investigating:

- whether or not similar results exist for upward planar graphs;
- a good approximation algorithm for solving MAGVD; and
- symmetric drawing issues.

References

1. J. A. Bondy and U. S. R. Murty, *Graph Theory with Application*, The Macmilan Press LTD, 1977.
2. G. Di Battista, P. D. Eades, R. Tamassia, and I. Tollis, Algorithms for Automatic Graph Drawing: An Annotated Bibliography, *Computational Geometry: Theory and Application*, 4, 235-282, 1994.
3. G. Di Battista and E. Nardelli, Hierarchies and Planarity Theory, *IEEE Tran. Sys. Man Cybern.* 18(6), 1035-1046, 1988.
4. G. Di Battista and R. Tamassia, Algorithms for Plane Representations of Acyclic Digraphs, *Theoretical Computer Science*, 61, 175-198, 1988.
5. G. Di Battista and R. Tamassia, On-Line Planarity Testing, *SIAM Journal on Computing*, 25(5), 1996.
6. G. Di Battista, R. Tamassia, and I. G. Tollis, Area Requirement and Symmetry Display of Planar Upward Drawings, *Discrete & Computational Geometry*, 7(381-401), 1992.
7. P. Eades, Q. Feng, X. Lin, and H. Nagamocha, Straight-Line Drawing Algorithms for Hierarchical Graphs and Clustered Graphs, to appear in *Algorithmic*.
8. P. Eades, X. Lin and R. Tamassia, An Algorithm for Drawing a Hierarchical Graph, *International Journal of Computational Geometry and Applications*, 6(2), 145-155, 1996.
9. M. R. Garey and D. S. Johnson, *Computers and Intractability - a guide to the theory of NP-Completeness*, Freeman, 1979.
10. A. Garg and R. Tamassia, On the Computational Complexity of Upward and Rectilinear Planarity Testing, *Graph Drawing'94*, LNCS, Springer-Verlag, 286-297, 1995.
11. L. S. Heath and S. V. Pemmaraju, Recognizing Leveled-Planar Dags in Linear Time, *Graph Drawing'95*, LNCS 1027, Springer-Verlag, 1995.
12. X. Lin, *Analysis of Algorithms for Drawing Graphs*, PHD thesis, University of Queensland, 1992.
13. X. Lin and P. Eades, Area Requirements for Drawing Hierarchically Planar Graphs, *Graph Drawing'97*, LNCS, Springer-Verlag, 1997.

14. X. Lin and P. Eades, Towards Area Requirements for Drawing Hierarchically Planar Graphs, *Manuscript*, UNSW, 1998.

15. R. Tamassia, On Embedding a Graph in the Grid with the Minimum Number of Bends, *SIAM J. Computing* 16(3), 421-444, 1987.

16. R. Tamassia and I. G. Tollis, Tesselation Representations of Planar Graphs, *Proc. 27th Annual Allerton Conf*, 1989.

Layout Problems on Lattice Graphs *

Josep Díaz[1], Mathew D. Penrose[2], Jordi Petit[1], and María Serna[1]

[1] Departament de Llenguatges i Sistemes Informàtics. Universitat Politècnica de Catalunya. Campus Nord C6. c/ Jordi Girona 1-3. 08034 Barcelona (Spain).
{diaz,jpetit,mjserna}@lsi.upc.es
[2] Department of Mathematical Sciences, University of Durham, South Road, Durham DH1 3LE, England. Mathew.Penrose@durham.ac.uk

Abstract. This work deals with bounds on the cost of layout problems for lattice graphs and random lattice graphs. Our main result in this paper is a convergence theorem for the optimal cost of the Minimum Linear Arrangement problem and the Minimum Sum Cut problem, for the case where the underlying graph is obtained through a subcritical site percolation process. This result can be viewed as an analogue of the Beardwood, Halton and Hammersley theorem for the Euclidian TSP. Finally we estimate empirically the value for the constant in the mentioned theorem.

1 Introduction

Several well-known optimization problems on graphs can be formulated as Layout Problems. Their goal is to find a layout (linear ordering) of the nodes of an input graph such that a certain function is minimized. For the problems we consider below, finding an optimal layout is **NP**-hard in general, and therefore it is natural to develop and analyze techniques to obtain tight bounds on restricted instances [1,7,5]. Graphs encoding circuits or grids are typical instances of linear arrangement problems. We consider these instances as sparse graphs that have clustering and geometric properties. For these classes of graphs, not much is known. In this paper, we are concerned with lattice graphs, and random instances of lattice graphs. For most of the layout problems it is an open problem to find exact or approximated polynomial time algorithms for lattice graphs which are non square [10,9].

A graph is said to be a *lattice graph* if it is a node-induced subgraph of the infinite lattice, that is, its vertex set is a subset of $\mathbb{Z}^2$ and two vertices are connected whenever they are at distance 1. *Percolation theory* provides a framework to study lattice graphs in a probabilistic setting. We consider site percolation, where nodes from the infinite lattice are selected with some probability p (selected nodes are called "open"). Let $\tilde{C}_0$ be the set of all open nodes connected to the origin. A basic question in percolation theory is whether or not $\tilde{C}_0$ can be infinite.

---

* This research was partially supported by ESPRIT LTR Project no. 20244 — ALCOM-IT, CICYT Project TIC97-1475-CE, and CIRIT project 1997SGR-00366.

T. Asano et al. (Eds.): COCOON'99, LNCS 1627, pp. 103–112, 1999.
© Springer-Verlag Berlin Heidelberg 1999

Let $\vartheta(p)$ denote the probability that $|\widetilde{C}_0| = \infty$, and set $p_c = \inf\{p : \vartheta(p) > 0\}$, the *critical value* of p. It is well-known [8] that $p_c \in (0.5, 1)$. In this paper, we consider only *subcritical* limiting regimes $p \in (0, p_c)$ in which all components are almost surely finite. Results for *supercritical* regimes are derived in [4,13]. In order to deal with bounded graphs, we introduce the class of *random lattice graphs* with parameters m and p denoted by $\mathcal{L}_{m,p}$ that corresponds to the lattice graphs whose set of vertices are obtained through the random selection (independently and uniformly) of elements from $\{0, ..., m - 1\}^2$ with probability p.

Our layout problems are formally defined as follows. A *layout* φ on a graph $G = (V, E)$ is a one-to-one function $\varphi : V \to \{1, ..., n\}$ with $n = |V|$. Given a graph G and a layout φ on G, let us define:

$$L(i, \varphi, G) = \{u \in V(G) \; : \; \varphi(u) \le i\} \quad R(i, \varphi, G) = \{u \in V(G) \; : \; \varphi(u) > i\}$$
$$\theta(i, \varphi, G) = \{uv \in E(G) \; : \; u \in L(i, \varphi, G) \wedge v \in R(i, \varphi, G)\}$$
$$\delta(i, \varphi, G) = \{u \in L(i, \varphi, G) \; : \; \exists v \in R(i, \varphi, G) : uv \in E(G)\}$$
$$\lambda(uv, \varphi, G) = |\varphi(u) - \varphi(v)| \qquad \text{where } uv \in E(G).$$

The problems we consider are:

- Bandwidth (BANDWIDTH): Given a graph $G = (V, E)$, find
  $\text{MINBW}(G) = \min_\varphi \text{BW}(\varphi, G)$ where $\text{BW}(\varphi, G) = \max_{uv \in E} \lambda(uv, \varphi, G)$.
- Minimum Linear Arrangement (MINLA): Given a graph $G = (V, E)$, find
  $\text{MINLA}(G) = \min_\varphi \text{LA}(\varphi, G)$, where $\text{LA}(\varphi, G) = \sum_{uv \in E} \lambda(uv, \varphi, G)$
  $= \sum_{i=1}^n |\theta(i, \varphi, G)|$.
- Minimum Cut Width (MINCUT): Given a graph $G = (V, E)$, find
  $\text{MINCUT}(G) = \min_\varphi \text{CUT}(\varphi, G)$ where $\text{CUT}(\varphi, G) = \max_{i=1}^n |\theta(i, \varphi, G)|$.
- Vertex Separation (VERTSEP): Given a graph $G = (V, E)$, find
  $\text{MINVS}(G) = \min_\varphi \text{VS}(\varphi, G)$ where $\text{VS}(\varphi, G) = \max_{i=1}^n |\delta(i, \varphi, G)|$.
- Minimum Sum Cut (MINSUMCUT): Given a graph $G = (V, E)$, find
  $\text{MINSC}(G) = \min_\varphi \text{SC}(\varphi, G)$ where $\text{SC}(\varphi, G) = \sum_{i=1}^n |\delta(i, \varphi, G)|$.
- Bisection (BISECTION): Given a graph $G = (V, E)$, find
  $\text{MINBIS}(G) = \min_\varphi \text{BIS}(\varphi, G)$ where $\text{BIS}(\varphi, G) = |\theta(\lfloor n/2 \rfloor, \varphi, G)|$.

Only a few results are known for those problems when restricted to lattice graphs. For the particular case of square or rectangular lattices, it is known that BANDWIDTH can be solved optimally using a diagonal mapping and that MINLA can also be solved with a non-trivial ordering [9,10]. Results for the case of d-dimensional c-ary arrays (a generalization of square lattices) on the BISECTION, MINCUT and MINLA problems are presented in [11]. On the other hand, [12] presents a dynamic programming algorithm to solve BISECTION on lattice graphs without holes. With regard to their complexity, BANDWIDTH, MINCUT and VERTSEP remain **NP**-complete even when restricted to lattice graphs [4]. For the remaining layout problems, the complexity on lattice graphs is open.

In the first part of this paper (Section 2), we deal with general lattice graphs. The results of this section are upper bounds for several layout problems on any lattice graph.

In the second part of the paper (Section 3), we move to a randomized setting where we deal with random lattice graphs. The main result in that section is to prove that for $p \in (0, p_c)$, there exist two nonzero finite constants $\beta_{LA}(p)$ and $\beta_{SC}(p)$ such that $\text{MINLA}(\mathcal{L}_{m,p})/m^2$ converges in probability to $\beta_{LA}(p)$ and $\text{MINSC}(\mathcal{L}_{m,p})/m^2$ converges in probability to $\beta_{SC}(p)$, as $m \to \infty$. This result can be viewed as an analogue of the celebrated Beardwood, Halton and Hammersley theorem on the cost for the traveling salesman problem (TSP) on random points distributed in $[0, 1]^d$,

BHH Theorem [2]. Let $X = \{X_i\}$ be a sequence of independent and uniformly distributed points in $[0, 1]^d$. Let $\text{MINTSP}(n)$ denote the length of the optimal solution of the TSP among the first n points of X. Then, there exists a constant $\beta_{TSP}(d)$ such that $\text{MINTSP}(n)/n^{(d-1)/d}$ converges to $\beta_{TSP}(d)$ almost surely as $n \to \infty$.

A key property to prove BHH-like results is geometric subadditivity [16, Chapter 3]. This property does not hold for our layout problems, therefore we take a completely different approach using percolation theory.

Finally, in the third part of this paper (Section 4), we report results on computational experiments in order to get an empirical estimate of the value of $\beta_{LA}(p)$.

2   Bounds for lattice graphs

In this section we give some deterministic upper bounds on the costs MINBIS, MINCUT and MINVS on subsets of the square integer lattice $\mathbb{Z}^2$. Each subset L of vertices in $\mathbb{Z}^2$ is identified with a lattice graph, namely the maximal subgraph of the 2-dimensional integer lattice with vertex set L.

Lemma 1. *For any lattice graph L with n vertices, and $m \in \{1, 2, \ldots, n\}$, there is a layout φ on L such that $|\theta(m, \varphi, L)| \leq 2^{3/2}\sqrt{n} + 1$.*

Proof. We are looking for a subset S of L consisting of m vertices, such that there are at most $2^{3/2}\sqrt{n} + 1$ edges between S and $L \backslash S$. Let $\alpha > 0$ be a constant, to be chosen later. For $x \in \mathbb{Z}$ let $S_x = \{y \in \mathbb{Z} : (x, y) \in L\}$. Let $V = \{x \in \mathbb{Z} : |S_x| \geq \alpha\sqrt{n}\}$.

For $i \in \mathbb{Z}$, let H_i denote the half-space $(-\infty, i] \times \mathbb{R}$. Set

$$i_0 = \min\{i \in \mathbb{Z} : |L \cap H_i| \geq m\}.$$

Consider the case $i_0 \notin V$. Then define S to be a set of the form

$$S = L \cap (H_{i_0-1} \cup (\{i_0\} \times (-\infty, j]))$$

with j chosen so that S has precisely m elements.

With this definition of S for $i_0 \notin V$, the number of horizontal edges between S and $L \backslash S$ is at most $|S_{i_0}|$, and hence is at most $\alpha\sqrt{n}$. There is at most one vertical edge between S and $L\backslash S$, so the number of edges from S to $L\backslash S$ is at most $\alpha\sqrt{n}+1$ when $i_0 \notin V$.

Now consider the other case $i_0 \in V$. Let $I = [i_1, i_2]$ be the largest integer interval which includes i_0 and is contained in V. Then $i_1 - 1 \notin V$, and $i_2 + 1 \notin V$. Also, as $|V| \leq \alpha^{-1}\sqrt{n}$, $i_2 - i_1 + 1 \leq \alpha^{-1}\sqrt{n}$. We have

$$|L \cap H_{i_1-1}| < m \leq |L \cap H_{i_2}|.$$

For $j \in \mathbb{Z}$ let $T_j = [i_1, i_2] \times (-\infty, j]$. Choose j_0 so that

$$|L \cap (H_{i_1-1} \cup T_{j_0-1})| < m \leq |L \cap (H_{i_1-1} \cup T_{j_0})|,$$

and let S be $L \cap (H_{i_1-1} \cup T_{j_0-1} \cup ([i_1, i_3] \times \{j_0\}))$, with $i_3 \in [i_1, i_2]$ chosen so that S has precisely m elements.

We estimate the number of edges between S and $L\backslash S$ for the case $i_0 \in V$. Since $i_1 - 1 \notin V$, and $i_2 + 1 \notin V$, the number of horizontal edges between S and $L\backslash S$ is at most $2\alpha\sqrt{n}+1$. Also, since $i_2 - i_1 + 1 \leq \alpha^{-1}\sqrt{n}$, the number of vertical edges between S and $L\backslash S$ is at most $\alpha^{-1}\sqrt{n}$. Combining these estimates we find that there are at most $(2\alpha + \alpha^{-1})\sqrt{n}+1$ edges between S and $L\backslash S$, whether or not $i_0 \in V$.

The minimum value of $2\alpha + \alpha^{-1}$ (achieved at $\alpha = 2^{-1/2}$) is $2\sqrt{2}$. Setting $\alpha = 2^{-1/2}$ in the above definition, we have the partition required.   □

Using Lemma 1 and taking $m = \lfloor n/2 \rfloor$ we get the following result.

Theorem 1. *For any lattice graph L with n vertices* MINBIS$(L) \leq 2^{3/2}\sqrt{n}+1$.

For the MINCUT problem the bound changes in the constant:

Theorem 2. *For any lattice graph L with n vertices,* MINCUT$(S) \leq 14\sqrt{n}$.

Proof. First suppose we have $n = 2^m$ for an integer m. The proof is based on repeated bisection, with the size guaranteed by lemma 1. Partition L into complementary subsets S_0, S_1 of size 2^{m-1}, with at most $2^{3/2}2^{m/2}+1$ edges between them. Recursively for each i, $i = 1, \ldots, m-1$, partition a set S_j at level i into disjoint subsets S_{j0}, S_{j1} of size 2^{m-i-1}, with at most $2^{3/2}2^{(m-i)/2}+1$ edges between them, for $j = 0^i, \ldots, 1^i$ (binary sequences of length i).

The decomposition of L into subsets S_α generated in this way can be represented, in the obvious way, as vertices in a complete binary tree of depth m. At level m, one ends up with 2^m singletons, each represented by a binary sequence of length m. Given $x \in S$, let α_x be the binary sequence such that $\{x\} = S_{\alpha_x}$, define $\varphi(x)$ as the position of α_x in the lexicographic ordering on binary sequences of length m. Then $\{x\}$ is a leaf of the binary tree, and has one ancestor set at each of levels $0, 1, 2, \ldots, m-1$. Therefore $|\theta(\varphi(x), \varphi, L)|$ is the sum of the numbers of edges in the bisection of the m ancestors of x in the tree. Thus,

$$|\theta(\varphi(x), \varphi, L)| \leq \sum_{j=1}^{m}(2^{3/2}2^{j/2}+1) = 4\frac{2^{m/2}-1}{2^{1/2}-1} + m.$$

Therefore we have,

$$\text{MINCUT}(L) \leq \left(\frac{4}{2^{1/2} - 1} \right) (2^{m/2} - 1) + m,$$

We can drop the assumption that $n = 2^m$, by taking m so that $n \leq 2^m < 2n$, and adding extra points until one has a set of size 2^m. By monotonicity this process does not reduce the MINCUT cost.

$$\text{MINCUT}(L) \leq \left(\frac{2^{5/2}}{2^{1/2} - 1} \right) \sqrt{n} + (\log_2(n) + 1) - \frac{4}{2^{1/2} - 1}$$

$$\leq 13.657\sqrt{n} + \log_2(n) - 8.$$

But notice that for any $x > 0$ we have $(\log_2(x) - 8)/\sqrt{n} < 0.067$, therefore the above bound for $\text{MINCUT}(L)$ is at most $14\sqrt{n}$ for all n.   □

As a consequence of the previous theorem, and the fact that for any graph G, $\text{MINLA}(G) \leq n\,\text{MINCUT}(G)$ and $\text{MINSC}(G) \leq n\,\text{MINVS}(G)$, we can extend the previous result to the remaining problems.

Corollary 1. *For any lattice graph L with n vertices, $\text{MINVS}(L) \leq 14\sqrt{n}$, $\text{MINLA}(L) \leq 14n\sqrt{n}$ and $\text{MINSC}(L) \leq 14n\sqrt{n}$.*

Comparing with the known results on square lattice graphs, we can see the previous bounds are equal up to a constant.

3   Stochastic results

Let us describe some basic concepts of site percolation for the lattice L_m with vertex set $V_m = ([0, m) \cap \mathbb{Z})^2$. Given $p \in (0, 1)$, *site percolation with parameter p on L_m* is obtained by taking a random set of *open* vertices of V_m with each vertex being open with probability p independently of the others. Let $\mathcal{L}_{m,p}$ be the subgraph of L_m obtained by taking all edges between open vertices. We say that $\mathcal{L}_{m,p}$ is a *random lattice graph*. Denote by $\mathbf{Pr}_p$ and $\mathbf{E}_p$ the probability and expectation with respect to the described process of site percolation with parameter p. By a *cluster* we mean the set of vertices in any connected component of $\mathcal{L}_{m,p}$. Let C_0 denote the cluster in $\mathcal{L}_{m,p}$ that includes $(0,0)$ (possibly the empty set), and write $|C_0|$ for the cardinality of C_0.

A similar site percolation process can be generated analogously on the infinite lattice with vertex set $\mathbb{Z}^2$ and edges between nearest neighbors. In the same way we can extend $\mathbf{Pr}_p$ and $\mathbf{E}_p$ to this infinite process. Let us denote by $\tilde{C}_0$ the cluster including the origin for site percolation on $\mathbb{Z}^2$ and let $|\tilde{C}_0|$ denote its cardinality. It may be the case that $\tilde{C}_0$ is empty. Notice that we can view the random lattice graph as generated by a site percolation process on $\mathbb{Z}^2$ and taking the open vertices in V_m.

In this section we consider random lattice graphs generated by *subcritical limiting regimes* ($p < p_c$), in which all clusters are almost surely finite. We begin by giving bounds for the MINCUT and VERTSEP problems on the subcritical percolation process on the lattice L_m.

Theorem 3. *Assume $0 < p < p_c$; then there exists constants $0 < c_1 < c_2$ such that*

$$\lim_{m \to \infty} \mathbf{Pr} \left[c_1 \le \frac{\mathrm{MINVS}(\mathcal{L}_{m,p})}{\sqrt{\log m}} \le \frac{\mathrm{MINCUT}(\mathcal{L}_{m,p})}{\sqrt{\log m}} \le c_2 \right] = 1.$$

Proof. Recall that for any graph G, $\mathrm{MINVS}(G) \le \mathrm{MINCUT}(G)$. As the MINCUT of a disconnected graph is the maximum of the MINCUTs of its connected components, then $\mathrm{MINCUT}(\mathcal{L}_{m,p}) = \max_{x \in V_m} \mathrm{MINCUT}(C_x)$. Hence by Theorem 2, for any positive constant c_2,

$$\mathbf{Pr} \left[\mathrm{MINCUT}(\mathcal{L}_{m,p}) \ge c_2 \sqrt{\log m} \right] = \mathbf{Pr} \left[\cup_{x \in V_m} \{ \mathrm{MINCUT}(C_x) \ge c_2 \sqrt{\log m} \} \right].$$

By the site percolation version of Theorem 5.75 of [8], there exists $\alpha > 0$ such that $\mathbf{Pr} \left[|\tilde{C}_0| \ge n \right] \le e^{-\alpha n}$, therefore by Theorem 2

$$\mathbf{Pr} \left[\mathrm{MINCUT}(\mathcal{L}_{m,p}) \ge c_2 \sqrt{\log m} \right] \le \mathbf{Pr} \left[\cup_{x \in V_m} \{ |C_x| \ge (c_2/14)^2 \log m \} \right]$$

$$\le m^2 \exp(-\alpha (c_2/14)^2 \log m).$$

Choosing $c_2 > 14\sqrt{2/\alpha}$ we get $\mathbf{Pr} \left[\mathrm{MINCUT}(\mathcal{L}_{m,p}) \ge c_2 \sqrt{\log m} \right] \to 0$.

To get a lower bound for $\mathrm{MINVS}(\mathcal{L}_{m,p})$, let $\delta > 0$ and let $T_1, \ldots, T_{j(m)}$ be disjoint lattice subsquares of L_m, each of side $\lfloor (\delta \log m)^{1/2} \rfloor$, where

$$j(m) = \lfloor m / \lfloor (\delta \log m)^{1/2} \rfloor \rfloor^2.$$

Set $\gamma = \log(1/p)$ so that $p = e^{-\gamma}$. Let A_j be the event that all sites in T_j are open. Then

$$\mathbf{Pr}[A_j] = \exp(-\gamma \lfloor (\delta \log m)^{1/2} \rfloor^2) \ge m^{-\gamma\delta}.$$

Hence, $\mathbf{Pr} \left[\cap_{j=1}^{j(m)} A_j^c \right] \le (1 - m^{-\gamma\delta})^{j(m)} \le \exp(-m^{-\gamma\delta} j(m))$ which tends to zero provided δ is chosen so that $\gamma\delta < 2$. But as $\mathrm{MINVS}(L_m) \ge m$, then

$$\cup_{j=1}^{j(m)} A_j \subset \{ \mathrm{MINVS}(\mathcal{L}_{m,p}) \ge (\delta \log m)^{1/2} \},$$

and taking $c_1 = \sqrt{\delta}$ we obtain the lower bound. $\square$

Notice that the above theorem only gives an order of magnitude result for the minimal cost and we do not have a convergence result. The order of magnitude is $\Theta(\log m)$, which contrast with the supercritical case $p > p_c$, for which $\mathrm{MINVS}(\mathcal{L}_{m,p})$ and $\mathrm{MINCUT}(\mathcal{L}_{m,p})$ are $\Theta(m)$ [13].

In the next lemma we prove that for a subcritical site percolation with parameter p, the expected ratio of the value of the $\mathrm{MINLA}(\tilde{C}_0)$ and $|\tilde{C}_0|$ is finite. We also give a similar result for the MinSumCut problem. To cover the case $\tilde{C}_0 = \emptyset$, we use the convention $0/0 = 0$ throughout the remainder of the paper.

Lemma 2. *For any $p \in (0, p_c)$,*

$$\mathbf{E}_p \left[\frac{\mathrm{MINLA}(\tilde{C}_0)}{|\tilde{C}_0|} \right] \in (0, \infty) \quad and \quad \mathbf{E}_p \left[\frac{\mathrm{MINSC}(\tilde{C}_0)}{|\tilde{C}_0|} \right] \in (0, \infty).$$

Proof. Let $R_0 = \min\{n : \tilde{C}_0 \subset [-n,n]^2\}$; then by considering the lexicographic ordering of vertices one sees that $\mathrm{MINSC}(L_m) \le m^3$ and $\mathrm{MINLA}(L_m) \le m^3$, which together with monotonicity gives us that $\mathrm{MINSC}(\tilde{C}_0) \le (2R_0+1)^3$ and $\mathrm{MINLA}(\tilde{C}_0) \le (2R_0+1)^3$. The statement of the lemma follows from the fact that $\mathbf{Pr}_p[R_0 > n]$ decays exponentially in n [8, Chapter 3]. □

We use this lemma to state one of our main results, namely that the value of MINLA on random lattices, divided by m^2, converges in probability to a constant. In section 4 we give some empirical bounds for the value of the constant for MINLA. The theorem also includes a similar result for MINSC. Recall [3], that if $\{X_n\}$ is a sequence of random variables and let X be a random variable, X_n converges *in probability* to X ($X_n \xrightarrow{\mathrm{Pr}} X$) if, for every $\epsilon > 0$ we have $\lim_{n\to\infty} \mathbf{Pr}[|X_n - X| > \epsilon] = 0$, and X_n converges to X *almost surely* (a.s) if as $n \to \infty$, $\mathbf{Pr}[\limsup X_n = X = \liminf X_n] = 1$

Theorem 4. *Assume* $0 < p < p_c$; *then as* $m \to \infty$,

$$\frac{\mathrm{MINLA}(\mathcal{L}_{m,p})}{m^2} \xrightarrow{\mathrm{Pr}} \mathbf{E}_p\left[\frac{\mathrm{MINLA}(\tilde{C}_0)}{|\tilde{C}_0|}\right] \quad and \quad \frac{\mathrm{MINSC}(\mathcal{L}_{m,p})}{m^2} \xrightarrow{\mathrm{Pr}} \mathbf{E}_p\left[\frac{\mathrm{MINSC}(\tilde{C}_0)}{|\tilde{C}_0|}\right].$$

Proof. Recall C_x is the cluster including x for $\mathcal{L}_{m,p}$. Think of $\mathcal{L}_{m,p}$ as being embedded in a site percolation process on the infinite lattice $\mathbb{Z}^2$, with clusters in this latter process denoted $\tilde{C}_x$.

$$\frac{\mathrm{MINLA}(\mathcal{L}_{m,p})}{m^2} = m^{-2} \sum_{x \in V_m} \frac{\mathrm{MINLA}(C_x)}{|C_x|}$$

$$= m^{-2} \sum_{x \in V_m} \frac{\mathrm{MINLA}(\tilde{C}_x)}{|\tilde{C}_x|} + m^{-2} \sum_{x \in V_m} \left(\frac{\mathrm{MINLA}(C_x)}{|C_x|} - \frac{\mathrm{MINLA}(\tilde{C}_x)}{|\tilde{C}_x|}\right). \quad (1)$$

Using theorem VII.6.9 from [6] and the Kolmogorov zero-one law,

$$m^{-2} \sum_{x \in V_m} \frac{\mathrm{MINLA}(\tilde{C}_x)}{|\tilde{C}_x|} \xrightarrow{\mathrm{Pr}} \mathbf{E}_p\left[\frac{\mathrm{MINLA}(\tilde{C}_0)}{|\tilde{C}_0|}\right].$$

Writing ∂V_m for the set of $x \in V_m$ with lattice neighbors in $\mathbb{Z}^2 \backslash V_m$, we get

$$m^{-2} \sum_{x \in V_m} \left|\left(\frac{\mathrm{MINLA}(C_x)}{|C_x|} - \frac{\mathrm{MINLA}(\tilde{C}_x)}{|\tilde{C}_x|}\right)\right| \le 2m^{-2} \sum_{x \in V_m, \tilde{C}_x \ne C_x} \frac{\mathrm{MINLA}(\tilde{C}_x)}{|C_x|}$$

$$\le 2m^{-2} \sum_{y \in \partial V_m} \mathrm{MINLA}(\tilde{C}_y).$$

By the proof of Lemma 2, $\mathbf{E}_p[\mathrm{MINLA}(\tilde{C}_y)]$ is finite and does not depend on y. Hence the mean of the above expression tends to zero. The result for MINLA then follows from (2), and the proof for MINSC is just the same. □

4   Empirical estimate of the limiting constant

Let us focus now on the MINLA problem. Our BHH-like Theorem 4 shows that for graphs in $\mathcal{L}_{m,p}$ with $p \in (0, p_c)$, there exists a nonzero finite constant $\beta_{\text{LA}}(p)$ such that as m goes to infinity, $\text{MINLA}(\mathcal{L}_{m,p})/m^2$ converges in probability to $\beta_{\text{LA}}(p)$, where $\beta_{\text{LA}}(p) = \mathbf{E}_p[\text{MINLA}(\tilde{C}_0)/|\tilde{C}_0|]$. This result is of importance in order to predict the expected value of a random lattice graph L sampled from $\mathcal{L}_{m,p}$, because for big values of m, we will have $\text{MINLA}(L) \sim \beta_{\text{LA}}(p)m^2$. The purpose of this section is to estimate the value of $\beta_{\text{LA}}(p)$ (for some values of p) through a computational experiment.

The setting of the basic experiment for a given value of p is obtained through the following procedure:

1. Generate at random a connected component $\tilde{C}_0$ centered at the origin where the probability of a node to be open is p.
2. If $|\tilde{C}_0|$ is *small enough*,
   a) Compute $\text{MINLA}(\tilde{C}_0)$.
3. Else
   a) Compute a lower bound of $\text{MINLA}(\tilde{C}_0)$.
   b) Compute an upper bound of $\text{MINLA}(\tilde{C}_0)$.

To perform step 1, we have applied the following algorithm. With probability $1 - p$, $\tilde{C}_0$ is empty. In order to generate at random a non empty $\tilde{C}_0$ component, we start by including node $(0,0)$ in $\tilde{C}_0$, marking it as "alive" and marking the rest of nodes as "waiting". Then, for each "alive" node, we process each of its "waiting" neighbors. Tossing a coin, with probability p, a "waiting" node is included in $\tilde{C}_0$ and its mark changes to "alive"; with probability $1 - p$ its mark changes to "dead". This procedure is iterated until no "alive" nodes exist with "waiting" neighbors. Observe that this procedure correctly generates random connected grid graphs, because for each node only one coin is tossed. Moreover, the procedure ends, because we are dealing with the subcritical phase.

When $|\tilde{C}_0| \leq 15$, we are able to compute the exact value of $\text{MINLA}(\tilde{C}_0)$ (step 2.1) in reasonable time. Otherwise we use the Degree method to obtain a lower bound, and Randomized Successive Augmentation algorithms to compute an upper bound. Both techniques are described and analyzed in [15].

This basic experiment (generation, lower bound and upper bound computation) has been repeated 10000 times, for each $p = 0,025i$, $4 \leq i \leq 18$. In this way, for each value of p, we have obtained an average lower bound and an average upper bound of $\mathbf{E}_p[\text{MINLA}(\tilde{C}_0)/|\tilde{C}_0|]$.

The obtained results are shown in Figure 1. As it can be seen, the estimation of $\beta_{\text{LA}}(p)$ degrades as p increases, but for $p = 0.45$, the factor is less than 3. However for the highest values of p the standard deviation measured is quite big, so the obtained results may not be close to the expected value.

We conjecture that the actual value of $\beta_{\text{LA}}(p)$ for high values of p is closer to the upper bound than to the lower bound. There are two reasons for this. On the one hand, the upper bounds we have obtained using another upper bounding

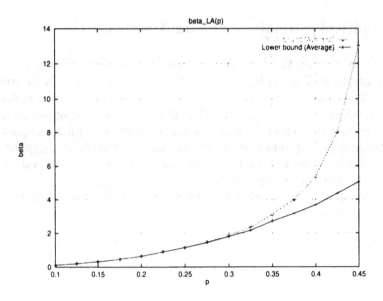

Fig. 1. Lower and upper bounds for the value of $\beta_{LA}(p)$.

technique (Spectral Sequencing with Simulated Annealing, SS+SA [14]) are only slightly lower than the ones obtained by multiple runs of Successive Augmentation algorithms. SS+SA is considered to be a powerful heuristic to approximate the MINLA problem on geometric instances, but requires very long runs. On the other hand, our experience is that even if the Degree method is the method that delivers the highest available lower bounds, these are usually far from the optimal values. Table 1 reproduces a result of [15] that gives strength to this conjecture, by showing upper bounds and lower bounds obtained with all these methods for a 33 × 33 mesh.

	Upper bound		Lower bound	
Method	Succ. Augm.	SS+SA	Degree	Optimal
mesh33	35456	34515	3135	31680

Table 1. Results for a 33 × 33 mesh

5   Conclusions

For the sake of clarity we have presented our results in section 3 for $\mathbb{Z}^2$, but they can be generalized easily to random lattice graphs in higher dimensions, subsets of $\mathbb{Z}^d$ for $d > 2$. Moreover the convergence in Theorem 4 holds almost surely by a straightforward (but messy) extension of our argument given here. It remains as an open problem to find better lower bounding techniques for connected lattice graphs; this would allow us a more accurate estimation of the constant in Theorem 4.

References

1. S. Arora, A. Frieze, and H. Kaplan. A new rounding procedure for the assignment problem with applications to dense graphs arrangements. In *37th IEEE Symposium on Foundations of Computer Science*, 1996.
2. J. Beardwood, J. Halton, and J.M. Hammersley. The shortest path through many points. *Proceedings of the Cambridge Philos. Society.*, 55:299–327, 1959.
3. K.L. Chung. *A Course in Probability Theory*. Academic Press, New York, 1974.
4. J. Díaz, M. D. Penrose, J. Petit, and M. Serna. Linear orderings of random geometric graphs. Technical report, Departament de Llenguatges i Sistemes Informàtics, UPC, http://www.lsi.upc.es/~jpetit/Publications, 1999.
5. J. Díaz. The δ-operator. In L. Budach, editor, *Fundamentals of Computation Theory*, pages 105–111. Akademie-Verlag, 1979.
6. N. Dunford and J. Schwartz. *Linear Operators. Part I: General Theory*. Interscience Publisher., New York, 1958.
7. S. Even and Y. Shiloach. NP-completeness of several arrangements problems. Technical report, TR-43 The Technion, Haifa, 1978.
8. G. Grimmett. *Percolation. (2nd edition)*. Springer-Verlag, Heidelberg, 1999.
9. G. Mitchison and R. Durbin. Optimal numberings of an $n \times n$ array. *SIAM Journal on Discrete Mathematics*, 7(4):571–582, 1986.
10. D.O. Muradyan and T.E. Piliposjan. Minimal numberings of vertices of a rectangular lattice. *Akad. Nauk. Armjan. SRR*, 1(70):21–27, 1980. In Russian.
11. K. Nakano. Linear layouts of generalized hypercubes. In J. van Leewen, editor, *Graph-theoretic concepts in computer science*, volume 790 of *Lecture Notes in Computer Science*, pages 364–375. Springer-Verlag, 1993.
12. C. H. Papadimitriou and M. Sideri. The bisection width of grid graphs. In *First ACM-SIAM Symp. on Discrete Algorithms.*, pages 405–410, San Francisco, 1990.
13. M.D. Penrose. Vertex ordering and partition problems for random spatial graphs. Technical report, University of Durham, 1999.
14. J. Petit i Silvestre. Combining Spectral Sequencing with Simulated Annealing for the MINLA Problem: Sequential and Parallel Heuristics. Technical Report LSI-97-46-R, Departament de Llenguatges i Sistemes Informàtics, Universitat Politècnica de Catalunya, 1997.
15. J. Petit i Silvestre. Approximation Heuristics and Benchmarkings for the MINLA Problem. In Roberto Battiti and Alan A Bertossi, editors, *Alex '98 — Building bridges between theory and applications* , http://www.lsi.upc.es/~jpetit/MinLA/Benchmark, February 1998. Università di Trento.
16. J.M. Steele. *Probability theory and Combinatorial Optimization*. SIAM CBMS-NSF Regional Conference Series in Applied Mathematics., 1997.

A New Transference Theorem in the Geometry of Numbers

Jin-Yi Cai[*]

Department of Computer Science, State University of New York at Buffalo, Buffalo, NY 14260. cai@cs.buffalo.edu.

Abstract. We prove a new transference theorem in the geometry of numbers, giving optimal bounds relating the successive minima of a lattice with the minimal length of generating vectors of its dual. It generalizes the transference theorem due to Banaszczyk. The theorem is motivated by our efforts to improve Ajtai's connection factors in the connection of average-case to worst-case complexity of the shortest lattice vector problem. Our proofs are non-constructive, based on methods from harmonic analysis.

1   Introduction

Stimulated by a recent breakthrough of Ajtai [1], there has been great interest in the complexity and structure of n-dimensional lattice problems lately. In [1] Ajtai established the first explicit connection between, in a certain technical sense, the worst-case and the average-case complexity of the problem of finding the shortest lattice vector or approximating its length, up to a large polynomial factor n^c. This paper stems from efforts to improve Ajtai's connection factor n^c.

We first give some general definitions and background. A lattice in $\mathbf{R}^n$ is the set of all integral linear combinations of a fixed set of linearly independent vectors over $\mathbf{R}$. Such a generating set of vectors is called a basis of the lattice. Equivalently a lattice can be defined as a discrete additive subgroup Γ of $\mathbf{R}^n$. The equivalence means that for any discrete additive subgroup Γ one can always find a basis in the above sense. The basis of a lattice is not unique; any two bases are related to each other by unimodular transformations. The characterization and the complexity of finding a good basis that consists of short vectors is a central problem in the algorithmic study of geometry of numbers.

The rank or the dimension of a lattice L, denoted by $\dim L$, is the dimension of the linear subspace it spans. The length of the shortest non-zero lattice vector is denoted by $\lambda_1(L)$. More generally, Minkowski's *successive minima* $\lambda_i(L)$ are defined as follows: for $1 \leq i \leq \dim L$, $\lambda_i(L) = \min_{v_1,\ldots,v_i \in L} \max_{1 \leq j \leq i} \|v_j\|$, where the sequence of vectors $v_1, \ldots, v_i \in L$ ranges over all i linearly independent lattice vectors. (In this paper, we consider exclusively the Euclidean 2-norm $\|\cdot\|$.) It is perhaps the first indication of the intricacies of higher dimensiyonal lattices that,

---

[*] Research supported in part by NSF grant CCR-9634665 and a J. S. Guggenheim Fellowship.

except for dimensions up to 3, the shortest vectors represented by the successive minima *do not* necessarily form a basis of the lattice [22].

The algorithmic complexity of the shortest lattice vector problem is a fascinating subject. Lovász in 1982 gave a polynomial time algorithm that finds a short vector within a factor of $\sqrt{2^n}$ of the shortest nonzero vector, which led to a number of breakthroughs [25,26,31]. The complexity of finding the shortest lattice vector or its length $\lambda_1(L)$ in l_2-norm has been a long standing open problem, and was finally settled recently by Ajtai [2]. Ajtai showed that finding or even approximating $\lambda_1(L)$ up to a factor of $1 + \frac{1}{2^{n^k}}$ is NP-hard under randomized reductions. The Ajtai connection [1] of worst-case to average-case complexity for lattice problems has been improved by Cai and Nerurkar [11]. They [12] also improve the NP-hardness result of Ajtai [2] to show that the problem of approximating the shortest vector length up to a factor of $1 + \frac{1}{n^\epsilon}$, for any $\epsilon > 0$, is also NP-hard. This improvement also works for all l_p-norms, for $1 \le p < \infty$. Micciancio [29] has improved this further to a constant factor $\sqrt{2} - \epsilon$. Prior to that, it was known that the shortest lattice vector problem is NP-hard for the l_∞-norm, and the nearest lattice vector problem, which is to find the closest lattice vector given an arbitrary vector, is NP-hard under all l_p-norms, $p \ge 1$ [23,33]. Arora et. al. showed that even finding an approximate solution to within any constant factor for the nearest vector problem for any l_p-norm is NP-hard [4]. This has been improved further by Dinur et. al. [14].

Regarding the limit of NP-hardness of approximation factors in these lattice problems, Lagarias, Lenstra and Schnorr [24] showed, as a consequence of their *transference theorem*, that the approximation problem (in l_2-norm) within a factor of $O(n)$ *cannot* be NP-hard, unless NP = coNP. Goldreich and Goldwasser showed that approximating the shortest lattice vector within a factor of $O(\sqrt{n/\log n})$ is not NP-hard assuming the polynomial time hierarchy does not collapse [15].

We present a new *transference theorem*. To describe our result, we need a few additional definitions. Define the generating radius $g(L)$ to be the minimum r such that a ball centered ar 0 $B(0;r) = \{x \in \mathbf{R}^n \mid ||x|| \le r\}$ contains a set of lattice vectors generating L. More generally, we say a sublattice $L' \subset L$ is a *saturated sublattice* if $L' = L \cap \Pi$, where Π is the linear subspace of $\mathbf{R}^n$ spanned by L'. Then we define $g_i(L)$ to be the minimum r such that the sublattice generated by $L \cap B(0;r)$ contains an i-dimensional saturated sublattice L', where $1 \le i \le \dim L$. Clearly for $d = \dim L$, $g(L) = g_d(L)$ and $\lambda_i(L) \le g_i(L)$, for all $1 \le i \le d$.

The dual lattice L^* of a lattice L of dimension n in $\mathbf{R}^n$ is defined as those vectors u, such that $\langle u, v \rangle \in \mathbf{Z}$, for all $v \in L$. It consists of all integral linear combinations of the dual basis vectors $b_1^*, \ldots, b_n^*$, where $\langle b_i^*, b_j \rangle = \delta_{ij}$. In particular $\det(L^*) = 1/\det(L)$, and $L^{**} = L$. For a lattice (embedded in $\mathbf{R}^n$) with dimension less than n, its dual is defined within its own linear span. There is a long history in geometry of numbers to study relationships between various quantities such as the successive minima associated with the primal and dual lattices, L and L^*. Such theorems are called transference theorems. Our main theorem is the following upper bound

$$g_i(L) \cdot \lambda_{n-i+1}(L^*) \le Cn, \tag{1}$$

for some universal constant C, and for all i, $1 \le i \le n$. This is an improvement of previously the best transference theorem of this type due to Banaszczyk [6], who showed that $\lambda_i(L)\lambda_{n-i+1}(L^*) \le C'n$, for some universal constant C'.

The estimate for this latter product has a long history: Mahler [28] proved that the upper bound $(n!)^2$ holds for all lattices. This was improved by Cassels [13] to $n!$. The first polynomial upper bound was obtained by Lagarias, Lenstra and Schnorr [24] where a bound of $n^2/6$ was shown for all $n \ge 7$. The Banaszczyk bound for the product $\lambda_i(L)\lambda_{n-i+1}(L^*)$ is optimal up to a constant, for Conway and Thompson (see [30]) showed that there exists a self-dual lattice family $\{L_n\}$ with $\lambda_1(L_n) = \Omega(\sqrt{n})$, i.e., bounded below asymptotically by $\sqrt{n}$. Since $g_i(L) \ge \lambda_i(L)$ for all i and for all L, our bound (1) is an even stronger bound than Banaszczyk's bound, and is also optimal up to a constant by the same Conway-Thompson construction. For a number of other related results see [5,20,6,9].

A special class of lattices where each lattice possesses an n^ϵ-unique shortest vector plays an important role in the recent breakthrough by Ajtai [1] on the connection between the average-case and the worst-case complexity of the shortest lattice vector problem, and the Ajtai-Dwork public-key cryptosystem [3]. In a paper which is to appear in the *Complexity Conference 99* [10], we will extend our proof techniques here to obtain near optimal results for such lattices. These results lead to a further improvement to previously the best connection factor proved in [11]. In this paper we will concentrate on the structural aspects, and prove the general transference theorems, which are of independent interests.

Finally we point out that although our work is mostly motivated by algorithmic complexity considerations, our proofs are non-constructive. We build on the work of Banaszczyk [6], using methods from harmonic analysis. It is also possible to extend our work to other norms using the results from [7] and [8].

2   Preliminaries

The main tools of our proofs are Gaussian-like measures on a lattice, and their Fourier transforms. For a given lattice L of dimension n in $\mathbf{R}^n$, $v \in L$, we define

$$\sigma_L(\{v\}) = \frac{e^{-\pi||v||^2}}{\sum_{x \in L} e^{-\pi||x||^2}}. \tag{2}$$

The Fourier transform of σ_L is

$$\widehat{\sigma_L}(u) = \int_{x \in \mathbf{R}^n} e^{2\pi i \langle u, x \rangle} d\sigma_L = \sum_{v \in L} e^{2\pi i \langle u, v \rangle} \sigma_L(\{v\}), \tag{3}$$

where $u \in \mathbf{R}^n$. Note that σ_L is an even function, so that

$$\widehat{\sigma_L}(u) = \sum_{v \in L} \sigma_L(\{v\}) \cos(2\pi \langle u, v \rangle) = \frac{\sum_{v \in L} e^{-\pi||v||^2} \cos(2\pi \langle u, v \rangle)}{\sum_{x \in L} e^{-\pi||x||^2}}.$$

Define

$$\tau_L(u) = \frac{\sum_{y \in L+u} e^{-\pi ||y||^2}}{\sum_{x \in L} e^{-\pi ||x||^2}}. \tag{4}$$

Then the following identity holds

Lemma 1.

$$\widehat{\sigma_L}(u) = \tau_{L^*}(u), \tag{5}$$

where L^* is the dual lattice of L.

The proof of Lemma 1 uses Poisson summation formula, see [21]. The following lemma is proved in [6] (Lemma 1.5) and is crucial to our proof (in the following \ denotes set difference):

Lemma 2. For each $c \geq 1/\sqrt{2\pi}$,

$$\sigma_L(L \backslash B(0; c\sqrt{n})) = \frac{\sum_{v \in L \backslash B(0; c\sqrt{n})} e^{-\pi ||v||^2}}{\sum_{x \in L} e^{-\pi ||x||^2}} < \left(c\sqrt{2\pi e} e^{-\pi c^2}\right)^n; \tag{6}$$

and for all $u \in \mathbf{R}^n$,

$$\frac{\sum_{v \in (L+u) \backslash B(0; c\sqrt{n})} e^{-\pi ||v||^2}}{\sum_{x \in L} e^{-\pi ||x||^2}} < 2 \left(c\sqrt{2\pi e} e^{-\pi c^2}\right)^n. \tag{7}$$

3   The Main Theorem

We now present a general inequality which relates the quantities $g_i(L)$ and $\lambda_{n-i+1}(L^*)$. This theorem generalizes the result of Banaszczyk [6].

Theorem 3. For every constant $c > 3/2\pi$, there exists an n_0, such that

$$g_{n-i+1}(L) \cdot \lambda_i(L^*) \leq cn,$$

for every lattice L of dimension $n \geq n_0$, and every $1 \leq i \leq n$.

We remark that it is known that the product $\lambda_i(L)\lambda_{n-i+1}(L^*)$ is at least 1 for all L and all $1 \leq i \leq n$. Since $g_i(L) \geq \lambda_i(L)$, we also have $g_i(L)\lambda_{n-i+1}(L^*) \geq 1$ for all L and i, and this lower bound is easily achievable, for example by the Gaussian lattice $\mathbf{Z}^n$.

The main idea will be a double estimate for the Fourier transforms of Gaussian-like measures on the lattice and a proper sublattice.

Suppose the inequality does not hold. Let c_1 and c_2 be two constants, such that $c_1 c_2 = c$ and $c_1 > 1/\sqrt{2\pi}$ and $c_2 > 3/\sqrt{2\pi}$. By a suitable scaling factor, we may assume that both

$$g_{n-i+1}(L) > c_1\sqrt{n}, \quad \text{and} \quad \lambda_i(L^*) > c_2\sqrt{n}.$$

Let $u_1, \ldots, u_{i-1}$ be a set of linearly independent lattice vectors in L^* attaining the successive minima $\lambda_1(L^*), \ldots, \lambda_{i-1}(L^*)$, respectively. Let S be the linear subspace of $\mathbf{R}^n$ spanned by $u_1, \ldots, u_{i-1}$. Thus $\dim S = i - 1$ and there are no vectors in $L^* \backslash S$ with norm less than $\lambda_i(L^*)$. This is clear if $\lambda_{i-1}(L^*) < \lambda_i(L^*)$, for in this case, any $u \in L^* \backslash S$ with $\|u\| < \lambda_i(L^*)$ would give us i linearly independent vectors $u_1, \ldots, u_{i-1}, u$ in L^* with norm all strictly less than $\lambda_i(L^*)$. In the general case, let $u \in L^* \backslash S$, For suppose there were some $u \in L^* \backslash S$, $\|u\| < \lambda_i(L^*)$. Let $j = \min\{k \mid \|u\| < \lambda_k(L^*)\}$. Then $1 \leq j \leq i$. Let $C = \{u_1, \ldots, u_{j-1}, u\}$. It is clear that C is a set of j linearly independent vectors in L^*, with $\max\{\|x\| \mid x \in C\} = \|u\| < \lambda_j(L^*)$. A contradiction. In particular there are no vectors in $L^* \backslash S$ with norm less than or equal to $c_2 \sqrt{n}$.

$L^* \cap S$ is a lattice of dimension $i - 1$. Let $v_1^*, \ldots, v_{i-1}^*$ be a basis of $L^* \cap S$, and this can be extended to a basis $v_1^*, \ldots, v_{i-1}^*, v_i^*, \ldots, v_n^*$ for the lattice L^*. Let $v_1, \ldots, v_n$ be its dual basis for L. We note that $v_1^*, \ldots, v_{i-1}^*$ is also a vector space basis for S, while $v_i, \ldots, v_n$ is a vector space basis for $S^\perp$ as well as a lattice basis for the sublattice $L \cap S^\perp$ of L.

We now define two projections from $\mathbf{R}^n$ to $S^\perp$:

$$\pi: \qquad \mathbf{R}^n \qquad \longrightarrow \quad S^\perp \tag{8}$$

$$\sum_{j=1}^{i-1} x_j v_j^* + \sum_{j=i}^{n} x_j v_j \mapsto \sum_{j=i}^{n} x_j v_j, \tag{9}$$

is the orthogonal projection onto $S^\perp$, and

$$\varphi: \qquad \mathbf{R}^n \quad \longrightarrow \quad S^\perp : \tag{10}$$

$$\sum_{j=1}^{n} x_j v_j \mapsto \sum_{j=i}^{n} x_j v_j, \tag{11}$$

is the projection "modulo $v_1, \ldots, v_{i-1}$ in terms of the basis $v_1, \ldots, v_n$". We note that both projections are well defined, and

$$\varphi(L) = L \cap S^\perp = \{\sum_{j=i}^{n} x_j v_j \mid x_j \in \mathbf{Z}\}.$$

We need several technical lemmas.

Lemma 4.

$$\pi(L^*) = (L \cap S^\perp)^*.$$

Proof. We have noted already that $L \cap S^\perp$ is a sublattice of L of dimension $n - i + 1$ with lattice basis $v_i, \ldots, v_n$. Thus its dual lattice $(L \cap S^\perp)^*$ is also an $(n - i + 1)$-dimensional lattice in the space $S^\perp$.

For every $v \in \pi(L^*)$, there exists a $v' \in L^*$, such that $v = \pi(v')$, i.e., $v - v' \in S$. Hence, for all $x \in S^\perp$,

$$\langle v, x \rangle = \langle v', x \rangle.$$

In particular, for all $x \in L \cap S^{\perp}$, $\langle v, x \rangle$ is an integer. Thus, $v \in (L \cap S^{\perp})^*$.
Conversely, for every $v \in (L \cap S^{\perp})^*$, let

$$v' = v - \sum_{j=1}^{i-1} \langle v, v_j \rangle v_j^*.$$

We claim that $\pi(v') = v$, and $v' \in L^*$, thus $v \in \pi(L^*)$. It is clear that $\pi(v') = v$, since $\sum_{j=1}^{i-1} \langle v, v_j \rangle v_j^* \in S$. It is also clear that for each basis vector v_j of L, if $j \leq i-1$ then $\langle v', v_j \rangle = 0$, and if $j \geq i$ then $\langle v', v_j \rangle = \langle v, v_j \rangle \in \mathbf{Z}$, since $v_j \in L \cap S^{\perp}$ for $j \geq i$.   □

Let L' be the sublattice of L generated by all lattice vectors of L with length at most $c_1 \sqrt{n}$, namely $L \cap B(0; c_1 \sqrt{n})$. As $g_{n-i+1}(L) > c_1 \sqrt{n}$, L' does not contain any $(n - i + 1)$-dimensional saturated sublattice of L.

Lemma 5. $\varphi(L')$ is a proper sublattice of $L \cap S^{\perp}$.

Proof. Clearly $\varphi(L') \subseteq \varphi(L) = L \cap S^{\perp}$. To show that it is a proper sublattice let's assume $\varphi(L') = L \cap S^{\perp}$. Then $v_i, \ldots, v_n \in \varphi(L')$. It follows that there are vectors $w_i, \ldots, w_n \in L'$, $v_j = \varphi(w_j)$, $j = i, \ldots, n$. Thus, each $w_j = v_j + \sum_{k=1}^{i-1} x_{jk} v_k$ for some integers x_{jk}, $j = i, \ldots, n$ and $k = 1, \ldots, i-1$.

Let T be the linear subspace generated by $w_i, \ldots, w_n$. Clearly $w_i, \ldots, w_n$ are linearly independent, so $\dim T = n - i + 1$, and $L \cap T$ is a saturated $(n - i + 1)$-dimensional sublattice of L. By the definition of L', L' does not contain $L \cap T$.

However we claim that $L' \cap T = L \cap T$. This would be a contradiction which would prove the lemma.

To show that $L' \cap T = L \cap T$, let any $u \in L \cap T$. $u \in T$ implies that there exist real numbers $r_i, \ldots, r_n$ such that $u = \sum_{j=i}^n r_j w_j$. $u \in L$ implies that in the above expression, when expressed in terms of $v_1, \ldots, v_n$, all coefficients in $v_1, \ldots, v_n$ are integers. In particular, the coefficients of $v_i, \ldots, v_n$, namely $r_i, \ldots, r_n$, are all integers. Thus, u belongs to the integral span of $w_i, \ldots, w_n \in L'$, and thus $u \in L'$.   □

We now replace $\varphi(L')$ by a full ranked proper sublattice $\tilde{L}$ of $L \cap S^{\perp}$, which contains $\varphi(L')$ (if $\varphi(L')$ is not already one). If $\dim(\varphi(L')) = n - i + 1$, then $\varphi(L')$ is already full ranked, we simply let $\tilde{L} = \varphi(L')$. If however, $\dim(\varphi(L')) < n - i + 1$, then we let $\tilde{L}$ be any proper sublattice of $L \cap S^{\perp}$, which is of dimension $n - i + 1$ and contains $\varphi(L')$. This can be accomplished as follows, for example. Let $k = \dim(\varphi(L')) \leq n - i$, and let $b_1, \ldots, b_k$ be a lattice basis of $L \cap \mathrm{span}(\varphi(L'))$. This can be extended to a lattice basis of $L \cap S^{\perp}$, say $b_1, \ldots, b_k, \ldots, b_{n-i+1}$. Then we may let $\tilde{L}$ be the integral span of $b_1, \ldots, b_k, \ldots, 2b_{n-i+1}$, say.

Summarizing,

Lemma 6.   *1.* $\varphi(L') \subseteq \tilde{L}$;
   2. $\dim \tilde{L} = n - i + 1$; *and*
   3. $\tilde{L}$ *is a proper sublattice of* $L \cap S^{\perp}$.

Now we let
$$L'' = \tilde{L} \oplus \langle v_1 \rangle \oplus \cdots \oplus \langle v_{i-1} \rangle.$$

Lemma 7. *1. $L' \subseteq L''$;*
2. $\dim L'' = n$; and
3. L'' is a proper sublattice of L.

Proof. For any $v \in L'$,

$$v = \varphi(v) + \sum_{k=1}^{i-1} x_k v_k,$$

for some integers x_k. Since $\varphi(v) \in \varphi(L') \subseteq \tilde{L}$, it follows that $v \in L''$.

$\dim L'' = n$ follows directly by the definition of L'' and $\dim(\tilde{L}) = n - i + 1$.

Finally, we show that L'' is a proper sublattice of L. That L'' is a sublattice of L is trivial. Moreover, $\varphi(L'') = \tilde{L}$ is a proper sublattice of $L \cap S^\perp = \varphi(L)$, hence L'' is a proper sublattice of L. □

Corollary 8. L^* *is a proper sublattice of* $(L'')^*$.

Lemma 9. *There exists a vector $x \in (L'')^* \backslash (L^* + S)$.*

Proof. Since $\tilde{L}$ is a proper sublattice of $L \cap S^\perp$ of full rank in the linear space $S^\perp$, $(\tilde{L})^*$ is an $(n - i + 1)$-dimensional lattice in $S^\perp$ properly containing $(L \cap S^\perp)^* = \pi(L^*)$. In particular there exists a $y \in (\tilde{L})^* \backslash \pi(L^*)$. Let

$$x = y - \sum_{k=1}^{i-1} \langle y, v_k \rangle v_k^*.$$

Then $\pi(x) = y$. This implies that $x \notin L^* + S$, for otherwise, $y = \pi(x) \in \pi(L^*)$.

We show next that $x \in (L'')^*$. Since $\dim L'' = n$ all we need to show is that for every $w \in L''$, $\langle x, w \rangle \in \mathbf{Z}$. Take any $w \in L''$,

$$w = \varphi(w) + \sum_{\ell=1}^{i-1} y_\ell v_\ell,$$

for some integral y_ℓ. By the definition of L'', $\varphi(w) \in \tilde{L}$. Then it is easy to verify that
$$\langle x, w \rangle = \langle y, \varphi(w) \rangle,$$
which belongs to $\mathbf{Z}$ since $y \in (\tilde{L})^*$ and $\varphi(w) \in \tilde{L}$. □

Now we come to the crucial combinatorial lemma:

Lemma 10. *There exists $x \in (L'')^*$, such that*

$$\min_{y \in L^*} \|x - y\| \geq \frac{\lambda_i(L^*)}{3}.$$

Proof. Suppose not. Then for every $x \in (L'')^*$, there exists a $y \in L^*$ such that

$$||x - y|| < \frac{\lambda_i(L^*)}{3}.$$

In particular we may choose our $x \in (L'')^* \setminus (L^* + S)$ by Lemma 9. Let $y \in L^*$ be the corresponding point in L^* as above. Denote $x - y$ by u, then we still have $u \in (L'')^* \setminus (L^* + S)$, for clearly $u \in (L'')^*$ by Corollary 8; and $u \notin L^* + S$, otherwise $x \in L^* + S$ as well. In particular $u \notin S$. Moreover, $||u|| < \frac{\lambda_i(L^*)}{3}$.

Consider the set of points $\{ku \mid k = 1, 2, \ldots\}$. Each ku is associated with a point in L^* of distance less than $\lambda_i(L^*)/3$. Since $u \notin S$, for sufficiently large k, the associated point of L^* cannot be in S. Let k_0 be the first such k, then $k_0 > 1$. Let $z \in L^* \setminus S$ be the point associated with $k_0 u$ and $z' \in L^* \cap S$ be the point associated with $(k_0 - 1)u$. Then $z - z' \in L^* \setminus S$. Furthermore,

$$||z - z'|| = ||(z - k_0 u) + u + ((k_0 - 1)u - z')|| < \lambda_i(L^*).$$

This contradicts the definition of S and $\lambda_i(L^*)$. □

We are now ready to prove Theorem 3. We will pick u to be the x promised in Lemma 10. Then,

$$\widehat{\sigma_L}(u) = \sum_{v \in L} \sigma_L(\{v\}) \cos(2\pi \langle u, v \rangle)$$

$$= \sum_{v \in L''} \sigma_{L''}(\{v\}) \cos(2\pi \langle u, v \rangle) + \sum_{v \in L''} (\sigma_L(\{v\}) - \sigma_{L''}(\{v\})) \cos(2\pi \langle u, v \rangle)$$

$$+ \sum_{v \in L \setminus L''} \sigma_L(\{v\}) \cos(2\pi \langle u, v \rangle)$$

$$= \widehat{\sigma_{L''}}(u) + A + B, \quad \text{say.}$$

Since $L \cap B(0; c_1 \sqrt{n}) \subset L' \subseteq L''$, the last term

$$|B| \leq \sum_{v \in L \setminus B(0; c_1 \sqrt{n})} \sigma_L(\{v\}) < \left(c_1 \sqrt{2\pi e} e^{-\pi c_1^2}\right)^n = \epsilon_1^n,$$

by Lemma 2 inequality (6).

Similarly for the other error term A, we can show that $|A| < \epsilon_1^n$. Hence

$$\widehat{\sigma_L}(u) > \widehat{\sigma_{L''}}(u) - 2\epsilon_1^n.$$

By Lemma 1 $\widehat{\sigma_L}(u) = \tau_{L^*}(u)$ and $\widehat{\sigma_{L''}}(u) = \tau_{(L'')^*}(u)$. Since $u \in (L'')^*$, $(L'')^* + u = (L'')^*$ so that $\tau_{(L'')^*}(u) = 1$.

On the other hand, since

$$\min_{p \in L^*} ||u - p|| \geq \frac{\lambda_1(L^*)}{3} > \frac{c_2}{3} \sqrt{n},$$

we note that every point in $L^* + u$ has norm greater than $\frac{c_2}{3}\sqrt{n}$, and so, it can be shown that

$$\tau_{L^*}(u) < 2\left(\frac{c_2}{3}\sqrt{2\pi e} e^{-\pi(\frac{c_2}{3})^2}\right)^n = 2\epsilon_2^n,$$

by Lemma 2 inequality (7). Since both c_1 and $c_2/3 > 1/\sqrt{2\pi}$, we have both ϵ_1 and $\epsilon_2 < 1$ by an elementary estimate. Thus

$$2\epsilon_2^n > 1 - 2\epsilon_1^n,$$

which is a contradiction for large n. The proof of Theorem 3 is complete.   □

We remark that the inequality in Theorem 3 can be made to hold for all n, and not just for sufficiently large n, with a moderate constant c. For example, $c = 2$ will do, with $c_1 = \sqrt{2/3}$ and $c_2 = \sqrt{6}$.

Acknowledgements

I thank Ajay Nerurkar for valuable discussions and comments.

References

1. M. Ajtai. Generating hard instances of lattice problems. In *Proc. 28th ACM Symposium on the Theory of Computing*, 1996, 99–108. Full version available from ECCC as TR96-007.
2. M. Ajtai. The shortest vector problem in L_2 is NP-hard for randomized reductions. In *Proc. 30th ACM Symposium on the Theory of Computing*, 1998, 10–19. Full version available from ECCC as TR97-047.
3. M. Ajtai and C. Dwork. A public-key cryptosystem with worst-case/average-case equivalence. In *Proc. 29th ACM Symposium on the Theory of Computing*, 1997, 284–293. Full version available from ECCC as TR96-065.
4. S. Arora, L. Babai, J. Stern, and Z. Sweedyk. The hardness of approximate optima in lattices, codes, and systems of linear equations. In *Proc. 34th IEEE Symposium on Foundations of Computer Science (FOCS)*, 1993, 724-733.
5. L. Babai. On Lovász' lattice reduction and the nearest lattice point problem. *Combinatorica*, 6:1–13, 1986.
6. W. Banaszczyk. New Bounds in Some Transference Theorems in the Geometry of Numbers. *Mathematische Annalen*, 296, pages 625–635, 1993.
7. W. Banaszczyk. Inequalities for Convex Bodies and Polar Reciprocal Lattices in $\mathbf{R}^n$, *Discrete and Computational Geometry*, 13:217–231, 1995.
8. W. Banaszczyk. Inequalities for Convex Bodies and Polar Reciprocal Lattices in $\mathbf{R}^n$ II: Application of K-Convexity. *Discrete and Computational Geometry*, 16:305–311, 1996.
9. J-Y. Cai. A Relation of Primal-Dual Lattices and the Complexity of Shortest Lattice Vector Problem. *Theoretical Computer Science* (207), 1998, pp 105–116.
10. J-Y. Cai. Applications of a New Transference Theorem to Ajtai's Connection Factor. To appear in the *14th Annual IEEE Conference on Computational Complexity* 1999.
11. J-Y. Cai and A. Nerurkar. An Improved Worst-Case to Average-Case Connection for Lattice Problems. In *Proc. 38th IEEE Symposium on Foundations of Computer Science (FOCS)*, 1997, 468–477.
12. J-Y. Cai and A. Nerurkar. Approximating the SVP to within a factor $\left(1 + \frac{1}{\dim^\epsilon}\right)$ is NP-hard under randomized reductions. In *Proc. of the 13th Annual IEEE Conference on Computational Complexity*, 1998, 46–55.

13. J. W. S. Cassels. *An Introduction to the Geometry of Numbers*. Berlin Göttingen Heidelberg: Springer 1959.
14. I. Dinur, G. Kindler and S. Safra. Approximating CVP to within almost-polynomial factors is NP-hard. In *Proc. 39th IEEE Symposium on Foundations of Computer Science*, 1998, 99–109.
15. O. Goldreich and S. Goldwasser. On the Limits of Non-Approximability of Lattice Problems. In *Proc. 30th ACM Symposium on Theory of Computing*, 1998, 1–9.
16. O. Goldreich, S. Goldwasser, and S. Halevi. Public-key cryptosystems from lattice reduction problems. In *Advances in Cryptology - CRYPTO '97*, Burton S. Kaliski Jr. (Ed.), Lecture Notes in Computer Science, 1294:112-131, Springer-Verlag, 1997.
17. M. Grötschel, L. Lovász, and A. Schrijver. *Geometric Algorithms and Combinatorial Optimization*. Springer Verlag, 1988.
18. P. M. Gruber. *Handbook of Convex Geometry*. Elsevier Science Publishers B.V., 1993.
19. P. M. Gruber and C. G. Lekkerkerker. *Geometry of Numbers*. North-Holland, 1987.
20. J. Håstad. Dual Vectors and Lower Bounds for the Nearest Lattice Point Problem. *Combinatorica*, Vol. 8, 1988, pages 75–81.
21. E. Hewitt and K. A. Ross. *Abstract Harmonic Analysis*, Vol II. Berlin Göttingen Heidelberg: Springer 1970.
22. A. Korkin and G. Zolotarev. Sur les formes quadratiques positives quaternaires. *Mathematische Annalen*, **5**, 581–583, 1872.
23. J. C. Lagarias. The computational complexity of simultaneous diophantine approximation problems. *SIAM Journal of Computing*, Volume 14, page 196–209, 1985.
24. J. C. Lagarias, H. W. Lenstra, and C. P. Schnorr. Korkin-Zolotarev Bases and Successive Minima of a Lattice and its Reciprocal Lattice. *Combinatorica*, 10:(4), 1990, 333-348.
25. A. K. Lenstra, H. W. Lenstra, and L. Lovász. Factoring polynomials with rational coefficients. *Mathematische Annalen*, 261:515-534, 1982.
26. H. W. Lenstra, Jr. Integer programming with a fixed number of variables. *Mathematics of Operations Research*, 8:538-548, 1983.
27. L. Lovász. *An Algorithmic Theory of Numbers, Graphs and Convexity*. SIAM, Philadelphia, 1986.
28. K. Mahler. Ein Übertragungsprinzip für konvexe Körper. *Čas. Pěstování Mat. Fys.* **68**, pages 93–102, 1939.
29. D. Micciancio. The Shortest Vector in a Lattice is Hard to Approximate to within Some Constant. In *Proc. 39th IEEE Symposium on Foundations of Computer Science*, 1998, 92–98.
30. J. Milnor and D. Husemoller. *Symmetric Bilinear Forms*. Berlin Heidelberg New York: Springer 1973.
31. A. Odlyzko and H.J.J. te Riele. Disproof of the Mertens conjecture. *Journal für die Reine und Angewandte Mathematik*, 357:138-160, 1985.
32. C. P. Schnorr. A hierarchy of polynomial time basis reduction algorithms. *Theory of Algorithms*, pages 375-386, 1985.
33. P. van Emde Boas. Another NP-complete partition problem and the complexity of computing short vectors in lattices. Technical Report 81-04, Mathematics Department, University of Amsterdam, 1981.

On Covering and Rank Problems for Boolean Matrices and Their Applications

Carsten Damm[1]*, Ki Hang Kim[2], and Fred Roush[2]

[1] Fachbereich IV — Informatik, Universität Trier, D-54286 Trier, Germany.
damm@uni-trier.de
[2] Mathematics Research Group, Alabama State University, Montgomery AL
36101-0271, U.S.A. **kkim@asu.alasu.edu, froush@asu.alasu.edu**

Abstract. We consider combinatorial properties of Boolean matrices and their application to two-party communication complexity. Let A be a binary $n \times n$ matrix and let K be a field. *Rectangles* are sets of entries defined by collections of rows and columns. We denote by $\mathrm{rank}_B(A)$ ($\mathrm{rank}_K(A)$, resp.) the least size of a family of rectangles whose union (sum, resp.) equals A.
We prove the following:
- With probability approaching 1, for a random Boolean matrix A the following holds:
$$\mathrm{rank}_B(A) \geq n(1 - o(1)).$$
- For finite K and fixed $\varepsilon > 0$ the following holds: If A is a Boolean matrix with $\mathrm{rank}_B(A) \leq t$ then there is some matrix $A' \leq A$ such that $A - A'$ has at most $\varepsilon \cdot n^2$ non-zero entries and $\mathrm{rank}_K(A') \leq t^{O(1)}$.

As applications we mention some improvements of earlier results: (1) With probability approaching 1 a random n-variable Boolean function has nondeterministic communication complexity n, (2) functions with *nondeterministic* communication complexity l can be *approximated* by functions with *parity* communication complexity $O(l)$. The latter complements a result saying that nondeterministic and parity communication protocols cannot efficiently *simulate* each other. Another consequence is: (3) matrices with small Boolean rank have small matrix rigidity over any field.

1 Introduction and Definitions

Matrices with Boolean entries—so-called *Boolean matrices*—appear in every branch of mathematics with a Boolean context, i.e., with AND, OR, and NOT as natural basic primitives. They have many applications such as to switching circuits, voting methods, applied logic (see [9] for these and other facts about Boolean matrices). Another area of applications, namely communication complexity of Boolean functions, will be mentioned more detailed later in this paper.

The combinatorial and algebraic structure of Boolean matrices is an appealing object of study. Although there are a lot of similarities to linear algebra over

---

* Partially supported by DFG grant Me 1077/14-1

T. Asano et al. (Eds.): COCOON'99, LNCS 1627, pp. 123–133, 1999.
© Springer-Verlag Berlin Heidelberg 1999

fields, Boolean matrix theory is much less structured. On the other hand the Boolean entries 0 and 1 have a meaning in every field. It would, therefore, be desirable to use this embedding (rather than the embedding into Boolean algebra) in order to have the rich arsenal of tools from linear algebra at one's disposal.

Though in general most of the Boolean properties will get lost during this translation we will show that one important combinatorial parameter will to some extent survive on this way. The notion we will focus on in this paper is the notion of the *rank* of a matrix which can be defined in a Boolean sense and with respect to a field.

In the first part of the paper we will show that Boolean and field rank have similarities in that most Boolean matrices have almost full Boolean rank and over any field most matrices have full field rank. With respect to Boolean rank we improve an earlier observation of [20] and give a new proof with strengthened meaning of "almost full Boolean rank". In the next part we show that matrices with small Boolean rank are "close to having" small field rank. In the remaining sections we apply these results to two-party communication complexity and matrix rigidity.

The necessary definitions on matrices and communication complexity will be given in the rest of this section.

1.1   Boolean Rank and Field Rank

The rank of an $n \times n$ matrix A over a field K is well-defined and has many equivalent characterizations: e.g., as the minimal number of rank 1 matrices summing up over K to A. Equivalently, the field rank of A is the least t such that A is the product over K of an $n \times t$ matrix and a $t \times n$ matrix. We denote the rank of A over K by $\mathrm{rank}_K(A)$.

A matrix A with Boolean entries 0 and 1 is called a *Boolean matrix*. Several rank concepts can be defined: the row (column) rank is the minimal size of a set of rows (columns) whose Boolean linear combinations (i.e., component-wise disjunction) include all rows (columns). For $n \geq 4$ the row rank of A may not equal its column rank. A third concept is *Boolean rank* of an $n \times n$ Boolean matrix A which is the least number of *rectangles* whose union is the set of 1-entries of A. Here, a rectangle is a set E of matrix entries for which there are sets R and C of (not necessarily consecutive) rows and columns such that entry (i, j) belongs to E if and only if $i \in R$ and $j \in C$. Hence, we look for minimal *exact covers* of the 1-entries in A by rectangles. Equivalently, the Boolean rank of a square Boolean matrix A is the least t such that A is the *Boolean* product of an $n \times t$ matrix and a $t \times n$ matrix. We denote the Boolean rank of A by $\mathrm{rank}_B(A)$. Boolean rank is much harder to deal with than field rank. E.g., Boolean rank can differ from both row and column rank (but it is always less than or equal to them).

The difficulty in dealing with Boolean rank can also be seen from the fact that determining Boolean rank is NP-complete (see, e.g., [14]), while determining field ranks can be done in deterministic polynomial time by the Gaussian algorithm or by very efficient parallel algorithms (see [1] for a survey).

Without special mentioning we will consider only square matrices.

We will often use the terms *random matrix* and *almost all matrices* in this paper. In case of $n \times n$ matrices with Boolean entries or $n \times n$ matrices with entries from a finite field this refers to the uniform distribution. In particular, we say that *almost all* matrices from a specified finite sample space have a certain property P if for all $\delta > 0$ there is some integer n_0 such that for all $n > n_0$ at least a fraction $(1 - \delta)$ of all order n square matrices of the sample space have property P.

Matrices over infinite fields have to be dealt with differently. We will give an explanation in the place where it is needed.

1.2   Communication Complexity

Communication complexity of Boolean functions has proved to be a useful and strong tool to derive lower bounds on the complexity of computations for many computational models. Basic among the various variants of communication complexity considered in the literature is the following one, which was introduced by A.C. Yao in [20]. Alice and Bob want to jointly compute the value of a Boolean function $f : \{0,1\}^s \times \{0,1\}^s \rightarrow \{0,1\}$ on input (x,y), where $x \in \{0,1\}^s$ are Alice's input bits and $y \in \{0,1\}^s$ are Bob's. Both know the function table but neither one has access to the bits of the other. Therefore they agree on a deterministic protocol P that governs their communication. The protocol is a set of rules that for each party specifies the bits to be sent in dependence of the given part of the input and the string of bits communicated before. It is required that the very last bit sent (we denote it $P(x,y)$) is the function value $f(x,y)$. We say P *computes* f if the protocol satisfies this requirement. The length of P is the number of communicated bits on the worst case input. The deterministic communication complexity $C_D(f)$ of f is the minimal length of a deterministic protocol that computes f. This model and its extensions have many interesting applications in lower bound proofs ranging from VLSI circuit models to branching programs and data structures for Boolean functions (see [12]).

The nondeterministic extension of the above model is the following: A nondeterministic two-party communication protocol for f is a family $\mathcal{P} = (P_i)$ of deterministic protocols, such that for all inputs x, y holds $f(x, y) = 1$ if and only if $\exists i : P_i(x, y) = 1$. In this case we say $\mathcal{P}$ *computes* f. The length of $\mathcal{P}$ is the maximum over all lengths of the P_i and the nondeterministic communication complexity of f is the minimal length of a nondeterministic protocol computing f. It is denoted $C_N(f)$.

The existential predicate underlying the definition of a nondeterministic protocol can be generalized to $f(x, y) = 1$ if and only if the number of P_i for which $P_i(x, y) = 1$ is not divisible by the positive integer m. Length of the protocol is defined as above. The minimal length of a protocol computing f in that sense is called MOD_m-communication complexity of f and is denoted $C_{MOD_m}(f)$. The special case of $m = 2$ is called parity communication complexity and is mentioned in the abstract.

Nondeterministic and MOD_m communication complexity has been used to derive lower bounds for various branching program and circuit models (see [3,11]. The basic tool to estimate these communication complexities is given below for the case of a prime $m = p$. To formulate it, we assign a $2^s \times 2^s$ matrix $M(f)$ whose rows and columns are indexed by $x, y \in \{0, 1\}^s$ and with entry $f(x, y)$ in row x and column y. Matrix $M(f)$ can be considered as Boolean matrix or as matrix over any field. For a matrix A let $\mathrm{rank}_p(A)$ denote $\mathrm{rank}_{GF(p)}(A)$, where $GF(p)$ is the Galois field of characteristic p.

Lemma 1 ([20,3]). *For $f : \{0, 1\}^s \times \{0, 1\}^s \to \{0, 1\}$ holds*

$$C_N(f) = \lceil \log \mathrm{rank}_B(M(f)) \rceil$$

and

$$C_{MOD_p}(f) = \lceil \log \mathrm{rank}_p(M(f)) \rceil.$$

2   Boolean and Field Rank of Random Matrices

Let A be a random element of the set B_n of all $n \times n$ Boolean matrices. We wish to find a lower bound on the Boolean rank of almost all Boolean matrices in the sense that the probability tends to 0 with n that the bound is violated. The best bound we know of is by the methods of A.C. Yao [20], which yield $n/(2 + \varepsilon)$ although he stated only $n/4$. Here we find the best asymptotic bound, namely n. We use the Erdős-Spencer method [6], and show that in general we cannot expect rank 1 submatrices to exist which cover asymptotically more than $n/2$ ones. But the total number of ones is almost always $n^2/2$ by the binomial distribution. We use a rather precise estimate on the central limit theorem due to Feller [7].

Lemma 2. *The expected number of $a \times b$ rectangles forming a rank 1 submatrix entirely of 1 in a random $n \times n$ matrix is*

$$2^{-ab} \binom{n}{a} \binom{n}{b}.$$

This number tends to zero whenever $a, b > 4 \log n$ at a rate faster than any negative power of n. The probability of having at least one submatrix consisting entirely of ones whose area is greater than $(n/2)(1 + 2n^{-5/12})$ tends to zero as $n \to \infty$.

Proof. This expected number is the probability that any fixed $a \times b$ block has this form times the number of such blocks. The number of all $a \times b$ matrices is 2^{ab} and the probability is its reciprocal. The number of all $a \times b$ submatrices (not necessarily consecutive rows/columns) in an $n \times n$ matrix is $\binom{n}{a}\binom{n}{b}$. That proves the first statement. In the case of the 2nd statement,

$$2^{-ab} \binom{n}{a} \binom{n}{b} \leq 2^{-ab} n^{a+b}$$

$$\leq 2^{-16 \log^2 n + 8 \log^2 n}$$

which tends to zero at a rate faster than any negative power of n. For the last statement, we must use a different method not involving expected value, but rather a better approximation to the normal distribution. For a given set of c rows, the number of ones in their intersection (entry-wise product) is binomially distributed with probability $p = 1/2^c$ and n trials which are the various entries of the intersection vector. This yields a mean of $n/2^c$ and a variance of $n(2^c - 1)/(2^{2c})$. The probability of being at least $n^{1/12}$ standard deviations from the mean by Feller [7], Corollary to Theorem XVI.7.1(p.552), is at most

$$(1 + O(n^{-3/4}))(1 - N(n^{1/12}))$$

which tends to zero at a rate at least

$$e^{-n^{1/6}/2}$$

since for $x > 2$

$$1 - N(x) < \frac{2}{x\sqrt{2\pi}}e^{-x^2/2}$$

by comparing initial values and derivatives. This quantity will outweigh all polynomial order factors including the choices of the shapes, and will, assuming at least one side is at most $4 \log n$ in size using the second statement, also outweigh the $\binom{n}{c} < n^c < n^{4 \log n}$ factor resulting from being able to choose the c rows in any way.

If this number of ones is less than $n^{1/12}$ standard deviations from the mean then the area is at most

$$c(n/2^c + n^{1/12}\sqrt{n(2^c - 1)/2^{2c}}).$$

For $c > 2$, the initial term makes this asymptotically less than or equal to $3n/8$. For $c = 1, 2$ its values are at most

$$n/2 + n^{7/12}\sqrt{1/4}$$

$$2(n/4 + n^{7/12}\sqrt{3/16})$$

which agree with the stated value.

Theorem 1. *Almost all Boolean matrices of order $n \times n$ have Boolean rank $n(1 - o(1))$.*

Proof. We prove that for any positive ε, the probability tends to 1 that a random matrix from B_n has Boolean rank at least $\frac{n}{(1+\varepsilon)}$.

Suppose it has lower Boolean rank. Then all ones in the matrix are contained in a union of rank 1 matrices. By facts about the binomial distribution, with probability approaching 1, the total number of ones in a random $n \times n$ matrix is $\frac{n^2}{2}(1 - o(1))$ with probability approaching 1. By Lemma 2, with probability approaching 1, the areas of all submatrices consisting entirely of 1 are at most $n(1 + o(1))/2$ in size, and hence it will take $n(1 - o(1))$ rectangles to do the covering.

Analogues to Theorem 1 are well-known also for rank over fields. The following statement for finite fields is somewhat stronger compared to Theorem 1. For completeness we give a short proof.

Theorem 2. *For any sequence $f(n)$ tending to infinity, the probability that a random $n \times n$ matrix over the fixed finite field $GF(q)$ has rank at least $n - f(n)$ tends to 1, in fact the probability of rank at least $n - k$ is at least*

$$exp(-\frac{q^{-k+1}}{(1-q)^2}).$$

Proof. We construct a large family of matrices with rank at least $n-k$ as follows. There are q^n possible row vectors; any subspace of rank k has q^k vectors. Its complement has cardinality $q^n - q^k$. Choose the first $n - k$ rows, each in turn as an element in the complement of the subspace generated by the previous rows. This can be done in $(q^n - 1)(q^n - q) \cdots (q^n - q^{n-k-1})$ ways. Then choose the remaining row vectors arbitrarily with $q^n \cdots q^n = q^{kn}$ choices. There are q^{n^2} $n \times n$ matrices in all. The probability is then at least

$$P(n,k) = (1 - q^{-k+1})(1 - q^{-k-2}) \cdots (1 - q^{-n})$$

We derive a lower bound on $P(n,k)$ looking at its natural logarithm, and using the Taylor series for $\ln(1 - x) = -\sum_n x^n/n$. The last line is a case of the

$$\begin{aligned}
\log P(n,k) &= \log(1 - q^{-k-1}) + \ldots + \log(1 - q^{-n}) \\
&\geq -q^{-k-1} - q^{-2k-2}/2 - \ldots - q^{-k-2} - q^{-2k-4}/2 - \ldots - q^{-n} - q^{-2n}/2 - \ldots \\
&\geq -q^{-k-1} - 2q^{-k-2} - 3q^{-k-3} - \ldots \\
&= -q^{-k-1}(1 + 2q^{-1} + 3q^{-2} + \ldots) \\
&= -q^{-k-1}\frac{1}{(1 - q^{-1})^2}.
\end{aligned}$$

binomial theorem for exponent -2. Taking exponentials gives the stated formula, and we observe that as k tends to infinity in any way, the exponent tends to zero and the estimate tends to 1.

In case of an infinite field K there is no straight-forward analogue to "almost all". However a satisfying formalization can be given using polynomial identities: We say that *almost all* order n square matrices over K (i.e., points from $K^{(n^2)}$) have a certain property P, if $P \cap K^{(n^2)}$ is an ZARISKI-open set, i.e., it is the complement of the solution set of a system of algebraic equations. In that sense, the following statement follows from the remark that a matrix is singular over a field if it's determinant (a polynomial in the matrix entries) vanishes over this field.

Theorem 3. *Let K be an infinite field. Then almost all matrices of order $n \times n$ have rank n over K.*

Finally, we mention the following result of [10]:

Theorem 4. *Let K be an infinite field. Then almost all Boolean matrices of order $n \times n$ have rank n over K.*

Here the notion *almost all* is used in the counting sense as defined at the end of Section 1.1.

3   Approximating Low Boolean Rank by Low Field Rank

In this section we prove, that over any field K a Boolean matrix with small Boolean rank can be approximated by a matrix with small K-rank. We prove this for the special case of the field $K = GF(2)$ and explain the generalizations later. First we put the statement into a more general context.

Let Ω be a finite set and $\mathcal{F} \subseteq 2^{\Omega}$ be a family of *base sets* that contains Ω and is *closed* under intersection. We call such families *closed families*. We recall also the following: the *symmetric sum of sets* $A_1, \ldots, A_t$ (denoted $A_1 \oplus \ldots \oplus A_t$) is the set of all elements a for which the number of A_i that contain a is odd. We use $\oplus$ also to denote addition modulo 2. Further we use the following convenient notions: a *t-union* (*t-intersection, symmetric t-sum*) over a set family is the union (intersection, symmetric sum) of at most t sets of the family. Finally, for $\varepsilon > 0$ a set A' is called an *ε-approximation* to $A \subseteq \Omega$ if $|A \oplus A'| \leq \varepsilon \cdot |\Omega|$.

Given that $A = E_1 \cup \ldots \cup E_t$, where the E_i are base sets. Then we have

$$A = \Omega \oplus \bigcap_{i=1}^{t} (\Omega \oplus E_i) = \bigoplus_{\emptyset \subset S \subseteq [t]} \bigcap_{i \in S} E_i.$$

Hence, a t-union over a closed family is a symmetric T-sum over this family for some $T \leq 2^t - 1$ which is easily seen to be optimal for some families. The lemma below says, that we can do much better, if we allow ε-approximations.

Lemma 3. *Let $A = E_1 \cup \ldots \cup E_t$ be a t-union over a closed family $\mathcal{F}$ and $\varepsilon > 0$. Then there is some ε-approximation $A' \subseteq \Omega$ for A that is a symmetric T-sum over $\mathcal{F}$, where $T \leq t^{O(\log(1/\varepsilon))}$.*

Proof. Let $A = E_1 \cup \ldots \cup E_t$ for base sets $E_1, \ldots, E_t$. Let $d = \lceil \log(1/\varepsilon) \rceil$ and let $w : [t] \rightarrow GF(2)^t$ be a random weight function. For any $a \in A$ let the weight of a be the sum of weights of all sets E_j containing a: $w(a) = \sum_{a \in E_j} w(j)$. Then for all elements $a \in A$

$$\mathbf{Pr}_w[w(a) = 0] \leq 1/2^d \leq \varepsilon.$$

Hence, by a standard counting argument, there is some w^* such that $\mathbf{Pr}_a[w^*(a) = 0 | a \in A] \leq \varepsilon$. Therefore $B = \{a \in A | w^*(a) \neq 0\} = \bigcup_{i=1}^{d} \{a \in A | w_i^*(a) = 1\} \subseteq A$ is an ε-approximation to A.

Observe that $w_i^*(a) = 1$ if and only if $\bigoplus_{a \in E_j} w_i^*(j) = 1$ if and only $a \in \bigoplus_{a \in E_j} E_j^i$, where

$$E_j^i = \begin{cases} E_j & \text{if} \quad w_i^*(j) = 1 \\ \emptyset & \text{else.} \end{cases}$$

On the other hand from $a \in \bigoplus_{a \in E_j} E_j^i$ follows $a \in \bigoplus_{j=1}^t E_j^i =: A^i$. Hence, we have $B \subseteq A' \subseteq A$ for $A' := \bigcup_{i=1}^d A^i = \Omega \oplus \bigoplus_{i=1}^d (\Omega \oplus A^i)$. Since each A^i is the symmetric sum of some E_j, by expansion the right hand side is a symmetric T-sum of (intersections of) base sets.

It can be proved, that Lemma 3 is basically equivalent to a lemma from [16].

Symmetric sum, union, and intersection, respectively, of subsets of an n-set Ω correspond to vector sum over $GF(2)^n$, component-wise OR, and AND of vectors, respectively. To generalize Lemma 3 to arbitrary field arithmetic we therefore consider $\mathcal{F}$ as a closed set of 0-1-vectors (i.e, it contains $(1, \ldots, 1) \in \{0,1\}^n$ and is closed under AND) and speak of linear t-combinations of vectors over $\mathcal{F}$ (i.e., linear combinations of at most t vectors from $\mathcal{F}$) instead of symmetric t-sums.

Lemma 4. *Let K be a field and $\mathcal{F} \subseteq \{0,1\}^n$ be a closed set of vectors. If $A = E_1 \vee \ldots \vee E_t$ is a t-union over $\mathcal{F}$ and $\varepsilon > 0$ then there is some ε-approximation A' for A that is a linear T-combination over $\mathcal{F}$, where $T \leq t^{O(\log(1/\varepsilon))}$ if K is finite and $T \leq t^{O(\log(1/\varepsilon) \log t)}$ if K is infinite.*

The generalization can be proved along similar lines as above, where in the case of an infinite field an approximation result of [18] is used. In a more special setting this was also exploited in [13]. For finite fields the above Lemma is basically equivalent to a Lemma of [17].

We observe that rectangles of a matrix form a closed family over the set Ω of entries of this matrix and contains the set Ω itself. By the definition of Boolean and field ranks, we can conclude the main result of this section:

Theorem 5. *Let A be a Boolean matrix and $\mathrm{rank}_B(A) \leq t$. Then for any field K and $\varepsilon > 0$ there is some matrix A' that differs from A in at most an ε-fraction of its 1-entries and such that $\mathrm{rank}_K(A') \leq T$, where $T \leq t^{O(\log(1/\varepsilon))}$ if K is finite and $T \leq t^{O(\log(1/\varepsilon) \log t)}$ if K is infinite.*

4   Applications to Communication Complexity and Matrix Rigidity

In this section we apply our results from Sections 2 and 3 to communication complexity and rigidity of matrices.

A direct consequence of Theorem 1 is the following:

Corollary 1. *Almost all Boolean functions $f : \{0,1\}^s \times \{0,1\}^s \to \{0,1\}$ have nondeterministic communication complexity s.* $\square$

This improves the asymptotic lower bound $s - 1$ given in [20] to the optimal value s (it is obvious that every function as above can be computed by a nondeterministic protocol of length s). For *deterministic* communication complexity this was observed in [5] on base of Theorem 4.

From Theorem 5 we can conclude the following

Corollary 2. *Let* $f : \{0,1\}^s \times \{0,1\}^s \to \{0,1\}$. *For any prime* p *and any* $\varepsilon > 0$ *there is a function* f' *that differs from* f *in at most an* ε-*fraction of its 1-values such that*

$$C_{MOD_p}(f) \le \log(1/\varepsilon) \cdot C_N(f).$$

$\square$

This is an improvement over a result in [2] where a weaker statement was proved for the special case $p = 2$ using different methods. It is also interesting to note, that it is not possible to simulate nondeterministic communication protocols by MOD_p-communication protocols and vice versa: in [4] examples of functions where constructed that have exponential higher complexity in one of the models compared to the other.

Theorem 5 can also be applied in a different context. The *rigidity* $\mathcal{R}_M^k(r)$ of an $n \times n$-matrix M *with respect to a field* K is the minimal number of entries of M that must be changed in order to reduce the rank of the resulting matrix over K to at most r. This notion was introduced by Valiant [19] for the study of algebraic circuit complexity. He showed that almost all matrices have "high" rigidity with respect to any field. However, proving large lower bounds on rigidity of *explicit* matrices remains a fundamental open problem. A solution would imply corresponding lower bounds for a variety of computational models [13]. The best bounds we have so far are of order $\Omega(\frac{n^2}{r})$ for infinite fields (e.g., [15]) and $\Omega(\frac{n^2}{r} \log \frac{n}{r})$ for finite fields [8].

As application of Theorem 5 we establish an upper bound on rigidity of a Boolean matrix M. In the statement we use the notation $\overline{M}$ to denote the Boolean complement of a Boolean matrix M (i.e., the matrix obtained from M by complementing each entry).

Corollary 3. *Let* M *be a binary* $n \times n$-*matrix and* K *a finite field. Then there is some* $c > 0$ *such that for any* $r \ge 1$ *and* $t = \max\{rank_B(M), rank_B(\overline{M})\}$ *holds*

$$\mathcal{R}_K^M(r) \le \frac{N^2}{2^{c \cdot \log r / \log t}}.$$

In case K *is infinite the upper bound holds with* $\log t$ *replaced by* $(\log t)^2$.

Proof. We consider only the case of a finite field, the other is similar.

Let $\mathcal{F}$ be the family of rectangles of an $n \times n$ matrix M. $\mathcal{F}$ is clearly closed under intersection. Considering a Boolean matrix as the set of its 1-entries we can write $M = R_1 \cup \ldots \cup R_t$, where each $R_i \in \mathcal{F}$. By Lemma 3 for fixed $\varepsilon > 0$ there is an ε-approximation $M' \subseteq M$ such that M' is a linear combination of at most $t^{O(\log 1/\varepsilon)}$ rectangles. Now, each rectangle — as a matrix of order $N \times N$ —

has rank 1 over K. Hence, the rank of its sum is at most $r_\varepsilon = t^{O(\log 1/\varepsilon)}$. In other words, to reduce the rank of M to at most r_ε, it is sufficient to change at most an ε-fraction of its entries. Setting $\varepsilon = \frac{1}{2^{c \cdot \log r / \log t}}$ for a suitable c guarantees that r_ε achieves the target rank r.

The upper bound in terms of $\mathrm{rank}_K(\overline{M})$ follows by considering the complement of M.

5   Conclusion

We conclude with mentioning two open problems:

1. Prove or disprove that almost all Boolean $n \times n$ matrices have Boolean rank n (this would be the full Boolean analogue of Theorem 4). To improve Theorem 1 it seems necessary to somehow use the fact that almost all of the rank 1 matrices are $1 \times n, 2 \times n, n \times 1, n \times 2$ matrices or to estimate the minimum overlap of these rectangles.
2. Prove that the converse of Theorem 5 is not true, i.e., present a sequence of $(0,1)$-matrices with small rank over some field, such that each approximating Boolean matrix has large Boolean rank. We believe that SYLVESTER-matrices are good candidates for such properties.

References

1. Stephen A. Cook. A taxonomy of problems with fast parallel algorithms. *Information and Control*, 64:2–22, 1985.
2. Carsten Damm. On Boolean vs. modular arithmetic for circuits and communication protocols. Forschungsbericht 98-06, Universität Trier, 1998.
3. Carsten Damm, Matthias Krause, Christoph Meinel, and Stephan Waack. Separating oblivious linear length MOD_p-branching program classes. *Journal of Information Processing and Cybernetics EIK*, 30:63–75, 1994.
4. Carsten Damm, Matthias Krause, Christoph Meinel, and Stephan Waack. On relations between counting communication complexity classes. *Journal of computer and System Sciences*, (to appear), 1998.
5. M. Dietzfelbinger, J. Hromkovič, and G. Schnitger. A comparison of two lower bound methods for communication complexity. In *Proc. of MFCS'94*, volume 841 of *Lecture Notes in Computer Science*, pages 326–336. Springer-Verlag, 1994.
6. P. Erdős and J. Spencer. *Probabilistic Methods in Combinatorics*. Academic Press, 1974.
7. W. Feller. *An Introduction to Probability Theory and its Applications, Vol.II*. Wiley, 1971.
8. J. Friedman. A note on matrix rigidity. *Combinatorica*, 13:235–239, 1993.
9. Ki Hang Kim. *Boolean Matrix Theory and Applications*. Marcel Dekker, New York, 1982.
10. J. Komlós. On the determinant of $(0,1)$-matrices. *Studia Sci. Math. Hungar.*, 2:7–21, 1965.
11. Matthias Krause and Stephan Waack. Variation ranks of communication matrices and lower bounds for depth-two circuits having nearly symmetric gates with unbounded fan-in. *Mathematical Systems Theory*, 28:553–564, 1995.

12. E. Kushilevitz and Noam Nisan. *Communication Complexity*. Cambridge University Press, 1996.
13. Satyanarayana V. Lokam. Spectral methods for matrix rigidity with applications to size-depth tradeoffs and communication complexity. In *FOCS*, pages 6–15, 1995.
14. George Markowsky. Ordering d-classes and computing schein rank is hard. *Semigroup Forum*, 44:373–375, 1992.
15. A. Razborov. On rigid matrices. manuscript, June 1989.
16. Alexander A. Razborov. Lower bounds for the size of bounded depth with basis $\{\wedge, \oplus\}$. *Mathematical Notes of the Academy of Sciences of the USSR*, 41:598–607, 1987.
17. Roman Smolensky. Algebraic methods in the theory of lower bounds for Boolean circuit complexity. In *19th ACM STOC*, pages 77–82, 1987.
18. Jun Tarui. Probabilistic polynomials, AC^0 functions, and the polynomial-time hierarchy. *Theoretical Computer Science*, 113:167–183, 1993.
19. Leslie Valiant. The complexity of computing the permanent. *TCS*, 8:189–201, 1979.
20. Andy Yao. Some complexity questions related to distributive computing. In *Proceedings of the 11th Annual ACM Symposium on the Theory of Computing*, 1979.

A Combinatorial Algorithm for Pfaffians

Meena Mahajan[1], P.R. Subramanya[2], and V. Vinay[2]

[1] Institute of Mathematical Sciences, Chennai, India. meena@imsc.ernet.in
[2] Dept. of Computer Science & Automation, Indian Institute of Science,
Bangalore - 560012, India. [subbu,vinay]@csa.iisc.ernet.in

Abstract. The Pfaffian of a graph is closely linked to Perfect Matching. It is also naturally related to the determinant of an appropriately defined matrix. This relation between Pfaffian and determinant is usually exploited to give a fast algorithm for computing Pfaffians.

We present the first completely combinatorial algorithm for computing the Pfaffian in polynomial time. In fact, we show that it can be computed in the complexity class GapL; this result was not known before. Our proof techniques generalize the recent combinatorial characterization of determinant [MV97] in novel ways.

As a corollary, we show that under reasonable encodings of a planar graph, Kasteleyn's algorithm for counting the number of perfect matchings in a planar graph is also in GapL. The combinatorial characterization of Pfaffian also makes it possible to directly establish several algorithmic and complexity theoretic results on Perfect Matching which otherwise use determinants in a roundabout way.

1 Introduction

The main result of this paper is a combinatorial algorithm for computing the Pfaffian of a graph. This is similar in spirit to a recent result of Mahajan & Vinay [MV97] who give a combinatorial algorithm for computing the determinant of a matrix. In complexity theoretic terms, we establish that computing the Pfaffian of a graph is in the class GapL.

GapL is the class of functions that are logspace reducible to computing the integer determinant of a matrix. It is known that computing the determinant of a matrix is equivalent to taking the difference of two #L functions [Vin91]. In other words, GapL is the class of functions that can be expressed as the difference in the number of accepting paths of two nondeterministic logspace machines. They define a space analog of the important counting classes, GapP and #P.

Pfaffians are intimately connected to determinants. For example, it is known that the square of the Pfaffian of a graph is equal to the determinant of a related matrix. This is, however, not adequate to imply our GapL algorithm as we do not know if GapL is closed under square roots (of positive integers).

One of the motivations for this work is to understand the complexity of Perfect Matching. Perfect Matching is not known to be in NC, but known to be in RNC [MVV87]. This result has been recently improved by Allender & Reinhardt

T. Asano et al. (Eds.): COCOON'99, LNCS 1627, pp. 134–143, 1999.

[AlRe98] who show that Perfect Matching is in the class SPL (non-uniformly). (SPL is a subclass of GapL, wherein functions take the value 0 or 1. Please refer to [AlRe98] for details of the definition.) Interestingly, Allender & Reinhardt make use of the Mahajan-Vinay clow sequences [MV97] critically to establish their result. We hope that our combinatorial characterization of Pfaffian will be a key in resolving the vexed question of the complexity of Perfect Matching. This is indeed our main motivation for this work.

Pfaffians arise naturally in the study of matchings; the pfaffian of a graph is just the sum over all possible perfect matchings except that each matching has an associated *sign* as well. This gives it a flavour similar to that of a determinant. In the absence of the sign, they would calculate the number of perfect matchings in a graph, a problem that is well-known to be complete for $\sharp P$ [Val79]. Also, in the case of special graphs, it is known that the graph may be oriented in such a way that all the terms of the pfaffian turn out to be positive. This obviously means there would be no cancellation and hence the pfaffian would count the number of perfect matchings in the underlying graph. Such orientations of graphs are called Pfaffian orientations.

A celebrated result of Kasteleyn [Kas67] proves that all planar graphs admit a pfaffian orientation. This result has since been improved by Little [Lit74] who has shown that all $K_{3,3}$-free graphs admit a pfaffian orientation (not all graphs admit of a pfaffian orientation). Finding such an orientation was shown to be in NC by Vazirani [Vaz89]. In this paper, we show that under reasonable encodings of planar graphs, the problem of counting the number of perfect matchings in a planar graph is in GapL as well. The problem of extending our result to $K_{3,3}$-free graphs remains to be investigated.

In section 2, a few preliminaries and definitions are stated. In section 3, we set up the combinatorial framework for pfaffians. Section 4 focuses on the combinatorial algorithm for computing Pfaffians. We indicate in section 5 that counting the number of perfect matchings in a planar graph is in GapL.

2   Preliminaries & Definitions

Let D be a $n \times n$ matrix. S_n is the permutation group on $\{1, 2, \ldots, n\}$ (denoted $[n]$). The permanent and determinant of D, $per(D)$ and $det(D)$, are defined as,

$$per(D) = \sum_{\sigma \in S_n} \prod_i d_{i\sigma(i)} \qquad\qquad det(D) = \sum_{\sigma \in S_n} sgn(\sigma) \prod_i d_{i\sigma(i)}$$

where $sgn(\sigma)$ is -1 if σ has an odd number of inversions, $+1$ otherwise. An equivalent definition of the sign of a permutation is in terms of the number of cycles in its cycle decomposition.

Every permutation $\sigma \in S_n$ can be decomposed into a set of cycles in the associated graph[1], G_D, of the matrix D. The cycles are non-intersecting (i.e.

---

[1] The graph G_D associated with a $n \times n$ matrix D is the complete directed graph K_n on n vertices (with self-loops) having the matrix elements as edge weights.

simple), disjoint and they cover every vertex in the graph, i.e. these are **cycle covers**. The sign of a cycle cover is defined in terms of the number of even length cycles in it. The sign is $+1$ if there are an even number of them, else it is -1.

A **clow** in G_D is a walk that starts at some vertex (called *head*), visits vertices larger than the head any number of times and returns to the head. This cycle in G_D is not always a simple cycle. Formally,

Definition 1. [MV97]

1. *A clow is an ordered sequence of edges $C = \langle e_1, e_2, \ldots, e_m \rangle$ such that $e_i = \langle v_i, v_{i+1} \rangle$ and $e_m = \langle v_m, v_1 \rangle$, $v_1 \neq v_j$ for $j \in \{2, 3, \ldots, m\}$ and $v_1 = min\{v_1, \ldots, v_m\}$. The vertex v_1 is called the head of the clow and denoted $h(C)$. The length of the clow is $|C| = m$, and the weight of the clow is $wt(C) = \prod_{i=1}^{m} wt(e_i)$. [Note: $C = \langle e \rangle$ where $e = \langle v, v \rangle$, i.e. a self-loop, is also a clow, of length one.]*

2. *An l-clow sequence is an ordered sequence of clows $C = (C_1, \ldots, C_k)$ such that $h(C_1) < \ldots < h(C_k)$ and $\sum_{i=1}^{k} |C_i| = l$.*

Pfaffians were introduced by Kasteleyn [Kas67] to count the number of dimer coverings of a lattice graph. We define matchings and Pfaffians more formally.

Definition 2. *Given an undirected graph $G = (V, E)$ with $V = \{1, 2, \ldots, n\}$, we define*

1. *A **matching** $\mathcal{M}$, is a subset of the edges of G such that no two edges have a vertex in common. That is, $\mathcal{M} \subseteq E(G)$ such that $e_1, e_2 \in \mathcal{M}$, $e_1 = \langle i_1, j_1 \rangle, e_2 = \langle i_2, j_2 \rangle$ and $i_1 = i_2 \Leftrightarrow j_1 = j_2$.*

2. *A matching $\mathcal{M}$ is a **perfect matching** if every vertex $i \in V(G)$ occurs as the end-point of some edge in $\mathcal{M}$.*

Let G be a graph on n vertices. With each edge $(i, j) \in G$, associate an indeterminate x_{ij} such that $x_{ij} = -x_{ji}$. For even n, let $\mathcal{M} = |\ i_1 i_2\ |\ i_3 i_4\ | \ldots |\ i_{n-1} i_n\ |$ be a partition of the vertices of G into $\frac{n}{2}$ unordered pairs. $\mathcal{M}$ corresponds in a natural way to a possible perfect matching in G. We define two functions on $\mathcal{M}$, $wt(\mathcal{M})$ and $sgn(\mathcal{M})$. $wt(\mathcal{M})$ should be non-zero if and only if $\mathcal{M}$ corresponds to a feasible perfect matching in G (i.e. every pair in $\mathcal{M}$ is indeed an edge in G). $sgn(\mathcal{M})$ is the sign of the permutation $\sigma_\mathcal{M}$ associated[2] with $\mathcal{M}$; $\sigma_\mathcal{M} = \begin{pmatrix} 1 & 2 & 3 & 4 & \ldots & n-1 & n \\ i_1 & i_2 & i_3 & i_4 & \ldots & i_{n-1} & i_n \end{pmatrix}$. That is, $sgn(\mathcal{M}) = sgn(\sigma_\mathcal{M})$.

The *Tutte matrix* for G is an $n \times n$ matrix $D = (d_{ij}) = \begin{cases} x_{ij} & (i, j) \in G, i < j \\ -x_{ji} & (i, j) \in G, i > j \\ 0 & \text{otherwise} \end{cases}$

Note that, D is a skew symmetric matrix. The Pfaffian of G, defined in terms of the *Tutte Matrix D* for G is,

$$Pf(G) = Pf(D) = \sum_{\mathcal{M}: \text{ partition}} sgn(\mathcal{M}) \cdot d_{i_1 i_2} d_{i_3 i_4} \ldots d_{i_{n-1} i_n}$$

---

[2] A partition of $[1, n]$ into $\frac{n}{2}$ pairs has $2^{\frac{n}{2}} \cdot (\frac{n}{2})!$ permutations that can yield it. Irrespective of which permutation one ascribes to a partition, the contribution to the Pfaffian is the same.

where the summation is over all partitions. In the rest of the paper, when we talk about the Pfaffian of a matrix D, we are actually referring to the Pfaffian of the underlying graph.

Using Linear Algebra we can prove the following properties of skew symmetric matrices D.

- If D has an odd number of rows, then $det(D) = 0$.
- If D has an even number of rows, then $det(D) = (\mathrm{Pf}(D))^2$.

Let G be a graph with three special vertices s, t_+ and t_-. Consider a function f defined as,

$$f = \sum_{\rho:\ s \mapsto t_+} wt(\rho) - \sum_{\eta:\ s \mapsto t_-} wt(\eta)$$

where ρ iterates over all paths from s to t_+ and η is over all s to t_- paths. GapL is precisely the class of functions that can be formulated in this fashion. Stated somewhat differently, GapL consists of those functions that are the difference of two $\sharp$L functions, where $\sharp$L is the counting class for NL. As stated earlier, since GapL is the class of languages logspace reducible to computing the integer determinant, the Mahajan & Vinay GapL algorithm for the determinant [MV97] can be used to compute $det(D)$, viz. $[\mathrm{Pf}(D)]^2$. This fails to yield a GapL algorithm for Pfaffian because GapL is not known to be closed under square roots.

3   A Combinatorial Setting for Pfaffians

In this section, we build the combinatorial framework for pfaffians using a variant of *clow sequences*. We also provide a new characterization for the sign of a pfaffian term that we utilize in our combinatorial algorithm for computing pfaffians.

Lemma 3. [LovPlu86] *Let G be an arbitrary orientation of an undirected graph G. Let F_1 and F_2 be two* perfect matchings *of G. Let k be the number of evenly oriented alternating cycles in $F_1 \cup F_2$. Then, $sgn(F_1) \cdot sgn(F_2) = (-1)^k$.*

We can fix a permutation (i.e. matching) and compute the signs of all other matchings with respect to it. This is a direct outcome of Lemma 1. Let us fix the *identity permutation* as our base matching. As the sign of the identity permutation is $+1$, the sign of any matching will be the number of evenly oriented cycles in its superposition with the identity permutation.

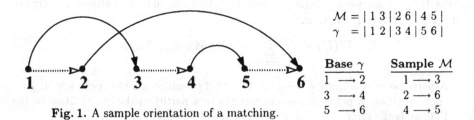

$$\mathcal{M} = |\,1\,3\,|\,2\,6\,|\,4\,5\,|$$
$$\gamma = |\,1\,2\,|\,3\,4\,|\,5\,6\,|$$

Base γ	Sample $\mathcal{M}$
$1 \longrightarrow 2$	$1 \longrightarrow 3$
$3 \longrightarrow 4$	$2 \longrightarrow 6$
$5 \longrightarrow 6$	$4 \longrightarrow 5$

Fig. 1. A sample orientation of a matching.

Consider a partition (i.e., a matching) $\mathcal{M} = |\ i_1\ i_2\ |\ i_3\ i_4\ |\ \ldots\ |\ i_{n-1}\ i_n\ |$. We know that there are several permutations that we can associate with $\mathcal{M}$. Each of these permutations will yield the same pfaffian term, $p_{\mathcal{M}} = sgn(\mathcal{M}) \cdot wt(\mathcal{M})$. This is so because, using a finite number of edge reversal and vertex reordering operations, we can arrive at any permutation associated with $\mathcal{M}$ from any other. Under each individual operation, the pfaffian term remains unchanged. Therefore, we have freedom in choosing a permutation to associate with $\mathcal{M}$.

The permutation, $\sigma_{\mathcal{M}}$, we choose to associate with $\mathcal{M}$ has forward-going arcs. That is, for any pairing $|\ i_j\ i_{j+1}\ |$ in $\mathcal{M}$, orient the edge as $\langle i_j, i_{j+1}\rangle$ ensuring that $i_j < i_{j+1}$, i.e. the arc is forward going. Fig 1 illustrates this convention. The dotted edges in Fig 1 are our base matching, while the normal edges constitute $\mathcal{M}$. The matching edges are ordered in increasing order of the edge source.

Our convention for orienting the graph G can be summarized as follows:
- The identity permutation, γ, is the base matching.
- The permutation, $\sigma_{\mathcal{M}}$, associated with $\mathcal{M}$ has forward going arcs only.
- γ and $\sigma_{\mathcal{M}}$ together impose an orientation G on G.
- Compute $sgn(\mathcal{M})$ with respect to G using Lemma 2.

Lemma 4. *The sign of a partition $\mathcal{M}$ in the pfaffian can be defined as,*

$$sgn(\mathcal{M}) = (-1)^{|\{i\ :\ \langle i,2j\rangle \in C,\ i<2j\}|\ +\ |\{i\ :\ \langle i,2j-1\rangle \in C,\ i>(2j-1)\}|\ +\ k}$$

where k is the number of cycles in the superposition of $\mathcal{M}$ with the identity permutation γ, and C is any cycle in the superposition.

Proof: Route each cycle $C \in (\mathcal{M} \cup \gamma)$ so that starting from the smallest vertex in C we take its matched edge in $\mathcal{M}$. The sign associated with C is -1 if the number of wrongly oriented edges is even, and $+1$ otherwise. The wrongly oriented edges on C can be grouped as follows (recall, edges of $\mathcal{M}$ are forward oriented)
- $\langle 2j-1, i\rangle \in \mathcal{M}$ is wrongly traversed.
- $\langle i, 2j\rangle \in \mathcal{M}$ is correctly traversed, but forces the γ-edge $\langle 2j, 2j-1\rangle$ to be traversed wrongly.
- $\langle 2j, i\rangle \in \mathcal{M}$ and $\langle 2j, 2j-1\rangle \in \gamma$ are traversed wrongly, so the net effect is nullified.

Let r be the total contribution from above. The sign of C is $(-1)^{r+1}$. (The additional 1 in the exponent is to get a -1 on an evenly oriented cycle.) We now need the parity of the number of even oriented cycle, which is the product of the signs of each of the cycles. $\square$

We need a variant of clows called **pclows** for pfaffians.

Definition 5.

- *A pair of edges $E = (e_1, e_2)$ is a **p-edge** if for some $i \in [1, n]$ either,*
  1. *$e_1 = \langle i, 2j\rangle$ and $e_2 = \langle 2j, 2j\text{-}1\rangle$, or*
  2. *$e_1 = \langle i, 2j\text{-}1\rangle$ and $e_2 = \langle 2j\text{-}1, 2j\rangle$.*
- *A **pclow** is a clow with its ordered sequence of edges being $P = \langle E_1, E_2, \ldots, E_m\rangle$ where each E_i is a p-edge.*

- *A l-pclow sequence is an ordered sequence of pclows, $\mathcal{P} = \langle P_1, \ldots, P_k \rangle$.*
- *We need to generalize the sign of a pclow sequence which is consistent with the sign of a perfect matching. Define the sign of a pclow sequence to be the parity of the number of evenly oriented pclows. The results of Lemma 2 generalizes to pclow sequences as well.*
- *The weight of a p-edge $E = (e_1, e_2)$ is the weight of the edge e_1. The contribution to the weight of the p-edge E by the second edge, e_2, is always 1. The weight of a pclow is the product of the p-edge weights.*

Fig. 2. Selecting $\langle i, 2j - 1 \rangle$ from i.   **Fig. 3.** Selecting $\langle i, 2j \rangle$ from i.

Fig 2 & 3 indicate the cases when a pair of consecutive edges are p-edges. Using pclow sequences, we prove a novel and powerful characterization of the Pfaffian.

Theorem 6.

$$\mathsf{Pf}(D) = \sum_{W:\ pclow\ sequence} sgn(\mathcal{W}) \cdot wt(\mathcal{W})$$

Proof: Pclow sequences that are cycle covers are the superposition of the base matching with a perfect matching. We need to show that pclow sequences that are not cycle covers do not contribute to the summation. We establish an involution on the set of pclow sequences. Non-cycle covers get mapped onto non-cycle covers of opposite signs. The fixed points of the involution are the cycle covers.

Our technique would be to pair a pclow sequence with another having the same set of edges but with an opposite sign. Consequently, they cancel each other's contribution to the summation. Note that, all pclows in any given pclow sequence are even in length.

However, a given sequence could have an odd length simple cycle in a pclow as shown in Fig 4 & 5. To pair such sequences, pick the pclow with the smallest head that has an odd simple sub-cycle. Walk down this pclow from its head, until you realize that you have gone around an odd cycle. Simply reverse the orientation of all the edges in this cycle. This defines a new pclow sequence. Conversely, starting with the new sequence, our mapping will consider the same (sub-)cycle and reversing its edges will give us the old pclow sequence; so they pair. Their total contribution is zero, since reversing an odd number of edges contributes a negative sign. The pclows in Fig 4 & 5 are an example of the above bijection.

We are left with pclow sequences all of which have pclow(s) with even sub-cycles. Let $\mathcal{P} = \langle P_1, \ldots, P_k \rangle$ be such a pclow sequence. Pick the smallest i such

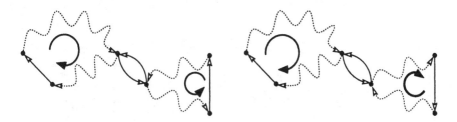

Fig. 4. Pclow with odd sub-cycle   **Fig. 5.** Pclow with odd sub-cycle reversed

that P_{i+1} to P_k are disjoint simple cycles. If $i = 0$, then $\mathcal{P}$ is a cycle cover and $\mathcal{P}$ maps onto itself. Else, traverse P_i till one of the following happens,

1. We hit a vertex that meets one of P_{i+1} to P_k.
2. We hit a vertex that completes an even length simple cycle in P_i.

Let v be this vertex. Note that, these two conditions are mutually exclusive because of the way we have traversed P_i. We never hit a vertex that simultaneously satisfies both the conditions.

Case 1: Suppose v touches P_j. The successor edge of v in P_i must be from the base matching (read ahead to see why). Let w be this vertex. w is either $v + 1$ or $v - 1$. Either the predecessor or the successor of v in P_j has to be w. (If w had been the predecessor of v in P_i, then we would have stopped our traversal at w itself.) The orientation of the (v, w) edge in P_j gives rise to two cases.

1. *If the edge in P_j is from v to w:* (v,w) is identically oriented in P_i and P_j. We simply stick P_j into P_i at v. Formally, map $\mathcal{P}$ to a pclow sequence $\mathcal{P}' = \langle P_1, \ldots, P_{i-1}, P_i', P_{i+1}, \ldots, P_{j-1}, P_{j+1}, \ldots P_k \rangle$. P_i' is obtained from P_i by inserting into it the simple cycle P_j at the first occurrence of v. Figure 6 illustrates this case.
2. *If the edge in P_j is from w to v:* (v,w) has opposite orientations in P_i and P_j. Stick P_j into P_i at v but only after reversing the orientation of all edges in P_j. Figure 7 shows the mapping.

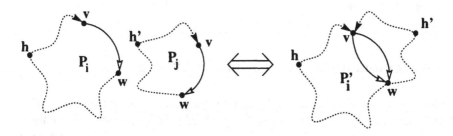

Fig. 6. P_i and P_j have (v,w) oriented the same way.

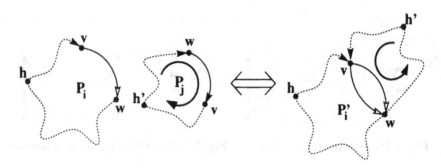

Fig. 7. P_i and P_j have (v,w) oriented differently.

Case 2: Suppose v completes a simple cycle P in P_i. P must be disjoint from all the later cycles. We modify the pclow sequence $\mathcal{P}$ by plucking out P from P_i and introducing it as a new pclow. P's position will be to the right of P_i as P_i's head would be smaller than P's. However, one additional change may be necessary. The definition of a p-edge demands that the out-going edge from the head be a matched edge. This *may not* be so with P. If so, it would not even be a valid pclow. In which case, we reverse the orientations of all the edges in P before inserting it into an appropriate position.

We need to argue the correctness of these mappings. It should be clear that the new sequences map back to the original sequences and hence, the mapping is an involution. We now show that the mapped pclow sequences have opposing signs, and as their weights are identical, they cancel each others contribution.

In *case 1*, when both P_i and P_j are oddly oriented, the resulting pclow, P_i', is evenly oriented. When P_i and P_j are both evenly oriented, so is P_i'. In the other two cases, P_i' will be oddly oriented. Therefore, in *case 1*, the number of even-oriented pclows in the two sequences differ by one and hence, they have different parity. Note that, the reversal of an even length pclow preserves its orientation parity. The analysis of the second case is similar.

Pclow sequences arising from the superposition of the identity permutation with some perfect matching, map onto themselves. These are the sole survivors. □

4   A Combinatorial Algorithm for Pfaffians

In this section we describe a combinatorial algorithm for computing the Pfaffian. We construct a layered directed acyclic graph H_D with three special vertices s, t_+ and t_-. We show that $\mathrm{Pf}(D) = \sum_{\rho:\, s\,\mapsto\, t_+} wt(\rho) \ - \ \sum_{\eta:\, s\,\mapsto\, t_-} wt(\eta)$. In this model of computation, all $s \mapsto t_+$ ($s \mapsto t_-$) paths of positive (negative) sign are in 1-1 correspondence with pclow sequences of positive (negative) sign.

H_D has the vertex set, $\{s, t_+, t_-\} \cup \{[p, h, u, i] | p \in \{0, 1\}, h, u \in [1, n], i \in \{0, \ldots, n-1\}\,\}$. A path from s to $[p, h, u, i]$ indicates that in the pclow sequence being constructed along this path, p is the parity of the pclow sequence, h is the head of the current pclow, u is the current vertex on the pclow and i is the number of edges seen so far. A $s \mapsto t_+$ ($s \mapsto t_-$) path corresponds to a n-pclow sequence having a positive (negative) sign.

H_D has n layers and layer i has vertices of the form $[\_, \_, \_, i]$. The edges from layer $(2j\text{-}1)$ to layer $2j$ are fixed and independent of G_D. The edges in H_D are,

1. $\langle s, [0, h, h, 0] \rangle$ for $h = 2i - 1$, where $i \in [1, \frac{n}{2}]$; edge weight is 1.
2. $\langle [p, h, u, 2i], [p, h, v, 2i + 1] \rangle$ if $\langle u, v \rangle \in G_D$, $v > h$, $v > u$, $i \in [0, \frac{n}{2} - 1]$; edge weight is d_{uv}.
3. $\langle [p, h, u, 2i], [\bar{p}, h, v, 2i + 1] \rangle$ if $\langle u, v \rangle \in G_D$, $v > h$, $v < u$, $i \in [0, \frac{n}{2} - 1]$; edge weight is d_{vu}.
4. $\langle [p, h, 2j - 1, 2i - 1], [p, h, 2j, 2i] \rangle$ if $2j - 1 > h$, $2i < n$; edge weight is 1.
5. $\langle [p, h, 2j, 2i - 1], [\bar{p}, h, 2j - 1, 2i] \rangle$ if $2j - 1 > h$, $2i < n$; edge weight is 1.
6. $\langle [p, h, h + 1, 2i - 1], [p, h', h', 2i] \rangle$ if $h' > h$, h' is odd, $2i < n$; edge weight is 1.
7. $\langle [0, h, h + 1, n - 1], t_+ \rangle$ and $\langle [1, h, h + 1, n - 1], t_- \rangle$ if $h = 2i - 1$, $i \in [1, \frac{n}{2}]$; edge weight is 1.

Theorem 7. *Given a $n \times n$ skew symmetric matrix D, let H_D be the graph described above. Then,*

$$\mathsf{Pf}(D) = \sum_{\rho:\; s \,\mapsto\, t_+} wt(\rho) - \sum_{\eta:\; s \,\mapsto\, t_-} wt(\eta)$$

The full version of the paper contains a proof of the theorem.

Using simple dynamic programming techniques we can evaluate $\mathsf{Pf}(D)$ in polynomial time. The algorithm proceeds in n stages, where in the ith stage we compute the sum of the weighted paths from s to any vertex x in layer i. Layer n has vertices t_+ and t_-, and we compute the difference of the weighted paths from s to t_+ and t_-.

The above algorithm looks at an edge in H_D once and hence, is a polynomial-time algorithm. It is clear that we can parallelize the above computation and evidently, computing pfaffian is in NC.

5 Finding Admissible Orientations for Planar Graphs

Counting the perfect matchings in a graph requires finding an admissible orientation and then, computing the Pfaffian of the associated matrix. We know how to compute Pfaffians in GapL. We know that the general problem of counting the perfect matchings in a graph is $\sharp$P-Complete. If we restrict ourselves to planar graphs, then finding admissible orientations to them is in GapL. This means that counting perfect matchings in planar graphs is in GapL.

In this section, we shall indicate these results and leave the proofs for the full version of the paper. The definitions of *orientation parity* and *admissible orientations* appear in [Kas67] and [VY89].

Lemma 8. *Finding an admissible orientation of a planar graph is logspace reducible to the problem of evaluating a* **parity tree**.

Lemma 9. *Parity Tree Evaluation can be done in logspace.*

Proof: Parity is associative and commutative. Evaluating the parity of a sequence of elements requires one to remember only the parity of the elements seen so far. Hence, the parity tree can be collapsed. The parity of the leaves is therefore that of the tree. Systematically finding these leaves can be done by a logspace machine. □

Combining the results of the previous sections, we have,

Theorem 10. *Given a planar graph G, counting the number of perfect matchings in G can be done within* GapL.

References

AlRe98.   E. Allender and K. Reinhardt, *Isolation, Matching, and Counting*, in IEEE Conference on Computational Complexity, pp 92-100, 1998.

Kas67.   P. W. Kasteleyn, *Graph Theory and Crystal Physics*, in Graph Theory and Theoretical Physics, ed: F. Harary, Academic Press, pp 43-110, 1967.

Lit74.   C. H. C. Little, *An extension of Kasteleyn's method of enumerating the 1-factors of planar graphs*, in Combinatorial Mathematics, Proc. 2nd Australian Conference, ed: D. Holton, *Lecture Notes in Mathematics*, 403, pp 63-72, 1974.

LovPlu86.   L. Lovasz and M. Plummer, *Matching Theory*, Annals of Discrete Mathematics 29, North-Holland, 1986.

Lyn77.   N. Lynch, *Log Space Recognition and Translation of Parenthesis Languages*, Journal of the ACM, Vol. 24(4), pp 583-590, 1977.

MV97.   M. Mahajan and V. Vinay, *Determinants: Combinatorics, algorithms, and complexity*, Chicago Journal of Theoretical Computer Science, vol. 5, 1997.

MRST97.   W. McCuaig, N. Robertson, P. D. Seymour and R. Thomas, *Permanents, Pfaffian Orientations, and Even Directed Circuits*, Proc. of IEEE Symposium on Theory of Computing, pp 402-405, 1997.

MVV87.   K. Mulmuley, U. V. Vazirani and V. V. Vazirani, *Matching is as easy as Matrix Inversion*, Combinatorica, vol. 7, pp 105-113, 1987.

Reif84.   J. Reif, *Symmetric Complementation*, Journal of the ACM, vol. 31(2), pp 401-421, 1984.

Val79.   L. G. Valiant, *The complexity of computing the permanent*, Theoretical Computer Science, vol. 8, pp 189-201, 1979.

Vaz89.   V. V. Vazirani, *NC algorithms for computing the number of perfect matchings in $K_{3,3}$-free graphs and related problems*, Information and Computation, vol. 80(2), pp 152 - 164, 1989.

VY89.   V. V. Vazirani and M. Yannakakis, *Pfaffian Orientations, 0-1 permanents and even cycles in directed graphs*, Discrete Applied Mathematics, vol. 25, pp 179 - 190, 1989.

Vin91.   V. Vinay, *Counting auxiliary pushdown automata and semi-unbounded arithmetic circuits*, Proc. 6th IEEE Structure in Complexity Theory Conference, pp 270-284, 1991.

How to Swap a Failing Edge
of a Single Source Shortest Paths Tree[*]

Enrico Nardelli[1,2], Guido Proietti[1], and Peter Widmayer[3]

[1] Dipartimento di Matematica Pura ed Applicata, Università di L'Aquila, Via Vetoio, 67010 L'Aquila, Italy. {nardelli,proietti}@univaq.it.
[2] Ist. di Analisi dei Sistemi e Informatica, CNR, V.le Manzoni 30, 00185 Roma, Italy.
[3] Institut für Theoretische Informatik, ETH Zentrum, 8092 Zürich, Switzerland. widmayer@inf.ethz.ch.

Abstract. In this paper we introduce the notion of *best swap* for a failing edge of a single source shortest paths tree (SPT) $S(r)$ rooted in r in a weighted graph $G = (V, E)$. Given an edge $e \in S(r)$, an edge $e' \in E \setminus \{e\}$ is a *swap edge* if the *swap tree* $S_{e/e'}(r)$ obtained by swapping e with e' in $S(r)$ is a spanning tree of G. A *best swap edge* for a given edge e is a swap edge minimizing some distance functional between r and the set of nodes disconnected from the root after the edge e is removed. A *swap algorithm* with respect to some distance functional computes a best swap edge for every edge in $S(r)$. We show that there exist fast swap algorithms (much faster than recomputing from scratch a new SPT) which also preserve the functionality of the affected SPT.

1 Introduction

Survivability of a communication network means the ability of the network to remain operational even if individual network components (such as a link or even a node) fail. In the past few years, several survivability problems have been studied intensely [4], mainly as a consequence of the advent of sparse fiber optic networks. For example, a classic survivability problem is that of maintaining the shortest path between two specified nodes in the network under the assumption that at most k arcs or nodes along the original shortest path are removed [2].

In the extreme, a network might be designed as a spanning tree of some underlying graph of all possible links. A sparse network, however, is more likely to react catastrophically to failures, especially of links, since the density of traffic through each link is very high. Therefore, it is important for sparse networks to take survivability into account from the very beginning. Assuming that damaged links can be restored quickly, the likelihood of having failures that overlap in time is quite small. Therefore, it makes sense to study the problem of dealing with the failure of each single link in the network, since we can expect that sooner or later each link will fail. Moreover, for several practical motivations [6], whenever

---

[*] This research was partially supported by the CHOROCHRONOS TMR Program of the European Community. The work of the third author was partially supported by grant "Combinatorics and Geometry" of the Swiss National Science Foundation.

T. Asano et al. (Eds.): COCOON'99, LNCS 1627, pp. 144–153, 1999.
© Springer-Verlag Berlin Heidelberg 1999

a link fails, it is important that the number of replacing edges is small. If the network is a spanning tree, the optimum would be replacing a link with a single link reconnecting the network. The question is: *which is the link that has to be chosen?*

Coping with a failure of a link in a network having the topology of a tree means, on the theoretical side, to define an interesting family of problems on graphs. Let $G = (V, E)$ be a biconnected, undirected graph, where V is the set of vertices and $E \subseteq V \times V$ is the set of edges, with a nonnegative real length $|e|$ associated with each edge $e \in E$. Let n and m denote the number of vertices and the number of edges, respectively. Let $T = (V, E_T)$, with $E_T \subseteq E$ be a spanning tree of G. A *swap edge* for an edge $e = (u, v) \in E_T$ is an edge $e' = (u', v') \in E \setminus \{e\}$ reconnecting the two subtrees created by the removal of e. Let in the following S_e denote the set of all swap edges for e and let $T_{e/e'}$ be the tree obtained by swapping e with e'. Let $F[T_{e/e'}]$ be a functional of $T_{e/e'}$. A *best swap* for an edge e is an edge $f \in S_e$ such that $F[T_{e/f}] \leq F[T_{e/e'}]$, for any $e' \in S_e$. A *swap algorithm* finds a best swap for every edge $e \in E_T$.

An interesting example arises when T is a minimum spanning tree (MST). Here, the natural functional to be defined is $F[T_{e/e'}] = \sum_{g \in E_{T_{e/e'}}} |g|$. In this case, it is easy to see that f is a swap edge of minimum length. Moreover, $T_{e/f}$ coincides with T', the MST of $G - e = (V, E \setminus \{e\})$. The fastest solution for finding a best swap for each edge in T runs in $O(m \cdot \alpha(m, n))$ time [9], where $\alpha(m, n)$ is the functional inverse of Ackermann's function [8].

Another interesting problem arises when T is a minimum diameter spanning tree (MDST). In this case, the natural functional is the diameter of $T_{e/e'}$, and a best swap is a swap edge which makes the diameter of the new spanning tree as low as possible. The problem of finding a best swap for each edge in T has been solved by Nardelli et al. in $O(n\sqrt{m})$ time [7]. They also showed that the diameter of $T_{e/f}$ is at most $5/2$ times the diameter of T', the MDST of $G - e$.

However, many network architectures are based on a single source shortest paths tree (SPT) rooted at a given node r, say $S(r) = (V, E_S)$ [1]. This is especially true for centralized network, where there exists a privileged node broadcasting messages to all the other nodes. In this case, a best swap policy for a failing edge is not unique as for the MST and the MDST, and depends on whatever is of interest from a network management point of view. Therefore, a rigorous approach to the problem requires the definition of the objective functional $F[S_{e/e'}(r)]$ to be minimized, and the complexity of finding a best swap for every edge in $S(r)$ will depend on the selected functional. In this paper, we propose efficient swap algorithms for three functionals of primary importance in SPTs. Let S_v denote the set of nodes disconnected from the root after the removal of $e = (u, v) \in S(r)$ and reconnected via the swap edge e' (see Figure 1).

Moreover, let $d(v_1, v_2)$ (resp., $d_{e/e'}(v_1, v_2)$) denote the distance between v_1 and v_2 in $S(r)$ (resp., $S_{e/e'}(r)$), for any $v_1, v_2 \in V$. The following functionals are considered, due to their primary importance in network applications:

1. $F[S_{e/e'}(r)] = \sum_{t \in S_v} d_{e/e'}(r, t);$

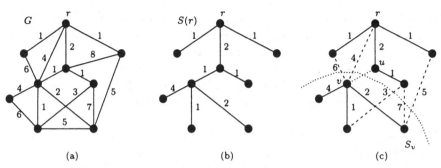

Fig. 1. (a) A weighted graph $G = (V, E)$; (b) a SPT $S(r)$ rooted in r; (c) edge $e = (u, v)$ is removed from $S(r)$: dashed edges are swap edges.

2. $F[S_{e/e'}(r)] = \max\limits_{t \in S_v}\{d_{e/e'}(r, t)\};$

3. $F[S_{e/e'}(r)] = \max\limits_{t \in S_v}\{d_{e/e'}(r, t) - d(r, t)\}.$

These functionals focus on S_v, since we are interested in studying how the nodes disconnected from the root are affected by the failure. The swap problems induced by these functionals will be named the *sum-problem*, the *height-problem* and the *increase-problem*, respectively. Our analysis shows that there exist fast swap algorithms for these functionals. These algorithms are much faster than recomputing from scratch for every edge $e \in S(r)$ the SPT $S'(r)$ rooted in r of $G - e$, (we say recomputing from scratch since no dynamic solution for this problem is known which is asymptotically better [3]). Moreover, we compare $S_{e/f}(r)$ with $S'(r)$, on the basis of the various studied functionals, showing that with respect to the chosen functionals, $S_{e/f}(r)$ is worse than $S'(r)$ only by a small constant factor, that is, $S_{e/f}(r)$ is *functionally* similar to $S'(r)$, in terms of the studied functionals. Therefore, we conclude that swapping in a SPT is cheap and effective.

The paper is organized as follows: Section 2 proposes the algorithms for solving the problems. In Section 3, we present a comparison between the solutions computed by the various swap algorithms and the respective exact solutions. Finally, in Section 4, we give conclusions and list some open problems.

2   The swap algorithms

2.1   Solving the sum-problem

Remember that the sum-problem asks for a swap edge minimizing the sum of the lengths of all the paths from r to each node in S_v. A brute force solution will consider all the edges in S_e, for any $e \in S(r)$. Since each edge can be a swap edge for $O(n)$ edges, and the above sum can be computed in $O(n)$ time, it follows that such an approach will cost $O(n^2m)$ time.

We now propose a more efficient solution. A high-level description of our algorithm is the following. We consider all the edges $e = (u, v) \in S(r)$ in any arbitrary postorder. Let us now fix such an edge (u, v). For each node t in S_v we do the following. We select a swap edge leading from r to t in $G - e$ on a path as short as possible, and we compute the corresponding sum of all the paths starting from r, going to t and leading to all the nodes in S_v. To do so efficiently, during the postorder traversal we keep track of the total length of all paths from t that stay within S_v and of the number of these paths. Finally, we select the minimum over these values for all nodes t in S_v and we return the corresponding best swap edge.

Let us now present a more detailed description of the algorithm. We make use of the following auxiliary data, for any $v \in V$ (see Figure 2):

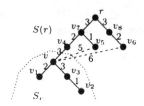

Fig. 2. A weighted graph and a SPT $S(r)$ (solid edges); non-tree edges are dashed. $S_v = \{v, v_1, v_3, v_2\}$, **son**$(v) = \{v_1, v_3\}$, $size(v) = 4$, $down(v) = 9$, $up(v, v_4) = 3$, $up(v, v_7) = 14$, $min\_path(v, v) = 9$, $min\_path(v, v_4) = 9$, $min\_path(v, v_7) = 11$, $all\_paths(v, v) = 45$, $all\_paths(v, v_4) = 57$, $all\_paths(v, v_7) = 100$.

- **son**(v): list of sons of v in $S(r)$;
- $size(v)$: number of nodes in S_v;
- $down(v)$: $\sum_{t \in S_v} d(v, t)$;
- $up(v, w)$: $\sum_{t \in S_w \setminus S_v} d(v, t)$, where w is an ancestor of v in $S(r)$, except r;
- $min\_path(v, w)$: $\min_{(e' = (u', v) \in E \setminus E_S) \wedge (u' \notin S_w)} \{d(r, u') + |e'|\}$, where w is either v

  or an ancestor of v in $S(r)$, except r; if no such edge e' exists, set $min\_path(v, w) = +\infty$;

The algorithm consists of the following steps:

Algorithm SUM_PROBLEM$(G, S(r))$;
Input: A weighted graph $G = (V, E)$ and a SPT $S(r) = (V, E_S)$;

Output: $\forall e = (u, v) \in E_S$, a swap edge $f | \sum_{t \in S_v} d_{e/f}(r, t) = \min_{e' \in S_e} \left\{ \sum_{t \in S_v} d_{e/e'}(r, t) \right\}$.

Step 1: For each node $v \in S(r)$ as considered by any postorder visit
Step 2:   Compute $son(v), size(v), down(v)$;
Step 3:   For each ancestor $w \neq r$ of v (including v) compute $min\_path(v, w)$;
Step 4: For each edge $e = (u, v) \in S(r)$ as considered by any postorder visit
Step 5:   For each node $v_i \in son(v)$
Step 6: $up(v_i, v) = down(v) - down(v_i) + [size(v) - size(v_i) - 1] \cdot |(v_i, v)|$;
Step 7:   For each node $t \in S_{v_i}$
Step 8: $up(t, v) = up(t, v_i) + [size(v) - size(v_i)] \cdot d(v_i, t) + up(v_i, v)$;
Step 9:   For each node $t \in S_v$
Step 10: $all\_paths(t, v) = down(t) + up(t, v) + min\_path(t, v) \cdot size(v)$;
Step 11: Compute t_{min}, where $all\_paths(t_{min}) = \min_{t \in S_v} \{all\_paths(t, v)\}$;
Step 12: Output the edge associated with $min\_path(t_{min}, v)$.

Theorem 1. *Given a graph $G = (V, E)$ with n vertices and m edges, with positive real edge lengths and a SPT $S(r)$ rooted in $r \in V$, the swap algorithm $SUM\_PROBLEM(G, S(r))$ solves the sum-problem in $O(n^2)$ time and space.*

Proof. The correctness of the algorithm is a consequence of the fact that it considers exhaustively at each step all the possible best swap edges. To establish the time and space complexity of the algorithm, let us analyze it step by step.

Concerning Step 2, we can compute $son(v)$, $size(v)$ and $down(v)$ in $O(n)$ time and space for each node (and $O(1)$ amortized over all the nodes). Therefore Step 2 can be accomplished in $O(n)$ time and space for all the nodes.

Step 3 can be accomplished in $O(n)$ time and space for each node in the following way: let $\langle r \equiv w_0, w_1, \ldots, w_k \equiv v \rangle$ be the path in $S(r)$ joining r and v. We start by bucketing the non-tree edges adjacent to v with respect to their nearest common ancestors. This can be done in $O(1)$ time for each edge [5], that is, it will cost $O(n)$ time for each node and $O(m)$ time for all the nodes. Let $b(v, w_i)$ be the bucket containing the edges associated with $w_i, i = 0, \ldots, k - 1$. We initially search for the edge in $b(v, w_0)$ minimizing the path from r to v. This can be done in time and space proportional to the number of elements in $b(v, w_0)$. This value defines $min\_path(v, w_1)$. Afterwards, we repeat the step for $b(v, w_1)$: if the found value is less than $min\_path(v, w_1)$, then this becomes $min\_path(v, w_2)$, otherwise we set $min\_path(v, w_2) = min\_path(v, w_1)$. The process goes on iteratively, up to $b(v, w_{k-1})$. In this way we spend $O(n)$ time and space for each node, and $O(n^2)$ time and space for all the nodes.

Step 6 can be accomplished in $O(1)$ time for each node and in $O(n)$ total time for all the nodes. Steps 7-11 can be executed in $O(n)$ time, and therefore require a total $O(n^2)$ time for all the nodes. Finally, Step 12 costs $O(1)$ time per node. Therefore, the overall time and space complexity is $O(n^2)$.   □

2.2   Solving the height-problem

Remember that the height-problem asks for a swap edge minimizing the length of a longest path starting from r and ending in S_v. The following can be proved:

Theorem 2. *Given a graph $G = (V, E)$ with n vertices and m edges, with positive real edge lengths and a SPT $S(r)$ rooted in $r \in V$, there exists a swap algorithm solving the height-problem in $O(n\sqrt{m})$ time and $O(m)$ space.*

Proof. This problem can be solved by slightly modifying the approach used in [7], where the problem of computing all the best swaps for a minimum diameter spanning tree has been solved. In fact, as a subroutine of the main algorithm, the length of a longest path starting from $t \in S_v$ and staying within S_v (which is exactly what we need, once that we add $d_{e/e'}(r, t)$) is there computed, and this costs $O(n\sqrt{m})$ time and $O(m)$ space.   □

2.3 Solving the increase-problem

Remember that the increase-problem asks for a swap edge minimizing the maximum increase of the distance from r to any node in S_v. The following can be proved:

Theorem 3. *Given a graph $G = (V, E)$ with n vertices and m edges, with positive real edge lengths and a SPT $S(r)$ rooted in $r \in V$, there exists a swap algorithm solving the increase-problem in $O(m \cdot \alpha(m, n))$ time and space.*

Proof. A best swap by definition is an edge f such that

$$\max_{t \in S_v}\{d_{e/f}(r, t) - d(r, t)\} = \min_{e' = (u', v') \in S_r}\left\{\max_{t \in S_v}\{d_{e/e'}(r, t) - d(r, t)\}\right\}. \tag{1}$$

For any swap edge e' and for any node $t \in S_v$ we have

$$d_{e/e'}(r, t) - d(r, t) \leq d_{e/e'}(r, v) + d_{e/e'}(v, t) - d(r, t) =$$

$$= d_{e/e'}(r, v) + d(v, t) - d(r, t) = d_{e/e'}(r, v) - d(r, v).$$

Then, (1) becomes $d_{e/f}(r, v) - d(r, v) = \min_{e' = (u', v') \in S_r}\{d_{e/e'}(r, v) - d(r, v)\}$, and therefore

$$d_{e/f}(r, v) = \min_{e' = (u', v') \in S_r}\{d_{e/e'}(r, v)\}.$$

Hence, to solve the increase-problem it suffices to get a swap edge minimizing the distance from r to v. To do that efficiently, we make use of a *transmuter* [9]. A transmuter $D_G(T)$ is a directed acyclic graph that represents the set of fundamental cycles of a graph G with respect to a spanning tree T. Basically, $D_G(T)$ contains for each tree edge e a source node $s(e)$, and for each non-tree edge e' a sink node $t(e')$, plus a certain number of additional nodes. The fundamental property of a transmuter is that there is a path from a given source $s(e)$ to a given sink $t(e')$ if and only if e and e' form a cycle in T. It is clear that in $S(r)$, all and only the edges belonging to S_e form a cycle with e. Therefore, we can build a transmuter having as source nodes all the edges belonging to $S(r)$ and as sink nodes all the non-tree edges. This can be done in $O(m \cdot \alpha(m, n))$ time and space [9]. To associate e with its best swap f, it remains to establish the value that has to be given to a sink node. If we associate in $O(1)$ time with e' the length of the (not simple) cycle in $S(r)$ starting from r, passing through e' and going back to r, that is

$$c(e') = d(r, u') + |e'| + d(r, v'),$$

we will have $d_{e/e'}(r, v) = c(e') - d(r, v)$ for any edge $e' \in S_e$ and therefore, a shortest cycle is associated with a best swap, and vice-versa. Finally, we can solve the increase-problem by processing the nodes of the transmuter in reverse topological order in $O(m \cdot \alpha(m, n))$ time. This completes the proof. □

3   Swapping versus recomputing from scratch

Since swapping a single edge for a failed one is fast and involves very few changes in the underlying network (e.g., as to routing information), it is interesting to see how the tree obtained from swapping compares with a true SPT that does not use the failed edge. In this section we address this task, comparing the SPT $S_{e/f}(r)$, obtained by swapping the failed edge e with a best swap edge f, and a true SPT $S'(r)$ of $G - e$. While it is natural to study each of the three quality criteria (functionals) for the algorithms that optimize the corresponding swap, we go one step further: We also study the effect that a swap algorithm has on the other criteria (that it does not aim at). Let $d'(v_1, v_2)$ denote the distance between v_1 and v_2 in $S'(r)$. For each swap algorithm, the following ratios in the two trees will be studied:

$$\sigma = \frac{\sum_{t \in S_v} d_{e/f}(r,t)}{\sum_{t \in S_v} d'(r,t)} ; \quad \rho = \frac{\max_{t \in S_v}\{d_{e/f}(r,t)\}}{\max_{t \in S_v}\{d'(r,t)\}} ; \quad \delta = \frac{d_{e/f}(r,v)}{d'(r,v)} . \tag{2}$$

In the following, $h(S_v)$ will denote the *height* of S_v, that is the length of a longest path between v and any node in S_v, while ℓ will denote the height of $S(r)$ *restricted* to S_v, that is the length of a longest path in $S(r)$ between r and any node in S_v. Similarly, ℓ' and $\ell_{e/f}$ will denote the heights of $S'(r)$ and $S_{e/f}(r)$ restricted to S_v. Note that $h(S_v) \leq \ell$, $h(S_v) \leq \ell'$ and $h(S_v) \leq \ell_{e/f}$.

3.1   Ratios of the swap algorithm for the sum-problem

We can prove the following result:

Theorem 4. *For the swap algorithm solving the sum-problem, we have $\sigma \leq 3$, $\rho \leq 4$ and δ unbounded. The bounds are tight.*

Proof. Let $f = (x, y)$ be a best swap edge and let $f' = (x', y')$ be the (only) swap edge such that $f' \in S'(r)$ and f' is on the shortest path from r to v in $S'(r)$. Concerning σ, let $\overline{\ell_{e/f}}$ and $\overline{\ell'}$ denote the average length of a path from r to $t \in S_v$ in $S_{e/f}(r)$ and $S'(r)$, respectively. Of course, $\sigma = \overline{\ell_{e/f}}/\overline{\ell'}$. We have

$$\overline{\ell_{e/f}} \leq \overline{\ell_{e/f'}} \leq d_{e/f'}(r,y') + d_{e/f'}(y',v) + \frac{\sum_{t \in S_v} d(v,t)}{size(S_v)}$$

and given that $d_{e/f'}(r,y') = d'(r,y')$ and $d_{e/f'}(y',v) = d(y',v) = d'(y',v)$ (since the old path from v to y' was a shortest path), it follows

$$\overline{\ell_{e/f}} \leq d'(r,v) + \frac{\sum_{t \in S_v} d(v,t)}{size(S_v)} .$$

Moreover, we have that for any $t \in S_v$, $d'(r, v) \leq d'(r, t) + d(t, v) \leq d'(r, t) + d(r, t) \leq 2d'(r, t)$, from which, for any node t in S_v, it follows that $d'(r, v) \leq 2\overline{\ell'}$. Furthermore

$$\frac{\displaystyle\sum_{t \in S_v} d(v, t)}{size(S_v)} \leq \frac{\displaystyle\sum_{t \in S_v} d(r, t)}{size(S_v)} \leq \frac{\displaystyle\sum_{t \in S_v} d'(r, t)}{size(S_v)} = \overline{\ell'}.$$

Therefore, we have that $\overline{\ell_{e/f}} \leq 3\overline{\ell'}$, that is $\sigma \leq 3$. The bound is tight as shown in Figure 3a.

Concerning ρ, we have $\ell_{e/f} \leq d_{e/f}(r, y) + 2h(S_v) \leq d_{e/f}(r, y) + 2\ell'$. Moreover, $d_{e/f}(r, y) \leq d_{e/f'}(r, y) \leq d_{e/f'}(r, v) + d_{e/f'}(v, y)$, and given that $d_{e/f'}(r, v) = d_{e/f'}(r, y') + d_{e/f'}(y', v) = d'(r, v)$ (since f' is on the shortest path from r to v in $S'(r)$, and this path contains the old shortest path from v to y') , it follows

$$d_{e/f}(r, y) \leq d'(r, v) + d(v, y) \leq \ell' + h(S_v) \leq 2\ell'$$

that is, $\ell_{e/f} \leq 4\ell'$ or $\rho \leq 4$. The bound is tight as shown in Figure 3b.

Finally, concerning δ, it is unbounded as shown in Figure 3c. $\square$

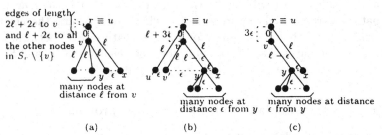

(a)   (b)   (c)

Fig. 3. In all the pictures, $S(r)$ (solid edges) with the removed edge (u, v); non-tree edges are dashed and the best swap is $f = (x, y)$. (a) In $S_{e/f}(r)$, $\overline{\ell_{e/f}} = 3\ell + \epsilon$, while $\overline{\ell'} = \ell + 2\epsilon, \epsilon \geq 0$, since in $S'(r)$ we have all the edges of length $\ell + 2\epsilon$ from r to $S_v \setminus \{v, y\}$; then, $\sigma = 3$. (b) The distance to node v' in $S_{e/f}(r)$ (i.e., the height) is $4\ell - \epsilon$, while in $S'(r)$ the height is $\ell + 3\epsilon, \epsilon \geq 0$, from which $\rho = 4$. (c) The distance to node v in $S_{e/f}(r)$ is 2ℓ, while in $S'(r)$ is $3\epsilon, \epsilon \geq 0$, from which δ is unbounded.

3.2   Ratios of the swap algorithm for the height-problem

We can prove the following result:

Theorem 5. *For the swap algorithm solving the height-problem, we have σ unbounded, $\rho \leq 2$ and δ unbounded. The bounds are tight.*

Proof. Let f and f' be defined as for Theorem 4. Concerning σ and δ, they are unbounded as shown in Figure 4a. Concerning ρ, we have

$$\ell_{e/f} \leq \ell_{e/f'} \leq d_{e/f'}(r, v) + h(S_v) = d'(r, v) + h(S_v) \leq \ell' + h(S_v) \leq 2\ell'$$

from which $\rho \leq 2$. The bound is tight as shown in Figure 4b. $\square$

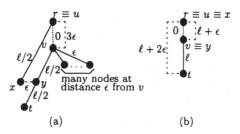

(a)   (b)

Fig. 4. In all the pictures, $S(r)$ (solid edges) with the removed edge (u, v); non-tree edges are dashed and the best swap is $f = (x, y)$. (a) $\overline{\ell}_{e/f} = \ell + 2\epsilon$, while $\overline{\ell}' = 4\epsilon$, and the distance to node v in $S_{e/f}(r)$ is $\ell + \epsilon$, while in $S'(r)$ is $3\epsilon, \epsilon \geq 0$, from which σ and δ are unbounded. (b) The distance to node t in $S_{e/f}(r)$ (i.e., the height) is $2\ell + \epsilon$, while in $S'(r)$ the height is $\ell + 2\epsilon, \epsilon \geq 0$, from which $\rho = 2$.

3.3   Ratios of the swap algorithm for the increase-problem

We can prove the following result:

Theorem 6. *For the swap algorithm solving the increase-problem, we have* $\sigma \leq 3$, $\rho \leq 2$ *and* $\delta = 1$. *The bounds are tight.*

Proof. Let f be a best swap edge and let t be any node in S_v. Concerning σ, from the fact that f is on the shortest path in $G - e$ from r to v, and then $S_{e/f}(r)$ and $S'(r)$ share that path, we have that $d_{e/f}(r, t) \leq d_{e/f}(r, v) + d(v, t) = d'(r, v) + d(v, t)$ and with $d'(r, v) \leq d'(r, t) + d(v, t)$ and $d(v, t) \leq d(r, t) \leq d'(r, t)$, it follows that $d_{e/f}(r, t) \leq 3d'(r, t)$, that is $\sigma \leq 3$. The bound is tight as shown in Figure 5a.

Concerning ρ, we have that $\ell_{e/f} \leq d_{e/f}(r, v) + h(S_v) = d'(r, v) + h(S_v) \leq \ell' + \ell' = 2\ell'$, that is $\rho \leq 2$. The bound is tight as shown in Figure 5b.

Finally, concerning δ, since $d_{e/f}(r, v) = d'(r, v)$, it follows that $\delta = 1$. $\square$

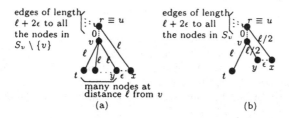

(a)   (b)

Fig. 5. In all the pictures, $S(r)$ (solid edges) with the removed edge (u, v); non-tree edges are dashed and the best swap is $f = (x, y)$. (a) $\overline{\ell}_{e/f} = 3\ell + \epsilon$, while $\overline{\ell}' = \ell + 2\epsilon, \epsilon \geq 0$, since in $S'(r)$ we have all the edges of length $\ell + 2\epsilon$ from r to $S_v \setminus \{v\}$, from which $\sigma = 3$. (b) The distance to node t in $S_{e/f}(r)$ (i.e., the height) is $2\ell + \epsilon$, while in $S'(r)$ the distance to all the nodes is $\ell + 2\epsilon, \epsilon \geq 0$, from which $\rho = 2$.

4   Summary and conclusions

Table 1 summarizes the bounds of the various algorithms for which we have performed comparisons between $S_{e/f}(r)$ and $S'(r)$. Interestingly, the algorithm for the increase-problem, which is the cheapest in terms of time complexity, is also the best with respect to the measures of quality we have defined. Our interpretation is that choosing as a best swap edge that one belonging to the new shortest path from r to v (as does the swap algorithm for the increase-problem) produces a swap tree topologically similar to the old SPT.

Measure	Algorithm		
	sum-problem	height-problem	increase-problem
Time	$O(n^2)$	$O(n\sqrt{m})$	$O(m \cdot \alpha(m,n))$
Space	$O(n^2)$	$O(m)$	$O(m \cdot \alpha(m,n))$
σ	3	unbounded	3
ρ	4	2	2
δ	unbounded	unbounded	1

Table 1. Time and space complexity and ratios for the studied swap algorithms

After we introduced in this paper the notion of *best swap edge* for a SPT, a large amount of work remains to be done. For example, different functionals $F[S_{e/e'}(r)]$ could be defined. Another open problem is the case of managing multiple simultaneous link failures. Clearly, it also remains to establish whether the algorithms we have proposed are optimal or not. Finally, the case of transient node failures deserves further studies.

Acknowledgements – The authors would like to thank Samir Khuller for helpful discussions on the topic.

References

1. R.K. Ahuia, T.L. Magnanti and J.B Orlin. *Network Flows: Theory, Algorithms and Applications*, Prentice Hall, Englewood Cliffs, NJ (1993).
2. A. Bar-Noy, S. Khuller and B. Schieber, The complexity of finding most vital arcs and nodes, CS-TR-3539, Dept. of Computer Science, Univ. of Maryland, 1995.
3. D. Frigioni, A. Marchetti-Spaccamela and U. Nanni, Fully dynamic output bounded single source shortest path problem, in *Proc. ACM-SIAM Symposium on Discrete Algorithms (SODA'96)*, 1996, 212--221.
4. M. Grötschel, C.L. Monma and M. Stoer, Design of survivable networks, in: *Handbooks in OR and MS, Vol. 7*, Elsevier (1995) 617–672.
5. D. Harel and R.E. Tarjan, Fast algorithms for finding nearest common ancestors, *SIAM J. Comput.*, **13**(2) (1984) 338–355.
6. G.F. Italiano and R. Ramaswami, Maintaining spanning trees of small diameter, *Proc. 21st Int. Coll. on Automata, Languages and Programming (ICALP'94)*, 1994, Lecture Notes in Computer Science, Vol. 820, 212–223.
7. E. Nardelli, G. Proietti and P.Widmayer, Finding all the best swaps of a minimum diameter spanning tree under transient edge failures, *Proc. 6th European Symp. on Algorithms (ESA'98)*, 1998, Lecture Notes in Computer Science, Vol. 1461, 55–66.
8. R.E. Tarjan. Efficiency of a good but not linear set union algorithm, *Journal of the ACM*, **22** (1975) 215–225.
9. R.E. Tarjan, Applications of path compression on balanced trees, *Journal of the ACM*, **26** (1979) 690–715.

On Bounds for the k-Partitioning of Graphs*

S.L. Bezrukov[1], R. Elsässer[2], and U.-P. Schroeder[2]

[1] Department of Mathematics and Computer Science, University of Wisconsin -
Superior, Superior, WI 54880-4500, USA,
[2] Department of Mathematics and Computer Science, University of Paderborn,
D-33102 Paderborn, Germany.

Abstract. Let an edge cut partition the vertex set of a graph into k disjoint subsets $A_1, \ldots, A_k$ with $||A_i| - |A_j|| \leq 1$. We consider the problem of determining the minimal size of such a cut for a given graph. For this we introduce a new lower bound method which is based on the solution of an extremal set problem and present bounds for some graph classes based on Hamming graphs.

1 Introduction

Many applied computational problems involve the partitioning of some associated graphs into subgraphs as a subproblem. For example, the approach to solve differential equations based on the finite elements method on a multiprocessor computing system requires a partitioning of the area into some simple figures (e.g. triangles or rectangles) and then an assignment of the nodes of the obtained graph to the processors of the computing system. In order not to affect the speed of computations, the assignment of the nodes should be possibly uniform, and the data exchange between the processors should be minimized. These demands cause a problem of partitioning the vertex set of the underlying graph into subsets of (almost) equal size by cutting a minimum number of edges.

Most graphs appearing in applications are highly irregular, which makes the evaluation of the partition quality difficult. Many papers in the literature deal with the design of partition algorithms. Consequently there exist a lot of libraries where these algorithms are implemented [8,16,9,13,12]. For evaluation of the quality of partition algorithms it is helpful to know exact results concerning the partitioning of some graph classes, or at least good lower bounds for the parameters of a partition. Such results are known in the literature for partitioning graphs into two parts [11,14] and for some special graphs [6,15]. Furthermore, in the classical paper [7] there is shown a lower bound for the edge cut of a partitioning of arbitrary graphs into k disjoint subsets. This bound is based on the spectrum of the Laplacian matrix of the graph. Let $A(G) = \{a_{ij}\}$ be the adjacency matrix of a graph G. Then we denote by $L(G)$ its Laplacian matrix

* This work is partially supported by the German Research Association (DFG) within the SFB 376 "Massive Parallelität: Algorithmen, Entwurfsmethoden, Anwendungen" and by the EC ESPRIT Long Term Research Project 20244 "ALCOM-IT".

T. Asano et al. (Eds.): COCOON'99, LNCS 1627, pp. 154–163, 1999.
© Springer-Verlag Berlin Heidelberg 1999

obtained from $A(G)$ by multiplying all its elements with -1 and replacing $a_{ii} = 0$ with the degree of the ith vertex of G. It is well known that $L(G)$ is positive semidefinite, all eigenvalues of $L(G)$ are real and non-negative, and the smallest eigenvalue is equal to 0 iff G is connected. However, even if we know all the eigenvalues of $L(G)$ it is not clear whether the spectral lower bound of [7] is quite good.

In our paper, we introduce a new lower bound method for the edge cut of the k-partitioning problem. Our approach is based on the solution of an extremal set problem which is called the edge-isoperimetric problem. In the literature many different names are used for various versions of isoperimetric problems such as the conductance of a graph, the isoperimetric number, etc. and have been proven to be particularly useful in a number of areas (cf. [2,3,5]).

Exemplarily, we apply our method to the partitioning of Hamming graphs, i.e. the Cartesian products of complete graphs. By the n-th Cartesian product G^n with $n \geq 2$ of a graph $G = (V, E)$, we mean the graph on the vertex set $V \times \cdots \times V$ where two tuples $(v_1, \ldots, v_n)$ and $(u_1, \ldots, u_n)$ form an edge iff they agree in some $n - 1$ entries and for the remaining entry, say i, holds $(v_i, u_i) \in E$. Throughout this paper let $[m, n]$ denote the set of integer numbers between m and n. Denote for $a \in \mathbb{N}$

$$V^n = \{(x_1, \ldots, x_n) \mid x_i \in [0, a - 1], i \in [1, n]\}.$$

For $u, v \in V^n$ the Hamming distance $\rho(u, v)$ is defined as the number of entries where u and v differ. The Hamming graph H_a^n is a graph on the vertex set V^n, where two vertices u, v are adjacent iff $\rho(u, v) = 1$.

These graphs are well studied with respect to the edge-isoperimetric problem [10], which we apply to get lower bounds for our problem. The results presented here extend the results of [1], where the partitioning of hypercubes is analyzed.

Moreover, we apply our method to the partitioning of graphs which are built by the Cartesian product of complete graphs without a certain number of perfect matchings. Let $a \in \mathbb{N}$ be even and $l \in [0, \frac{a}{2}]$, $V_1 = \{0, \ldots, \frac{a}{2} - 1\}$, and $V_2 = \{\frac{a}{2}, \ldots, a - 1\}$. We define the graph $H_{a,l} = (V, E)$ as follows: $V = V_1 \cup V_2$ and with $j = 0, \ldots, l - 1$ we have $E = \{(x, y) \mid x, y \in V_1 \lor x, y \in V_2\} \cup \{(x, y) \mid x \in V_1 \land y \in V_2 \land y \neq a - 1 - (x + j) \bmod a/2\}$.

These graphs are regular of degree $a - l - 1$ and are of special interest due to the fact that if we have a lower bound for an edge cut of a k-partitioning of these graphs we get immediately a lower bound for an edge cut of a k-partitioning of any regular graph of order a and degree $a - l - 1$. This statement holds also if we build the n-th Cartesian product of $H_{a,l}$ and a corresponding regular graph [3].

For a graph $G = (V_G, E_G)$ and a fixed k, consider a partition of V_G into the parts A_i, $i \in [1, k]$, satisfying $\|A_i\| - |A_j\| \leq 1$, or in other words

$$\left\lfloor \frac{|V_G|}{k} \right\rfloor \leq |A_i| \leq \left\lceil \frac{|V_G|}{k} \right\rceil. \tag{1}$$

For such a partition $\mathcal{A}_G = \{A_1, \ldots, A_k\}$ denote

$$\nabla \mathcal{A}_G = \{(u, v) \in E_G \mid u \in A_i, v \in A_j, i \neq j\}.$$

We say that a partition $\mathcal{A}_G$ is *minimal* (with respect to a given k) if $|\nabla \mathcal{A}_G|$ is minimal among all partitions satisfying (1) and denote by $\nabla_G(k)$ the size of the minimal partition.

We deal with the problem of finding the function $\nabla_{H_a^n}(k)$ for the graph H_a^n, respectively $\nabla_{H_{a,l}^n}(k)$ for the graph $H_{a,l}^n$. The paper is organized as follows. In the next section we present a general approach to get lower bounds for the k-partitioning problem. In Section 3 we determine lower and upper bounds for $\nabla_G(k)$ for general values of k for the mentioned graphs $G \in \{H_a^n, H_{a,l}^n\}$. Section 4 is devoted to the asymptotic of $\nabla_{H_a^n}(k)$ in the case of a fixed a and growing n where k is a constant. Some extended results for related graphs and possible directions for further research conclude the paper in Section 5.

2   Derivation of a Lower Bound

Let $G = (V_G, E_G)$ be a graph and $A \subseteq V_G$. Denote

$$\partial_G A = \{(u, v) \in E_G \mid u \in A, v \notin A\},$$
$$\gamma_G(m) = \min_{\substack{A \subseteq V_G \\ |A| = m}} |\partial_G A|.$$

By the edge-isoperimetric problem for the graph G we mean the problem to estimate the function γ_G. We call a set $A \subseteq V_G$ optimal if $|\partial_G A| = \gamma_G(|A|)$.

Lemma 1. *Let $\mathcal{A}_G = \{A_1, \ldots, A_k\}$ be a partition of $G = (V_G, E_G)$, satisfying (1). Then*

$$\nabla_G(k) \geq \frac{k}{2} \min \left\{ \gamma_G\left(\left\lfloor \frac{|V_G|}{k} \right\rfloor\right), \gamma_G\left(\left\lceil \frac{|V_G|}{k} \right\rceil\right) \right\}. \tag{2}$$

Proof.
For $i \neq j$ denote $c_{i,j} = |\{(u, v) \in E_G \mid u \in A_i, v \in A_j\}|$ and put $c_{i,i} = 0$, $i \in [1, k]$. Considering $\partial_G A_i$ one has $\sum_{j=1}^k c_{i,j} = |\partial_G A_i| \geq \gamma_G(|A_i|)$. Now we summarize it for $i \in [1, k]$. Since $c_{i,j} = c_{j,i}$, we get $\sum_{i=1}^k \sum_{j=1}^k c_{i,j} = 2|\nabla \mathcal{A}_G| \geq \sum_{i=1}^k \gamma_G(|A_i|)$. The lemma follows by taking into account (1) and that $\gamma_G(|A_i|)$ is not less than the minimum in (2). $\qquad \square$

Corollary 1. *Let $\mathcal{A}_G = \{A_1, \ldots, A_k\}$ be a partition of $G = (V_G, E_G)$, satisfying (1). If each subset A_i is optimal, then the partition $\mathcal{A}_G$ is minimal.*

3 Application to Special Graph Classes

3.1 Bounds for $\nabla_{H_a^n}(k)$

We introduce the *lexicographic number* $\ell(u)$ of a vertex $u = (x_1, \ldots, x_n) \in V^n$ defined by $\ell(u) = \sum_{i=1}^{n} x_i a^{n-i}$. Denote

$$L_m^n = \{u \in V^n \mid 0 \le \ell(u) < m\}.$$

We say that subsets $A, B \subseteq V_G$ are *congruent* (denotation $A \cong B$) if B is the image of A in some automorphism of G. We call a subset of vertices $F \subseteq V^n$ a *face* of H_a^n of dimension p ($0 \le p \le n$) if the subsets F and $L_{a^p}^n$ are congruent.

Lemma 2. (Lindsey [10]) L_m^n *is an optimal set in* H_a^n *for any n and* $m \in [1, a^n]$.

Thus, each face of the graph H_a^n is an optimal set and we get the following Corollaries.

Corollary 2. *Let k divide a. Then*

$$\nabla_{H_a^n}(k) = \frac{k-1}{2k} \cdot a^{n+1}.$$

Corollary 3. *Let p be a constant with $0 < p \le n$. Then*

$$\nabla_{H_a^n}(a^p) = \frac{(a-1)p}{2} \cdot a^n.$$

Theorem 1. *Let $n > 2$, $a \ge 2$, $a^{p-1} < k < a^p$ and $n > 2(p-1)$. Then*

$$\frac{(a-1)(p-1)}{2} \le \frac{\nabla_{H_a^n}(k)}{a^n} \le \frac{3ap - 2p + 2a}{4}.$$

Proof.
To prove the lower bound we apply Lemma 1 and estimate the minimum in (2). Let $m = \lfloor a^n/k \rfloor$ and partition V^n into faces of dimension $n - p + 1$. Now L_m^n and L_{m+1}^n are proper subsets of one such face, say F. Furthermore, let $E_F = \{(u, v) \in F \times F \mid \rho(u, v) = 1\}$.

$$\min\{\gamma(m), \gamma(m+1)\} \ge m(a-1)(p-1) + \min\{|\partial L_m^n \cap E_F|, |\partial L_{m+1}^n \cap E_F|\}$$
$$\ge (m(p-1) + n - p + 1)(a-1).$$

Since $m \ge a^n/k - 1$ then

$$\nabla_{H_a^n}(k) \ge \frac{k}{2}(m(p-1) + (n - (p-1)))(a-1)$$

$$\ge \frac{k}{2} \cdot \frac{a^n}{k}(p-1)(a-1) + \frac{k}{2}(a-1)(n - 2(p-1))$$

$$\ge \frac{(a-1)(p-1)}{2} \cdot a^n.$$

To show the upper bound we first partition V^n into the faces $F_1, \ldots, F_a$ of dimension $n-1$ and then isomorphically partition each F_i into k optimal subsets $\{A_1^i, \ldots, A_k^i\}$, $i \in [1, a]$. Let $A_j = \cup_{i=1}^a A_j^i$ for $j \in [1, k]$. Obviously, all these sets A_j are optimal sets and we get a partition $\mathcal{A} = \{A_1, \ldots, A_k\}$ of H_a^n. Since k is not a power of a, then $|A_j^i| \in \{m, m+1\}$ with $m = \lfloor \frac{a^{n-1}}{k} \rfloor$, hence $A_j \in \{am, am+a\}$.

Such a partition can be balanced by deleting the vertices from larger parts and adding them to smaller parts to obtain a partition satisfying (1). We claim that at most $ak/4$ vertices of the initial partition will be reassigned. To see this denote by m_1 and m_2 the number of subsets of size am and $am+a$ in this partition. We move from each partition having $am+a$ vertices, $a - m_2a/k$ to the partitions which have only a vertices, where we assume w.l.o.g. that we operate with an integer value. Therefore, there will be moved $m_2(a - \frac{m_2a}{k}) \leq \frac{ak}{4}$ vertices. The balancing results in increasing of $|\nabla \mathcal{A}|$ at most on $\frac{ak}{4}(a-1)n$ and

$$\nabla_{H_a^n}(k) \leq a \cdot \nabla_{H_a^{n-1}}(k) + a(a-1)kn/4,$$

from where for $t \geq p$ follows

$$\nabla_{H_a^n}(k) \leq a^{n-t} \cdot \nabla_{H_a^t}(k) + \frac{a(a-1)k}{4}\left(n + (n-1)a + \cdots + (t+1)a^{n-t-1}\right). \quad (3)$$

Taking into account

$$\sum_{i=1}^r i\, a^i = \frac{r\, a^{r+2} - (r+1)\, a^{r+1} + a}{(a-1)^2}$$

recursion (3) implies

$$\nabla_{H_a^n}(k) \leq a^{n-t} \cdot \nabla_{H_a^t}(k) + \frac{ak(at+a-t)}{4a^t(a-1)}\, a^n. \quad (4)$$

We apply (4) with $t = p$. To compute $\nabla_{H_a^p}(k)$ note that for $\frac{a^p}{2} < k < a^p$ in any minimal k-partition of V^p exactly $2k - a^p$ parts consist of just one vertex, and each of the remaining $a^p - k$ parts consists of two vertices. In this case the size of the cut is larger than in the case $a^{p-1} < k \leq \frac{a^p}{2}$, because in the last case then each part consists at least of two vertices, and, thus, has at least one inner edge. Therefore,

$$\nabla_{H_a^p}(k) \leq \frac{a(a-1)p}{2} \cdot a^{p-1} - (a^p - k) = \frac{ap - p}{2} \cdot a^p - a^p + k. \quad (5)$$

Substituting (5) into (4) and taking into account $\frac{k}{a^p} < 1$ and $\frac{a^2}{a-1} \leq a + 2$, we get

$$\frac{\nabla_{H_a^n}(k)}{a^n} \leq \frac{ap - p - 2}{2} + \left(\frac{k}{a^p} + \frac{k}{a^p} \cdot \frac{a}{4(a-1)}\, (a(p+1) - p)\right)$$

$$\leq \frac{3ap - 2p + 2a}{4}. \quad \square$$

The spectral lower bound of [7] for $\nabla_{H_a^n}(k)$ is given by $\frac{1}{2} \cdot \frac{a^n}{k} \cdot (\lambda_1 + \ldots + \lambda_k)$ where $\lambda_1 \leq \ldots \leq \lambda_i$ are the i smallest eigenvalues of the Laplacian matrix of H_a^n. Therefore, for all reasonable values for k (by this we mean $a^2 < k < n \cdot (a-1)$) our lower bound for $\nabla_{H_a^n}(k)$ is better than the spectral lower bound.

3.2   Bounds for $\nabla_{H_{a,l}^n}(k)$

In [3] we show that each initial segment of the lexicographic order is an optimal subset in $H_{a,l}^n$ for $l \leq \lfloor \frac{a}{4} \rfloor$. Therefore, we get with similar arguments as in the section before the following results where we have omit the proofs because of space limitation.

Corollary 4. *Let k divide a and $l \leq \lfloor \frac{a}{4} \rfloor$. Then*

$$\nabla_{H_{a,l}^n(k)} = \frac{(k-1)a - lk}{2k} \cdot a^n.$$

Corollary 5. *Let p be a constant with $0 < p \leq n$ and $l \leq \lfloor \frac{a}{4} \rfloor$. Then*

$$\nabla_{H_{a,l}^n(a^p)} = \frac{(a-l-1)p}{2} \cdot a^n.$$

Theorem 2. *Let $n > 2$, $a \geq 2$, $a^{p-1} < k < a^p$, $l \leq \lfloor \frac{a}{4} \rfloor$, and $n > 2(p-1)$. Then*

$$\frac{(a-l-1)(p-1)}{2} \leq \frac{\nabla_{H_{a,l}^n(k)}}{a^n} \leq \frac{3ap - 2p - 2lp + 2a}{4}.$$

If we compare this lower bound with the spectral lower bound of [7] for the graphs $H_{a,l}^n$ we can conclude with the same arguments as in the previous section that for k and n large enough our lower bound is better than the spectral one.

4   Asymtotics for $\nabla_{H_a^n}(k)$

In this section we consider the case where a and k are constants and $n \to \infty$.

As soon as we are interested in the asymptotic of $\nabla_{H_a^n}(k)$ only, it is convenient to operate with partitions, where each set may not be optimal, but is in a sense close to be optimal. To be more exact, let $A \subseteq V^n$, $|A| = m$ and let c be some constant. We say that A is a *quasioptimal* set (with respect to the constant c), if there exists an optimal subset $B \cong L_m^n$, such that $|A \triangle B| \leq c$, where $\triangle$ denotes the symmetric difference.

In the following, we construct partitions $\mathcal{A} = \{A_1, \ldots, A_k\}$ of V^n into k quasioptimal subsets A_i of cardinality (1). Assuming that n is growing we show that for each subset A_i there exists a set B with $B \cong L_{|A_i|}^n$ (w.r.t. a constant c) such that $|A_i \triangle B| \leq c'$, where the constant c' depends on c and k, but not on n. We will not specify the constants c and c' exactly in our constructions, but will merely ensure that such constants do exist.

For each quasioptimal set $A_i \subseteq V^n$ of cardinality (1) it holds $|\partial_{H_a^n} A_i| - \gamma_{H_a^n}(|A_i|) \leq c'an$. Accordingly, we call the partition $\mathcal{A}_{H_a^n}$ quasiminimal, if $|\nabla \mathcal{A}_{H_a^n}| - \frac{k}{2}\gamma_{H_a^n}(\lfloor|V^n|/k\rfloor) \leq c''n$, where the constant c'' depends on k, c', a, but not on n.

Corollary 6. *Let k be fixed and $\mathcal{A} = \{A_1, \ldots, A_k\}$ be a partition of V^n, satisfying (1). If each subset A_i is quasioptimal in H_a^n, then the partition $\mathcal{A}$ is quasiminimal.*

Indeed, similarly as in the proof of Lemma 1 we get

$$2|\nabla \mathcal{A}_{H_a^n}| = \sum_{i=1}^{k} |\partial_{H_a^n} A_i| \leq \sum_{i=1}^{k} \left(\gamma_{H_a^n}(|A_i|) + c'an\right)$$

$$\leq k\left(\gamma_{H_a^n}(\lfloor|V^n|/k\rfloor) + an\right) + aknc' = k\gamma_{H_a^n}(\lfloor|V^n|/k\rfloor) + akn(c'+1).$$

Obviously, $\gamma_{H_a^n}(\lfloor|V^n|/k\rfloor)$ is exponential on n if k and a are fixed. Therefore, for $n \to \infty$

$$\nabla_{H_a^n}(k) \sim \frac{k}{2}\gamma(\lfloor|V^n|/k\rfloor), \qquad (6)$$

if V^n admits a partitioning into k quasioptimal subsets.

Lemma 3. *Let H_a^n be partitioned into k quasioptimal subsets. Then, for any constant q and $n \to \infty$*

$$\nabla_{H_a^{n+q}}(a^q \cdot k) \sim \frac{q(a-1)}{2} \cdot a^{n+q} + a^q \cdot \nabla_{H_a^n}(k).$$

Proof.
Given a partition of V^n into k quasioptimal subsets, we construct a partition of V^{n+q} into $a^q \cdot k$ quasioptimal subsets. To do this, we first partition V^{n+q} into a^q faces of dimension n. By Corollary 3 this cut has the cardinality $\frac{q(a-1)}{2} \cdot a^{n+q}$. After that we partition each face into k quasioptimal subsets, assuming that the partitions of all the faces are isomorphic. It remains to note that if a set A is quasioptimal in H_a^n, then it is also quasioptimal in H_a^{n+q}. $\qquad \square$

Theorem 3. *There exists a partition of V^n into $k = a^p + 1$ quasioptimal subsets satisfying (1).*

Proof.
Let $n \geq 2a + 1$ and consider first a partition of V^n into $k - 1 = a^p$ faces $F_1, \ldots, F_{k-1}$ of dimension $n - p$. Now, for H_a^n let $A_i' \subseteq F_i$ (for $i \in [1, k-1]$) with $A_i' \cong L_m^{n-p}$ and $m = \lfloor\frac{a^{n-p}}{k}\rfloor$. We construct a partition $\{A_1, \ldots, A_k\}$ of V^n with $A_i = F_i \setminus A_i'$ for $i \in [1, k-1]$ and $A_k = A_1' \cup \cdots \cup A_{k-1}'$.

All the subsets A_i are optimal, but the resulting partition may not satisfy (1). Since a and p are constants, we can reassign a constant number of vertices (depending on a, p only) between the parts A_k and A_i with $i < k$, so that the obtained subsets $\tilde{A}_i$, $i \in [1, k]$, will be quasioptimal and therefore the partition $\{\tilde{A}_i, \, i \in [1, k]\}$ is quasiminimal by Corollary 6. □

In order to compute $\nabla_{H_a^n}(k)$ for $k = a^p + 1$ we first compute $g_n = \gamma(m)$ for $m = \lfloor a^n/k \rfloor$. For that consider again a partition of V^n into $k - 1 = a^p$ faces of dimension $n - p$. Note, that the number of vertices in each such a face exceeds m. Thus, the set L_m^n is a proper subset just of one of these faces and the numbers g_n satisfy the recursion $g_n = g_{n-p} + \frac{p(a-1)}{k} a^n + O(n)$, where the term $O(n)$ occurs due to omitting the integer parts. This recursion provides

$$g_n \sim \frac{(a-1)(k-1)p}{k(k-2)} a^n \tag{7}$$

as $n \to \infty$. Substituting (7) into (6) we get

$$\nabla_{H_a^n}(a^p + 1) \sim \frac{(a-1)p\,a^p}{2(a^p - 1)} a^n.$$

With $a^p + a^q = a^q(a^{p-q} + 1)$ and applying Theorem 3 and Lemma 3, we get

Corollary 7. *Let $p > q \geq 0$. If $n \to \infty$, then*

$$\nabla_{H_a^n}(a^p + a^q) \sim \frac{pa^p - qa^q}{a^p - a^q} \cdot \frac{a-1}{2} \cdot a^n.$$

Now we switch our attention to the case $k = a^p - a^q$ with $p > q \geq 0$. In the sequel we introduce some special faces of V^n and represent them by n-dimensional characteristic vectors over the alphabet $\{0, 1, \ldots, a - 1, *\}$. If some entry of the characteristic vector is "$*$", then the corresponding face contains all vertices of H_a^n obtained by replacing this entry with any element of the set $\{0, 1, \ldots, a - 1\}$.

Theorem 4. *There exists a partition of V^n with $n \geq 2p - 1$ into $k = a^p - 1$ quasioptimal subsets satisfying (1).*

Proof.
We omitted the very technical proof because of space limitation and refer the interested reader to [4] where the full proof is given. □

In order to compute $\nabla_{H_a^n}(k)$ for $k = a^p - 1$ we first compute $g_n = \gamma(\lfloor a^n/k \rfloor)$. Let us partition H_a^n into a^{p-1} faces of dimension $n - p + 1$ with the characteristic vectors $(x_1, \ldots, x_{p-1}, *, \ldots, *)$, $x_1, \ldots, x_{p-1} \in [0, a-1]$. Clearly, the set $A = L_m^n$ with $m = \lfloor a^n/k \rfloor$ is a subset of the face with the characteristic vector $(\underbrace{0, \ldots, 0}_{p-1}, *, \ldots, *)$. Denote this face by F.

Note that $|A| = a^{n-p} + \lfloor \frac{1}{k} a^{n-p} \rfloor$. Thus, if one partitions the face F into subfaces F_i of dimension $n - p$ with the characteristic vectors of the form $(\underbrace{0, \ldots, 0}_{n-p}, i, *, \ldots, *)$, $i \in [0, a-1]$, then the set A has a nonempty intersection just with two such subfaces F_0 and F_1. These subfaces partition A into two parts $A_0 = F_0$ and $A_1 \subseteq F_1$ with $|A_1| = \left\lfloor \frac{a^{n-p}}{k} \right\rfloor$. Therefore, ∂A consists of four parts: $\partial^{n-p} A_1$ (in F_1), $\{(u,v) \in E \mid u \in F_0,\ v \in F_1 \setminus A_1\}$, $\{(u,v) \in E \mid u \in A,\ v \in F_i \text{ with } i \geq 2\}$, and $\{(u,v) \in E \mid u \in A,\ v \notin F\}$. Computing the cardinalities of these parts leads us to a recursion

$$g_n = g_{n-p} + \frac{k-1}{k} \cdot a^{n-p} + (a-2) \cdot a^{n-p} + \frac{a-2}{k} \cdot a^{n-p} + \frac{(a-1)(p-1)}{k} \cdot a^n + O(n).$$

The term $O(n)$ in this recursion occurs due to omitting the integer parts. Taking into account that a is a constant, the recursion provides

$$g_n \sim \frac{p(a-1)(k+1) - 2}{k^2} \cdot a^n$$

as $n \to \infty$. Applying Lemma 1 we get

$$\nabla_{H_a^n}(a^p - 1) \sim \frac{p(a-1)a^p - 2}{2(a^p - 1)} \cdot a^n.$$

This formula combined with Lemma 3 provides

Corollary 8. *Let $p > q \geq 0$. If $n \to \infty$, then*

$$\nabla_{H_a^n}(a^p - a^q) \sim \frac{p(a-1)a^p - q(a-1)a^q - 2a^q}{2(a^p - a^q)} \cdot a^n.$$

5   Concluding Remarks

In this paper we have presented a new approach to determine a lower bound for the edge cut of the k-partitioning problem for a given graph. Using this new strategy we showed some bounds for the k-partitioning problem for Hamming graphs and a related graph class. The importance of these results raise with the observation that a lower bound for an edge cut of a k-partitioning of these graphs is also a lower bound for the n-th Cartesian product of any regular graph of the same order and the same degree.

We can obtain similar results for some related graph families. For example, for the graphs B_a^n defined as the Cartesian products of n complete bipartite graphs with $2a$ vertices each and also for the graphs $B_{a,l}^n$ as the Cartesian product of n complete bipartite graphs without a certain number of perfect matchings.

Finally, let us consider an important particular case of the above results for the graphs H_2^n and B_2^{n-1}. Clearly, both graphs are isomorphic to the binary hypercube Q^n. Theorems 3 and 4 imply that for the series of k mentioned there the limit $\lim_{n \to \infty} \nabla_{H_2^n}(k)/2^n$ exists. Denoting this limit by c_k we have the table listed below.

The entries for $k = 2, 4, 8, 16$ follow from Corollary 3, c_7, c_{14}, and c_{15} are given by Corollary 8. Constructions for $k = 3, 5, 6, 9, 10, 12, 17, 18$ are provided by Corollary 7. The values for $k = 11, 13, 19$ are presently unknown. It is interesting to note that the sequence c_k is not monotone.

Table 1. Some asymptotic results

k	2	3	4	5	6	7	8	9	10	11	12	13	14	15	16	17	18	19	20
c_k	$\frac{1}{2}$	1	1	$\frac{4}{3}$	$\frac{3}{2}$	$\frac{11}{7}$	$\frac{3}{2}$	$\frac{12}{7}$	$\frac{11}{6}$	?	2	?	$\frac{29}{14}$	$\frac{31}{15}$	2	$\frac{32}{15}$	$\frac{31}{14}$	?	$\frac{7}{3}$

References

1. Bezrukov S.L.: On k-partitioning the n-cube. In: *Proceedings Graph-Theoretic Concepts in Computer Science WG'96*, LNCS 1197, Springer Verlag (1997), 44–55.
2. Bezrukov S.L.: Isoperimetric Problems in Discrete Spaces. In: *Extremal Problems for Finite Set*, Bolyai Soc. Math. Stud. 3, ed.: P. Frankl, Z. Füredi, G. Katona, D. Miklos, Budapest (1994), 59–91.
3. Bezrukov S.L.: Edge isoperimetric problems on graphs (a survey). To appear in *Graph Theory and Combinatorial Biology*, Bolyai Soc. Math. Stud. 7, ed.: L. Lovasz, A. Gyarfas, G. Katona, A. Recski, L. Szekely, Budapest.
4. Bezrukov S.L., Elsässer R., Schroeder U.-P.: On k-Partitioning of Hamming Graphs. To appear in *Discrete Applied Mathematics*.
5. Chung F.R.K.: Spectral Graph Theory. *American Mathematical Society*, Regional Conference Series in Mathematics **92** (1997).
6. Cypher R.: Theoretical Aspects of VLSI Pin Limitations. *SIAM J. Comput.* **22** (1993), 356–378.
7. Donath W.E., Hoffman A.J.: Lower Bounds for the Partitioning of Graphs. *IBM J. Res. Develop.* **17** (1973), 420–425.
8. Hendrickson B., Leland R.: The chaco user's guide: Version 2.0. *Technical Report SAND94-2692*, Sandia National Laboratories, Albuquerque, NM, (1994).
9. Karypis G., Kumar V.: A fast and high quality multilevel scheme for partitioning irregular graphs. *Technical Report 95-035*, University of Minnesota, (1995).
10. Lindsey II J.H.: Assignment of numbers to vertices. *Amer. Math. Monthly* **7** (1964), 508–516.
11. Lipton R.J., Tarjan R.E.: A Separator Theorem for Planar Graphs. *SIAM J. Appl. Math.* **36** (1979), 177–189.
12. Preis R., Diekmann R.: PARTY - A Software Library for Graph Partitioning. *Advances in Computational Mechanics with Parallel and Distributed Processing*, ed.: B.H.V. Topping, Civil-Comp Press (1997), 63–71.
13. Pellegrini F., Roman J.: Scotch: A software package for static mapping by dual recursive bipartitioning of process and architecture graphs. In: *Proc. of HPCN'96*, LNCS 1067, Springer Verlag (1996), 493–498.
14. Rolim J., Sýkora O., Vrto I.: Optimal Cutwidth and Bisection Width of 2- and 3-Dimensional Meshes. In: *Proc. of WG'95*, LNCS 1017, Springer Verlag (1995), 252–264.
15. Schwabe E.J.: Optimality of a VLSI Decomposition Scheme for the De-Bruijn Graph. *Parallel Process. Lett.* **3** (1993), 261–265.
16. Walshaw C., Cross M., Everett M.G.: A localised algorithm for optimising unstructured mesh partitions. *Int. J. Supercomputer Appl.* **9**(4) (1995), 280–295.

A Faster Algorithm for Computing Minimum 5-Way and 6-Way Cuts in Graphs*

Hiroshi Nagamochi, Shigeki Katayama, and Toshihide Ibaraki

Kyoto University, Kyoto, Japan 606-8501
{naga, ibaraki}@kuamp.kyoto-u.ac.jp

Abstract. For an edge-weighted graph G with n vertices and m edges, the minimum k-way cut problem is to find a partition of the vertex set into k non-empty subsets so that the weight sum of edges between different subsets is minimized. For this problem with $k = 5$ and 6, we present a deterministic algorithm that runs in $O(n^{k-2}(nF(n,m) + C_2(n,m) + n^2)) = O(mn^k \log(n^2/m))$ time, where $F(n,m)$ and $C_2(n,m)$ denote respectively the time bounds required to solve the maximum flow problem and the minimum 2-way cut problem in G. The bounds $\tilde{O}(mn^5)$ for $k = 5$ and $\tilde{O}(mn^6)$ for $k = 6$ improve the previous best randomized bounds $\tilde{O}(n^8)$ and $\tilde{O}(n^{10})$, respectively.

1 Introduction

Let $G = (V, E)$ stand for a simple undirected graph with a set V of vertices and a set E of edges, which are weighted by non-negative real numbers. Let n and m denote the numbers of vertices and edges, respectively. For an integer $k \geq 2$, a *k-way cut* is a partition $\{V_1, V_2, \ldots, V_k\}$ of V consisting of k non-empty subsets. The problem of partitioning V into k non-empty subsets so as to minimize the weight sum of the edges between different subsets is called *the minimum k-way cut problem*. The problem has several important applications such as cutting planes in the traveling salesman problem, VLSI design, task allocation in distributed computing systems and network reliability. The 2-way cut problem (i.e., the problem of computing edge-connectivity) can be solved by $\tilde{O}(nm)$ time deterministic algorithms [6,17] and by $\tilde{O}(n^2)$ and $\tilde{O}(m)$ time randomized algorithms [12,13,14]. For an unweighted planar graph, Hochbaum and Shmoys [8] proved that the minimum 3-way cut problem can be solved in $O(n^2)$ time. However, the complexity status of the problem for general $k \geq 3$ in an arbitrary graph G has been open for several years. Goldschmidt and Hochbaum proved that the problem is NP-hard if k is an input parameter [5]. In the same article [5], they presented an $O(n^{k^2/2-3k/2+4}F(n,m))$ time algorithm for solving the minimum k-way cut problem, where $F(n,m)$ denotes the time required to find a minimum (s,t)-cut (i.e., a minimum 2-way cut that separates

---

* This research was partially supported by the Scientific Grant-in-Aid from Ministry of Education, Science, Sports and Culture of Japan, and the subsidy from the Inamori Foundation.

two specified vertices s and t) in an edge-weighted graph with n vertices and m edges. A minimum (s,t)-cut can be obtained by applying a maximum flow algorithm. This running time is polynomial for any fixed k. Afterwards, Karger and Stein [14] proposed a randomized algorithm that solves the minimum k-way cut problem with high probability in $O(n^{2(k-1)}(\log n)^3)$ time, and a deterministic $O(n^{2k-3}m)$ time algorithm is claimed in [10] (where no full proof is given).

For $k = 3$, Kapoor [11] and Kamidoi et al. [9] showed that the problem can be solved in $O(n^3 F(n,m))$ time, which was then improved to $\tilde{O}(mn^3)$ by Burlet and Goldschmidt [1]. For $k = 4$, Kamidoi et al. [9] gave an $O(n^4 F(n,m)) = \tilde{O}(mn^5)$ time algorithm.

Let us call a non-empty and proper subset X of V a cut. If we can identify the first cut V_1 in a minimum k-way cut $\{V_1, \ldots, V_k\}$, then the rest of cuts $V_2, \ldots, V_k$ can be obtained by solving the minimum $(k-1)$-way cut problem in the graph induced by $V - V_1$. For $k = 3$, Burlet and Goldschmidt [1] succeeded to characterize a set of $O(n^2)$ number of cuts which contains at least one such cut V_1. Thus, by solving $O(n^2)$ minimum $(k-1)$-way cut problems, a minimum k-way cut can be computed. They also showed that such $O(n^2)$ number of cuts can be enumerated in $\tilde{O}(mn^3)$ time, yielding an $\tilde{O}(mn^3)$ time algorithm for the minimum 3-way cut problem. Recently, the article [18] gave a new characterization of a set of $O(n)$ number of such candidate cuts for the minimum k-way cut problems with $k = 3$ and 4, and showed that those $O(n)$ cuts can be obtained in $O(n^2 F(n,m))$ time by using Vazirani and Yannakakis's algorithm for enumerating small cuts [19]. This yields an $O(mn^k \log(n^2/m))$ time minimum k-way cut algorithm for $k = 3$ and 4. However, it was left open to extend the argument in [18] to obtain an $\tilde{O}(mn^k)$ time minimum k-way cut algorithm for $k \geq 5$.

In this paper, by examining the structure of three cuts crossing each other, we show that there is a set of $O(n)$ number of candidate cuts for $k = 5$ and 6, and that those $O(n)$ cuts again can be computed in $O(n^2 F(n,m))$ time. Therefore, we can find a minimum 5-way cut in $O(mn^5 \log(n^2/m))$ time and a minimum 6-way cut in $O(mn^6 \log(n^2/m))$ time. The bounds $\tilde{O}(mn^5)$ for $k = 5$ and $\tilde{O}(mn^6)$ for $k = 6$ improve upon the previous best randomized bounds $\tilde{O}(n^8)$ and $\tilde{O}(n^{10})$ [14], respectively. Also, it is shown that our algorithm runs in $O(n^k)$ time for $k = 5$ and 6 if G is an edge-weighted planar graph.

2 Preliminaries

2.1 Notations and definitions

A singleton set $\{x\}$ may be simply written as x. For a finite set V, a cut is defined as a non-empty and proper subset X of V. For two disjoint subsets $S, T \subset V$, we say that a cut X separates S and T if $S \subseteq X \subseteq V - T$ or $T \subseteq X \subseteq V - S$ holds. A cut X intersects another cut Y if $X - Y \neq \emptyset$, $Y - X \neq \emptyset$ and $X \cap Y \neq \emptyset$ hold, and X crosses Y if, in addition, $V - (X \cup Y) \neq \emptyset$ holds. Let $\mathcal{X} = \{X_1, X_2, \cdots, X_p\}$ be a set of cuts, where $\mathcal{X}$ is not necessarily a partition of V. We call a subset $V' \subseteq V$ a cell generated by $\mathcal{X}$ if it is a maximal subset V' such that no two vertices $x, y \in V'$ are separated by any cut $X \in \mathcal{X}$. Let $\gamma[\mathcal{X}]$ denote the set of all cells generated by $\mathcal{X}$. A partition $\pi_0 = \{V_1, \ldots, V_k\}$ of V is called *compatible* with $\mathcal{X}$ if each subset V_i is the union of some cells generated by $\mathcal{X}$.

Let $G = (V, E)$ be a simple undirected graph with a set V of vertices and a set E of edges weighted by non-negative reals. The weight of an edge $e \in E$ is denoted by $c(e)$. For a non-empty subset $X \subseteq V$, let $G[X]$ denote the graph induced from G by X. For $X, Y \subseteq V$, let $E_G(X, Y)$ be the set of edges between X and Y. We denote $\sum_{e \in E_G(X,Y)} c(e)$ by $c_G(X, Y)$. In particular, $E_G(X, V - X)$ and $c_G(X, V - X)$ may also be denoted by $E_G(X)$ and $c_G(X)$, respectively. We denote a k-way cut as a partition $\pi = \{V_1, \cdots, V_k\}$ of V with k non-empty subsets, and its weight in G is defined by

$$\Psi_G(\pi) = \frac{1}{2}(c_G(V_1) + c_G(V_2) + \cdots + c_G(V_k)),$$

i.e., the weight sum of the edges between different cuts in π. A k-way cut π is called *optimal* in G if its weight $\Psi_G(\pi)$ is minimum over all k-way cuts. We call a cut X a *k-base cut* in G if there is an optimal k-way cut $\pi = \{V_1, \cdots, V_k\}$ such that $V_1 = X$ (or $V_1 = V - X$) and $c_G(V_1) = \min_{V_i \in \pi} c_G(V_i)$.

We use the next easy lemma to analyze the running time of our algorithm in section 4.

Lemma 1. *Let $H = (V, E)$ be a simple undirected graph. If $|E| > 2|V| - 2$, then there is a cycle of length at least 5 in H, and it can be obtained in $O(nm)$ time.* □

2.2   Enumerating all 2-way cuts

In [19], Vazirani and Yannakakis presented an algorithm that finds all the 2-way cuts in G in the order of non-decreasing weights. The algorithm repeatedly finds the next 2-way cut by solving at most $2n - 3$ maximum flow problems. This algorithm will be used in section 2.3 to design a minimum k-way cut algorithm.

Theorem 2. [19] *For an edge-weighted graph $G = (V, E)$ with $n = |V|$ and $m = |E|$, 2-way cuts in G can be enumerated in the order of non-decreasing weights with $O(nF(n, m))$ time delay between two consecutive outputs, where $F(n, m)$ is the time required to find a minimum 2-way cut that two separates specified disjoint subsets $S, T \subset V$ in G.* □

For an edge-weighted graph G, it is well known that a minimum 2-way cut that separates specified two disjoint subsets $S, T \subset V$ can be found by solving a single maximum flow problem. Therefore $F(n, m) = O(mn \log(n^2/m))$ holds if we use the maximum flow algorithm of [4], where n and m are the numbers of vertices and edges in G, respectively. In the planar graph case, we can enumerates cuts more efficiently by making use of the shortest path algorithm in dual graphs.

Corollary 3. [18] *For an edge-weighted planar graph $G = (V, E)$ with $n = |V|$, 2-way cuts in G can be enumerated in the order of non-decreasing weights with $O(n)$ time delay between two consecutive outputs, where the first 2-way cut can be found in $O(n^2)$ time.* □

2.3 An Algorithm for the minimum k-way cut problem

In this subsection, we describe a minimum k-way cut algorithm for a general k. The algorithm first finds a k-base cut X (= V_1, or $V - V_1$) of an optimal k-way cut $\pi^* = \{V_1, \cdots, V_k\}$, from which we can find the remaining $(k - 1)$ cuts $V_2, \ldots, V_{k-1}$ of π^* by solving the minimum $(k - 1)$-way cut problem in the induced subgraph $G[V - X]$ (or $G[X]$ if $X = V - V_1$). To find a k-base cut X, we enumerate 2-way cuts $X_1, X_2, \ldots$ in G in the order of non-decreasing weights (i.e., $c_G(X_1) \leq c_G(X_2) \leq \cdots$), while it is guaranteed that the sequence contains such a cut X. More precisely, the algorithm is described as follows.

MULTIWAY
Input: A positive integer k, and an edge-weighted graph $G = (V, E)$ with $|V| \geq k$.
Output: A minimum k-way cut π in G.

1. Enumerate 2-way cuts $X_1, X_2, \ldots$ in G in the order of non-decreasing weights, until the set $\mathcal{X} = \{X_1, X_2, \ldots, X_r\}$ of all cuts enumerated so far contains a subset of cuts $\mathcal{Y}$ such that $|\gamma[\mathcal{Y}]| \geq k$ and
   (A): there is a k-way cut $\pi_0 = \{V_1, V_2, \ldots, V_k\}$, which is compatible with $\mathcal{Y}$ and satisfies

$$\Psi_G(\pi_0) \leq \frac{k}{2} \max_{X \in \mathcal{Y}} c_G(X).$$

2. For each cut $X_i \in \mathcal{X}$, $i = 1, \ldots, r$, with $|V - X_i| \geq k$, compute a k-way cut $\pi_i^+ = \{X_i, V_2, \ldots, V_k\}$ with the minimum weight over all k-way cuts π with $X_i \in \pi$ (such π_i^+ can be obtained by finding a minimum $(k - 1)$-way cut $\{V_2, \ldots, V_k\}$ in $G[V - X_i]$). Similarly, for each $i = 1, \ldots, r$, with $|X_i| \geq k$, compute a minimum k-way cut π_i^- in the form of $\{V - X_i, V_2, \ldots, V_k\}$.
3. Output a k-way cut $\pi \in \{\pi_0, \pi_1^+, \pi_1^-, \ldots, \pi_r^+, \pi_r^-\}$ with the minimum weight as an optimal k-way cut. □

The correctness of this algorithm can be observed as follows. Let $\mathcal{X} = \{X_1, X_2, \ldots, X_r\}$ be the set of the first r smallest cuts, and let $\mathcal{Y} \subseteq \mathcal{X}$ and π_0 be a set of cuts and a k-way cut satisfying condition (A). Assume that π_0 is not optimal (otherwise we are done). That is, $\Psi_G(\pi_0) > \Psi_G(\pi^*)$ holds for an optimal k-way cut $\pi^* = \{V_1', \ldots, V_k'\}$, where we assume $c_G(V_1') \leq c_G(V_2') \leq \cdots \leq c_G(V_k')$ without loss of generality. In this case, by (A) and $\mathcal{Y} \subseteq \mathcal{X}$, we have

$$\Psi_G(\pi^*) < \Psi_G(\pi_0) \leq \frac{k}{2} \max_{X \in \mathcal{Y}} c_G(X) \leq \frac{k}{2} \max_{X \in \mathcal{X}} c_G(X),$$

from which it holds $c_G(V_1') < \max_{X \in \mathcal{X}} c_G(X_i) = c_G(X_r)$. Thus the k-base cut $X = V_1'$ of π^* has strictly smaller weight than $c_G(X_r)$. Hence, such a cut X must have been enumerated as one of the $X_1, X_2, \ldots, X_{r-1}$. In other words, an optimal k-way cut π^* has been obtained in Step. 2 as one of the k-way cuts $\pi_1^+, \pi_1^+, \pi_1^-, \ldots, \pi_r^+, \pi_r^-$. This proves the correctness of MULTIWAY.

Now we consider the running time of MULTIWAY. For a given k and a graph G with n vertices, let us define $r(k, n)$ to be the smallest number r such that any set $\mathcal{X}$ of r smallest cuts contains a subset $\mathcal{Y}$ that satisfies condition (A). For an edge-weighted graph G with n vertices and m edges, let $C_k(n, m)$ $(k \geq 2)$ denote the time required to find an optimal k-way cut, and $D(n, m)$ denote the time delay required to compute the i-th minimum 2-way cut after enumerating the first $(i - 1)$ minimum 2-way cuts. Then, the running time of MULTIWAY becomes

$$C_k(n, m) = O(r(k, n)(D(n, m) + C_{k-1}(n, m) + T_1(k, n, m)) + T_2(k, n, m)),$$

where $T_1(k, n, m)$ and $T_2(k, n, m)$ denote the time to check whether $r > O(r(k, n))$ holds or not and the time to find a k-way cut π_0 in Step. 1, respectively.

The remaining task is to estimate $r(k, n)$ (or its upper bound) and to construct an efficient procedure for finding a subset $\mathcal{Y}$ and a k-way cut π_0 satisfying (A). For $k = 3$ and 4, the article [18] proved that any subset $\mathcal{Y}$ consisting of two crossing cuts $X_i, X_j \in \mathcal{X}$ satisfies (A), and that one of the k-way cuts compatible with $\mathcal{Y} = \{X_i, X_j\}$ becomes a desired k-way cut π_0. From this observation, $r(3, n), r(4, n) \leq 3n - 5$ follows and such π_0 can be computed in $O(n^2)$ time.

In sections 3 and 4 of this paper, we shall prove that $r(5, n), r(6, n) \leq 15n - 60$ holds. Based on the proof, we also give an efficient procedure for finding a subset $\mathcal{Y}$ and a k-way cut π_0 in condition (A).

3   Structure of a Set of Cuts

Now to estimate $r(5, n)$ and $r(6, n)$, we investigate the structure of a set of cuts.

3.1   Structure of Minimum k-way Cuts

In this subsection, we discuss some methods of constructing a k-way cut π with a small weight from a given set $\mathcal{Y}$ of cuts. The next property is observed by Kamidoi et al. [10] (under a slightly different statement).

Lemma 4. [10] *Given an edge-weighted graph $G = (V, E)$ and an integer k with $2 \leq k \leq |V|$, let $\mathcal{Y}$ be a set of cuts such that*

$$|\mathcal{Y}| \leq \lceil \tfrac{k}{2} \rceil \ and \ |\gamma[\mathcal{Y}]| \geq 2\lceil \tfrac{k}{2} \rceil. \tag{3.1}$$

Then there exists a k-way cut π_0 that is compatible with $\mathcal{Y}$ and satisfies $\Psi_G(\pi_0) \leq \tfrac{k}{2} \max_{X \in \mathcal{Y}} c_G(X)$, and such π_0 can be obtained in $O(kn+m)$ time, where $n = |V|$ and $m = |E|$. $\square$

Notice that a set $\mathcal{Y}$ satisfying (3.1) also satisfies the condition (A) in algorithm MULTIWAY in subsection 2.3. However, the characterization of $\mathcal{Y}$ by (3.1) is not enough to prove $r(k, n) = O(n)$ for $k \geq 5$. Let us call a set $\mathcal{X}$ of cuts *graphic* if $|X| = 2$ for all cuts $X \in \mathcal{X}$. It is not difficult to observe that no graphic set $\mathcal{X}$ contains a subset $\mathcal{Y}$ (which is also graphic) satisfying condition (3.1) for $k \geq 5$. Since there exits a graphic set $\mathcal{X}$ with $|\mathcal{X}| = \Omega(n^2)$, we need a different characterization of $\mathcal{Y}$ to show $r(k, n) = O(n)$ for $k = 5, 6$. For this, we define some notations. For a given set $\mathcal{Y}$ of cuts in V, let $p = |\mathcal{Y}|$ and $\gamma[\mathcal{Y}] = \{V_1, \ldots, V_h\}$. For each integer i with $0 \leq i \leq p$, we denote by $\mathcal{E}_i$ the set of pairs $\{V_j, V_{j'}\}$ of cells in $\gamma[\mathcal{Y}]$ such that there are exactly i cuts $X \in \mathcal{Y}$ which separate V_j and $V_{j'}$. Clearly, $\{\mathcal{E}_1, \mathcal{E}_2, \ldots, \mathcal{E}_p\}$ is a partition of the set $\{\{V_j, V_{j'}\} \mid 1 \leq j < j' \leq h\}$ (where $\mathcal{E}_i$ may be empty for some i). Clearly $\Psi_G(\gamma[\mathcal{Y}]) = \sum_{1 \leq i \leq p} cost(\mathcal{E}_i)$ and $\sum_{X \in \mathcal{Y}} c_G(X) = \sum_{1 \leq i \leq p} i \cdot cost(\mathcal{E}_i)$ hold.

Lemma 5. *Given an edge-weighted graph $G = (V, E)$ and an integer k with $2 \leq k \leq |V|$, let $\mathcal{Y}$ be a set of cuts such that*

$$|\mathcal{Y}| \leq k, \ |\gamma[\mathcal{Y}]| \geq k + a \ \text{and} \ |\mathcal{E}_1| \leq 2a \ \text{hold for some integer } a \geq 0. \tag{3.2}$$

Then there exists a k-way cut π_0 that is compatible with $\mathcal{Y}$ and satisfies $\Psi_G(\pi_0) \leq \frac{k}{2} \max_{X \in \mathcal{Y}} c_G(X)$, and such π_0 can be obtained in $O(kn + m + |\gamma[\mathcal{Y}]|)$ time, where $n = |V|$ and $m = |E|$.

Proof. Assume that a set $\mathcal{Y}$ of cuts satisfying (3.2) is given, and denote $\mathcal{E}_1 = \{e_1, e_2, \cdots, e_{|\mathcal{E}_1|}\}$, where each e_i is a pair $\{V_j, V_j'\}$ of cells. By assumption $|\mathcal{E}_1| \leq 2a$ holds. For a set $\mathcal{E}'$ of pairs of cells in $\gamma[\mathcal{Y}]$ we denote $cost(\mathcal{E}') = \sum_{\{V_j, V_{j'}\} \in \mathcal{E}'} c_G(V_j, V_{j'})$. We consider the following two cases.

(1) $|\mathcal{E}_1| < a$: Note that $k + a - |\mathcal{E}_1| > k$. In the partition $\gamma[\mathcal{Y}]$, we choose all pairs $\{V_j, V'\}$ in $\mathcal{E}_1$, and merge each pair of cells into a single cell. The resulting partition π^* has at least $|\gamma[\mathcal{Y}]| - |\mathcal{E}_1| \geq k + a - |\mathcal{E}_1| > k$ cells (because merging two cells in a pair $\{V_j, V_{j'}\}$ decreases the number of cells at most by one). The weight of π^* is

$$\Psi_G(\pi^*) = cost(\mathcal{E}_2) + cost(\mathcal{E}_3) + \cdots + cost(\mathcal{E}_p)$$
$$\leq \frac{1}{2}\Big[cost(\mathcal{E}_1) + 2cost(\mathcal{E}_2) + 3cost(\mathcal{E}_3) + \cdots + p \cdot cost(\mathcal{E}_p)\Big]$$
$$= \frac{1}{2} \sum_{X \in \mathcal{Y}} c_G(X) \leq \frac{|\mathcal{Y}|}{2} \max_{X \in \mathcal{Y}} c_G(X) \leq \frac{k}{2} \max_{X \in \mathcal{Y}} c_G(X).$$

(2) $a \leq |\mathcal{E}_1|$: Let $\mathcal{E}_1' = \{e_1, \cdots, e_a\}$ and $\mathcal{E}_1'' = \{e_{|\mathcal{E}_1| - a + 1}, \cdots, e_{|\mathcal{E}_1|}\}$ which may not be disjoint. In $\gamma[\mathcal{Y}]$, choose all pairs $\{V_j, V'\}$ in $\mathcal{E}_1'$, and merge each pair of cells into a single cell. Then the resulting partition π' contains at least k cells by $|\gamma[\mathcal{Y}]| \geq k + a$. The weight of π' is $\Psi_G(\pi_0') = cost(\mathcal{E}_1) + cost(\mathcal{E}_2) + \cdots + cost(\mathcal{E}_{|\mathcal{Y}|}) - cost(\mathcal{E}_1')$. Analogously construct the partition π'' from $\mathcal{E}_1''$. Then $\Psi_G(\pi'') = cost(\mathcal{E}_1) + cost(\mathcal{E}_2) + \cdots + cost(\mathcal{E}_p) - cost(\mathcal{E}_1'')$. By summing up these weights, we have

$$\Psi_G(\pi') + \Psi_G(\pi'') = (cost(\mathcal{E}_1) - cost(\mathcal{E}_1' \cap \mathcal{E}_1'')) + 2\Big[cost(\mathcal{E}_2) + cost(\mathcal{E}_3) + \cdots + cost(\mathcal{E}_p)\Big]$$
$$\leq cost(\mathcal{E}_1) + 2cost(\mathcal{E}_2) + 3cost(\mathcal{E}_3) + \cdots + p \cdot cost(\mathcal{E}_p)$$
$$= \sum_{X \in \mathcal{Y}} c_G(X) \leq k \max_{X \in \mathcal{Y}} c_G(X).$$

This implies that one of π' and π'' (say π^*) has weight equal to or less than $\frac{k}{2} \max_{X \in \mathcal{Y}} c_G(X)$.

In both cases, since $|\pi^*| \geq k$, there is a k-way cut π_0 that is compatible with π^*, and satisfies $\Psi_G(\pi_0) \leq \Psi_G(\pi^*)$. It is easy to see that such π_0 can be obtained from $\mathcal{Y}$ in $O(|\mathcal{Y}|n + |\gamma[\mathcal{Y}]| + m) = O(kn + m + |\gamma[\mathcal{Y}]|)$ time.   □

Now we specialize this lemma to the case in which a given set $\mathcal{Y}$ of cuts is graphic. Associated with a graphic set $\mathcal{X}$ of cuts in a finite set V, we define a graph $\mathcal{G}(\mathcal{X}) = (\mathcal{V}, \mathcal{E})$ as follows: the vertex set $\mathcal{V}$ is given by the set $\gamma[\mathcal{X}]$ of cells generated by $\mathcal{X}$, and it has an edge $e = (V', V'') \in \mathcal{E}$ if and only if $\mathcal{X}$ contains a cut $X = V' \cup V''$.

Lemma 6. *Given an edge-weighted graph $G = (V, E)$ and an integer k with $2 \leq k \leq |V|$, let $\mathcal{Y}$ be a set of cuts such that*

$$\mathcal{Y} \text{ is graphic, and } \mathcal{G}(\mathcal{Y}) \text{ is a cycle of length at least } k - 1. \qquad (3.3)$$

Then there exists a k-way cut π_0 that is compatible with $\mathcal{Y}$ and satisfies $\Psi_G(\pi_0) \leq \frac{k}{2} \max_{X \in \mathcal{Y}} c_G(X)$, and such π_0 can be obtained in $O(kn+m)$ time, where $n = |V|$ and $m = |E|$.

Proof. Since $\mathcal{G}(\mathcal{Y})$ is a cycle, $\mathcal{Y}$ can be given by $\{Y_1, Y_2, \ldots, Y_p\}$ ($p = |\mathcal{Y}|$) so that $Y_i \cap Y_{i+1}$ is a cell in $\gamma[\mathcal{Y}]$ for each $i = 1, 2, \ldots, p$ (where Y_{i+1} with $i = p$ means Y_1). In this case, any two cells in $\gamma[\mathcal{Y}]$ are separated by at least two cuts in $\mathcal{Y}$. Therefore, $|\mathcal{E}_1| = 0$ holds for this $\mathcal{Y}$. If $|\mathcal{Y}| \leq k$ (i.e., $|\mathcal{Y}| = k$ or $k - 1$), then $|\gamma[\mathcal{Y}]| \geq k$ holds (by $|V| \geq k$ even if $|\mathcal{Y}| = k - 1$). Thus, $\mathcal{Y}$ satisfies (3.2) with $a = 0$, and the lemma follows from Lemma 3. If $|\mathcal{Y}| \geq k + 1$, then we choose a subset $\mathcal{Y}' = \{Y_1, Y_2, \ldots, Y_k\} \subset \mathcal{Y}$, so that the graph $\mathcal{G}(\mathcal{Y}')$ becomes a path of length k. Clearly, $|\gamma[\mathcal{Y}']| = k + 2$, where one cell in $\gamma[\mathcal{Y}']$ is $\overline{Y} = V - (Y_1 \cup Y_2 \cup \cdots \cup Y_k)$. Now $\mathcal{E}_1$ for $\mathcal{Y}'$ is given by $\{\{\overline{Y}, Y_1 - Y_2\}, \{Y_1 - Y_2, Y_1 \cap Y_2\}, \{Y_{k-1} \cap Y_k, Y_k - Y_{k-1}\}, \{Y_k - Y_{k-1}, \overline{Y}\}\}$. By $|\mathcal{E}_1| = 4$, $\mathcal{Y}'$ satisfies (3.2) with $a = 2$, and the lemma follows from Lemma 3.   □

4   Minimum 5-Way and 6-Way Cuts

We call a set $\mathcal{X}$ of cuts *admissible* if there is a subset $\mathcal{Y} \subseteq \mathcal{X}$ which satisfies one of the conditions (3.1) and (3.3).

4.1   Upper bounds on $r(k, n)$ for $k = 5, 6$

In this subsection, we derive an upper bound on the size $|\mathcal{X}|$ of a non-admissible set $\mathcal{X}$ of cuts. A set $\mathcal{X}$ of cuts in V is called *simple* if, for each subset $X \subseteq V$, at most one of X and $V - X$ belongs to $\mathcal{X}$.

Lemma 7. *Let V be a finite set with $n = |V| \geq 5$, and $\mathcal{X}$ be a simple set of cuts in V. If $\mathcal{X}$ is not admissible, then $|\mathcal{X}| \leq 15n - 60$.*

Proof. We proceed by induction on n. For $n = 5$, a simple set $\mathcal{X}$ has at most $2^4 - 1 = 15$ cuts, and hence $|\mathcal{X}| \leq 15 \leq 15n - 60$ holds. Let $n > 5$.

Case-1. There is a cut $X \in \mathcal{X}$ such that $|X| \geq 3$ and $|V - X| \geq 3$: Choose such a cut X.

(a) No cut $Y \in \mathcal{X}$ crosses X: Without loss of generality we assume that $Y \subset X$ or $Y \subset V - X$ for all cuts $Y \in \mathcal{X} - X$ (by replacing Y with $V - Y$ if necessary). Let $\mathcal{X}_1$ (resp., $\mathcal{X}_2$) be the set of all cuts $Y \in \mathcal{X}$ such that $Y \subset X$ (resp., $Y \subset V - X$), where $\mathcal{X} = \mathcal{X}_1 \cup \mathcal{X}_2$. Clearly, both sets $\mathcal{X}_1$ and $\mathcal{X}_2$ are simple and non-admissible. By contracting the elements in X (resp., in $V - X$) into a single element x (resp., $\bar{x}$), $\mathcal{X}_2$ (resp., $\mathcal{X}_1$) remains a simple and non-admissible set of cuts in the resulting set $V_2 = (V - X) \cup \{x\}$ (resp., $V_1 = X \cup \{\bar{x}\}$). By $\min\{|X|, |V - X|\} \geq 3$, it holds $|V_i| < n$ for $i = 1, 2$. Hence, by induction hypothesis, we obtain $|\mathcal{X}_1| \leq 15(|V - X| + 1) - 60$ if $|V - X| + 1 \geq 5$ and $|\mathcal{X}_1| \leq 7$ if $|V - X| + 1 = 4$. Similarly, we have $|\mathcal{X}_2| \leq 15(|X| + 1) - 60$ if $|X| + 1 \geq 5$ and $|\mathcal{X}_2| \leq 7$ if $|X| + 1 = 4$. From these, it holds $|\mathcal{X}| \leq |\mathcal{X}_1| + |\mathcal{X}_2| \leq \max\{15|V| - 90, 14, 15(|V| - 2) - 53\} \leq 15|V| - 60$ for $|V| > 5$.

(b) There exists a cut $Y \in \mathcal{X}$ that crosses X. Let $\gamma[\{X, Y\}] = \{V_1, V_2, V_3, V_4\}$. By $|X| \geq 3$ and $|V - X| \geq 3$, there are at least two cells in $\{V_1, V_2, V_3, V_4\}$ which contain more than one element; $|V_1|, |V_2| \geq 2$ is assumed without loss of generality. We say that a cut Z *divides* a subset V_1 if $Z \cap V_1 \neq \emptyset \neq V_1 - Z$ holds. We denote a subset $\mathcal{Z} \subseteq \mathcal{X}$ of all cuts which divide V_1. Since $\mathcal{X}$ is not admissible, any cut $Z \in \mathcal{Z}$, which divides V_1, cannot divide the other cells in $\{V_2, V_3, V_4\}$ (otherwise, if Z divides V_i, then the set $\{X, Y, Z\}$ generating at least 6 cells would satisfy (3.1)). Thus, $Z - V_1$ is the union of some cells in $\{V_2, V_3, V_4\}$. In particular, either $V_i \cap Z = \emptyset$ or $V_i \subset Z$ holds for all cuts $Z \in \mathcal{Z}$ and $i \in \{2, 3, 4\}$. Hence, $\mathcal{Z}$ remains simple and non-admissible in the set $V' = V_1 \cup \{v_2, v_3, v_4\}$ which is obtained from V by contracting V_2, V_3, V_4 into v_2, v_3, v_4, respectively. Since $|V'| < |V|$ holds by $|V_2| \geq 2$, we have by induction hypothesis that $|\mathcal{Z}| \leq 15(|V_1| + 3) - 60$. On the other hand, no cut $X' \in \mathcal{X} - \mathcal{Z}$ divides V_1, and $\mathcal{X} - \mathcal{Z}$ is simple and non-admissible in the set $V'' = (V - V_1) \cup \{v_1\}$ obtained from V by contracting V_1 into an element v_1. By $|V''| < |V|$ (from $|V_1| \geq 2$), we obtain by induction hypothesis that $|\mathcal{X} - \mathcal{Z}| \leq 15(|V| - |V_1| + 1) - 60$. Therefore, we have $|\mathcal{X}| = |\mathcal{Z}| + |\mathcal{X} - \mathcal{Z}| \leq 15n - 60$.

Case-2. For all cuts $X \in \mathcal{X}$, $|X| \leq 2$ holds: Let $\mathcal{X}_1$ be the set of cuts $X \in \mathcal{X}$ with $|X| = 1$. Clearly, $|\mathcal{X}_1| \leq |V|$. Then the set $\mathcal{X}_2 = \mathcal{X} - \mathcal{X}_1$ is graphic, and remains non-admissible. By Lemma 4, the graph $\mathcal{G}(\mathcal{X}_2) = (V, \mathcal{E})$ associated with $\mathcal{X}_2$ has no cycle of length at least 5 ($\geq k - 1$ for $k = 5, 6$). Therefore, by Lemma 1, we have $|\mathcal{X}_2| = |\mathcal{E}| \leq 2|V| - 2 = 2|\gamma[\mathcal{X}_2]| - 2 \leq 2|V| - 2$. Thus, $|\mathcal{X}| = |\mathcal{X}_1| + |\mathcal{X}_2| \leq 3|V| - 2 \leq 15|V| - 60$ for $|V| \geq 5$.

This proves the theorem.   □

Consequently, if $|\mathcal{X}| > 15n - 60$, then $\mathcal{X}$ is always admissible, establishing the next result.

Theorem 8. $r(5, n) \leq 15n - 60$ *and* $r(6, n) \leq 15n - 60$ *holds for all* $n \geq 5$.   □

4.2   Running time of MULTIWAY for $k=5,6$

Finally, we estimate the running time of MULTIWAY for $k = 5, 6$. In Step 1 of MULTIWAY, we can stop enumerating 2-way cuts when $r = 15n - 60$ cuts have been enumerated. Thus, we can test whether $|\mathcal{X}| > O(r(k, n))$ holds or not in $T_1(k, n, m) = O(1)$ time. Given an admissible $\mathcal{X}$, based on the inductive proof of Lemma 5, we can find in $O(n^2|\mathcal{X}|)$ time a subset $\mathcal{Y}$ that satisfies (3.1) or (3.2). From such a set $\mathcal{Y}$ of cuts, we can construct a desired k-way cut π_0 in $O(kn + m)$ time by Lemmas 2 and 4. Thus, the time to find a π_0 from a set $\mathcal{X}$ with $|\mathcal{X}| = 15n - 60$ is $T_2(k, n, m) = O(n^2 r(k, n)) + O(kn + m) = O(n^3)$. Therefore, from the argument in section 2.3 and by Theorem 1, the time to find minimum k-way cut is

$$C_6(n, m) = O(n^2 F(n, m) + n C_5(n, m) + n^3), \text{ for } k = 6,$$

$$C_5(n, m) = O(n^2 F(n, m) + n C_4(n, m) + n^3), \text{ for } k = 5.$$

Since $C_4(n, m) = O(n^3 F(n, m) + n^2 C_2(n, m))$ is known in [18], we establish the following result.

Theorem 9. *For an edge-weighted graph $G = (V, E)$, where $n = |V| \geq 5$ and $m = |E|$, a minimum 5-way cut and a minimum 6-way cut can be computed in $O(n^4 F(n, m) + n^3 C_2(n, m))$ and $O(n^5 F(n, m) + n^4 C_2(n, m))$ time, respectively.* □

Currently, $F(n, m) = O(mn \log(n^2/m))$ [4] and $O(\min\{n^{2/3}, m^{1/2}\}m \log(n^2/m) \log U)$ (where U is the maximum edge weight in the case where weights are all integers) [3] are among the fastest, and $C_2(n, m) = O(mn \log(n^2/m))$ [6] is known for weighted graphs. Thus our bounds become

$$C_5(n, m) = O(mn^5 \log(n^2/m)) \text{ or } O(\min\{n^{2/3}, m^{1/2}\}mn^3 \log(n^2/m) \log U),$$

$$C_6(n, m) = O(mn^6 \log(n^2/m)) \text{ or } O(\min\{n^{2/3}, m^{1/2}\}mn^4 \log(n^2/m) \log U).$$

For unweighted graphs, these becomes $C_5(n, m) = O(\min\{mn^{2/3}, m^{3/2}\}n^5)$ and $C_6(n, m) = O(\min\{mn^{2/3}, m^{3/2}\}n^6)$, since $F(n, m) = O(\min\{mn^{2/3}, m^{3/2}\})$ is known for unweighted graphs [2,15].

For a planar graph G, we can obtain better bounds by applying Corollary 1.

Corollary 10. *For an edge-weighted planar graph $G = (V, E)$, where $n = |V| \geq 5$, a minimum 5-way cut and a minimum 6-way cut can be computed in $O(n^5)$ and $O(n^6)$ time, respectively.* □

5   Concluding Remark

In this paper, we proposed an algorithm MULTIWAY for computing a minimum k-way cut in an edge-weighted graph, for general k. The key part in analyzing the time complexity of MULTIWAY is to estimate the number $r(k, n)$ (defined in subsection 2.3). In this paper and [18], $r(k, n) = O(n)$ has been shown for $k \leq 6$. It is left open to derive an upper bound on $r(k, n)$ for $k \geq 7$. We conjecture that $r(k, n) = O(f(k)n)$ holds, where $f(k)$ is a function of k (such as $f(k) = 2^k$).

References

1. M. Burlet and O. Goldschmidt, *A new and improved algorithm for the 3-cut problem*, Operations Research Letters, vol.21, (1997), pp. 225–227.
2. S. Even and R. E. Tarjan, *Network flow and testing graph connectivity*, SIAM J. Computing, vol.4, (1975), pp. 507–518.
3. A. V. Goldberg and S. Rao, *Beyond the flow decomposition barrier*, Proc. 38th IEEE Annual Symp. on Foundations of Computer Science, (1997), pp. 2–11.
4. A. V. Goldberg and R. E. Tarjan, *A new approach to the maximum flow problem*, J. ACM, vol.35, (1988), pp. 921–940.
5. O. Goldschmidt and D. S. Hochbaum, *A polynomial algorithm for the k-cut problem for fixed k*, Mathematics of Operations Research. vol.19, (1994), pp. 24–37.
6. J. Hao and J. Orlin, *A faster algorithm for finding the minimum cut in a directed graph*, J. Algorithms, vol. 17, (1994), pp. 424–446.
7. M. R. Henzinger, P. Klein, S. Rao and D. Williamson, *Faster shortest-path algorithms for planar graphs*, J. Comp. Syst. Sc., vol. 53, (1997), pp. 2-23.
8. D. S. Hochbaum and D. B. Shmoys, *An $O(|V|^2)$ algorithm for the planar 3-cut problem*, SIAM J. Algebraic Discrete Methods, vol.6, (1985), pp. 707–712.
9. Y. Kamidoi, S. Wakabayashi and N. Yoshida. *Faster algorithms for finding a minimum k-way cut in a weighted graph*, Proc. IEEE International Symposium on Circuits and Systems, (1997), pp. 1009–1012.
10. Y. Kamidoi, S. Wakabayashi and N. Yoshida. *A new approach to the minimum k-way partition problem for weighted graphs*, Technical Report of Inst. Electron. Inform. Comm. Eng., COMP97-25, (1997), pp. 25–32.
11. S. Kapoor, *On minimum 3-cuts and approximating k-cuts using cut trees*, Lecture Notes in Computer Science 1084, Springer-Verlag, (1996), pp. 132–146.
12. D. R. Karger, *Minimum cuts in near-linear time*, Proceedings 28th ACM Symposium on Theory of Computing. (1996), pp. 56–63.
13. D. R. Karger and C. Stein, *An $\tilde{O}(n^2)$ algorithm for minimum cuts*, Proceedings 25th ACM Symposium on Theory of Computing, (1993), pp. 757–765.
14. D. R. Karger and C. Stein, *A new approach to the minimum cut problems*, J. ACM, vol.43, no.4, (1996), pp. 601–640.
15. A. V. Karzanov, *O nakhozhdenii maksimal'nogo potoka v setyakh spetsial'nogo vida i nekotorykh prilozheniyakh*, In Mathematicheskie Voprosy Upravleniya Proizvodstvom, volume 5, Moscow State University Press, Moscow, (1973). In Russian; title translation: On finding maximum flows in networks with special structure and some applications.
16. N. Katoh. T. Ibaraki and H. Mine, *An efficient algorithm for K shortest simple paths*, Networks. vol.12, (1982). pp. 441–427.
17. H. Nagamochi and T. Ibaraki, *Computing the edge-connectivity of multigraphs and capacitated graphs*, SIAM J. Discrete Mathematics, vol.5, (1992), pp. 54–66.
18. H. Nagamochi and T. Ibaraki, *A fast algorithm for computing minimum 3-way and 4-way cuts*, Lecture Notes in Computer Science, Springer-Verlag, 7th Conference on Integer Programming and Combinatorial Optimization. June 9-11, Graz, Austria, (1999) (to appear).
19. V. V. Vazirani and M. Yannakakis, *Suboptimal cuts: Their enumeration, weight, and number*, Lecture Notes in Computer Science. 623, Springer-Verlag, (1992), pp. 366–377.

Probabilities to Accept Languages by Quantum Finite Automata

Andris Ambainis[1], Richard Bonner[2], Rūsiņš Freivalds[3], and Arnolds Ķikusts[3]

[1] Computer Science Division, University of California, Berkeley, CA 94720-2320[†]
[2] Department of Mathematics and Physics, Mälardalens University
[3] Institute of Mathematics and Computer Science, University of Latvia,
Raiņa bulv. 29, Riga, Latvia[‡]

Abstract. We construct a hierarchy of regular languages such that the current language in the hierarchy can be accepted by 1-way quantum finite automata with a probability smaller than the corresponding probability for the preceding language in the hierarchy. These probabilities converge to $\frac{1}{2}$.

1 Introduction

Quantum computation is a most challenging project involving research both by physicists and computer scientists. The principles of quantum computation differ from the principles of classical computation very much. The classical computation is based on classical mechanics while quantum computation attempts to exploit phenomena specific to quantum physics.

One of features of quantum mechanics is that a quantum process can be in a combination (called *superposition*) of several states and these several states can interact one with another. A computer scientist would call this *a massive parallelism*. This possibility of massive parallelism is very important for Computer Science. In 1982, Nobel prize winner physicist Richard Feynman (1918-1988) asked what effects the principles of quantum mechanics can have on computation[Fe 82]. An exact simulation of quantum processes demands exponential running time. Therefore, there may be other computations which are performed nowadays by classical computers but might be simulated by quantum processes in much less time.

R.Feynman's influence was (and is) so high that rather soon this possibility was explored both theoretically and practically. David Deutsch[De 89] introduced quantum Turing machines, quantum physical counterparts of probabilistic Turing machines. He conjectured that they may be more efficient that classical Turing machines. He also showed the existence of a universal quantum Turing machine. This construction was subsequently improved by Bernstein and Vazirani [BV 97] and Yao [Ya 93].

[†] Supported by Berkeley Fellowship for Graduate Studies.
[‡] Research supported by Grant No.96.0282 from the Latvian Council of Science

T. Asano et al. (Eds.): COCOON'99, LNCS 1627, pp. 174–183, 1999.
© Springer-Verlag Berlin Heidelberg 1999

Quantum Turing machines might have remained relatively unknown but two events caused a drastical change. First, Peter Shor [Sh 97] invented surprising polynomial-time quantum algorithms for computation of discrete logarithms and for factorization of integers. Second, joint research of physicists and computer people have led to a dramatic breakthrough: all the unusual quantum circuits having no classical counterparts (such as quantum bit teleportation) have been physically implemented. Hence universal quantum computers are to come soon. Moreover, since the modern public-key cryptography is based on intractability of discrete logarithms and factorization of integers, building a quantum computer implies building a code-breaking machine.

In this paper, we consider quantum finite automata (QFAs), a different model of quantum computation. This is a simpler model than quantum Turing machines and and it may be simpler to implement.

Quantum finite automata have been studied in [AF 98,BP 99,KW 97,MC 97]. Surprisingly, QFAs do not generalize deterministic finite automata. Their capabilities are incomparable. QFAs can be exponentially more space-efficient[AF 98]. However, there are regular languages that cannot be recognized by quantum finite automata[KW 97].

This weakness is caused by reversibility. Any quantum computation is performed by means of unitary operators. One of the simplest properties of these operators shows that such a computation is reversible. The result always determines the input uniquely. It may seem to be a very strong limitation. Luckily, for unrestricted quantum algorithms (for instance, for quantum Turing machines) this is not so. It is possible to embed any irreversible computation in an appropriate environment which makes it reversible[Be 89]. For instance, the computing agent could keep the inputs of previous calculations in successive order. Quantum finite automata are more sensitive to the reversibility requirement.

If the probability with which a QFA is required to be correct decreases, the set of languages that can be recognized increases. In particular[AF 98], there are languages that can be recognized with probability 0.68 but not with probability 7/9. In this paper, we extend this result by constructing a hierarchy of languages in which each next language can be recognized with a smaller probability than the previous one.

2   Preliminaries

2.1   Basics of quantum computation

To explain the difference between classical and quantum mechanical world, we first consider one-bit systems. A classical bit is in one of two classical states *true* and *false*. A *probabilistic* counterpart of the classical bit can be *true* with a probability α and *false* with probability β, where $\alpha + \beta = 1$. A *quantum bit* (*qubit*) is very much like to it with the following distinction. For a *qubit* α and β can be arbitrary complex numbers with the property $\|\alpha\|^2 + \|\beta\|^2 = 1$. If we observe a qubit, we get *true* with probability $\|\alpha\|^2$ and *false* with probability

$\|\beta\|^2$, just like in probabilistic case. However, if we modify a quantum system without observing it (we will explain what this means), the set of transformations that one can perform is larger than in the probabilistic case. This is where the power of quantum computation comes from.

More generally, we consider quantum systems with m basis states. We denote the basis states $|q_1\rangle$, $|q_2\rangle$, ..., $|q_m\rangle$. Let ψ be a linear combination of them with complex coefficients

$$\psi = \alpha_1 |q_1\rangle + \alpha_2 |q_2\rangle + \ldots + \alpha_m |q_m\rangle.$$

The l_2 norm of ψ is

$$\|\psi\| = \sqrt{|\alpha_1|^2 + |\alpha_2|^2 + \ldots + |\alpha_m|^2}.$$

The state of a quantum system can be any ψ with $\|\psi\| = 1$. ψ is called a *superposition* of $|q_1\rangle$, ..., $|q_m\rangle$. α_1, ..., α_m are called *amplitudes* of $|q_1\rangle$, ..., $|q_m\rangle$. We use $l_2(Q)$ to denote the vector space consisting of all linear combinations of $|q_1\rangle$, ..., $|q_m\rangle$.

Allowing arbitrary complex amplitudes is essential for physics. However, it is not important for quantum computation. Anything that can be computed with complex amplitudes can be done with only real amplitudes as well. This was shown for quantum Turing machines in [BV 93][1] and the same proof works for QFAs. However, it is important that *negative* amplitudes are allowed. For this reason, we assume that all amplitudes are (possibly negative) reals.

There are two types of transformations that can be performed on a quantum system. The first type are unitary transformations. A unitary transformation is a linear transformation U on $l_2(Q)$ that preserves l_2 norm. (This means that any ψ with $\|\psi\| = 1$ is mapped to ψ' with $\|\psi'\| = 1$.)

Second, there are measurements. The simplest measurement is observing $\psi = \alpha_1 |q_1\rangle + \alpha_2 |q_2\rangle + \ldots + \alpha_m |q_m\rangle$ in the basis $|q_1\rangle, \ldots, |q_m\rangle$. It gives $|q_i\rangle$ with probability α_i^2. ($\|\psi\| = 1$ guarantees that probabilities of different outcomes sum to 1.) After the measurement, the state of the system changes to $|q_i\rangle$ and repeating the measurement gives the same state $|q_i\rangle$.

In this paper, we also use *partial measurements*. Let $Q_1, \ldots, Q_k$ be pairwise disjoint subsets of Q such that $Q_1 \cup Q_2 \cup \ldots \cup Q_k = Q$. Let E_j, for $j \in \{1, \ldots, k\}$, denote the subspace of $l_2(Q)$ spanned by $|q_j\rangle$, $j \in Q_i$. Then, a *partial measurement* w.r.t. $E_1, \ldots, E_k$ gives the answer $\psi \in E_j$ with probability $\sum_{i \in Q_j} \alpha_i^2$. After that, the state of the system collapses to the projection of ψ to E_j. This projection is $\psi_j = \sum_{i \in Q_j} \alpha_i |q_i\rangle$.

2.2   Quantum finite automata

Quantum finite automata were introduced twice. First this was done by C. Moore and J.P. Crutchfield [MC 97]. Later in a different and non-equivalent way these automata were introduced by A. Kondacs and J. Watrous [KW 97].

---

[1] For unknown reason, this proof does not appear in [BV 97].

The first definition just mimics the definition of 1-way probabilistic finite automata only substituting *stochastic* matrices by *unitary* ones. We use a more elaborated definition [KW 97].

A QFA is a tuple $M = (Q; \Sigma; V; q_0; Q_{acc}; Q_{rej})$ where Q is a finite set of states, Σ is an input alphabet, V is a transition function, $q_0 \in Q$ is a starting state, and $Q_{acc} \subset Q$ and $Q_{rej} \subset Q$ are sets of accepting and rejecting states. The states in Q_{acc} and Q_{rej} are called *halting states* and the states in $Q_{non} = Q - (Q_{acc} \cup Q_{rej})$ are called *non halting states*. κ and $\$$ are symbols that do not belong to Σ. We use κ and $\$$ as the left and the right endmarker, respectively. The *working alphabet* of M is $\Gamma = \Sigma \cup \{\kappa; \$\}$.

The transition function V is a mapping from $\Gamma \times l_2(Q)$ to $l_2(Q)$ such that, for every $a \in \Gamma$, the function $V_a : l_2(Q) \to l_2(Q)$ defined by $V_a(x) = V(a, x)$ is a unitary transformation.

The computation of a QFA starts in the superposition $|q_0\rangle$. Then transformations corresponding to the left endmarker κ, the letters of the input word x and the right endmarker $\$$ are applied. The transformation corresponding to $a \in \Gamma$ consists of two steps.

1. First, V_a is applied. The new superposition ψ' is $V_a(\psi)$ where ψ is the superposition before this step.

2. Then, ψ' is observed with respect to $E_{acc}, E_{rej}, E_{non}$ where $E_{acc} = span\{|q\rangle : q \in Q_{acc}\}$, $E_{rej} = span\{|q\rangle : q \in Q_{rej}\}$, $E_{non} = span\{|q\rangle : q \in Q_{non}\}$ (see section 2.1).

If we get $\psi' \in E_{acc}$, the input is accepted. If we get $\psi' \in E_{rej}$, the input is rejected. If we get $\psi' \in E_{non}$, the next transformation is applied.

We regard these two transformations as reading a letter a. We use V'_a to denote the transformation consisting of V_a followed by projection to E_{non}. This is the transformation mapping ψ to the non-halting part of $V_a(\psi)$. We use ψ_y to denote the non-halting part of QFA's state after reading the left endmarker κ and the word $y \in \Sigma^*$.

We compare QFAs with different probabilities of correct answer. This problem was first considered by A. Ambainis and R. Freivalds[AF 98]. The following theorems were proved there:

Theorem 1. *Let L be a language and M be its minimal automaton. Assume that there is a word x such that M contains states q_1, q_2 satisfying:*

1. *$q_1 \neq q_2$,*
2. *If M starts in the state q_1 and reads x. it passes to q_2,*
3. *If M starts in the state q_2 and reads x. it passes to q_2, and*
4. *q_2 is neither "all-accepting" state, nor "all-rejecting" state.*

Then L cannot be recognized by a 1-way quantum finite automaton with probability $7/9 + \epsilon$ for any fixed $\epsilon > 0$.

Theorem 2. *Let L be a language and M be its minimal automaton. If there is no q_1, q_2, x satisfying conditions of Theorem 1 then L can be recognized by a 1-way reversible finite automaton (i.e. L can be recognized by a 1-way quantum finite automaton with probability 1).*

Theorem 3. *The language a^*b^* can be recognized by a 1-way QFA with the probability of correct answer $p = 0.68...$ where p is the root of $p^3 + p = 1$.*

Corollary 1. *There is a language that can be recognized by a 1-QFA with probability 0.68... but not with probability $7/9 + \epsilon$.*

For probabilistic automata, the probability of correct answer can be increased arbitrarily and this property of probabilistic computation is considered as evident. Theorems above show that its counterpart is not true in the quantum world! The reason for that is that the model of QFAs mixes reversible (quantum computation) components with nonreversible (measurements after every step).

In this paper, we consider the best probabilities of acceptance by 1-way quantum finite automata the languages $a^*b^* \ldots z^*$. Since the reason why the language a^*b^* cannot be accepted by 1-way quantum finite automata is the property described in the Theorems 1 and 2, this new result provides an insight on what the hierarchy of languages with respect to the probabilities of their acceptance by 1-way quantum finite automata may be. We also show a generalization of Theorem 3 in a style similar to Theorem 2.

3   Main results

Lemma 1. *For arbitrary real $x_1 > 0$, $x_2 > 0$, ..., $x_n > 0$, there exists a unitary $n \times n$ matrix $M_n(x_1, x_2, ..., x_n)$ with elements m_{ij} such that*

$$m_{11} = \frac{x_1}{\sqrt{x_1^2 + ... + x_n^2}}, \quad m_{21} = \frac{x_2}{\sqrt{x_1^2 + ... + x_n^2}}, \quad ..., m_{n1} = \frac{x_n}{\sqrt{x_1^2 + ... + x_n^2}}.$$

$\square$

Let L_n be the language $a_1^* a_2^* ... a_n^*$.

Theorem 4. *The language L_n $(n > 1)$ can be recognized by a 1-way QFA with the probability of correct answer p where p is the root of $p^{\frac{n+1}{n-1}} + p = 1$ in the interval $[1/2, 1]$.*

Proof: Let m_{ij} be the elements of the matrix $M_k(x_1, x_2, ..., x_k)$ from Lemma 1. We construct a $k \times (k-1)$ matrix $T_k(x_1, x_2, ..., x_k)$ with elements $t_{ij} = m_{i,j+1}$. Let $R_k(x_1, x_2, ..., x_k)$ be a $k \times k$ matrix with elements $r_{ij} = \frac{x_i \cdot x_j}{x_1^2 + ... + x_k^2}$ and I_k be the $k \times k$ identity matrix.

For fixed n, let $p_n \in [1/2, 1]$ satisfy $p_n^{\frac{n+1}{n-1}} + p_n = 1$ and p_k $(1 \le k < n) = p_n^{\frac{k-1}{n-1}} - p_n^{\frac{k}{n-1}}$. It is easy to see that $p_1 + p_2 + ... + p_n = 1$ and

$$1 - \frac{p_n(p_k + ... + p_n)^2}{(p_{k-1} + ... + p_n)^2} = 1 - \frac{p_n p_n^{\frac{2(k-1)}{n-1}}}{p_n^{\frac{2(k-2)}{n-1}}} = 1 - p_n^{\frac{n+1}{n-1}} = p_n. \tag{1}$$

Now we describe a 1-way QFA accepting the language L_n.

The automaton has $2n$ states: q_1, q_2, ... q_n are non halting states, q_{n+1}, q_{n+2}, ... q_{2n-1} are rejecting states and q_{2n} is an accepting state. The transition function is defined by unitary block matrices

$$V_\kappa = \begin{pmatrix} M_n(\sqrt{p_1}, \sqrt{p_2}, ..., \sqrt{p_n}) & 0 \\ 0 & I_n \end{pmatrix},$$

$$V_{a_1} = \begin{pmatrix} R_n(\sqrt{p_1}, \sqrt{p_2}, ..., \sqrt{p_n}) & T_n(\sqrt{p_1}, \sqrt{p_2}, ..., \sqrt{p_n}) & 0 \\ T_n^T(\sqrt{p_1}, \sqrt{p_2}, ..., \sqrt{p_n}) & 0 & 0 \\ 0 & 0 & 1 \end{pmatrix},$$

$$V_{a_2} = \begin{pmatrix} 0 & 0 & 1 & 0 & 0 \\ 0 & R_{n-1}(\sqrt{p_2}, ..., \sqrt{p_n}) & 0 & T_{n-1}(\sqrt{p_2}, ..., \sqrt{p_n}) & 0 \\ 1 & 0 & 0 & 0 & 0 \\ 0 & T_{n-1}^T(\sqrt{p_2}, ..., \sqrt{p_n}) & 0 & 0 & 0 \\ 0 & 0 & 0 & 0 & 1 \end{pmatrix},$$

....

$$V_{a_k} = \begin{pmatrix} 0 & 0 & I_{k-1} & 0 & 0 \\ 0 & R_{n+1-k}(\sqrt{p_k}, ..., \sqrt{p_n}) & 0 & T_{n+1-k}(\sqrt{p_k}, ..., \sqrt{p_n}) & 0 \\ I_{k-1} & 0 & 0 & 0 & 0 \\ 0 & T_{n+1-k}^T(\sqrt{p_k}, ..., \sqrt{p_n}) & 0 & 0 & 0 \\ 0 & 0 & 0 & 0 & 1 \end{pmatrix},$$

...,

$$V_{a_n} = \begin{pmatrix} 0 & 0 & I_{n-1} & 0 \\ 0 & 1 & 0 & 0 \\ I_{n-1} & 0 & 0 & 0 \\ 0 & 0 & 0 & 1 \end{pmatrix},$$

$$V_\$ = \begin{pmatrix} 0 & I_n \\ I_n & 0 \end{pmatrix}.$$

Case 1. The input is $\kappa a_1^* a_2^* ... a_n^* \$$.

The starting superposition is $|q_1\rangle$. After reading the left endmarker the superposition becomes $\sqrt{p_1}|q_1\rangle + \sqrt{p_2}|q_2\rangle + ... + \sqrt{p_n}|q_n\rangle$ and after reading a_1^* the superposition remains the same.

If the input contains a_k then reading the first a_k changes the non-halting part of the superposition to $\sqrt{p_k}|q_k\rangle + ... + \sqrt{p_n}|q_n\rangle$ and after reading all the rest of a_k the non-halting part of the superposition remains the same.

Reading the right endmarker maps $|q_n\rangle$ to $|q_{2n}\rangle$. Therefore, the superposition after reading it contains $\sqrt{p_n}|q_{2n}\rangle$. This means that the automaton accepts with probability p_n because q_{2n} is an accepting state.

Case 2. The input is $\kappa a_1^* a_2^* ... a_k^* a_k a_m ... (k > m)$.

After reading the last a_k the non-halting part of the superposition is $\sqrt{p_k}|q_k\rangle + ... + \sqrt{p_n}|q_n\rangle$. Then reading a_m changes the non-halting part to

$\frac{\sqrt{p_m}(p_k+...+p_n)}{(p_m+...+p_n)}|q_m\rangle + ... + \frac{\sqrt{p_n}(p_k+...+p_n)}{(p_m+...+p_n)}|q_n\rangle$. This means that the automaton accepts with probability $\leq \frac{p_n(p_k+...+p_n)^2}{(p_m+...+p_n)^2}$ and rejects with probability at least

$$1 - \frac{p_n(p_k + ... + p_n)^2}{(p_m + ... + p_n)^2} \geq 1 - \frac{p_n(p_k + ... + p_n)^2}{(p_{k-1} + ... + p_n)^2} = p_n$$

that follows from (1). □

Corollary 2. *The language L_n can be recognized by a 1-way QFA with the probability of correct answer at least $\frac{1}{2} + \frac{c}{n}$, for a constant c.*

Proof: By resolving the equation $p^{\frac{n+1}{n-1}} + p = 1$, we get $p = \frac{1}{2} + \Theta(\frac{1}{n})$. □

Theorem 5. *The language L_n cannot be recognized by a 1-way QFA with probability greater than p where p is the root of*

$$(2p - 1) = \frac{2(1 - p)}{n - 1} + 4\sqrt{\frac{2(1 - p)}{n - 1}} \tag{2}$$

in the interval $[1/2, 1]$.

Proof: Assume we are given a 1-way QFA M. We show that, for any $\epsilon > 0$, there is a word such that the probability of correct answer is less than $p + \epsilon$.

Lemma 2. *[AF 98] Let $x \in \Sigma^+$. There are subspaces E_1, E_2 such that $E_{non} = E_1 \oplus E_2$ and*

(i) *If $\psi \in E_1$, then $V_x(\psi) \in E_1$,*
(ii) *If $\psi \in E_2$, then $\|V'_{x^k}(\psi)\| \to 0$ when $k \to \infty$.*

We use $n - 1$ such decompositions: for $x = a_2$, $x = a_3$, ..., $x = a_n$. The subspaces E_1, E_2 corresponding to $x = a_m$ are denoted $E_{m,1}$ and $E_{m,2}$.

Let $m \in \{2, ..., n\}$, $y \in a_1^* a_2^* ... a_{m-1}^*$. Remember that ψ_y denotes the superposition after reading y (with observations w.r.t. $E_{non} \oplus E_{acc} \oplus E_{rej}$ after every step). We express ψ_y as $\psi_y^1 + \psi_y^2$, $\psi_y^1 \in E_{m,1}$, $\psi_y^2 \in E_{m,2}$.

Case 1. $\|\psi_y^2\| \leq \sqrt{\frac{2(1-p)}{n-1}}$ for some $m \in \{2, ..., n\}$ and $y \in a_1^* ... a_{m-1}^*$.

Let $i > 0$. Then, $ya_{m-1} \in L_n$ but $ya_m^i a_{m-1} \notin L_n$. Consider the distributions of probabilities on M's answers "accept" and "reject" on ya_{m-1} and $ya_m^i a_{m-1}$. If M recognizes L_n with probability $p+\epsilon$, it must accept ya_{m-1} with probability at least $p+\epsilon$ and reject it with probability at most $1-p-\epsilon$. Also, $ya_m^i a_{m-1}$ must be rejected with probability at least $p + \epsilon$ and accepted with probability at most $1 - p - \epsilon$. Therefore, both the probabilities of accepting and the probabilities of rejecting must differ by at least

$$(p + \epsilon) - (1 - p - \epsilon) = 2p - 1 + 2\epsilon.$$

This means that the *variational distance* between two probability distributions (the sum of these two distances) must be at least $2(2p - 1) + 4\epsilon$. We show that it cannot be so large.

First, we select an appropriate i. Let k be so large that $\|V'_{a^k_m}(\psi^2_y)\| \leq \delta$ for $\delta = \epsilon/4$. $\psi^1_y, V'_{a_m}(\psi^1_y), V'_{a^2_m}(\psi^1_y), \ldots$ is a bounded sequence in a finite-dimensional space. Therefore, it has a limit point and there are i, j such that

$$\|V'_{a^j_m}(\psi^1_y) - V'_{a^{i+j}_m}(\psi^1_y)\| < \delta.$$

We choose i, j so that $i > k$.

The difference between the two probability distributions comes from two sources. The first source is the difference between ψ_y and $\psi_{ya^i_m}$ (the states of M before reading a_{m-1}). The second source is the possibility of M accepting while reading a^i_m (the only part that is different in the two words). We bound each of them.

The difference $\psi_y - \psi_{ya^i_m}$ can be partitioned into three parts.

$$\psi_y - \psi_{ya^i_m} = (\psi_y - \psi^1_y) + (\psi^1_y - V'_{a^i_m}(\psi^1_y)) + (V'_{a^i_m}(\psi^1_y) - \psi_{ya^i_m}). \tag{3}$$

The first part is $\psi_y - \psi^1_y = \psi^2_y$ and $\|\psi^2_y\| \leq \sqrt{\frac{2(1-p)}{n-1}}$. The second and the third parts are both small. For the second part, notice that V'_{a_m} is unitary on $E_{m,1}$ (because V_{a_m} is unitary and $V_{a_m}(\psi)$ does not contain halting components for $\psi \in E_{m,1}$). Hence, V'_{a_m} preserves distances on $E_{m,1}$ and

$$\|\psi^1_y - V'_{a^i_m}(\psi^1_y)\| = \|V'_{a^j_m}(\psi^1_y) - V'_{a^{i+j}_m}(\psi^1_y)\| < \delta$$

For the third part of (3), remember that $\psi_{ya^i_m} = V'_{a^i_m}(\psi_y)$. Therefore,

$$\psi_{ya^i_m} - V'_{a^i_m}(\psi^1_y) = V'_{a^i_m}(\psi_y) - V'_{a^i_m}(\psi^1_y) = V'_{a^i_m}(\psi_y - \psi^1_y) = V'_{a^i_m}(\psi^2_y)$$

and $\|\psi^2_{ya^i_m}\| \leq \delta$ because $i > k$. Putting all three parts together, we get

$$\|\psi_y - \psi_{ya^i_m}\| \leq \|\psi_y - \psi^1_y\| + \|\psi^1_y - \psi^1_{ya^i_m}\| + \|\psi^1_{ya^i_m} - \psi_{ya^i_m}\| \leq \sqrt{\frac{2(1-p)}{n-1}} + 2\delta.$$

Lemma 3. [BV 97] *Let ψ and ϕ be such that $\|\psi\| \leq 1$, $\|\phi\| \leq 1$ and $\|\psi - \phi\| \leq \epsilon$. Then the total variational distance resulting from measurements of ϕ and ψ is at most 4ϵ.*

This means that the difference between any probability distributions generated by ψ_y and $\psi_{ya^i_m}$ is at most

$$4\sqrt{\frac{2(1-p)}{n-1}} + 8\delta.$$

In particular, this is true for the probability distributions obtained by applying $V_{a_{m-1}}$, $V_\$$ and the corresponding measurements to ψ_y and $\psi_{ya^i_m}$.

The probability of M halting while reading a_m^i is at most $\|\psi_k^2\|^2 = \frac{2(1-p)}{n-1}$. Adding it increases the variational distance by at most $\frac{2(1-p)}{n-1}$. Hence, the total variational distance is at most

$$\frac{2(1-p)}{n-1} + 4\sqrt{\frac{2(1-p)}{n-1}} + 8\delta = \frac{2(1-p)}{n-1} + 4\sqrt{\frac{2(1-p)}{n-1}} + 2\epsilon.$$

By definition of p, this is the same as $(2p-1)+2\epsilon$. However, if M distinguishes y and ya_m^i correctly, the variational distance must be at least $(2p-1)+4\epsilon$. Hence, M does not recognize one of these words correctly.

Case 2. $\|\psi_y^2\| > \sqrt{\frac{2(1-p)}{n-1}}$ for every $m \in \{2,\dots,n\}$ and $y \in a_1^* \dots a_{m-1}^*$.

We define a sequence of words $y_1, y_2, \dots, y_m \in a_1^* \dots a_n^*$. Let $y_1 = a_1$ and $y_k = y_{k-1}a_k^{i_k}$ for $k \in \{2,\dots,n\}$ where i_k is such that

$$\|V'_{a_k^{i_k}}(\psi_{y_{k-1}}^2)\| \le \sqrt{\frac{\epsilon}{n-1}}.$$

The existence of i_k is guaranteed by (ii) of Lemma 2.

We consider the probability that M halts on $y_n = a_1 a_2^{i_2} a_3^{i_3} \dots a_n^{i_n}$ before seeing the right endmarker. Let $k \in \{2,\dots,n\}$. The probability of M halting while reading the $a_k^{i_k}$ part of y_n is at least

$$\|\psi_{y_{k-1}}^2\|^2 - \|V'_{a_k^{i_k}}(\psi_{y_{k-1}}^2)\|^2 > \frac{2(1-p)}{n-1} - \frac{\epsilon}{n-1}.$$

By summing over all $k \in \{2,\dots,n\}$, the probability that M halts on y_n is at least

$$(n-1)\left(\frac{2(1-p)}{n-1} - \frac{\epsilon}{n-1}\right) = 2(1-p) - \epsilon.$$

This is the sum of the probability of accepting and the probability of rejecting. Hence, one of these two probabilities must be at least $(1-p) - \epsilon/2$. Then, the probability of the opposite answer on any extension of y_n is at most $1 - (1-p - \epsilon/2) = p + \epsilon/2$. However, y_n has both extensions that are in L_n and extensions that are not. Hence, one of them is not recognized with probability $p + \epsilon$. □

By solving the equation (2), we get

Corollary 3. L_n cannot be recognized with probability greater than $\frac{1}{2} + \frac{3}{\sqrt{n-1}}$.

Let $n_1 = 2$ and $n_k = \frac{9n_{k-1}^2}{c^2} + 1$ for $k > 1$ (where c is the constant from Theorem 4). Also, define $p_k = \frac{1}{2} + \frac{c}{n_k}$. Then, Corollaries 2 and 3 imply

Theorem 6. For every $k > 1$, L_{n_k} can be recognized with by a 1-way QFA with the probability of correct answer p_k but cannot be recognized with the probability of correct answer p_{k-1}.

Thus, we have constructed a sequence of languages $L_{n_1}, L_{n_2}, \dots$ such that, for each L_{n_k}, the probability with which L_{n_k} can be recognized by a 1-way QFA is smaller than for $L_{n_{k-1}}$.

Our final theorem is a counterpart of Theorem 2. It generalizes Theorem 3.

Theorem 7. *Let L be a language and M be its minimal automaton. If there is no q_1, q_2, q_3, x, y such that*

1. *the states q_1, q_2, q_3 are pairwise different,*
2. *If M starts in the state q_1 and reads x, it passes to q_2,*
3. *If M starts in the state q_2 and reads x, it passes to q_2, and*
4. *If M starts in the state q_2 and reads y, it passes to q_3,*
5. *If M starts in the state q_3 and reads y, it passes to q_3,*
6. *both q_2 and q_3 are neither "all-accepting" state, nor "all-rejecting" state,*

then L can be recognized by a 1-way quantum finite automaton with probability $p = 0.68....$

References

AF 98.   Andris Ambainis and Rūsiņš Freivalds. 1-way quantum finite automata: strengths, weaknesses and generalizations. *Proc. 39th FOCS*, 1998, p. 332–341. $http://xxx.lanl.gov/abs/quant - ph/9802062$

Be 89.   Charles Bennett. Time-space tradeoffs for reversible computation. *SIAM J. Computing*, 18:766-776, 1989.

BP 99.   A. Brodsky, N. Pippenger. Characterizations of 1-way quantum finite automata. $http://xxx.lanl.gov/abs/quant - ph/9903014$

BV 93.   Ethan Bernstein, Umesh Vazirani, Quantum complexity theory. *Proceedings of STOC'93*, pp.1-10.

BV 97.   Ethan Bernstein, Umesh Vazirani, Quantum complexity theory. *SIAM Journal on Computing*, 26:1411-1473, 1997.

De 89.   David Deutsch. Quantum theory, the Church-Turing principle and the universal quantum computer. *Proc. Royal Society London, A400*, 1989. p. 96–117.

Fe 82.   Richard Feynman. Simulating physics with computers. *International Journal of Theoretical Physics*, 1982, vol. 21, No. 6/7, p. 467-488.

Fr 79.   Rūsiņš Freivalds. Fast probabilistic algorithms. *Lecture Notes in Computer Science*, 1979, vol. 74, p. 57-69.

Ki 98.   Arnolds Ķikusts. A small 1-way quantum finite automaton. $http://xxx.lanl.gov/abs/quant - ph/9810065$

KW 97.   Attila Kondacs and John Watrous. On the power of quantum finite state automata. In *Proc. 38th FOCS*, 1997, p. 66–75.

MC 97.   Christopher Moore, James P. Crutchfield. Quantum automata and quantum grammars. *Theoretical Computer Science*, to appear. Also available at $http://xxx.lanl.gov/abs/quant - ph/9707031$

Sh 97.   Peter Shor. Polynomial time quantum algorithms for prime factorization and discrete logarithms on a quantum computer. *SIAM Journal on Computing*, 1997, vol. 26, p. 1484-1509.

Ya 93.   Andrew Chi-Chih Yao. Quantum circuit complexity. In *Proc. 34th FOCS*, 1993, p. 352–361.

Distributionally-Hard Languages

Lance Fortnow[1]*, A. Pavan[2], and Alan L. Selman[3]**

[1] Department of Computer Science, University of Chicago, 1100 East 58th Street, Chicago, IL, 60637, fortnow@cs.uchicago.edu
[2] Department of Computer Science and Engineering, University at Buffalo, 226 Bell Hall, Buffalo, NY 14260, aduri@cs.buffalo.edu
[3] Department of Computer Science and Engineering, University at Buffalo, 226 Bell Hall, Buffalo, NY 14260, selman@cs.buffalo.edu

Abstract. Define a set L to be *distributionally-hard to recognize* if for every polynomial-time computable distribution μ with infinite support, L is not recognizable in polynomial time on the μ-average. Cai and Selman [5] defined a modification of Levin's notion of average polynomial time and proved that every P-bi-immune language is distributionally-hard. Pavan and Selman [23] proved that there exist distributionally-hard sets that are not P-bi-immune if and only P contains P-printable-immune sets. We extend this characterizion to include assertions about several traditional questions about immunity, about finding witnesses for NP-machines, and about existence of one-way functions. Similarly, we address the question of whether NP contains sets that are distributionally hard. Several of our results are implications for which we cannot prove whether or not their converse holds. In nearly all such cases we provide oracles relative to which the converse fails. We use the techniques of Kolmogorov complexity to describe our oracles and to simplify the technical arguments.

1 Introduction

Levin [18] was the first to advocate the general study of average-case complexity and he provided the central notions for its study. More recently, Cai and Selman [5] observed that Levin's definition of Average-P has limitations when applied to distributional problems with unreasonable distributions and when applied to exponential time-bounds. For example, for every language L in the complexity class E, (L, μ) is in Average-P, where μ is the flat distribution defined by $\mu'(x) = 4^{-|x|}$. However, E contains sets that are almost-everywhere complex. That is, E contains sets that require more than polynomial time to recognize for all but finitely many inputs. It is unnatural to consider such a set to have average polynomial-time complexity for any distribution with infinite support, and certainly not for any flat distribution. Cai and Selman modified Levin's definition to remove these limitations. In particular, letting AVP denote the class

* funded in part by NSF grant CCR97-32922
** The author performed part of this research while visiting the Department of Computer Science, University of Chicago

T. Asano et al. (Eds.): COCOON'99, LNCS 1627, pp. 184–193, 1999.

of all distributional problems that are polynomial on the μ-average according to the definition of Cai and Selman, if a language L is almost-everywhere complex, then (L, μ) does not belong to AVP for any distribution μ. (We will provide all formal definitions in the next section.) Nevertheless, the difference between AVP and Average-P is modest: Define a distribution to be *reasonable* if there exists a constant $s > 0$ such that $\mu(\{x \mid |x| \geq n\}) = \Omega(\frac{1}{n^s})$. The reason of course is that distributions that decrease too quickly give too much weight to small instances, and for this reason are unreasonable. All distributions of natural problems in the literature, including the distributions of all known DistNP-complete problems, are reasonable. The class AVP $\subseteq$ Average-P, and Cai and Selman showed that, for all reasonable distributions μ, $(L, \mu) \in$ AVP if and only if $(L, \mu) \in$ Average-P.

It is well-known that a set L is almost-everywhere complex if and only if it is P-bi-immune [2]. Thus, if L is P-bi-immune, then there is no distribution μ with infinite support such that (L, μ) belongs to AVP. We say that such languages are *distributionally-hard*.

In this paper we raise and address questions about distributionally-hard sets. For example, we extensively characterize the question of whether there exist distributionally-hard sets that are not P-bi-immune. Pavan and Selman [23] proved that such sets exist if and only if P contains sets that are P-printable-bi-immune. We provide several other equivalent characterizations. These include assertions about the inability to compute witnesses for sets in NP and an assertion about the existence of one-way functions that are hard to invert almost everywhere.

It is easy to see that every distributionally-hard set is P-printable-bi-immune. We will show that if there exist (1) P-printable-bi-immune sets that are not distributionally-hard, then (2) there are distributionally-hard sets that are not P-bi-immune. The question of whether there exist such sets may not be forthcoming. To wit, we obtain oracles relative to which the polynomial hierarchy is infinite and sets satisfying (2) exist, and we obtain oracles relative to which the polynomial hierarchy is infinite and such sets do not exist. We obtain oracles relative to which sets satisfying (1) hold, and relative to which (1) does not hold. We obtain an oracle relative to which there exist sets that satisfy (2), but there do not exist sets that satisfy (1).

Now consider the question of whether NP contains distributionally-hard sets, for, if so, then DistNP (the class of all distributional problems (L, μ), where $L \in$ NP, and mu is polynomial-time computable) and AVP are distinct. Suppose that NP contains a P-bi-immune set L. Then, L is a distributionally-hard set in NP. We will show that $L \cap 0^* \in$ NP is P-immune, from which it follows that P contains P-printable-immune sets, and therefore that there exist sets that are distributionally-hard but not P-bi-immune. Indeed, if NP contains a P-immune tally set, then NP contains a P-immune set, which implies that P contains P-printable-immune sets. (These are straightforward observations.)

Relative to a random oracle, Lutz's hypothesis that the p-measure of NP is not zero is true [16]. Thus, relative to a random oracle, NP contains P-bi-immune, distributionally-hard sets [21]. We will see from our results that relative to an oracle constructed by Impagliazzo and Tardos [15], P contains P-printable-immune sets but NP does not contain P-immune tally sets. Hemachandra and Jha [12] constructed an oracle relative to which NP contains P-immune sets but no P-immune tally sets. Here, we will construct an oracle relative to which P contains a P-printable-immune set but NP does not contain any P-immune set. Thus, relative to this oracle, there exist distributionally-hard sets that are not P-bi-immune, nevertheless, NP does not contain distributionally-hard sets. Relative to these oracles, the converses of the implications set forward in the previous paragraph do not hold.

We will use Kolmogorov complexity to describe the oracles that we construct. Because of this, both the ideas and the technical details of our proofs will be much easier than they would be otherwise. However, there is not room for proofs of these results in this conference proceedings.[1]

2   Preliminaries

We assume that all languages are subsets of $\Sigma^* = \{0,1\}^*$ and we assume that Σ^* is ordered by standard lexicographic ordering. We use standard notation for complexity theoretic notions. In particular, $E = \bigcup_{c>0} \text{DTIME}(2^{cn})$ and $NE = \bigcup_{c>0} \text{NTIME}(2^{cn})$. We will use complexity class names to abbreviate descriptions of Turing machines. For example, an E-*machine* is a deterministic 2^{cn}, $c \geq 0$, time-bounded Turing machine.

A *transducer* is a Turing machine with a read-only input tape, a write-only output tape, and accepting states in the usual manner. A transducer computes a value y on an input string x if there is an accepting computation on x for which y is the final contents of the output tape. In general, such transducers compute partial, multivalued functions. PF is the set all partial functions that are computed by deterministic polynomial time-bounded transducers. A function $f \in PF$ is *honest* if there is a polynomial q such that for every y in range(f), there exists x in dom(f) such that $f(x) = y$ and $|y| \leq q(|x|)$. A *one-way* function is an honest partial function in PF that cannot be inverted in polynomial time. Such functions exist if and only if $P \neq NP$ [25]. Define an honest partial function f in PF to be *almost-always one-way* if no polynomial-time Turing machine inverts f correctly on more than a finite subset of range(f).

A *distribution function* $\mu : \{0,1\}^* \to [0,1]$ is a nondecreasing function from strings to the closed interval $[0,1]$ that converges to one. The corresponding *density function* μ' is defined by $\mu'(0) = \mu(0)$ and $\mu'(x) = \mu(x) - \mu(x-1)$. Clearly, $\mu(x) = \sum_{y \leq x} \mu'(y)$. For any subset of strings S, we will denote by $\mu(S) = \sum_{x \in S} \mu'(x)$, the probability of the event S. Define $u_n = \mu(\{x \mid |x| = n\})$. For each n, let $\mu'_n(x)$ be the conditional probability of x in $\{x \mid |x| = n\}$. That is, $\mu'_n(x) = \mu'(x)/u_n$, if $u_n > 0$, and $\mu'_n(x) = 0$ for $x \in \{x \mid |x| = n\}$, if $u_n = 0$.

A function μ from Σ^* to $[0,1]$ is *computable in polynomial time* [17] if there is a polynomial time-bounded transducer M such that for every string x and every

---

[1] The reader may find a complete version of this paper, with proofs, at the Web page of the Networked Computer Science Technical Reference Library, www.ncstrl.org.

positive integer n, $|\mu(x) - M(x, 1^n)| < \frac{1}{2^n}$. Consistent with Levin's hypothesis that natural distributions are computable in polynomial time, we restrict our attention *entirely* to such distributions. If μ is computable in polynomial time, then the density function μ' is computable in polynomial time. (The converse is false unless P = NP [7].) All distributions are to have infinite support—we explicitly exclude from consideration distributions μ for which $\mu'(x) = 0$ for all but a finite number of strings x. Consideration of such distributions would allow every problem to be an essentially finite problem.

Levin [18] defines a function f from Σ^* to nonnegative reals to be *polynomial on μ-average* if there is an integer $k > 0$ such that

$$\sum_{|x| \geq 1} \mu'(x) \frac{(f(x))^{1/k}}{|x|} < \infty. \tag{1}$$

Average-P is the class of distributional problems (L, μ), where L is a language and μ is a polynomial-time computable distribution, such that L can be decided by some Turing machine M whose running time T_M is polynomial on μ-average.

For any time-constructible function T that is monotonically increasing, and hence invertible, Cai and Selman [5] define T on the μ-average as follows[2]: Let μ be a distribution on Σ^*, and let $W_n = \mu(\{x : |x| \geq n\})$. A function f is T on the μ-average if for all $n \geq 1$,

$$\sum_{|x| \geq n} \mu'(x) \cdot \frac{T^{-1}(f(x))}{|x|} \leq W_n. \tag{2}$$

Then, AVTIME$(T(n))$ denotes the class of distributional problems (L, μ), where L is a language and μ is a polynomial-time computable distribution, such that L can be decided by some Turing machine M whose running time T_M is T on the μ-average.

Define AVP $= \bigcup_{k \geq 1}$ AVTIME(n^k). Clearly, AVP $\subseteq$ Average-P.

A distribution μ is *reasonable* if there exists $s > 0$ such that $W_n = \Omega\left(\frac{1}{n^s}\right)$. We will require the following results of Cai and Selman [5] and Gurevich [7].

Proposition 1. 1. *If μ is a reasonable distribution, then (L, μ) belongs to Average-P (Levin's definition) if and only if (L, μ) belongs to AVP (Cai and Selman's definition).*

2. *If μ satisfies the stronger condition that there exists $s > 0$ such that $u_n = \Omega\left(\frac{1}{n^s}\right)$, then all of the following are equivalent:*

   (i) *(L, μ) belongs to Average-P;*
   (ii) *(L, μ) belongs to AVP;*

---

[2] Cai and Selman restricted their attention to functions that belong to Hardy's [8] class of logarithmico-exponential functions. We do not need to concern ourselves with this for the purpose of this paper.

(iii) There is a Turing machine M that accepts L and an integer $k > 0$ such that for all $n \geq 1$,

$$\sum_{|x|=n} \mu'(x) \frac{(T_M(x))^{1/k}}{|x|} \leq u_n. \tag{3}$$

Here we have given only the definitions and properties that we need for this paper; we refer the reader to the recent expositions by Impagliazzo [14] and Wang [26] for deeper understanding of average-case complexity.

2.1 Immunity

A language L is *immune* to a complexity class $\mathcal{C}$, or $\mathcal{C}$-*immune*, if L is infinite and no infinite subset of L belongs to $\mathcal{C}$. A language L is *bi-immune* to a complexity class $\mathcal{C}$, or $\mathcal{C}$-*bi-immune*, if L is infinite, no infinite subset of L belongs to $\mathcal{C}$, and no infinite subset of $\overline{L}$ belongs to $\mathcal{C}$. A language is DTIME$(T(n))$-*complex* if L does not belong to DTIME$(T(n))$ almost everywhere; that is, every Turing machine M that accepts L runs in time greater than $T(|x|)$, for all but finitely many words x. Balcázar and Schöning [2] proved that for every time-constructible function T, L is DTIME$(T(n))$-complex if and only if L is bi-immune to DTIME$(T(n))$.

Mayordomo [22] proved that the p-measure of the class of DTIME(2^n)-bi-immune sets is 1, and therefore, if the p-measure of NP is not 0, then NP contains a DTIME(2^n)-bi-immune set. Cai and Selman [5] proved, for all P-bi-immune sets L and for all polynomial-time computable distributions μ, that $(L, \mu) \notin$ AVP. Thus, if NP does not have p-measure 0, then there is a language L such that for every polynomial-time computable distribution μ, the distributional problem (L, μ) belongs to DistNP but does not belong to AVP. (Independently, Schuler and Yamakami [24] obtained a similar result.)

We define a language L to be *distributionally-hard to recognize* if for all polynomial-time computable distributions μ with infinite support, $(L, \mu) \notin$ AVP. As we noted, every P-bi-immune language is distributionally-hard to recognize.

A set L is *P-printable* if there exists $k \geq 1$ such that all the elements of L up to size n can be printed by a deterministic Turing machine in time $n^k + k$ [10,9]. Every P-printable set is sparse and belongs to P. A set A is *P-printable-immune* if A is infinite and no infinite subset of A is P-printable.

3 Distributional Hardness

The following is the central theorem of this paper. It completely characterizes the question of whether there exist languages that are distributionally-hard to recognize other than the P-bi-immune sets.

Theorem 1. *The following assertions are equivalent.*

1. *There exists a distributionally-hard set that is not* P-*bi-immune.*
2. *There exists a* P-*printable-bi-immune set that is not* P-*bi-immune.*

3. There exists a set that is P-printable-bi-immune, not P-bi-immune, but whose complement is P-immune.

4. P contains a P-printable-immune set.

5. NP contains a P-printable-immune set.

6. There is an infinite set S in NE and an NE-machine M that accepts S such that no E-machine correctly computes infinitely many accepting computations of M.

7. There is an infinite tally language L in NP and an NP-machine M that accepts L such that no P-machine correctly computes infinitely many accepting computations of M.

8. There is an infinite set S in NP and an NP-machine M that accepts S such that no P-machine correctly computes infinitely many accepting computations of M.

9. Almost-always one-way functions exist.

Theorem 1 shows that the question of whether there exist languages that are distributionally hard but not P-bi-immune is equivalent to several previously studied conjectures about immunity, about computing witnesses of nondeterministic machines, and about existence of strong one-way functions. Pavan and Selman [23] obtained the equivalence of items 1 and 4. The equivalence of items 4 and 5 is due to Russo [1]. Also, Theorem 1 strengthens a result of Hemaspaandra, Rothe, and Wechsung [13], which is that item 8 implies item 4.

Recall that every language in the class E (NE) is identifiable with a tally language in P (NP), respectively [4]. This observation yields the fact that items 6 and 7 are equivalent. Similarly, NP contains a P-immune tally language if and only if NE contains an E-immune set. One does not expect existence of a P-immune set in NP to imply existence of a P-immune tally language in NP, because in general such downward separations do not hold. Hemaspaandra and Jha [12] support this contention with an oracle relative to which NP contains a P-immune set but no P-immune tally language. Thus, the equivalence of items 7 and 8 is surprising and demonstrates that search problems and decision problems have different properties.

Theorem 2. *There exist recursive oracles A and B relative to which the polynomial hierarchy is infinite; relative to A the properties listed in Theorem 1 are true, and relative to B these properties are false.*

For the proof of Theorem 2, to construct A, begin with a random oracle and then apply techniques of Fortnow [6] with an oracle that makes the polynomial hierarchy infinite [27,11]. To construct B, begin with a sparse oracle relative to which $E^{NP} = E$ and then, in a similar manner, apply results of Long and Selman [20].

The equivalence of items 1 and 2 of Theorem 1 lead us to ask whether distributionally-hard and P-printable-bi-immune are equivalent. First, we note the following proposition.

Proposition 2. *Every distributionally-hard to recognize set is P-printable-bi-immune.*

The proof is clear: If either L or $\overline{L}$ has an infinite P-printable subset S, then define a polynomial-time computable distribution μ such that $\mu(S) = 1$. Then, it is easy to see that $(L, \mu) \in$ AVP.

Now, for any set L, we have the following implications:

$$L \text{ is P-bi-immune}$$
$$\Rightarrow \tag{4}$$
$$L \text{ is distributionally-hard}$$
$$\Rightarrow \tag{5}$$
$$L \text{ is P-printable-bi-immune.}$$

To study our question further, consider the following hypothesis:

Hypothesis 1. *There exists a P-printable-bi-immune-set that is not distributionally hard.*

Item 1 of Theorem 1 asserts that the implication 4 does not collapse. Hypothesis 1 asserts that the implication 5 does not collapse. Hypothesis 1 implies item 2 of Theorem 1. Thus, by Theorem 1, if 4 collapses, then 5 collapses.

Theorem 3. *There exists an oracle relative to which Hypothesis 1 is true, and there is an oracle relative to which the assertions in Theorem 1 hold but Hypothesis 1 is false.*

We construct these oracles using the sophisticated techniques of Kolmogorov complexity.

Now we turn to the question of whether NP contains distributionally-hard to recognize sets. The following for the most part are easy to prove and essentially known [23].

Theorem 4. *For any language L in NP,*

$$L \text{ is P-bi-immune}$$
$$\Rightarrow \tag{6}$$
$$L \text{ is distributionally-hard}$$
$$\Rightarrow \tag{7}$$
$$L \cap 0^* \text{ is P-immune}$$
$$\Leftrightarrow \tag{8}$$
$$L \cap 0^* \text{ is P-printable-immune}$$
$$\Rightarrow \tag{9}$$
$$\text{The assertions of Theorem 1 all hold.}$$

The equivalence (8) holds because a tally language is P-immune if and only if it is P-printable immune. Then, it is immediate that item 4 of Theorem 1 holds, which proves that (9) holds.

It follows immediately from Theorem 4 that NP has distributionally-hard sets relative to a random oracle (because NP has P-bi-immune sets), and that, relative to an oracle for which the conditions of Theorem 1 fails, NP does not have distributionally-hard sets.

If NP contains P-bi-immune sets, then NP contains sets having various combinations of immunity properties. For example, we have the following result.

Theorem 5. *If* NP *contains a* P-*printable-immune set, then* NP *contains sets that are*

1. P-*printable-immune and not* P-*immune, and*
2. P-*printable-bi-immune and not* P-*bi-immune.*

If the p-measure of NP is not 0, then NP contains a P-bi-immune set. The obvious corollaries hold, of which, the following are the most interesting.

Corollary 1. *(i) If the p-measure of* NP *is not* 0, *then there is a language that is distributionally-hard to recognize but not* P-*bi-immune.*
(ii) If the p_2-*measure of* NP *is not* 0, *then* NP *contains a language that is distributionally-hard to recognize but not* P-*bi-immune.*
(iii) If the p-measure of NP *is not* 0, *then* NP *contains sets that are*
   1. P-*printable-immune and not* P-*immune, and*
   2. P-*printable-bi-immune and not* P-*bi-immune.*

Parts (i) and (ii) are due to Pavan and Selman [23]. (The proof of Corollary 1(ii) is not obvious; it follows from a close analysis the proof of Theorem 1.) In light of these results, it is interesting to ask the following question: If NP contains a P-printable-immune set, does NP contain a P-immune set? Let us consider this question in the context of the following obvious implications:

$$\text{NP has a P-immune tally set} \qquad (10)$$

$$\Rightarrow$$

$$\text{NP has a P-immune set} \qquad (11)$$

$$\Rightarrow$$

$$\text{NP has a P-printable-immune set.} \qquad (12)$$

Item (12) is one of the equivalent assertions in Theorem 1. Item 6 of Theorem 1 was studied by Impagliazzo and Tardos [15], who obtained an oracle relative to which NE = E and this assertion holds. Thus, there is an oracle relative to which (12) holds and (10) does not hold.[3] Hemaspaandra and Jha [12] constructed an oracle relative to which (11) holds and (10) fails. However, it remains an open question as to whether there is an oracle relative to which (11) holds and NE = E. The following result completes our study of these assertions.

---

[3] Actually, the property studied by Impagliazzo and Tardos is somewhat weaker. Nevertheless, our claim is correct, as is our attribution.

Theorem 6. *There is a recursive oracle relative to which* NP *has a* P-*printable-immune set but* NP *has no* P-*immune set.*

In particular, neither the assertions listed in Theorem 1 (12) nor the existence of P-immune sets in NP (11) appear to be strong enough to ensure that NP contains distributionally-hard sets. We leave as open the questions of whether (10) implies that NP contains distributionally-hard sets, and whether existence of distributionally-hard sets in NP implies that NP contains P-bi-immune sets.

References

1. E. Allender and R. Rubinstein. P-printable sets. *SIAM Journal on Computing*, 17(6):1193–1202, 1988.
2. J. Balcázar and U. Schöning. Bi-immune sets for complexity classes. *Mathematical Systems Theory*, 18(1):1–10, June 1985.
3. S. Ben-David, B. Chor, O. Goldreich, and M. Luby. On the theory of average case complexity. *Journal of Computer and System Sciences*, 44(2):193–219, 1992.
4. R. Book. Tally languages and complexity classes. *Info. and Control*, 26:186–193, 1974.
5. J-Y. Cai and A. Selman. Fine separation of average time complexity classes. *SIAM Journal on Computing*, 28(4):1310–1325, 1999.
6. L. Fortnow. Relativized worlds with an infinite hierarchy. *Information Processing Letters*, 69(6):309–313, 1999.
7. Y. Gurevich. Average case completeness. *Journal of Computer and System Sciences*, 42:346–398, 1991.
8. G. Hardy. *Orders of Infinity, The 'infinitärcalcül' of Paul du Bois-Reymond*, volume 12 of *Cambridge Tracts in Mathematics and Mathematical Physics*. Cambridge University Press, London, 2nd edition, 1924.
9. J. Hartmanis, N. Immerman, and V. Sewelson. Sparse sets in NP-P: EXPTIME versus NEXPTIME. *Inf. Control*, 65:158–181, 1985.
10. J. Hartmanis and Y. Yesha. Computation times of NP sets of different densities. *Theoretical Computer Science*, 34:17–32, 1984.
11. J. Håstad. Almost optimal lower bounds for small depth circuits. In *Proceedings of the 18th Annual ACM Symposium on Theory of Computing*, pages 6–20, 1986.
12. L. Hemachandra and S. Jha. Defying upward and downward separation. In *Proceedings of the Tenth Annual Symposium on Theoretical Aspects of Computer Science*, volume 665 of *Lecture Notes in Computer Science*, pages 185–195. Springer-Verlag, Berlin, 1993.
13. L. Hemaspaandra, J. Rothe, and G. Wechsung. Easy sets and hard certificate schemes. *Acta Informatica*, 34(11):859–879, 1997.
14. R. Impagliazzo. A personal view of average-case complexity. In *Proceedings of the Tenth Annual IEEE Conference on Structure in Complexity Theory*, pages 134–147, 1995.
15. R. Impagliazzo and G. Tardos. Search versus decision in super-polynomial time. In *Proc. IEEE Foundations of Computer Science*, pages 222–227, 1991.
16. S. Kautz and P. Miltersen. Relative to a random oracle, NP is not small. In *Proceedings of the Ninth Annual IEEE Structure in Complexity Theory Conference*, pages 162–174, 1994.

17. K. Ko. On self-reducibility and weak P-selectivity. *Journal of Computer and System Sciences*, 26:209–211, 1983.
18. L. Levin. Average case complete problems. *SIAM Journal on Computing*, 15:285–286, 1986.
19. M. Li and P. Vitányi. *An Introduction to Kolmogorov Complexity and Its Applications*. Graduate Texts in Computer Science. Springer, New York, second edition, 1997.
20. T. Long and A. Selman. Relativizing complexity classes with sparse oracles. *Journal of the ACM*, 33(3):618–627, 1986.
21. J. Lutz and E. Mayordomo. Cook versus Karp-Levin: Separating completeness notions if NP is not small. In *Proceedings of the Eleventh Annual Symposium on Theoretical Aspects of Computer Science. Lecture Notes in Computer Science*, volume 755, pages 415–426, Berlin, 1994. Springer-Verlag.
22. E. Mayordomo. Almost every set in exponential time is P-bi-immune. *Theoretical Computer Science*, 136:487–506, 1994.
23. A. Pavan and A. Selman. Complete distributional problems, hard languages, and resource-bounded measure. *Theoretical Computer Science*, 1998. To appear.
24. R. Schuler and T. Yamakami. Sets computable in polynomial time on the average. In *Proceedings of the First Annual International Computing and Combinatorics Conference, Lecture Notes in Computer Science*, volume 959, pages 650–661. Springer-Verlag, Berlin, 1995.
25. A. Selman. A survey of one-way functions in complexity theory. *Mathematical Systems Theory*, 25:203–221, 1992.
26. J. Wang. Average-case computational complexity theory. In L. Hemaspaandra and A. Selman, editors, *Complexity Theory Retrospective II*, pages 295–328. Springer-Verlag, 1997.
27. A. Yao. Separating the polynomial-time hierarchy by oracles. In *Proceedings of the 26th IEEE Symposium on Foundations of Computer Science*, pages 1–10, 1985.

Circuits and Context-Free Languages

Pierre McKenzie[1], Klaus Reinhardt[2], and V. Vinay[3]

[1] Informatique et recherche opérationnelle, Université de Montréal, C.P. 6128, Succ. Centre-Ville, Montréal (Québec), H3C 3J7 Canada. Research performed while on leave at Universität Tübingen. mckenzie@iro.umontreal.ca

[2] Wilhelm-Schickard Institut für Informatik, Universität Tübingen, Sand 13, D-72076 Tübingen, Germany. reinhard@informatik.uni-tuebingen.de

[3] Department of Computer Science and Automation, Indian Institute of Science Bangalore-560 012 India. vinay@csa.iisc.ernet.in

Abstract. Simpler proofs that DAuxPDA-TIME(polynomial) equals LOG(DCFL) and that SAC^1 equals LOG(CFL) are given which avoid Sudborough's multi-head automata [Sud78]. The first characterization of LOGDCFL in terms of polynomial proof-tree-size is obtained, using circuits built from the multiplex select gates of [FLR96]. The classes L and NC^1 are also characterized by such polynomial size circuits: "self-similar" logarithmic depth captures L, and bounded width captures NC^1.

1 Introduction

The class LOGCFL (an NL machine equipped with an auxiliary pushdown) occupies a central place in the landscape of parallel complexity classes. LOGCFL sits between two interesting classes, NL and AC^1: NL is often viewed as the space analog of NP, and AC^1 characterizes the problems solvable on a PRAM in $O(\log n)$ time using a polynomial number of processors. As a class, LOGCFL has attracted a lot of attention due to its seemingly "richer structure" than that of NL. Sudburough [Sud78] showed that LOG(CFL) (logspace closure of CFL) was in fact LOGCFL by using multi-head pushdown automata. Then, Ruzzo [Ruz80] characterized LOGCFL alternating Turing machines. This was followed by Venkateswaran's surprising semi-unbounded circuits characterization [Ven91]: this clean circuit characterization of LOGCFL stands in contrast to its somewhat "messy" machine definitions. Then came a flurry of papers, for example showing closure of LOGCFL under complementation [BCD+88], characterizing LOGCFL in terms of groupoids [BLM93], linking it to depth reduction [Vin96,AJMV98], and studying the descriptive complexity of LOGCFL [LMSV99].

At the same time, progress was made in understanding LOGDCFL. The initial results of Sudburough ([Sud78], discussed below) were followed by Cook [Coo79] who showed that LOGDCFL is contained in deterministic space $O(\log^2 n)$. Dymond and Ruzzo [DR86] drew an important equivalence between between LOGDCFL and owner-write PRAM models. A variety of papers have since extended these results [FLR96,MRS93].

As mentioned, Sudburough proved that DAuxPDA-TIME(pol) $\subseteq$ LOG(DCFL) and that NAuxPDA-TIME(pol) $\subseteq$ LOG(CFL) by reducing the numbers of heads

T. Asano et al. (Eds.): COCOON'99, LNCS 1627, pp. 194–203, 1999.

[Iba73,Gal74] in the multi-head automata characterizing the pushdown classes D(N)AuxPDA-TIME(pol) [Har72]. In the light of the many advancements since, it is natural to wonder whether simpler proofs cannot be given. Indeed, we exhibit simpler and more intuitive proofs here.

In Section 3 we show how a standard nondeterministic pushdown automaton (with no auxiliary work tape) can evaluate a suitably represented semi-unbounded circuit. This establishes $SAC^1 \subseteq LOG(CFL)$ in a direct way. As in [Rei90], where alternating pushdown automata were considered, this direct route from circuits to AuxPDAs exploits the fact that the work of a PDA with an auxiliary work tape can be separated into a logspace transduction followed by a pure pushdown part.

In Section 4 we extend our strategy to circuits built from *multiplex select* gates. These gates (defined precisely in Section 2) have as inputs a sequence of *data* wire bundles and a single *selection* wire bundle whose setting identifies the input data bundle to be passed on as gate output.

Multiplex select gates were introduced to better reflect the difficulty of speeding up parallel computations [Rei97], and they turned up again in an incisive characterization of the power of Dymond and Ruzzo's owner-write PRAMs [FLR96]. Here we observe that a natural notion of *proof tree size* exists for multiplex circuits. Using this notion and further observations, we simplify the harder direction (common to [DR86] and [FLR96]) in the proof that CROW-PRAMs efficiently simulate AuxDPDAs. Coupled with an application of our strategy from Section 3, this implies that LOGDCFL is characterized by polynomial size multiplex circuits having polynomial size proof trees. Proof tree size characterizations are known for LOGCFL [Ruz80,Ven91], but this is the first such characterization of LOGDCFL.

Multiplex select gates are thus intriguing: designing a circuit from such gates, rather than from the more usual boolean gates, forces one to pay closer attention to *routing* aspects which (amazingly) precisely capture the owner-write restriction of a PRAM or the deterministic restriction of an AuxDPDA. So how does this routing aspect mesh with further restrictions on circuit resources? We already know that log depth and poly size yields LOGDCFL. In Section 5, we examine a restriction, called *self-similarity*, which allows log depth poly size circuits to capture the class L. Finally, we point out that bounded-width poly size multiplex circuits characterize NC^1: this follows from the known characterization of NC^1 by bounded-width poly size *boolean* circuits [Bar89], despite the fact that a multiplex gate in a bounded-width circuit can a priori access all the circuit inputs.

2   Preliminaries

We assume basic familiarity with complexity theory such as can be found in recent texts on the subject. We recall that LOG(CFL) (resp. LOG(DCFL)) is the class of languages logspace-reducible to a context-free (resp. deterministic context-free) language. A PDA is a pushdown automaton and an AuxPDA is a machine with a work tape, which alone is subjected to space bounds. along with an auxiliary pushdown. LOGCFL(resp. LOGDCFL) is the class of languages recognized by non-deterministic (resp deterministic) logspace auxiliary pushdown automata with polynomial running time.

Definition [FLR96]. A *multiplex* circuit is a circuit in the usual sense, whose only gates are *multiplex select* gates (see Figure 1). The inputs to a multiplex select gate are grouped into a bundle of $k \in O(\log n)$ steering bits and 2^k equal size bundles of $l \in O(\log n)$ data bits; the gate output is a single bundle of l data bits. One special gate in this circuit is the output gate, whose data bit bundles have size 1. Other special gates are the input gates $1, 2, \ldots, n$, also considered to have an output bundle of size 1. Bit bundles cannot be split or merged, that is, if the output of a gate A is input to a gate B, then the complete output bundle of A enters B as a indivisible and complete (steering or data) bundle; the only exception to this rule is that bundles can be extended by any (logarithmic) number of high order constant bits.

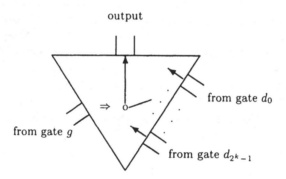

Fig. 1. A multiplex select gate

On input $x_1 x_2 \cdots x_n \in \{0, 1\}^*$, the multiplex circuit input gates respectively take on the values $x_1, x_2, \ldots, x_n$. Then each multiplex select gate having as in Figure 1 a bundle g of k steering bits eventually determines its output value by interpreting its steering bits as the binary encoding of a number j, $0 \le j < 2^k$, and by passing on its jth input bundle as output. The circuit accepts $x_1 x_2 \cdots x_n$ iff the output gate outputs 1.

For the purposes of this abstract, we use logspace uniformity for multiplex circuit families, although tighter uniformities work as well. Hence we say that a multiplex circuit family is uniform iff its direct connection language is logspace-computable, where the direct connection language consists of the sequence of encodings of each gate in the circuit (a gate as in Figure 1 is described by its number, and by the numbers of the gates $g, d_0, \ldots d_{2^k-1}$ combined, as the case may be, with high order constant bits).

The *size* and the *depth* of a multiplex circuit (or family) are defined in the usual way. A multiplex circuit can be unraveled into a tree like any other circuit (by duplicating nodes having outdegree greater than one, all the way down to, and including, the input gates). Given a multiplex circuit C, a *proof tree on input*

x is a pruning of the unraveled circuit which eliminates all the gates except the root of T and, for each non-input gate g kept, both the gate computing the steering bundle of g and the gate computing the data bundle selected by g. The *Prooftreesize* of C is the maximum, over all inputs x of the appropriate size, of the size of the proof tree on input x.

To define width for a multiplex circuit, we follow [Bar89] and assume that the circuit is leveled, i.e. that each gate g, except $1, 2, \ldots, n$, can be assigned a unique level given by the length of any one of its paths to the circuit output gate[1]. The *width* of a leveled multiplex circuit is the maximal number of wires connecting one level to the next. (Wires coming from an input of the circuit are not counted, and this is a standard practice to allow for sub-linear widths) The *gate-width* of a leveled multiplex circuit is the number of non-input gates in its largest level[2]. Observe that the number of circuit input gates accessed by a (gate-width-bounded or width-bounded gate) is not subject to the width bound; in particular, any single gate could even access all the circuit inputs.

Definition: MDepthSize(d, s) is the class of languages recognized by a uniform family of multiplex circuits having depth d and size s [FLR96]. Analogously, MProoftreesize-Size(p, s), MGatewidth-Size(g, s). MWidth-Size(w, s) are defined using proof tree size p. gate-width g and width w. Finally, SMDepthSize(d, s), where the S stands for self-similar, is defined from leveled circuits having the property that all gates across a level have the same sequence of data inputs, meaning that d_j for every $0 \leq j < 2^k$ is identical for every gate across the level.

Accordingly. Prooftreesize-Size(pol,pol) is defined using SAC-circuits, where a prooftree for an input x in the language is an accepting subcircuit (certificate).

3 Simpler proof for LOG(CFL)

Lemma 1. Prooftreesize-Size(pol,pol) $\subseteq$ LOG(CFL).

Proof: We must reduce, in logspace, an arbitrary $Y \in$ Prooftreesize-Size(pol,pol) to a language recognized by a standard NPDA. The main idea is to code each gate and its connections in the SAC circuit for Y in a clever way. Concatenating the descriptions of all the gates yields the encoding of the circuit. To get around the restriction that the NPDA has a one-way input head, we concatenate a polynomial number of copies of the circuit encoding, the polynomial being at least the size of an accepting proof tree in the circuit. This will be the output of the logspace transducer.

The code for a gate g is a tuple with (1) the label of g, (2) the label of g in reverse (this is motivated later), (3) the type of g, and (4) the labels in reverse of the gates that are inputs to g.

---

[1] As in the case of boolean circuits, this assumption can be enforced at the expense of inconsequential increases in size, since any wire bundle can be made to contribute to the depth by replacing this bundle with a gate having this same bundle duplicated as steering, multiple data, and output.

[2] Width is usually defined in the context of bounded fan-in circuits, and no distinction is necessary between width as defined here and gate-width in that case.

Initially, the PDA starts by pushing the label corresponding to the output gate. At any intermediate point, the top of the stack contains a gate (label); we wish to assert that this gate evaluates to a 1. On the input tape, we move the head forward until we are at a gate that matches the gate on top of stack. The semantics of the gate type now takes over. If it is an input gate, the PDA checks if its value is 1 and continues with the remaining gates (if any) on the stack; if its value is 0 the PDA rejects and halts. If the gate is an AND gate, the PDA places the two inputs to the AND gate on the stack, but since their labels are reversed on the input tape, they are pushed in the correct order onto the stack for future comparisons. If the gate is an OR gate, one of its inputs gates is non-deterministically pushed onto the stack.

To finish, we need to explain the necessity of component (2), in the code for a gate. In the previous paragraph, we did not explain how the PDA matches the gate on the stack with one on the input tape. Of course, we can do this with non-determinism: we can simply move the input head non-deterministically until it is magically under the right gate label. But we would like to do this deterministically as we use the idea for a later simulation. (And we all know that non-deterministic bits are costly!) So we check if the gate under the input head matches the gate on top of stack by popping identical bits. If the gates are different, we will notice it at some point. Let the input head see $pbq, q^R bp^R, \ldots$ and the PDA have $pc\ldots$ on its stack (where c is the complement of b). The machine will now match p and detect a difference on bit b. Skip this bit on the input tape and push the remaining bits of the label onto the stack; the input tape now looks like $q^R bp^R, \ldots$ and the stack looks like $q^R c\ldots$. Now match the input head and the stack until there is a difference, at which point we simply push the remaining bits of the label onto the stack to recreate the original gate label on the stack.

The language $L_{polproof}$, which is accepted by the PDA is contextfree, where we need not care about words which do not encode a circuit. Thus every language in Prooftreesize-Size(pol,pol) can be reduced to $L_{polproof}$ by a logspace transducer from input x using the uniformity for $C_{|x|}$.   □

Corollary 1. $SAC^1 \subseteq LOG(CFL)$.

4   New characterization of LOGDCFL

Consider a proof tree in a multiplex circuit. For a gate in a proof tree only two child gates have to be in the proof tree: the gate where the steering bundle comes from, and the gate where the selected bundle comes from. Thus we get the following:

Lemma 2. MDepth-Size(log,pol)⊆MProoftreesize-Size(pol,pol).

In the spirit of simplifying proofs, we can bypass the difficult direction in [DR86,FLR96] showing DAuxPDA-TIME(pol) ⊆ CROWTI-PR(log,pol) by the following naive algorithm simulating a DAuxPDA with only polynomial proof tree size. The function works on extended surface configurations containing the state, the top of the pushdown, the contents of the work tape, the position of the head on the input tape, the time and the current height of the pushdown. We may assume w.l.o.g. that every configuration either pushes or pops.

> **function** GO(x) :=
>   **if** Configuration x pushes Push(x) $\neq \lambda$
>   **if** going to Configuration y
>   **then** z :=GO(y); **return** GO(POP(z,Push(x)))
>   **else return** x

POP(z,a) is the configuration following z if z is a popping configuration and a is the symbol, which is popped from the pushdown.

The function outputs the last reachable surface configuration reached from x, without popping deeper symbols from the pushdown. W.l.o.g. GO(x_0) is the end configuration. The translation of this function to a multiplex-select circuit yields

Lemma 3. DAuxPDA-TIME(pol) $\subseteq$ MProoftreesize-Size(pol,pol)

Proof: A circuit calculating GO(x) using subcircuits for GO(y) and GO(POP(z, a)) is partially depicted on Figure 2. Each extended surface configuration and therefore each gate appears only once in a proof tree, which thus has polynomial size. $\square$

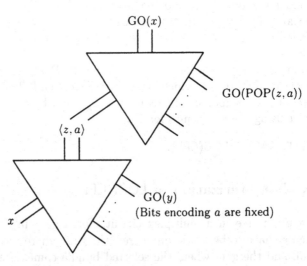

Fig. 2. Circuit used in proving Lemma 3

Lemma 4. MProoftreesize-Size(pol, pol) $\subseteq$ LOG(DCFL).

Proof: As in the proof of Lemma 1, we encode for every j-th data input of a gate g

$$w_j^g := (bin_j \# g\$ g^R \# bin_j^R \# f_j^R \# d_j^R \#)$$

where bin_j is the binary encoding of j, f_j is the fixed part of the j-th data input and d_j is the binary encoding of the gate to which the j-th data input bundle is connected (d_j is optional). For every gate g having a steering input from gate s_g, we encode

$$w_g := (g\$g^R\#s_g^R)w_1^g w_2^g...w_k^g$$

and let w_V be the concatenation of all w_g with $g \in V$. Let furthermore s be the proof tree size of the circuit. The logspace transducer generates $w := w_V^{2s}$ (which has polynomial length).

To evaluate a gate g, a pushdown automaton uses the same method as in Lemma 1 to find the right $(g\$g^R\#s_g^R)$ and pushes $g^R\#s_g^R$, which allows to evaluate the gate s, where the steering input came from and return with the value, which now becomes bin_j over g, which is still on the pushdown. Again the same method allows to find the right w_j^g, where $f_j^R\$d_j^R$ is pushed, which allows to evaluate the gate d_j and regard (because of the distinction between $\$$ and $\#$) f_j concatenated with the returned value as the new value, which is to return.   □

5   Multiplex circuits for L and for NC¹

Theorem 1. L = MGatewidth-Size$(1, pol)$ = MWidth-Size(log, pol).

Proof: Since the output-bundle of a gate consists of a logarithmic number of wires, it follows from the definition that MGatewidth-Size$(1, pol) \subseteq$ MWidth-Size(log, pol).

We get MWidth-Size$(log, pol) \subseteq$ L by evaluating a circuit level by level in polynomial time. Since only the outputs of the previous level have to be stored, logarithmic space is sufficient.

To show L $\subseteq$ MGatewidth-Size$(1, pol)$ we have to construct for a logspace machine a multiplex-select circuit, which consists of a single path of multiplex select gates, where the output bundle of a gate is connected to the steering input of the following gate. The value v of such a bundle encodes a surface configuration containing the state, the contents of the work tape, the position of the head on the input tape and on the lowest order bit the symbol (0 or 1) on the input tape at the current head position. The first gate gets on the steering input the start configuration. The v-th data input bundle of every (except the last) gate contains an encoding of the following configurations on v where the lowest order bit is connected to the corresponding bit on the input and all the other bits are given as constants[3]. The last gate has 1's on the data inputs whose numbers encode accepting configurations and 0's on the others.   □

Theorem 2. L = SMDepth-Size(log, pol).

---

[3] This means that the uniformity condition for the circuit has to be able to verify a single step of a logspace machine; the number of the input bit used is contained as head position in the encoding of the configuration.

Proof: $\subseteq$: Let l be the rounded up logarithm of the running time and k be the number of surface configurations. The first level consists of k gates of the same kind as in the proof of theorem 1 having the number $(c, 1)$ and getting the configuration c as (a constant) steering input. Thus their output will be the following configuration of c. The gates having the number (c, h) on level h with $1 < h \leq l$ get the output of gate $(c, h - 1)$ as steering input and the output of gate $(c', h - 1)$ as the c'-th data input, which means that the data input bundles on the h-th level are always connected the same way. By induction on h it follows that the output of gate number (c, h) will be the 2^h-th configuration following c. Thus the output of gate number (c_0, l) contains the end configuration [4] and is connected to the last gate, which, as in the proof of theorem 1 has 1's on the data inputs whose numbers encode accepting configurations and 0's on the others.

$\supseteq$: A logspace machine can evaluate a self-similar circuit in the following way: To evaluate a gate, it first recursively evaluates the gate where the steering input comes from; here it does not store the number of the gate but only the fact (one bit) that it is evaluating the steering input for a gate of that level. When it returns the value v for this steering input, it recursively evaluates the gate where the v-th data input comes from; because of self-similarity this is the same for all gates on this level thus the number of the exact gate is not necessary. Again only the fact (one bit) that it is evaluating a data input is stored for that level, telling the machine to return the value to the next level, if the value from the previous level is returned. The machine accepts if the evaluation of the last gate is 1 and rejects if it is 0. Since there is a logarithmic number of levels and the value always has a logarithmic number of bits, logarithmic space is sufficient. $\square$

One difference between bounded-width (bounded-fan-in) boolean circuits and bounded-width multiplex circuits is that any gate in the latter can access all the circuit inputs. However, this extra ability of multiplex gates is useless within a bounded-width circuit:

Theorem 3. MWidth-Size(constant,pol) $= NC^1$.

Proof: $\subseteq$: Consider a multiplex circuit C of bounded width w. The width bound implies that the steering input bundle to any gate in C can only take a constant number of different values. Even if there were n circuit inputs attached to the data portion of this gate, only a constant number of them can be selected. These selectable inputs can be identified at "compile time", and those not selectable can be set to constants (while preserving the circuit uniformity). The result is an equivalent multiplex circuit using at most a constant number of circuit inputs per level. Each gate in this circuit has constant fan-in, omitting constant bits. Now any constant fan-in gate (thus also a multiplex select gate) can be replaced by a constant size NC subcircuit. Replacing all multiplex gates in this way yields a bounded-width polynomial size boolean circuit in the sense of [Bar89, Section 5], who shows that such circuits characterize NC^1.

---

[4] W.l.o.g. the machine stays in its end configuration.

$\supseteq$: A binary multiplex select gate[5] (see Figure 1) simulates a NOT-gate if d_0 has the constant value 1 and d_1 has the constant value 0, it simulates an AND-gate if d_0 has the constant value 0, and it simulates an OR-gate if d_1 has the constant value 1. Thus any NC circuit is simulated within the same resource bounds by a multiplex circuit. In particular, the bounded-width polynomial size circuits of [Bar89] are simulated by bounded-width polynomial size multiplex circuits.   □

6   Conclusion

We have simplified proofs involving LOGCFL and LOGDCFL, and obtained new characterizations of LOGDCFL, L, and NC^1. We would like to characterize NL in a similar way, by restricting SAC^1 circuits. Now we observe that the SAC^1 circuits obtained in a canonical way by encoding the NL-complete reachability problem have the property that they are robust against a simplified NL evaluation algorithm, which would of course make mistakes when evaluating general SAC^1-circuits. An open problem is whether this robustness property has a nice[6] characterization.

Acknowledgment. The first author acknowledges useful discussions with Markus Holzer.

References

AJMV98. E. Allender, J. Jiao, M. Mahajan and V. Vinay, Non-commutative arithmetic circuits: depth reduction and size lower bounds, *Theoret. Comput. Sci.* 209, pp. 47–86, 1998.

Bar89.   D.A. Barrington, Bounded-width polynomial-size branching programs recognize exactly those languages in NC^1, *Journal of Computer and System Sciences*, 38(1):150–164, 1989.

BCD+88. A. Borodin, S. A. Cook, P. W. Dymond, W. L. Ruzzo, and M. Tompa. Two applications of complementation via inductive counting. In *Proc. of the 3rd IEEE Symp. on Structure in Complexity*, 1988.

BLM93.   F. Bédard, F. Lemieux, and P. McKenzie. Extensions to barrington's M-program model. *Theoret. Comput. Sci. A*, 107(1):33–61, 1993.

Coo79.   S. Cook, Deterministic CFL's are accepted simultaneously in polynomial time and log squared space, Proc. of the 11th ACM Symp. on Theory of Computing, 338–345, 1979.

DR86.   P. W. Dymond and W. L. Ruzzo. Parallel RAMs with owned global memory and deterministic language recognition. In *Proc. of the 13th Int. Conf. on Automata, Languages, and Programming*, LNCS226, pages 95–104. Springer-Verlag, 1986.

FLR96.   H. Fernau, K.-J. Lange, and K. Reinhardt. Advocating ownership. In V. Chandru, V. Vinay, editor, *Proc. of the 16th Conf. on Foundations of Software Technology and Theoret. Comp. Sci.*, vol. 1180 of *LNCS*, pages 286–297, Springer, 1996.

---

[5] having only 1 steering bit and 2 data bits
[6] by a condition which is in some way 'locally checkable'.

Gal74.   Z. Galil. Two–way deterministic pushdown automata and some open prob-
            lems in the theory of computation. In *Proc. 15th IEEE Symp. on Switching
            and Automata Theory*, pages 170–177, 1974.

Har72.   J. Hartmanis. On non-determinancy in simple computing devices. *Acta
            Informatica*, 1:336–344, 1972.

Iba73.   O. H. Ibarra. On two way multi-head automata. *Journal of Computer and
            System Sciences*, 7(1):28–36, 1973.

LMSV99.   C. Lautemann, P. McKenzie, H. Vollmer and T. Schwentick, The descriptive
            complexity approach to LOGCFL, *Proc. of the 16th Symp. on the Theoret.
            Aspects of Comp. Sci.*, 1999.

MRS93.   B. Monien, W. Rytter, and H. Schäpers. Fast recognition of deterministic
            cfl's with a smaller number of processors. *Theoret. Comput. Sci.*, 116:421–
            429, 1993. Corrigendum, 123:427, 1993.

Rei90.   K. Reinhardt. Hierarchies over the context-free languages. In J. Dassow and
            J. Kelemen, editors, *Proceedings of the 6th International Meeting of Young
            Computer Scientists*, number 464 in LNCS, pages 214–224. Springer-Verlag,
            1990.

Rei97.   K. Reinhardt. Strict sequential P–completeness. In R. Reischuk, editor,
            Proceedings of the 14th STACS, number 1200 in LNCS, pages 329-338,
            Lübeck, February 1997.

Ruz80.   W. Ruzzo, Tree-size bounded alternation, *Journal of Computer and System
            Sciences*, 21:218–235, 1980.

Sud78.   I. H. Sudborough. On the tape complexity of deterministic context-free
            languages. *Journal of the ACM*, 25:405–414, 1978.

Ven91.   H. Venkateswaran. Properties that characterize LOGCFL. *Journal of Com-
            puter and System Sciences*, 43:380-404, 1991.

Vin96.   V. Vinay. Hierarchies of circuit classes that are closed under complement.
            In *Proc. of the 11th IEEE Conf. on Computational Complexity (CCC-96)*,
            pp. 108–117, 1996.

On the Negation-Limited Circuit Complexity of Merging

Kazuyuki Amano[1], Akira Maruoka[1], and Jun Tarui[2]

[1] Graduate School of Information Sciences, Tohoku University
Sendai 980–8579, Japan
{ama|maruoka}@ecei.tohoku.ac.jp
[2] Department of Communications and Systems
University of Electro-Communications
Chofu, Tokyo 182–8585, Japan
jun@sw.cas.uec.ac.jp

Abstract. A negation-limited circuit is a combinational circuit that consists of AND, OR gates and a limited number of NOT gates. In this paper, we investigate the complexity of negation-limited circuits. The (n, n) merging function is a function that merges two presorted binary sequences $x_1 \leq \cdots \leq x_n$ and $y_1 \leq \cdots \leq y_n$ into a sequence $z_1 \leq \cdots \leq z_{2n}$. We prove that the size complexity of the (n, n) merging function with $t = (\log_2 \log_2 n - a)$ NOT gates is $\Theta(2^a n)$.

1 Introduction

To derive a strong lower bound on the size of a combinational circuit, that consists of AND, OR and NOT gates, for an explicitly defined function is one of the most challenging problems in theoretical computer science. But so far, only linear lower bounds have been known (See [7] for an excellent survey on the complexity of Boolean functions). In sharp contrast, there has been substantial progress in obtaining strong lower bounds on the size of a monotone circuit, that consists of AND and OR gates. Razborov[13] obtained superpolynomial lower bounds on the size of a monotone circuit for the clique function. Shortly thereafter, Andreev[1] and Alon and Boppana[3] obtained exponential lower bounds on the size of a monotone circuit for the clique function and for some other functions. In short, we have strong lower bounds on circuit size when we restrict ourselves to circuits without NOT gates, whereas we have very weak lower bounds when we consider circuits with an arbitrary number of NOT gates. So it is natural to consider the size of a circuit with a limited number of NOT gates, which we call the negation-limited circuit complexity.

There has been obtained many interesting results in this line of research. Markov[10] proved that $r = \lceil \log_2(n + 1) \rceil$ NOT gates are sufficient to compute any collection of Boolean functions on n variables and that there exists a function that requires r NOT gates to compute. Fisher[8] showed that restricting the number of available NOT gates to r entails only a polynomial blowup in circuit size. Beals, Nishino and Tanaka[6] strengthen this result by showing that, for

T. Asano et al. (Eds.): COCOON'99, LNCS 1627, pp. 204–209, 1999.
© Springer-Verlag Berlin Heidelberg 1999

any Boolean function f, the size of a smallest circuit with at most r NOT gates computing f is at most $2\text{size}(f) + O(n \log n)$, where $\text{size}(f)$ is the size of a smallest circuit with an arbitrary number of NOT gates computing f. Recently, Amano and Maruoka[4] proved that the clique function is not computable by a polynomial size circuit with at most $(1/6) \log \log n$ NOT gates.

Tardos[11] proved that there exists a function that can be computed by a circuit of polynomial size whereas it requires monotone circuits of exponential size. In contrast, Berkowitz[5] and Beals, Nishino and Tanaka[6] proved that NOT gates are almost powerless for slice functions. A function f is a *slice* function if $f(x_1, \ldots, x_n) = 0$ for any $\sum_{i=1}^{n} x_i < k$ and $f(x_1, \ldots, x_n) = 1$ for any $\sum_{i=1}^{n} x_i > k$ for some $0 \le k \le n$ (for input contains exactly k ones, the output of f is not constrained). They proved that for any slice function f, the monotone circuit complexity of f is at most $2\text{size}(f) + O(n \log n)$ where $\text{size}(f)$ is the circuit complexity of f.

The effect of NOT gates on circuit complexity looks like a mystery and better understanding of it is necessary to obtain a good lower bound for circuit complexity.

In this paper, we consider the problem : *how fast a complexity of a function deceases when the number of available NOT gates in a circuit increases?* That is one of the most interesting problems on negation-limited circuit complexity.

The merging function is a Boolean function that takes two presorted binary sequences each of length n as inputs, and merges them into a sorted sequence of length $2n$. It was shown that the merging function has circuit complexity $\Theta(n)$ and has monotone circuit complexity $\Theta(n \log n)$ (See e.g., [7], [9, Chap.5.3.4]). In this paper, we prove the nontrivial lower and upper bounds on the negation-limited circuit complexity of merging. Remarkably, both bounds we obtained are matched up to a constant factor for *any* values of the number of available NOT gates. Precisely, we prove

Theorem 1 (Main)   The size complexity for merging two presorted binary sequences each of length n with $t = (\log_2 \log_2 n - a)$ NOT gates is $\Theta(2^a n)$.

Roughly speaking, this theorem says that the size of a smallest circuit for the merging function is halved when the number of available NOT gates increases one, and if we can use $\log_2 \log_2 n$ NOT gates then the merging function can be computed by a circuit of linear size. In addition, the proof of the upper bound is constructive, so, for any number t of available NOT gates, we give a circuit for merging with t NOT gates whose size is optimal up to a constant factor.

The paper is organized as follows: First, in Section 2 we give some basic notations and definitions. In Section 3, we prove the lower bound of our main theorem by using the technique of partial assignment to the input variables. In Section 4, we prove the upper bound of our main theorem by giving the construction of a circuit for merging. Finally, we provide some open problems related our work in Section 5.

2 Preliminaries

A *circuit* is a combinational circuit that consists of AND gates of fan-in two, OR gates of fan-in two and NOT gates. In particular, a circuit without NOT gates is called *monotone*. A Boolean function f is monotone if $f(x_1, \ldots, x_n) \leq f(y_1, \ldots, y_n)$ whenever $x_i \leq y_i$ for all i. Monotone functions are exactly those computed by monotone circuits. The *size* of a circuit C is the number of gates in the circuit C.

Let F be a collection of Boolean functions $f_1, \ldots, f_m$. The *circuit complexity* (*monotone circuit complexity*, resp.) of F is the size of a smallest circuit (monotone circuit, resp.) that computes F, i.e., that outputs $f_1(x_1, \ldots, x_n), \ldots, f_m(x_1, \ldots, x_n)$ for the inputs $x_1, \ldots, x_n$. Following Beals, Nishino and Tanaka[6], we call a circuit including at most t NOT gates a *t-circuit*.

The (n, n) *merging function*, denoted by $\mathrm{MERGE}(n, n)$, is a collection of functions that merges two presorted binary sequences $x_1 \leq x_2 \leq \cdots \leq x_n$ and $y_1 \leq y_2 \leq \cdots \leq y_n$ into a sequence $z_1 \leq z_2 \leq \cdots \leq z_{2n}$, i.e., $z_i = 1$ if and only if the total number of 1's in the input sequences is not less than $2n - i + 1$, for all i. All logarithms in this paper are base two.

Now we restate our main theorem.

Theorem 1 (Main)   The size complexity of $\mathrm{MERGE}(n, n)$ with $t = (\log_2 \log_2 n - a)$ NOT gates is $\Theta(2^a n)$.

3 Lower Bound

In this section, we prove the lower bound of Theorem 1, that is, the size of any t-circuit, where $t = (\log_2 \log_2 n - a)$, computing $\mathrm{MERGE}(n, n)$ is $\Omega(2^a n)$.

Proof. The outline of the proof is as follows: First, we prove that if C is a t-circuit for $\mathrm{MERGE}(n, n)$ then C has a monotone subcircuit which computes $\mathrm{MERGE}(n/2^t, n/2^t)$. Combining this and the fact that the monotone circuit complexity of $\mathrm{MERGE}(n/2^t, n/2^t)$ is $\Omega((n/2^t) \log(n/2^t)) = \Omega(2^a n)$ to obtain the lower bound for the size of C.

Let C be an optimal t-circuit for $\mathrm{MERGE}(n, n)$. Without loss of generality, we may assume that $n = 2^r$ for some positive integer r. Let g_i, for $1 \leq i \leq t$, be the function computed by the input of i-th NOT gates counting from the inputs to the outputs. A *minterm* of a function f is a minimal assignment to some subset of the variables which determined f to be 1.

Let $X = \{x_1, \ldots, x_n\}$ and $Y = \{y_1, \ldots, y_n\}$ be the sets of input variables of $\mathrm{MERGE}(n, n)$. Let X^h (respectively, Y^h) be the second half of X (respectively, the second half of Y), i.e., $X^h = \{x_{n/2+1}, \ldots, x_n\}$ and $Y^h = \{y_{n/2+1}, \ldots, y_n\}$.

Now we define a partial assignment p_1 to $X \cup Y$ that makes g_1 a constant function. Note that g_1 is a monotone since g_1 is the input of the lowest NOT gates in C. If g_1 has a minterm such that all the variables appeared in it are contained in $X^h \cup Y^h$, then let p_1 be the partial assignment that assigns the value 1 to all the variables in $X^h \cup Y^h$. Otherwise, if g_1 does not have such minterms, then let p_1 be the partial assignment that assigns the value 0 to all the variables in $(X - X^h) \cup (Y - Y^h)$. For a Boolean function f and a partial assignment p, we denote the induced function of f under p by $f|_p$. It is clear that $g_1|_{p_1}$ is the constant 1 for the first case and is the constant 0 for the second case. It is also clear that $g_2|_{p_1}$ is a monotone.

Next, we define a partial assignment p_2 from p_1 by adding the assignments to half of the variables unassigned by p_1. Let $X_1 \subseteq X$ and $Y_1 \subseteq Y$ be the sets of all variables unassigned by p_1. Let X_1^h be the second half of X_1 and Y_1^h be the second half of Y_1. Note that $|X_1| = |Y_1| = n/2$ and $|X_1^h| = |Y_1^h| = n/4$. If $g_2|_{p_1}$ has a minterm such that all the variables appeared in it are contained in $X_1^h \cup Y_1^h$, then let p_2 be the partial assignment obtained from p_1 by adding the assignments of the value 1 to all the variables in $X_1^h \cup Y_1^h$. Otherwise, let p_2 be the partial assignment obtained from p_1 by adding the assignments of the value 0 to all the variables in $(X_1 - X_1^h) \cup (Y_1 - Y_1^h)$. It is clear that both $g_1|_{p_2}$ and $g_2|_{p_2}$ are constant functions and $g_3|_{p_2}$ is a monotone.

By continuing above procedure recursively for all the NOT gates in C, we obtain the partial assignment p_t such that (i) p_t leaves two sets of $n/2^t$ consecutive variables $\{x_{i+1}, x_{i+2}, \ldots, x_{i+n/2^t}\}$ in X and $\{y_{i+1}, y_{i+2}, \ldots, y_{i+n/2^t}\}$ in Y, for some i, (ii) p_t assigns the value 0 to $x_1, \ldots, x_i, y_1, \ldots, y_i$ and the value 1 to $x_{i+n/2^t+1}, \ldots, x_n, y_{i+n/2^t+1}, \ldots, y_n$, and (iii) p_t makes the inputs of all the NOT gates in C a constant.

Under p_t we obtain the induced circuit of C by eliminating all gates whose values become determined. By (i) and (ii), the sequence of the outputs $z_{2i+1}, \ldots, z_{2i+2n/2^t}$ of the induced circuit is equal to the merged sequence of $x_{i+1}, \ldots, x_{i+n/2^t}$ and $y_{i+1}, \ldots, y_{i+n/2^t}$. By (iii), the induced circuit has no NOT gates.

Thus the circuit C "contains" a monotone circuit for MERGE($n/2^t, n/2^t$) and the size of C is lower bounded by the monotone circuit complexity of MERGE($n/2^t, n/2^t$). Since the monotone circuit complexity of MERGE(N, N) is known to be $\Omega(N \log N)$ (See e.g., [7],[9]), the lower bound is $\Omega((n/2^t) \log(n/2^t)) = \Omega((n/2^t) \log n) = \Omega(2^a n)$ as desired. $\square$

4   Upper Bound

In this section, we prove the upper bound of Theorem 1, that is, there exists a t-circuit, where $t = (\log_2 \log_2 n - a)$, computing MERGE($n, n$) of size $O(2^a n)$.

Proof.   We describe below how to construct a circuit of size $O(n)$ with log log n NOT gates for bitonic sorting. We say that a binary sequence $z_1, \ldots, z_{2n}$ is *bitonic* if $z_1 \geq \cdots \geq z_k \leq \cdots \leq z_{2n}$ for some k, $1 \leq k \leq 2n$. A circuit that sorts for bitonic sequences can obviously compute MERGE(n, n) since a sequence $x_n, \ldots, x_1, y_1, \ldots, y_n$ is bitonic when $x_1 \leq \cdots \leq x_n$ and $y_1 \leq \cdots \leq y_n$. When only (log log $n - a$) NOT gates are available, exactly the same construction yields a circuit of size $O(2^a n)$.

Let $x_1, \ldots, x_{2n}$ be a bitonic sequence. The first layer of the standard bitonic sorter (with usual min/max gates expressed as AND/OR gates) produces the smaller half $y_1 = x_1 \wedge x_{n+1}, y_2 = x_2 \wedge x_{n+2}, \cdots, y_n = x_n \wedge x_{2n}$ and the larger half $z_1 = x_1 \vee x_{n+1}, \cdots, z_n = x_n \vee x_{2n}$. (One of y's or z's is an all 0 or all 1 sequence, and the other is a bitonic sequence.)

Let N be the number of 1's among x_i's. Now, put $b = y_1 \vee y_2 \vee \cdots \vee y_n$. Then, $b = 1$ if and only if N is equal to or greater than $n + 1$. Put $c = \neg(b)$, and put $w_1 = c \wedge z_1, \cdots, w_n = c \wedge z_n$, and $u_1 = w_1 \vee y_1, \cdots, u_n = w_n \vee y_n$.

If $b = 1$, all z's are 1's and since $c = 0$, all w's are set to 0, and u's $= y$'s. If $b = 0$, w's equal z's, and since all y's are 0's, u's $= z$'s.

Recursively construct a circuit for a bitonic sequence u of length n, repeating this procedure $t = \log \log n$ times. We obtain the most significant t bits b's of the binary representation of $N - 1$ (if $N = 0$, then all b's are 0's), and are left with the bitonic sequence u-last of length $2n/\log n$. We can sort u-last by AKS-sorting network[2] of size $O(n)$ into the sorted list S.

From t bits b's, their negation c's and the list S, a monotone circuit of linear size described below can produce the sorting of the original x's : Let us denote b's by $b_1^1, \ldots, b_t^1$, where b_i^1 is the i-th significant bit of b's and let $b_1^0, \ldots, b_t^0$ be their negations. Note that we have already computed b_i^0's as c's. For $0 \le i \le \log n$, put $l_i = \vee_{k=i}^{\log n - 1}(b_1^{k_1} \wedge \cdots \wedge b_t^{k_t})$, where k_j is the j-th significant bit of the binary representation of k. Note that, for $0 < i < \log n$, $l_i = 1$ if and only if N is greater than $i(2n/\log n)$, and that l_0 is always 1 and $l_{\log n}$ is always 0. Finally, for $1 \le i \le 2n$, output $l_{\lceil i/(2n/\log n) \rceil} \vee$ (the $\{(i-1) \bmod (2n/\log n)+1\}$-st largest bit of S) as the i-th largest bit of the sorted list of x's.

It is easy to check that the size of the entire circuit constructed above is $\Theta(n)$. This completes the proof. $\square$

5   Concluding Remarks and Open Problems

In the current paper, we established the asymptotic growth rate of the size of a smallest circuit with t negation gates for merging up to a constant factor, for any values of t. Actually, same result holds for bitonic sorting, which subsumes merging. It is an interesting open problem that whether we can extend our technique to obtain the negation-limited circuit complexity for (general) sorting.

The depth of our circuit for merging is $O((t + 1) \log n)$ when the number of available NOT gates is t. In particular, we obtained the linear size circuit with $\log \log n$ NOT gates of depth $O(\log n \log \log n)$. It is obvious that any circuit for merging has depth $\Omega(\log n)$. To see if there exists a linear size circuit of depth $O(\log n)$ with $\log \log n$ NOT gates for merging is an also interesting open problem.

References

1. A.E. Andreev, "On a Method for Obtaining Lower Bounds for the Complexity of Individual Monotone Functions", *Soviet Math. Dokl.*, Vol. 31, No. 3, pp. 530–534, 1985.
2. M. Ajtai, J. Komlós and E. Szmerédi, "An $O(n \log n)$ Sorting Network", *Combinatorica*, Vol. 3, No. 1, pp. 1–19, 1983.
3. N. Alon and R.B. Boppana, "The Monotone Circuit Complexity of Boolean Functions", *Combinatorica*, Vol. 7, No. 1, pp. 1–22, 1987.

4. K. Amano and A. Maruoka, "A Superpolynomial Lower Bound for a Circuit Computing the Clique Function with $(1/6)\log\log n$ Negation Gates", *Proc. 23rd MFCS*, *LNCS*, No. 1450, pp. 399–408, 1998.

5. S.J. Berkowitz, "On Some Relationships between Monotone and Non-monotone Circuit Complexity", *Tech. Rep. Univ. of Tronto*, 1982.

6. R. Beals, T. Nishino and K. Tanaka, "More on the Complexity of Negation-Limited Circuits", *Proc. 27th STOC*, pp. 585–595, 1995.

7. R.B. Boppana and M. Sipser, "The Complexity of Finite Functions", *Handbook of Theoretical Computer Science*, pp. 757–804, Elsevier Science, 1990.

8. M.J. Fischer, "The Complexity of Negation-Limited Networks–A Brief Survey", *LNCS*, No. 33, pp. 71–82, 1974.

9. D.E. Knuth, *"The Art of Computer Programming Vol. 3 : Sorting and Searching (Second Edition)"*, Addison Wesley, 1998.

10. A.A. Markov, "On the Inversion Complexity of a System of Functions", *J. ACM*, Vol. 5, pp. 331–334, 1958.

11. É. Tardos, "The Gap Between Monotone and Non-Monotone Circuit Complexity is Exponential", *Combinatorica*, Vol. 7, No. 4, pp. 141–142, 1987.

12. K. Tanaka and T. Nishino, "On the Complexity of Negation-Limited Boolean Networks", *Proc. 26th STOC*, pp. 38–47, 1994.

13. A.A. Razborov, "On the Method of Approximations", *Proc. 21st STOC*, pp. 167–176, 1989.

Super-Polynomial Versus Half-Exponential Circuit Size in the Exponential Hierarchy

Peter Bro Miltersen[1]*, N.V. Vinodchandran[1]**, and Osamu Watanabe[2]***

[1] BRICS, Department of Computer Science, University of Aarhus.
bromille,vinod@brics.dk
[2] Department of Mathematical and Computing Sciences, Tokyo Institute of
Technology. Meguro-ku Ookayama, Tokyo 152-8552. watanabe@is.titech.ac.jp

Abstract. Lower bounds on circuit size were previously established for functions in Σ_2^p, ZPP^{NP}, Σ_2^{exp}, $ZPEXP^{NP}$ and MA_{exp}. We investigate the general question: Given a time bound $f(n)$. What is the best circuit size lower bound that can be shown for the classes MA-TIME$[f]$, ZP-TIME$^{NP}[f]$,... using the techniques currently known? For the classes MA_{exp}, $ZPEXP^{NP}$ and Σ_2^{exp}, the answer we get is "half-exponential". Informally, a function f is said to be half-exponential if f composed with itself is exponential.

1 Introduction

One of the main issues of complexity theory is to investigate how powerful non-uniform (e.g. circuit based) computation is, compared to uniform (machine based) computation. In particular, a 64K dollar question is whether exponential time has polynomial size circuits. This being a challenging open question, a series of papers have looked at circuit size of functions further up the exponential hierarchy. In the early eighties, Kannan [10] established that there are languages in $\Sigma_2^{exp} \cap \Pi_2^{exp}$ which do not have circuits of polynomial size. Later, using methods from learning theory, in particular Bshouty et al [6], Köbler and Watanabe [12] improved Kannan's result and showed that in fact $ZPEXP^{NP}$ contains languages which do not have polynomial size circuits. Recently, Buhrman, Fortnow and Thierauf [7] improved it further to show that even the exponential version of the interactive class MA does not have polynomial size circuits. More precisely, they showed that $MA_{exp} \cap coMA_{exp} \not\subseteq P/poly$. This is an improvement since it follows from the result $MA \subseteq ZPP^{NP}$ (due to Arvind and Köbler [1], and independently, to Goldreich and Zuckerman [8]) using padding that $MA_{exp} \subseteq ZPEXP^{NP}$. In contrast to the previous results, their proof uses non-relativizable techniques.

* Supported by the ESPRIT Long Term Research Programme of the EU under project number 20244 (ALCOM-IT)
** Part of N.V. Vinodchandran's work was done while he was a Junior Research Fellow at the Institute of Mathematical Sciences, Chennai 600 113, India.
*** Supported in part by JSPS/NSF International Collaboration Grant.

T. Asano et al. (Eds.): COCOON'99, LNCS 1627, pp. 210–220, 1999.
© Springer-Verlag Berlin Heidelberg 1999

Some of the lower bounds mentioned above have analogous versions in the polynomial hierarchy. For instance Kannan showed that for any k, $\Sigma_2^p \cap \Pi_2^p$ contains a language that does not have circuits of size n^k. Köbler and Watanabe showed the same statement with $\Sigma_2^p \cap \Pi_2^p$ replaced with ZPP^{NP}. However, interestingly, it is not known if MA has linear sized circuits. So there the analogy seemingly stops. It seems interesting to understand the scenario better, so we consider the following question. *What is the smallest time bound for which* MA-TIME[f] *is known not to have linear sized circuits?* Also, it is natural to ask how far the techniques can be extended if one is not happy with just super-polynomial bounds. For example, do the techniques of Kannan in fact show that there are languages in $\Sigma_2^{\exp}$ that require exponential size circuits? In general, we can ask the following question: *Given a time bound f. What is the best circuit size lower bound we can show for functions in* MA-TIME[f], ZP-TIME^{NP}[f], Σ_2^f? One of the main objectives of this paper is to investigate these questions.

The complexity (in the uniform setting) of functions with *exponential* circuit complexity is of particular interest (note that a random function will have exponential circuit complexity). In this direction, the only previous work seems to be done by Kannan [10]. Indeed, Kannan shows that the third level of the exponential hierarchy, and in fact, $\Sigma_3^e \cap \Pi_3^e$, contains a function with *maximum* circuit size. As was shown by Shannon and Lupanov, this happens to be $\Theta(2^n/n)$.

In [10], the author also makes claim to the following statement (*) [10, Theorem 4, page 48], apparently answering one of the questions we asked in the beginning of this paper:

(*) *There is a universal constant l such that for any time-constructible function $f(\cdot)$ satisfying $n^l \leq f(n) \leq 2^{n/20} \forall n$, there is a language in $\Sigma_2^{f(n)} \cap \Pi_2^{f(n)}$ that does not have $O((f(n))^{1/l})$-size circuits.*

In particular, statement (*) implies that the second level of the exponential hierarchy does not have $2^{o(n)}$ sized circuits.

Though statement (*) is very likely to be a true, it is not clear to us that it was, in fact, given a proof in [10]. Nor do we know how to prove it. We suggest that the statement is reopened and considered an open problem. We analyse this issue in Section 3. In this context, we note that the bound of $\Sigma_3^e \cap \Pi_3^e$ for functions requiring maximum size, can be improved to Δ_3^e by using a binary search approach. This may be a folklore, but does not seem to be explicitly mentioned anywhere in the literature. Also, we need this improvement for some of the statements we prove later.

2 Notations, definitions and results

We assume the definitons of standard compelxity classes. Please refer to [5,14] for these and other standard complexity-theoretic definitions.

Let $f(n) \geq n$ be a time constructible function. Then the complexity classes of interest to us (with time bound f) are TIME[f], Σ_k^f, Π_k^f, Δ_k^f, ZP-TIME^{NP}[f]

(class of languages recognized by a randomized Turing machine with an NP-oracle, running in *expected* time $f(n)$ and always returning the right answer) and MA-TIME[f] (class of languages recognized by a Merlin-Arthur game, where the length of Merlin's proof and the time of Arthur's computation is bounded by $f(n)$). For any class of time bounds $\mathcal{F}$ and any of the the above-mentioned complexity class $\mathcal{C}$, we denote the class $\cup_{f \in \mathcal{F}} \mathcal{C}$ by $\mathcal{C}[\mathcal{F}]$. The polynomial and exponential versions of these classes are of special interest. For polynomial versions the notations are standard. The notations for the exponential versions that we use are EXP, $\Sigma_k^{\exp}$, $\Pi_k^{\exp}$, $\Delta_k^{\exp}$, ZPEXP^{NP} and MA$_{\exp}$, respectively . SIZE[f] denotes the class of languages accepted by circuit families of size bounded by $f(n)$ for sufficiently large n.

Furthermore, we let Σ_k^e (Π_k^e, Δ_k^e) denote the class $\cup_f \Sigma_k^f$ ($\cup_f \Pi_k^f, \cup_f \Delta_k^f$ respectively) where the union is over all $f \in 2^{O(n)}$.

2.1   Fractionally exponentially growing functions

First we motivate the definition of such functions. One of the main difficulties in answering the questions posed in the introduction is that the "best" lower bounds to be obtained are not easily expressible in terms of conventional mathematical notation. Usually, lower bounds in complexity theory can be described with expressions involving the operations $\{+, -, *, \exp, \log\}$ only. Growth rates so expressible are called *L-functions* by Hardy [9]. Unfortunately, the answers we get to the questions posed in the introduction involves functions that are not approximated well by any L-function. For instance, the best lower bound that can be shown using current techniques for the classes MA$_{\exp}$, ZPEXP^{NP}, or $\Sigma_2^{\exp}$ seems to be "half-exponential", i.e. it is a bound that, composed with itself, becomes exactly exponential. Any L-function with a smaller growth rate (making it a valid substitute in a lower bound statement) will have, in fact, much, much smaller growth rate, and thus make the statement much weaker.

Intuitively, the notions of $\frac{1}{2}$-exponentially, or even $\frac{1}{k}$-exponentially growing functions are clear. One naive approach of defining them is as follows: We say that a "nice" function f has at most, say, half-exponential growth if $f(f(n)) \leq 2^{p(n)}$ for some polynomial p and all n. With an appropriate interpretation of "nice", such a definition would be adequate for some of the statements we prove, such as "For any half-exponential function f, there is a problem g in MA$_{\exp}$, so that g does not have circuits of size f". However, it is not obvious how to generalize this approach to meaningfully express and prove statements such as "For any (7/8)-exponential function f, there is problem g in MA-TIME[$\exp_{14/8}$], so that g does not have circuits of size f". Also, suppose we want to express our lower bounds in terms of a statement such as "There are function in complexity class C requiring circuit size at least $f(n)$" for some *specific* function f. Then it is not obvious how to pick such function f in a way close to optimal, i.e., as fast growing as possible. This seems unsatisfactory: To get any feeling for what the lower bound really means, it seems desirable to have concrete examples of functions of *exactly*, say, half-exponential growth in mind.

With these issues in mind, we take the following approach (see [16] for a discussion). Let $e(x) = e^x - 1$, where e denotes the base of the natural logarithm. Consider the following functional equation (called Abel's equation in the mathematics literature).

$$A(e(x)) = A(x) + 1. \qquad (1)$$

Let $A : R_+ \to R_+$ be a fixed solution to this (R_+ denote the set of positive real numbers). Then, with respect to this solution we can define the α-*iterate* of $e(x)$ as; $e_\alpha(x) = A^{-1}(A(x) + \alpha)$. Now, define the class of time bounds

$$\exp_\alpha = \{f : \mathbf{N} \to \mathbf{N} \mid f \text{ is time constructible }, \exists k \forall n, f(n) \leq e_\alpha(n^k)\}.$$

In order to use these class of functions as complexity bounds, we would like the following robustness property to hold among these classes of functions. Let α, β rationals. Then for $f \in \exp_\alpha$ and $g \in \exp_\beta$, we would like $f(g(.))$ to be in $\exp_{\alpha+\beta}$. There exist solutions for Equation 1 which give rise to functions with this property. The one due to Szekeres [16] is an example (please refer to [15] for a proof this). We use this property in many of our proofs, often with out making any explicit reference to it.

Time constructibility of these functions is a more subtle issue. In [13], the authors give a numerical procedure for approximating $e_\alpha(x)$. We strongly believe that a rigorous analysis of the procedure given in [13] will give us the time constructibility of these functions also. This analysis may require finding out the rate of convergence of the procedure given.

2.2   New results in this paper

We first show the extension of the lower bounds shown in [12] and [7] to get a general lower bound result. More precisely, we show the following theorems.

Theorem 1. *For any rational value c, $0 \leq c < 1$, and any $k \geq 1$, there is a language $L_{c,k}$ in ZP-TIME$^{\mathrm{NP}}[\exp_{2c}]$, so that, for infinitely many n, $L_{c,k} \cap \{0,1\}^n$ is not computed by a circuit of size $e_c(n^k)$. This holds relative to any oracle.*

Theorem 2. *For any rational value c, $0 \leq c \leq \frac{1}{2}$, and any k, there is a language $L_{c,k}$ in MA-TIME$[\exp_{c+\frac{1}{2}}] \cap$ coMA-TIME$[\exp_{c+\frac{1}{2}}]$, and, for any $\frac{1}{2} \leq c < 1$, and any k, there is a language $L_{c,k}$ in MA-TIME$[\exp_{2c}] \cap$ coMA-TIME$[\exp_{2c}]$, so that, for infinitely many n, $L_{c,k} \cap \{0,1\}^n$ cannot be computed by a circuit of size $e_c(n^k)$.*

As Buhrman et al already established in [7] that there are oracles relative to which MA$_{\mathrm{exp}}$ has polynomial circuits, it is clear that Theorem 2 does not relativize.

It is interesting to compare the bounds for MA-TIME and ZP-TIME^{NP}. The lower bounds we can prove for MA$_{\mathrm{exp}}$ and ZPEXP^{NP} are essentially the same; namely, half-exponential. They are also the same for time bounds bigger than exponential. But as soon as we consider time bounds smaller than exponential, the lower bound for MA-TIME becomes weaker than the one for ZP-TIME^{NP}. In particular, addressing a question from the introduction, for any $k > 1$, we can prove that MA-TIME[$\exp_{\frac{1}{2}}$] does not have circuits of size n^k, but we are unable to prove the same lower bound for MA-TIME[$\exp_\sigma$] for any value of $\sigma < \frac{1}{2}$. In contrast, we know that ZPP^{NP} does not have circuits of size n^k. On the hand, we see that for any $\epsilon > 0$, MA-TIME[$\exp_{\frac{1}{2}+\epsilon}$] does not have polynomial sized circuits, thus improving the result of [7] stating that this is the case for MA$_{exp}$.

We are unable to improve these lower bounds without going to the complexity class $\Delta_3^{f(n)}$. In particular, we don't know if a super-half-exponential lower bounds can be proven for Σ_2^{exp}.

Next, we consider how well circuits can *approximate* members of the uniform classes. We say that a function f on input domain $\{0,1\}^n$ is approximated by a circuit C, if $C(x) = f(x)$ for at least a $\frac{1}{2} + \frac{1}{p(n)}$ fraction of the input domain[1], for some polynomial p. In particular we show the following.

Theorem 3. *For any rational value c, $0 < c \leq \frac{1}{2}$, and any k, there is a language $L_{c,k}$ in MA-TIME[$\exp_{c+\frac{1}{2}}$]$\cap$coMA-TIME[$\exp_{c+\frac{1}{2}}$], and, for any rational $\frac{1}{2} \leq c < 1$, and any k, there is a language $L_{c,k}$ in MA-TIME[$\exp_{2c}$]$\cap$coMA-TIME[$\exp_{2c}$], so that for infinitely many n, $L_{c,k} \cap \{0,1\}^n$ cannot be approximated by circuits of size $e_c(n^k)$.*

For proving this, we use the random-self reducibility properties of some classes of high complexity along with known (by now standard) techniques for increasing the hardness of functions. These techniques are heavily employed in the context of pseudorandom generator constructions.

The oracle constructed in [7] witnesses the fact that this theorem does not hold in all relativized world. However, by replacing MA-TIME $\cap$ coMA-TIME with ZP-TIME^{NP}, it is possible to prove a theorem that does hold relative to any oracle.

Note that, except for the fact that the case $c = 0$ is not covered, Theorem 3 improves Theorem 2 by replacing "computed" with "approximated". In contrast, the ZP-TIME^{NP}-version of the theorem does not strictly improve Theorem 1, as the lower bounds for subexponential time bounds become worse. It would be interesting to remedy this, and, in particular, to show that ZPP^{NP} cannot be approximated by linear sized circuits.

3   Kannan revisited

In [10], Kannan proves that $\Sigma_3^e \cap \Pi_3^e$ contains a function with maximum circuit complexity. The argument used in the proof actually gives that $\Sigma_3^f \cap \Pi_3^f$ contains functions of superpolynomial circuit complexity for any superpolynomial f.

---

[1] The setting of the desired level of approximation to inverse polynomial is somewhat arbitrary; it avoids a third parameter besides time and circuit size in the theorem, thus improving readability.

In the paper, statement (*) of the Introduction, i.e., Theorem 4, page 48, is also claimed.

In order to claim (*), the following lemma is proved (Lemma 4, page 47).

Lemma *If $f(n)$ is any increasing time-constructible super-polynomial function, then there is a language L in $\Sigma_2^{f(n)} \cap \Pi_2^{f(n)}$ that does not have small circuits.*

By *small circuits* are meant circuits of size $O(n^k)$ for some fixed k (page 41). The proof of the lemma proceeds as follows: If SAT (the problem of deciding the satisfiability of boolean formulae) does not have polynomial circuits, then the lemma is true. In the case of SAT having polynomial circuits, by Karp and Lipton's theorem [11], the polynomial hierarchy collapses to $\Sigma_2^p \cap \Pi_2^p$. This implies that Σ_3^f can be simulated in $\Sigma_2^{f^{O(1)}}$, and since it was already established that $\Sigma_3^f \cap \Sigma_3^f$ contains functions of superpolynomial circuit complexity for any superpolynomial f, the lemma follows.

After proving the above lemma, statement (*) is claimed. But if our description above is accurate, it seems that what has actually been proven is: Either SAT does not have polynomial sized circuits or statement (*) is true. But this does not seem to imply (*), as the fact that SAT does not have polynomial sized circuits does not imply that it has exponential sized circuits.

In order to get a provable statement (†) of the same syntactic form as (*), it seems necessary to make up a more "balanced" "Either A or B" statement so that A as well as B implies (†).

To make up such as statement, we note that Karp and Lipton's technique actually gives the following lemma:

Lemma 1. *Let $f(n) \geq n$ be any time constructible functions. There is a constant c, so that if SAT has circuits of size $f(n)$ then Σ_3^n is included in $\Sigma_2^{f(n^c)^c}$.*

With this in mind, we now construct the statement (†):

(†) *For any time-constructible function $f(n) \leq 2^n$, there is a language in $\Sigma_2^{f(f(n)^c)^c} \cap \Pi_2^{f(f(n)^c)^c}$ that on infinitely many n does not have $f(n)^{\frac{1}{2}}$-size circuits.*

We prove the statement (†). Let $f(n)$ be given. Suppose SAT does not have circuits of size $f(n)$, then (†) follows. Otherwise SAT has circuits of size $f(n)$. Then, from Lemma 1, it follows that Σ_3^n is included in $\Sigma_2^{f(n^c)^c}$. By padding, we conclude that $\Sigma_3^{f(n)}$ is included in $\Sigma_2^{f(f(n)^c)^c}$. Then $\Sigma_3^{f(n)} \cap \Pi_3^{f(n)}$ is included in $\Sigma_2^{f(f(n)^c)^c} \cap \Pi_2^{f(f(n)^c)^c}$. But since $\Sigma_3^{f(n)} \cap \Pi_3^{f(n)}$ contains a function that does not have circuits of size $f(n)^{\frac{1}{2}}$, we are done.

In particular, the second level of the exponential hierarchy does not have circuit size $f(n)$ for any fixed half-exponential function f. The statement (†) can certainly be improved a bit by polynomial fiddling, but we don't see how to get any *essential* improvement. In particular, we don't see how to establish that the second level of the exponential hierarchy does not have σ-exponential circuits for any $\sigma > \frac{1}{2}$.

In the next two sections, the simple idea above is extended to ZP-TIME^{NP} and MA-TIME, proving the first two theorems of the introduction.

Before we move on to proving the theorems stated in the introduction, we note that in fact Δ_3^{ϵ} contains functions with maximum circuit complexity. We give a proof of this fact here since we shall need this result to prove Theorem 1. Let $M(n)$ denote the maximum possible circuit complexity of a function on n variables.

Lemma 2. *There is a language $L \in \Delta_3^{\epsilon}$ so that, for all n, the circuit complexity of $L \cap \{0,1\}^n$ is $M(n)$.*

Proof. Let a *truth table* be a string over $\{0,1\}$ of length 2^n for some n - such a string can be interpreted as the truth table of a Boolean function on n variables.

Let L_1 be the language consisting of tuples $\langle x, 1^{2^n}, 1^s \rangle$ so that x is a Boolean string of length less than 2^n that is the prefix of some truth table of length 2^n, so that the corresponding Boolean function cannot be computed by a circuit of size of s. Clearly, $L_1 \in NP^{NP}$; we guess the truth table and verify that no small circuit computes the same function. The procedure in figure 1 now generates on input n the lexicographically first truth table of a Boolean function on n variables with maximum circuit complexity. The Δ_3^{ϵ} language L with maximum circuit complexity is $L = \{x | \text{hard}(|x|)_{\text{index}(x)} = 1\}$, where $\text{index}(x)$ is the lexicographic index of x in $\{0,1\}^{2^n}$.

> Procedure hard(int: n)
> $\quad s := 2^n$;
> $\quad$ repeat until $\langle \lambda, 1^{2^n}, 1^s \rangle \in L_1$
> $\qquad s := s - 1$;
> $\quad t := \lambda$;
> $\quad$ while $|t| < 2^n$ do
> $\qquad$ if $\langle t0, 1^{2^n}, 1^s \rangle \in L_1$ then $t := t0$ else $t := t1$ endif
> $\quad$ return t

Fig. 1. Generating the truth table of a hard function

Lemma 3. *For any time constructible function $n \leq f(n) \leq M(n)$, there is a language L_f in $\Delta_3^{f(n)^2}$ so that for all n, the circuit complexity of $L_f \cap \{0,1\}^n$ is at least $f(n)$.*

Proof. By Shannon's theorem, the Boolean function on $g(n) = \min\{n, \lceil 2\log f(n) \rceil\}$ variables with maximum circuit complexity has circuit complexity at least $f(n)$. Let L be the language of Lemma 2 and let $x \in L_f$ if and only if $x_{1...g(n)} \in L$.

4   Zero error algorithms with an oracle for NP

In this section, we prove Theorem 1. We shall follow the line of proof of Köbler and Watanabe, showing that ZPP^{NP} does not have linear circuits. The difference of their proof to Kannan's sketched in the previous section, is in using an improved collapse of the polynomial hierarchy on the assumption of SAT having small circuits. We state a result from [6,12] as a lemma, suitable for our application.

Lemma 4 ([6,12]). *Let $f(n) \geq n$ be any time constructible function. Assume that SAT has circuits of size $f(n)$. Then, there is a randomized Turing machine with access to an NP-oracle that on input 1^n runs in expected time polynomial in $f(n)$, halts with probability 1, and outputs a circuit for SAT of size $f(n)$ when it does.*

Proof of Theorem 1. According to Lemma 3, there is a language L in $\Delta_3^{\mathrm{exp}_c}$ which does not have circuits of size $e_c(n^k)$. Let M be the machine accepting this language, running in time f for some $f \in \mathrm{exp}_c$, using an $\mathrm{NP}^{\mathrm{NP}}$-oracle. The $\mathrm{NP}^{\mathrm{NP}}$-oracle can be simulated by a polynomial time non-deterministic machine M_1 with an oracle for SAT. On an input x, the machine M queries M_1 which queries its SAT oracle. The longest query to the SAT oracle, when M is given a input of length n, has length at most some polynomial in f, say f^d. Now, $f^d \in \mathrm{exp}_c$. We can assume, without loss of generality, that all queries have length f^d.

Suppose SAT does not have circuits of size $e_c(n^k)$. Then we are done. So let us assume that SAT has circuits of size $e_c(n^k)$. Then we will show how to simulate the machine M (accepting L) by a ZP-$\mathrm{TIME}^{\mathrm{NP}}[g]$ algorithm for some $g \in \mathrm{exp}_{2c}$. This simulation works in two steps.

The first step is to find a SAT circuit for instances of length f^d. By assumption, there is such a circuit of size at most h, for some $h \in \mathrm{exp}_{2c}$. Hence, according to Lemma 4, we can find the circuit by a ZP-$\mathrm{TIME}^{\mathrm{NP}}[\mathrm{exp}_{2c}]$ computation.

Now, having found the circuit, the second step is to simulate the machine M using M_1 as oracle, but with M_1 using the circuit in place of its SAT-oracle (We can easily modify M and M_1 so that the circuit is given as input). Since a circuit of size h has already been found, this simulation can be done in $\mathrm{TIME}^{\mathrm{NP}}[\mathrm{exp}_{2c}]$. The two steps together form a ZP-$\mathrm{TIME}^{\mathrm{NP}}[\mathrm{exp}_{2c}]$ computation.

Furthermore, it is easy to verify that the above proof and also Lemma 4 relativises. Hence we have the Theorem.

5   Merlin-Arthur games

In this section, we prove Theorem 2. In [7], for showing that $\mathrm{MA}_{\mathrm{exp}}$ contains languages without polynomial size circuits, the authors make use of a theorem due to Babai et al [4] which states: "$\mathrm{EXP} \subseteq \mathrm{P/poly} \Rightarrow \mathrm{EXP} = \mathrm{MA}$". We follow the same line of argument as in [7], but we need the following refinement of the result of Babai et al [4], stated with a more general range of time and size bounds. The result essentially follows from a theorem in [2] on *transparent* proofs.

Lemma 5. *Let $g(n) \geq n$ and $s(n) \geq n$ are increasing time constructible functions. Then there is a constant $c > 1$ so that the following holds. If $g(n) \leq 2^n$ then $\mathrm{TIME}[g(n)^c] \subseteq \mathrm{SIZE}[s(n)] \Rightarrow \mathrm{TIME}[g(n)] \subseteq \mathrm{MA\text{-}TIME}[s(3n)^c]$. If $g(n) > 2^n$ then $\mathrm{TIME}[2^{cn}] \subseteq \mathrm{SIZE}[s(n)] \Rightarrow \mathrm{TIME}[g(n)] \subseteq \mathrm{MA\text{-}TIME}[s(3\log g(n))^c]$.*

Proof. Given a Turing machine T operating in $g(n)$ steps, we can construct a Turing transducer T', operating in $g(n)^c$ steps which on input x outputs a *transparent* proof [2] of the fact that T accepts (or rejects) on input x. Also, we can make a Turing machine T'' operating in $g(|x|)^c$ steps which on input $\langle x, i \rangle$, outputs the i'th bit of the output string of T' on input x. We now construct an Arthur-Merlin game accepting the same language as T, as follows: Merlin sends Arthur the description of a circuit of size $s(n)$ computing the same function of as T''. By the assumption, such a circuit exists. Note that this circuit is a succinct representation of the transparent proof for the computation. Arthur, following the protocol in [2], now verifies that the circuit indeed is a succinct encoding of a string close to a transparent proof of the correct computation.

The statement of the theorem now follows; note that the two cases of the theorem corresponds to which part of the input $\langle x, i \rangle$ is the larger. The factor 3, occurring twice in the statement of the theorem, is an upper bound on the ratio of the length of $\langle x, y \rangle$ and the length of x or y for a reasonable pairing function $\langle x, y \rangle$.

Proof of Theorem 2. We divide the proof into two cases, according to whether $c \leq \frac{1}{2}$ or $c > \frac{1}{2}$.

Case 1 $c \leq \frac{1}{2}$: We need to show that MA-TIME[$\exp_{c+\frac{1}{2}}$] does not have circuits of size $e_c(n^k)$. Clearly, we can assume that TIME[$\exp_{c+\frac{1}{2}}$] *does* have circuits of size $e_c(n^k)$, otherwise we are done. But then according to Lemma 5, TIME[$\exp_{c+\frac{1}{2}}$] is included in MA-TIME[$\exp_c$]. By padding, TIME[$\exp_{c+1}$] is included in MA-TIME[$\exp_{c+\frac{1}{2}}$]. But TIME[$\exp_{c+1}$] contains a language that cannot be computed by circuits of size $e_c(n^k)$, and we are done.

Case 2 $c > \frac{1}{2}$: We need to show that MA-TIME[$\exp_{2c}$] does not have circuits of size $e_c(n^k)$. We can again assume that TIME[$\exp_{2c}$] *does* have circuits of size $e_c(n^k)$. In particular, this is the case for EXP, and by Lemma 5, we conclude that TIME[$\exp_{2c}$] is included in MA-TIME[$\exp_{c-1+2c}$]. By padding, TIME[$\exp_{2c+(1-c)}$] is included in MA-TIME[$\exp_{c-1+2c+(1-c)}$] which is the class MA-TIME[$\exp_{2c}$]. But TIME[$\exp_{2c+(1-c)}$] = TIME[$\exp_{c+1}$] contains a language that cannot be computed by circuits of size $e_c(n^k)$ and we are done.

6   Non-approximability

In this section, we show Theorem 3 of the introduction. We need a result from [4]. We state it as a lemma in a form convenient for our application.

Lemma 6. [4] *There is a constant $\epsilon > 0$ so that the following holds. There is a deterministic quasi-polynomial time procedure* harden, *taking as input the truth table of a Boolean function f on n variables, and outputting the truth table of a Boolean function* harden(f) *on n^2 variables, with the following property. Let $s > n^{1/\epsilon}$. If f cannot be computed by circuits of size s, then any circuit of size s^ϵ taking n^2 inputs, will agree with* harden(f) *on at most a $\frac{1}{2} + s^{-\epsilon}$ fraction of the input domain $\{0, 1\}^{n^2}$.*

Proof of Theorem 3. We divide the proof into two cases.

Case 1. $c \geq \frac{1}{2}$: According to Theorem 2, there is a language L in MA-TIME[$\exp_{2c}$] $\cap$ coMA-TIME[$\exp_{2c}$] that cannot be computed by circuits of size $(e_c(n^{2k}))^{1/\epsilon}$. We now define a new language fulfilling the requirements of the theorem. As the language only needs to satisfy the desired property on infinitely many input lengths, we shall only consider input lengths of the form n^2. The following computation defines the language. Since $2c \geq 1$, we can, on input x of length n^2, compute the truth table t of the characteristic function of $L \cap \{0,1\}^n$, without leaving the complexity class MA-TIME[$\exp_{2c}$]$\cap$coMA-TIME[$\exp_{2c}$], as this class is closed under exponential time Turing reductions. Now, compute the truth table t' =harden(t), and look up x in t'. This is the result of the computation. The language so defined full-fills the requirements of the theorem and we are done.

Case 2. $c < \frac{1}{2}$: Let $c = \frac{1}{2} - \delta$. We already know that MA$_{\exp}$∩coMA$_{\exp}$ contains a language that cannot be approximated by circuits of size $e_{\frac{1}{2}}(n^k)$. More than that, from the proof we see that we can ensure that any circuit of this size will actually err on at least a $\frac{1}{2} - e_{1/2}(n)^{-1}$ fraction of the input domain for infinitely many n. By padding, MA-TIME[$\exp_{1-\delta}$] $\cap$ coMA-TIME[$\exp_{1-\delta}$] contains a language, so that any circuit of size $e_{\frac{1}{2}-\delta}(n^k)$ has to err on at least a $\frac{1}{2} - e_{1/2}(e_{-\delta}(n))^{-1} = \frac{1}{2} - e_c(n)^{-1}$ fraction of the input domain. As $c > 0$, this means that no circuit of size $e_{\frac{1}{2}-\delta}(n^k)$ approximates the language, and we are done.

References

1. V. Arvind and J. Köbler. On resource-bounded measure and pseudorandomness. Proceedings of the *17th conference on the Foundations of Software Technology and Theoretical Computer Science*, LNCS Vol. 1346, (1997), pp. 235–249.

2. L. Babai, L. Fortnow, L. Levin and M. Szegedy. Checking computations in polylogarithmic time. Procedings of the *23rd ACM Symposium on the Theory of Computation*, (1991), pp. 21–31.

3. L. Babai, L. Fortnow, and C. Lund. Non-Deterministic Exponential Time has Two-Prover Interactive Protocols. *Computational Complexity*, 1, (1990), pp. 3–40.

4. L. Babai, L. Fortnow, N. Nisan and A. Wigdersen. BPP has subexponential time simulations unless EXPTIME has publishable proofs. *Computational Complexity*, 3, (1993), pp. 307–318.

5. J. L. Balcázar, J. Díaz and J. Gabarró. *Structural Complexity - I & II*. Springer Verlag, Berlin Heidelberg, 1988.

6. N.H. Bshouty, R. Cleve, R. Gavalda, S. Kannan, and C. Tamon. Oracles and queries that are sufficient for exact learning. *Journal of Computer and System Sciences*, 52, (1996), pp. 421-433 .

7. H. Buhrman, L. Fortnow and T. Thierauf. Nonrelativizing separations. Proceedings of the *13th IEEE conference on Computational Complexity*, (1998), pp. 8–12.

8. O. Goldreich and D. Zuckerman. Another proof that BPP ⊆ PH (and more). *ECCC TR97-045*, (1997). Available at http://www.eccc.uni-trier.de/eccc/.

9. G.H. Hardy. *Orders of Infinity*. Cambridge Tracts in Mathematics and Mathematical Physics, No. 12. Second edition. Cambridge University Press, Cambridge, 1924.

10. R. Kannan. Circuit-size lower bounds and non-reducibility to sparse sets. *Information and Control*, 55, (1982), pp. 40–56.

11. R. M. Karp and R. J. Lipton. Some connections between nonuniform and uniform complexity classes. Proceedings of the *12th ACM Symposium on Theory of Computing*, (1980), pp. 302–309.

12. J. Köbler and O. Watanabe. New collapse consequences of NP having small circuits. Proceedings of the *International Colloquium on Automata, Languages and Programming*, LNCS Vol. 944, (1995), pp. 196–207.

13. K.W. Morris and G. Szekeres. Tables of the logarithm of iteration of $e^x - 1$. *J. Australian Math. Soc.* 2, (1962), pp. 321-327.

14. C. Papadimitriou. *Computational Complexity*. Addison-Wesley Publishing Company, 1994.

15. P. B. Miltersen, N. V. Vinodchandran and O. Watanabe. Super-polynomial versus half-exponential circuit size in the exponential hierarchy. *Research Report c-130*, Dept. of Math. and Comput. Sc., Tokyo Inst. of Tech. Available at http://www.is.titech.ac.jp/research/research-report/C/, 1999.

16. G. Szekeres. Fractional iteration of exponentially growing functions. *J. Australian Math. Soc.* 2, (1962), 301-320.

Efficient Learning of Some Linear Matrix Languages

Henning Fernau

Wilhelm-Schickard-Institut für Informatik; Universität Tübingen
Sand 13; D-72076 Tübingen; Germany
fernau@informatik.uni-tuebingen.de

Abstract. We show that so-called deterministic even linear simple matrix grammars can be inferred in polynomial time using the query-based learner-teacher model proposed by Angluin for learning deterministic regular languages in [3]. In such a way, we extend the class of efficiently learnable languages both beyond the even linear languages and the even equal matrix languages proposed in [10,11,13,14,15,16,17].

1   Introduction

Since Angluin showed in [3] how to learn regular languages in polynomial time using a query-based learner-teacher model, it has been an important issue in learning theory to extend the class of languages learnable in polynomial time beyond the class of regular languages.

Several attempts have been made to go beyond regular languages:

Even linear languages (as introduced in [2]) have been proposed by Radhakrishnan, Nagaraja, Takada, Sempere and García [10,13,11] for learning-theoretic purposes.

Even equal matrix languages (a restriction of equal matrix languages introduced in [12]) were shown to be polynomial time learnable by Takada [14,15,16].

Control languages come up as a natural tool for transferring Angluin's learnability result to the classes listed in the first two items; indeed, the learning problem for those classes is reduced to the learning problem of (regular) control languages by using universal grammar normal forms. Obviously, one could now use, e.g., even linear languages as control language for universal even linear grammars, hence obtaining a whole hierarchy (similar to the one of Khabbaz [6]) of efficiently learnable language classes by iterating the above-sketched argument. Takada's papers [16,17] explore the learnability of levels of such hierarchies.

In this paper, we try to present a unified framework of efficiently learnable language classes, and moreover we extend the class of efficiently learnable languages further. In order to obtain our learnability result in Section 6, we use non-trivial arguments known from formal language theory.

T. Asano et al. (Eds.): COCOON'99, LNCS 1627, pp. 221-230, 1999.
© Springer-Verlag Berlin Heidelberg 1999

Moreover, the class of even linear simple matrix languages we propose to this end contains both typical "pushdown languages" and typical "queue languages", an observation which should be of definite interest to possible linguistic applications of learning theory, since natural languages typically consists both of parenthesis structures (as exemplified by relative clauses) and "copy-structures" (like in Swiss German). Finally, the Khabbaz hierarchy has been identified within a hierarchy akin to the tree-adjoining languages as done by Weir [18].

Our, i.e., Angluin's, learning model differs from others proposed for linguistic purposes in [1], where the learner has only a passive role when observing input.

2   Definitions and examples

In general, we use standard notions from formal language theory. So, $|x|$ denotes the length of a word x which is formally an element of the free monoid X^* generated by some finite alphabet X. The neutral element in X^* is called empty word and is denoted by λ. We use $\Sigma_m = \{a_1, \ldots, a_m\}$ to denote some fixed m-element alphabet.

A *linear simple matrix grammar of degree n*, $n \geq 1$, is (cf. [9]) an $(n + 3)$-tuple $G = (V_1, \ldots, V_n, \Sigma, M, S)$, where $\{S\}, V_1, \ldots, V_n, \Sigma$ are pairwise disjoint alphabets ($V_N = \bigcup_{i=1}^{n} V_i \cup \{S\}$ contains the nonterminals and Σ the terminals), and M is a finite set of matrices of the form

1. $(S \to A_1 \ldots A_n)$, for $A_i \in V_i$, $1 \leq i \leq n$, or
2. $(A_1 \to x_1, \ldots, A_n \to x_n)$, for $A_i \in V_i$, $x_i \in \Sigma^*$, $1 \leq i \leq n$, or
3. $(A_1 \to x_1 B_1 y_1, \ldots, A_n \to x_n B_n y_n)$, for $A_i, B_i \in V_i$, $x_i, y_i \in \Sigma^*$, $1 \leq i \leq n$.

We now define three restrictions on such grammars:

G is a *right-linear simple matrix grammar* if the last condition 3 on rules is replaced by

3'. $(A_1 \to x_1 B_1, \ldots, A_n \to x_n B_n)$, $A_i, B_i \in V_i$, $x_i \in \Sigma^*$, $1 \leq i \leq n$.

G is an *even linear simple matrix grammar* if the last condition on rules is replaced by

3''. $(A_1 \to x_1 B_1 y_1, \ldots, A_n \to x_n B_n y_n)$, for $A_i, B_i \in V_i$, $x_i, y_i \in \Sigma^*$ such that $|x_1| = |x_i| = |y_i|$ for all $1 \leq i \leq n$.

G is an *even right-linear simple matrix grammar* (or *even equal matrix grammar* as introduced by Takada [15]) if the last condition on rules is replaced by

3'''. $(A_1 \to x_1 B_1, \ldots, A_n \to x_n B_n)$, $A_i, B_i \in V_i$, $x_i \in \Sigma^*$ such that $|x_1| = |x_i|$ for all $1 \leq i \leq n$.

Let $V_G = V_N \cup \Sigma$. For $x, y \in V_G^*$, we write $x \Rightarrow y$ iff either (i) $x = S$, $(S \to y) \in M$, or (ii) $x = u_1 A_1 v_1 \ldots u_n A_n v_n$, $y = u_1 w_1 v_1 \ldots u_n w_n v_n$, and $(A_1 \to w_1, \ldots, A_n \to w_n) \in M$. In other words, all rules in a matrix are applied in parallel, hence replacing all nonterminals occurring in the sentential form at once. As usual, define $L(G) = \{x \in \Sigma^* \mid S \overset{*}{\Rightarrow} x\}$, where $\overset{*}{\Rightarrow}$ is the reflexive transitive closure of relation $\Rightarrow$.

The families of linear simple matrix grammars of degree n, right-linear simple matrix grammars, even linear simple matrix grammars, and even right-linear simple matrix grammars, respectively, as well as the corresponding language families are denoted by $SL(n)$, $SRL(n)$, $ESL(n)$ and $ESRL(n)$, respectively. Sometimes, we use notations like $ESL(n, m)$ in order to specify the terminal alphabet Σ_m explicitly.

In the following, we provide some examples to show the power of the mechanism even in the simplest case. Moreover, those examples are thought to be of linguistic relevance.

Example 1. Consider $G_1 = (\{A_1\}, \{A_2\}, \{a, b\}, P_1, S)$, where P_1 contains the following rules:

1. $(S \rightarrow A_1 A_2)$,
2. $(A_1 \rightarrow \lambda, A_2 \rightarrow \lambda)$,
3. $(A_1 \rightarrow aA_1, A_2 \rightarrow aA_2), (A_1 \rightarrow bA_1, A_2 \rightarrow bA_2)$.

G_1 is an $ESRL(2)$ grammar which generates $L(G_1) = \{ww \mid w \in \{a, b\}^*\}$.

Example 2. Take $G_2 = (\{A_1, B_1\}, \{A_2, B_2\}, \{a, b\}, P_2, S)$ with P_2 having:

1. $(S \rightarrow A_1 A_2)$,
2. $(A_1 \rightarrow \lambda, A_2 \rightarrow \lambda), (B_1 \rightarrow \lambda, B_2 \rightarrow \lambda)$,
3. $(A_1 \rightarrow aA_1, A_2 \rightarrow aA_2), (A_1 \rightarrow bB_1, A_2 \rightarrow bB_2)$,
   $(B_1 \rightarrow bB_1, B_2 \rightarrow bB_2)$.

$G_2 \in ESRL(2)$ generates $L(G_2) = \{a^n b^m a^n b^m \mid n, m \geq 0\}$.

Example 3. Take $G_3 = (\{A_1\}, \{A_2\}, \{A_3\}, \{a, b\}, P_3, S)$ with P_3 having:

1. $(S \rightarrow A_1 A_2 A_3)$,
2. $(A_1 \rightarrow \lambda, A_2 \rightarrow \lambda, A_3 \rightarrow \lambda)$,
3. $(A_1 \rightarrow aA_1, A_2 \rightarrow bA_2, A_3 \rightarrow aA_3)$.

G_3 is an $ESRL(3)$ grammar which generates $L(G_3) = \{a^n b^n a^n \mid n \geq 0\}$.

In the following, we restrict our considerations to (even) SL languages. The corresponding results for the (even) right-linear case can be obtained analogously, cf. also the works of Takada.

3   Control languages

Let $G = (V_1, \ldots, V_n, \Sigma, M, S) \in SL(n)$ be chosen arbitrarily. To every valid derivation sequence, there corresponds (at least) one sequence of matrices $\pi \in M^*$ (which have been applied in order to get that derivation sequence), and every sequence of matrices $\pi \in M^*$ determines (at most) one valid derivation sequence. As shorthand, we write $x \Rightarrow^\pi y$ if y is obtained from x by applying the matrices listed in the so-called *control word* π in that sequence. For $C \subseteq M^*$, define

$$L_C(G) = \{w \in \Sigma^* \mid \exists \pi \in C : S \Rightarrow^\pi w\}.$$

$L_C(G)$ is called the language generated by G with *control set* C.
In the following, we need two auxiliary notions: Let G be an SL grammar. G is said to have *terminal rule bound* $\ell_1 \geq 1$ if for all terminal rules $(A_1 \rightarrow x_1, \ldots, A_n \rightarrow x_n)$ in G, we have $|x_1| + \ldots |x_n| \leq \ell_1$. G is said to have *nonterminal rule bound* $\ell_2 \geq 1$ if for all nonterminal rules $(A_1 \rightarrow x_1 B_1 y_1, \ldots, A_n \rightarrow x_n B_n y_n)$ in G, we have $\max(|x_1 y_1|, \ldots, |x_n y_n|) \leq \ell_2$.

By putting the control within the first component of the grammar using a pairing construction, the following lemma is rather straightforward.

Lemma 4. *Let $n \geq 1$. For every SL(n) grammar G and every regular control set C, one can construct an SL(n) grammar G' generating $L_C(G)$ with the same terminal and nonterminal rule bounds.* □

So, regular control sets do not increase the descriptive power of simple linear matrix grammars. On the other hand, they may serve to simplify the notation of SL languages; to see this, we need another auxiliary notion.
Consider $G_{SL}(n, m, \ell_1, \ell_2) = (\{S_1\}, \ldots, \{S_n\}, \Sigma_m, M(n, m, \ell_1, \ell_2), S)$, where $M(n, m, \ell_1, \ell_2)$ contains the following rules:

1. $m_1 = (S \rightarrow S_1 \ldots S_n)$.
2. $m_{x_1, \ldots, x_n} = (S_1 \rightarrow x_1, \ldots, S_n \rightarrow x_n)$ for all $x_i \in \Sigma_m^*$
   such that $|x_1| + \ldots + |x_n| \leq \ell_1$.
3. $m_{x_1, \ldots, x_n; y_1, \ldots, y_n} = (S_1 \rightarrow x_1 S_1 y_1, \ldots, S_n \rightarrow x_n S_n y_n)$ for all $x_i, y_i \in \Sigma_m^*$
   such that $\max\{|x_1 y_1|, \ldots, |x_n y_n|\} \leq \ell_2$.

Let us term $G_{SL}(n, m, \ell_1, \ell_2)$ *standard SL(n) grammar*. With analogous restrictions, one can define standard ESL grammars. Clearly, $L(G_X(n, m, \ell_1, \ell_2)) = \Sigma_m^*$.

Lemma 5. *Let $n \geq 1$ and $X \in \{SL, ESL\}$. For every $X(n)$ grammar $G = (V_1, \ldots, V_n, \Sigma_m, M, S)$ with terminal and nonterminal rule bounds ℓ_1 and ℓ_2, one can construct a regular set C such that $L(G) = L_C(G_X(n, m, \ell_1, \ell_2))$.*

Proof. Consider $G = (V_1, \ldots, V_n, \Sigma_m, M, S)$. Define the right-linear grammar $G' = (V, M(n, m, \ell_1, \ell_2), P, S)$ by $V = V_1 \times \ldots \times V_n \cup \{S\}$, and let P contain the following rules:

1. $S \rightarrow m_1(A_1, \ldots, A_n)$ if $(S \rightarrow A_1 \ldots A_n) \in M$.
2. $(A_1, \ldots, A_n) \rightarrow m_{x_1, \ldots, x_n}$ if $(A_1 \rightarrow x_1, \ldots, A_n \rightarrow x_n) \in M$.
3. $(A_1, \ldots, A_n) \rightarrow m_{x_1, \ldots, x_n; y_1, \ldots, y_n}(B_1, \ldots B_n)$
   if $(A_1 \rightarrow x_1 B_1 y_1, \ldots, A_n \rightarrow x_n B_n y_n) \in M$.

The claim $L(G) = L_{L(G')}(G(n, m, \ell_1, \ell_2))$ can be shown by induction on the length of the control words. □

In order to show learnability of ESL(n), we have to restrict the standard grammars further, since we need unambiguous derivations in those grammars. This is done in the following section.

4   Universal grammars

Lemma 6. *Let $n \geq 1$. Every ESL(n) language can be generated by some standard grammar $G_{ESL}(n, m, \ell_1, 2)$ without unit productions (i.e., nonterminal rules are of the form $m_{a_1,\ldots,a_n;b_1,\ldots,b_n}$ with $a_i, b_i \in \Sigma_m$ only) with the help of a regular control set.*

Proof. By Lemma 2, we can assume that $L \in \text{ESL}(n)$, $L \subseteq \Sigma_m^*$, is generated as $L_C(G_{ESL}(n, m, \ell_1, \ell_2))$, where $C \subseteq M(n, m, \ell_1, \ell_2)^*$ is a regular set. Consider the morphism $h : M(n, m, \ell_1, \ell_2)^* \to M(n, m, \ell_1, 2)^*$ with:

1. $h(m_1) = m_1$;
2. $h(m_{x_1,\ldots,x_n}) = m_{x_1,\ldots,x_n}$;
3. $h(m_{\lambda,\ldots,\lambda;\lambda,\ldots,\lambda}) = \lambda$, and furthermore,
   $h(m_{x_1,\ldots,x_n;y_1,\ldots,y_n}) = m_{b_{11},\ldots,b_{n1};c_{11},\ldots,c_{n1}} \cdots m_{b_{1r},\ldots,b_{nr};c_{1r},\ldots,c_{nr}}$,
   where $x_i = b_{i1} \ldots b_{ir}$ and $y_i = c_{ir} \ldots c_{i1}$, $b_i, c_i \in \Sigma_m$ for $1 \leq i \leq r \leq \ell_2/2$.

Obviously, $L = L_{h(C)}(G_{ESL}(n, m, \ell_1, 2))$, and $h(C)$ is regular.   □

Fix $\Sigma_m = \{a_1, \ldots, a_m\}$ and consider the ESL(n, m) grammar

$$G(n, m) = (\{S_1\}, \ldots, \{S_n\}, \Sigma, M(n, m), S), \tag{1}$$

where $M(n, m) \subset M(n, m, 2n - 1, 2)$ contains the matrices

1. $m_1 = (S \to S_1 \ldots S_n)$ and
2. $m_x = (S_1 \to \lambda, \ldots, S_{n-1} \to \lambda, S_n \to x)$ with $x \in \Sigma_m^*$, $|x| < 2n$ and
3. $m_{b_1 \ldots b_n, c_1 \ldots c_n} = (S_1 \to b_1 S_1 c_1, \ldots, S_n \to b_n S_n c_n)$
   with $b_j, c_j \in \Sigma_m$ for $1 \leq j \leq n$.

We now prove our main theorem.

Theorem 7 (Main Theorem). *$G(n, m)$ is universal for ESL(n, m), i.e., for every ESL(n, m) grammar G, a regular language C can be algorithmically constructed such that $L(G) = L_C(G(n, m))$.*

Proof. Consider an arbitrary ESL(n) grammar G with $L(G) \subseteq \Sigma_m^*$. By Lemmas 2 and 3, from G one can construct a regular control set C such that $L(G) = L_C(G_{ESL}(n, m, \ell_1, 2))$, where ℓ_1 is the terminal rule bound of G. A typical control word in $M_{ESL}(n, m, \ell_1, 2)^*$ looks as follows:

$$\pi = m_1 m_{b_{11},\ldots,b_{n1};c_{11},\ldots,c_{n1}} \cdots m_{b_{1q},\ldots,b_{nq};c_{1q},\ldots,c_{nq}} m_{x_1,\ldots,x_n}. \tag{2}$$

(Here, $b_{ij}, c_{ij} \in \Sigma_m$.) Using π, $G_{ESL}(n, m, \ell_1, 2)$ generates

$$w = \prod_{\ell=1}^{n} \left(\prod_{j=1}^{q} b_{\ell,j} \cdot x_\ell \cdot \prod_{j=1}^{q} c_{\ell,q+1-j} \right),$$

Assume now $x_i \neq \lambda$, i.e., $x_i = x_i' a$ for some $a \in \Sigma_m$. Consider the control word $\pi_{r,i}$ defined as

$$m_1 \, m_{b_{11},\ldots,b_{i1}, \boxed{c_{i1}}, b_{i+2,1},\ldots,b_{n1};c_{11},\ldots,c_{i-1,1}, \boxed{c_{i,2}}, c_{i+1,1},\ldots,c_{n1}}$$
$$\cdot \prod_{j=2}^{q-1} m_{b_{1j},\ldots,b_{ij}, \boxed{b_{i+1,j-1}}, b_{i+2,j},\ldots,b_{nj};c_{1j},\ldots,c_{i-1,j}, \boxed{c_{i,j+1}}, c_{i+1,j},\ldots,c_{nj}}$$
$$\cdot m_{b_{1q},\ldots,b_{iq}, \boxed{b_{i+1,q-1}}, b_{i+2,q},\ldots,b_{nq};c_{1q},\ldots,c_{i-1,q}, \boxed{a}, c_{i+1,q},\ldots,c_{nq}}$$
$$\cdot m_{x_1,\ldots,x_{i-1}, \boxed{x_i'}, \boxed{b_{i+1,q}x_{i+1}}, x_{i+2},\ldots,x_n}$$

(We indicate the important changes from π to $\pi_{r,i}$ by using box frames.) Obviously, w can also be generated using that control word, since

$$w = \prod_{\ell=1}^{i-1}\left(\prod_{j=1}^{q} b_{\ell,j} \cdot x_\ell \cdot \prod_{j=1}^{q} c_{\ell,q+1-j}\right)$$
$$\cdot \left(\prod_{j=1}^{q} b_{i,j}\right) \cdot x_i' \cdot \left(a \cdot \prod_{j=1}^{q-1} c_{i,q+1-j}\right)$$
$$\cdot \left(c_{i,1} \cdot \prod_{j=1}^{q-1} b_{i+1,j}\right) \cdot \left(b_{i+1,q}x_{i+1}\right) \cdot \left(\prod_{j=1}^{q} c_{i+1,q+1-j}\right)$$
$$\cdot \prod_{\ell=i+2}^{n}\left(\prod_{j=1}^{q} b_{\ell,j} \cdot x_\ell \cdot \prod_{j=1}^{q} c_{\ell,q+1-j}\right).$$

Pictorially, passing from π to $\pi_{r,i}$ means a local "right shift" in the derivation tree structure.

For $1 \leq i < n$, consider now the rational transductions $\tau_{i \to i+1,d}$ mapping $M_{\mathrm{ESL}}(n,m,\ell_1,2)$ to itself which transforms π into $\pi_{r,i}$ as indicated above if $|x_i| > d$, and which just preserves π if $|x_i| \leq d$. (Since only control words of the form considered in Eq. (2) are of interest, $\tau_{i \to i+1,d}$ may map other words in $M_{\mathrm{ESL}}(n,m,\ell_1,2)^*$ into the empty set.)

Analogously, one could define a "left shift" in the derivation tree structure. Consider again w generated using π as defined above under the condition that $x_i \neq \lambda$, here, $x_i = \bar{a}x_i''$. Define $\pi_{\ell,i}$ to be

$$m_1 \, m_{b_{11},\ldots,b_{i-1,1}, \boxed{b_{i2}}, b_{i+1,1},\ldots,b_{n1};c_{11},\ldots,c_{i-2,1}, \boxed{b_{i,1}}, c_{i,1},\ldots,c_{n1}}$$
$$\cdot \prod_{j=2}^{q-1} m_{b_{1j},\ldots,b_{i-1,j}, \boxed{b_{i,j+1}}, b_{i+1,j},\ldots,b_{nj};c_{1j},\ldots,c_{i-2,j}, \boxed{c_{i-1,j-1}}, c_{i,j},\ldots,c_{nj}}$$
$$\cdot m_{b_{1q},\ldots,b_{i-1,q}, \boxed{\bar{a}}, b_{i+1,q},\ldots,b_{nq};c_{1q},\ldots,c_{i-2,q}, \boxed{c_{i-1,q-1}}, c_{i,q},\ldots,c_{nq}}$$
$$\cdot m_{x_1,\ldots,x_{i-2}, \boxed{x_{i-1}c_{i-1,q}}, \boxed{x_i''}, x_{i+2},\ldots,x_n}$$

For $1 \leq i < n$, define the rational transductions $\tau_{i \to i-1,d}$ by $\pi \mapsto \pi_{\ell,i}$ if $|x_i| > d$, and $\pi \mapsto \pi$ if $|x_i| \leq d$.

First, transform a control word π into

$$\pi' := \tau_1(\pi) := \tau_{n-1 \to n,0}^{\ell_1}(\ldots(\tau_{1 \to 2,0}^{\ell_1}(\pi))\ldots).$$

One easily verifies that $\pi' \in A(n,m) \cdot \{m_{\lambda,\ldots,\lambda,x_n} \mid |x_n| \leq \ell_1\}$, where $A(n,m) = \{m_1\}\{m_{b_1,\ldots,b_n;c_1,\ldots,c_n} \mid b_i, c_i \in \Sigma_m\}^*$. Further, transform π' into

$$\pi'' := \tau_2(\pi') := \tau_{n \to n-1,2s}^{2s}(\ldots(\tau_{n \to 1,2s}^{2s}(\pi'))\ldots),$$

1. output(m_1).
2. if $|w| < 2n$ then output(m_w); stop.
3. if $|w| \geq 2n$ then if w is decomposed into $w = b_1 w_1 c_1 \ldots b_n w_n c_n$, $b_i, c_i \in \Sigma_m$, $|w_1| = \ldots = |w_{n-1}|$, $|w_n| < |w_1| + 2n$ (cf. Eq. (3)), then
    a) output($m_{b_1 \ldots b_n . c_1 \ldots c_n}$);
    b) set $w = w_1 \ldots w_n$;
    c) goto 2.

Table 1. How to transform an input into a control word

where s is the truncated result of dividing ℓ_1 by $2n$ and r is the remainder of such division, so that $\ell_1 = 2sn + r$, and

$$\tau_{n \to j, d}(\bar{\pi}) := \tau_{j+1 \to j, d}(\ldots (\tau_{n \to n-1, d}(\bar{\pi})) \ldots)$$

results in a "left-shift" in the derivation tree of $n - j$ units. Now,

$$\pi'' \in A(n, m) \cdot \{m_{x_1, \ldots, x_n} \mid |x_1| = \ldots = |x_{n-1}| \text{ is even}, |x_n| < 2n + |x_1|\} . \quad (3)$$

Finally, consider the homomorphism g that acts as identity on all letters except from terminal rule matrices used in Eq. (3) which are transformed according to

$$g(m_{b_{11} \ldots b_{1s} c_{1s}, \ldots c_{11}, \ldots b_{n-1,1} \ldots b_{n-1,s} c_{n-1,s} \ldots c_{n-1,1}, b_{n1} \ldots b_{ns} d_1 \ldots d_r c_{ns} \ldots c_{n1}})$$

$$= m_{b_{11}, \ldots, b_{n1}; c_{11}, \ldots c_{n1}} \cdots m_{b_{1s}, \ldots, b_{ns}; c_{1s}, \ldots, c_{ns}} m_{\lambda, \ldots, \lambda, d_1 \ldots d_r} ,$$

where $b_{ij}, c_{ij}, d_j \in \Sigma_m$ $r < 2n$. So, we get a regular set $C' = g(\tau_2(\tau_1(C)))$ with $L(G) = L_{C'}(G(n, m))$.   □

Applying finally Lemma 1, we can easily derive that regularly-controlled $G(n, m)$ characterize $ESL(n, m)$. We now investigate $G(n, m)$ closer.

Proposition 8. $L(G(n, m)) = \Sigma_m^*$, and every word in Σ_m^* has a unique derivation in $G(n, m)$.

We omit a formal proof of the previous assertion; instead, the following proposition should give enough insight how a control word is obtained from the input.

Proposition 9. There exists a linear time algorithm which transforms an input word $w \in \Sigma_m^*$ to its control word $\pi \in M(n, m)^*$ and vice versa.

Here and in the following, time is always measured on a random access machine (RAM). An algorithm for transforming an input word into a control word is given in Table 1.

5   Formal language issues

Here, we concentrate on the classes $ESL(n)$ again. The results should be contrasted with the corresponding ones for $SL(n)$ and the right-linear cases contained in [4,9,15,16].

Using standard constructions and the main theorem 1, one can show:

Theorem 10 (Closure properties). *Each class* ESL(n) *is closed under the Boolean operations and letter-to-letter morphisms, but not under arbitrary (inverse) morphisms nor catenation nor Kleene star.*   □

Decidability results can be easily shown using the main theorem 1:

Theorem 11 (Decidability). *Let* $n \geq 1$ *be fixed. The equivalence, inclusion, emptiness and finiteness problems are decidable for* ESL(n) *grammars.*   □

In formal language theory, universal grammars have been used to derive Chomsky-Schützenberger-Stanley characterizations of language families. In this spirit, and similar to [5, Theorems 1 & 6], one can deduce:

Corollary 12. *For each* $n, m \geq 1$, *there exists a language* $L(n, m)$ *over some alphabet* $\Sigma(n, m)$ *and a letter-to-letter morphism* $h : \Sigma(n, m)^* \to \Sigma_m^*$, *so that for each language* $L \in$ ESL(n, m) *there exists a regular set* $R \subseteq \Sigma(n, m)^*$ *such that* $L = h(L(n, m) \cap R)$. *On the other hand, every language represented as* $h(L(n, m) \cap R)$, *where* h *is a letter-to-letter morphism mapping into* Σ_m^* *and* R *a regular set, is in* ESL(n, m).   □

With some additional effort, we can show a similar morphic characterization result for SL(n), namely by admitting arbitrary morphisms h. We just state a corollary of that assertion:

Corollary 13. *The full trio closure of* ESL(n) *equals* SL(n). *That trio is generated by* $\{(ww^R)^n \mid w \in \{a, b\}^*\}$.   □

Without proof, we mention the following non-trivial result:

Theorem 14 (Strict inclusions). *For all* $n \geq 1$, *we find that* ESRL$(n) \subset$ SRL$(n) \subset$ SL(n) *and* ESL$(n) \subset$ SL(n).   □

In fact, ESL$(n) \subseteq$ ESL(n') if and only if n divides n'.

Finally, let us mention without proof that building a Khabbaz/Takada-like hierarchy does not lead us outside the ESL-formalism in the following sense:

Theorem 15. *For every* ESL(n) *grammar* G *controlled by a* ESL(n') *language* C, $L_C(G) \in$ ESL$(4nn')$.   □

6   The learnability of ESL(n)

We now turn to the issue of learning of ESL(n) languages for a given $n \geq 1$. The model we use is the following: a learner L has the task to deduce an ESL(n) grammar for a certain ESL(n) language L which is known to its teacher T. At the beginning, L is just informed about the terminal alphabet Σ_m of L. L may ask T the following questions:

Membership query Is $w \in L$?
Equivalence query Does the ESL(n) grammar G generate L?

Teacher T reacts as follows to the questions:

1. To a membership-query, T answers either "yes" or "no".
2. To an equivalence-query, T answers either "yes" (here, the learning process may stop, since L has performed its task successfully) or "no, I show you a counterexample w."

(By Theorem 3, teacher T can answer equivalence-queries effectively.)

We now give a learning algorithm based on the ideas of Takada [13]. We assume there is a learner L' which can learn regular languages, e.g., the one described by Angluin [3].

Learner L transforms a counterexample w received from T using the algorithm given in Table 1. On the other hand, a query π of L' will be translated by L into the unique w determined by $S \Rightarrow^\pi w$ using the universal grammar $G(n, m)$, see (1). That w is passed to the teacher T.

Finally, if learner L' has a guess C of a possible regular (control) set, L transforms this guess into an ESL(n) grammar G' for $L_C(G(n, m))$ as in the proof of Theorem 1 and passes G' to teacher T.

So, L is mainly an interface between the ESL(n)-teacher T and the regular set learner L'. Our main theorem 1 ensures that in fact every ESL(n) languages is learnable.

Using Angluin's algorithm [3, pages 94–96] for learning regular languages, the above-sketched algorithm has running time polynomial in the size of the number of states of an deterministic finite automaton for the control set C and in the length of the longest counterexample and in the parameter n (which influences the size of the alphabet of the control set to be learnt).

So, we conclude to have a polynomial learning algorithm if we want to learn "deterministic" ESL grammars (which are formalizable as in [11]).

Theorem 16 (Learnability). *For each $n \geq 1$, the class of deterministic ESL grammars is inferrable in polynomial time using membership and equivalence queries within Angluin's learner/teacher model.*   □

7   Conclusions

Having in mind Theorems 4 and 5, we can conclude that by introducing ESL grammars we properly extended the learnability both of even linear and of even equal matrix languages and of the Khabbaz/Takada hierarchy mentioned in the introduction.

As a next step, learnability from positive examples similar to [7,8] will be explored.

We gratefully acknowledge discussions with E. Mäkinen, J. M. Sempere, F. Stephan, and Y. Takada.

References

1. N. Abe. Feasible learnability for formal grammars and the theory of natural language acquisition. In *Proceedings of the 12 th COLING*, pages 1–6, Budapest, 1988.

2. V. Amar and G. Putzolu. On a family of linear grammars. *Information and Control*, 7:283–291, 1964.

3. D. Angluin. Learning regular sets from queries and counterexamples. *Information and Computation*, 75:87–106, 1987.

4. J. Dassow and Gh. Păun. *Regulated Rewriting in Formal Language Theory*. Berlin: Springer, 1989.

5. S. Hirose and M. Nasu. Left universal context-free grammars and homomorphic characterizations of languages. *Information and Control*, 50:110–118, 1981.

6. N. A. Khabbaz. A geometric hierarchy of languages. *Journal of Computer and System Sciences*, 8:142–157, 1974.

7. T. Koshiba, E. Mäkinen, and Y. Takada. Learning deterministic even linear languages from positive examples. *Theoretical Computer Science*, 185(1):63–79, October 1997.

8. E. Mäkinen. A note on the grammatical inference problem for even linear languages. *Fundamenta Informaticae*, 25:175–181, 1996.

9. Gh. Păun. Linear simple matrix languages. *Elektronische Informationsverarbeitung und Kybernetik*, 14:377–384, 1978.

10. V. Radhakrishnan and G. Nagaraja. Inference of even linear grammars and its application to picture description languages. *Pattern Recognition*, 21:55–62, 1988.

11. J. M. Sempere and P. García. A characterization of even linear languages and its application to the learning problem. In *Proc. of the 2nd International Coll. on Grammatical Inference (ICGI-94)*, volume 862 of *LNAI/LNCS*, pages 38–44, 1994.

12. R. Siromoney. On equal matrix languages. *Information and Control*, 14:133–151, 1969.

13. Y. Takada. Grammatical inference of even linear languages based on control sets. *Information Processing Letters*, 28:193–199, 1988.

14. Y. Takada. Algorithmic learning theory of formal languages and its applications. Technical Report IIAS-RR-93-6E, IIAS-SIS: International Institute for Advanced Study of Social Information Sciences, Fujitsu Laboratories, 1992. Also PhD Thesis.

15. Y. Takada. Learning even equal matrix languages based on control sets. In *Parallel Image Analysis, ICPIA '92*, volume 652 of *LNCS*, pages 274–289, 1992.

16. Y. Takada. Learning formal languages based on control sets. In *Algorithmic Learning for Knowledge-Based Systems*, volume 961 of *LNAI/LNCS*, pages 317–339, 1995.

17. Y. Takada. A hierarchy of language families learnable by regular language learning. *Information and Computation*, 123:138–145, 1995.

18. D. J. Weir. A geometric hierarchy beyond context-free languages. *Theoretical Computer Science*, 104:235–261, 1992.

Minimizing Mean Response Time in Batch Processing System

Xiaotie Deng[1,*] and Yuzhong Zhang[2,**]

[1] Department of Computer Science
City University of Hong Kong
Hong Kong, P. R. China
Deng@cs.cityu.edu.hk,
[2] Institute of Operations Research
Qufu Normal University
Qufu, Shandong, P.R. China
yuzhong@dns1.qfnu.edu.cn

Abstract. We consider batch processing of jobs to minimize the mean response time. A batch processing machine can handle up to B jobs simultaneously. Each job is represented by an arrival time and a processing time. The processing time of a batch of several jobs is the maximum of their processing times. In batch processing, non-preemptive scheduling is usually required and we discuss this case. The batch processing problem reduces to the ordinary uni-processor system scheduling problem if $B = 1$. We focus on the case $B = +\infty$. Even for this seemingly simple extreme case, we are able to show that the problem is NP-hard. We further show that several important special cases of the problem can be solved in polynomial time.

1 Introduction

We are interested in minimizing the mean completion time in the batch processing system. Each job is associated with an arrival time and a processing time. The batch machine can process up to B jobs at the same time. The jobs are processed without preemption. That is, once the processing of a batch of jobs starts, the machine is occupied until the process is completed. The processing time of a batch of B jobs is the longest processing time of these jobs. That is, for a batch consisting of jobs with processing time $p_1 \leq p_2 \cdots \leq p_k$, their completion time will be p_k units after the batch machine starts to process them.

Let $r_1 \leq r_2 \leq \cdots \leq r_n$ be job arrival times. Let w_i be the weight of job i of length p_i, $i = 1, 2, \cdots, n$. We focus on the case $B \geq n$. Therefore, at any point in time, all the jobs already in the system can be processed at the same time if the batch machine is free. This is the same as setting $B = +\infty$. Jobs processed

---

[*] Research partially supported by a CERG grant of Hong Kong UGC and an SRG grant of City University of Hong Kong
[**] Research partially supported by Natural Science Foundation of China and Shandong

in a batch has the same completion time, that is, their common starting time plus the processing time of the longest job in the batch. The response time of a job i is $C_i - r_i$, its completion time minus its arrival time. The weighted total response time is defined to be $\sum_{i=1}^{n} w_i(C_i - r_i)$. The weighted mean response time is the weighted total response time divided by the sum of the weights of all n jobs. To minimize the weighted mean response time is equivalent to minimize the weighted total response time. The latter is, in turn, the same as to minimize the weighted total completion time $\sum_{i=1}^{n} w_i C_i$, since the weighted sum of job arrival times is a constant. W.l.o.g., we will consider the total completion time from now on. We denote our problem as $1|r_j, B| \sum w_j C_j$, applying the notations in Graham, et al., [4].

The batch processing problem has its root in various application areas, for example, the semiconductor manufacturing system. In very large scale integrated circuit manufacturing process, the last stage is the final testing to ensure that the defective products can not be passed on to the customer. This is done by using automated testing equipment to exam each integrated circuit and determine whether or not it is operating at the required specifications (See, e.g., [2]).

P. Brucker, et al., had a through discussion of the scheduling problem of batch machine under various constraints and objective functions [1]. To minimize the weighted total completion time, the discussion there was only for the case job arrival times are all zero. They pointed out that the case with unequal arrival times to be much more difficult even when $B = +\infty$. In this paper we pinpoint the difficulty by proving the problem to be NP-complete. More precisely, where $B = +\infty$ (or equivalently $B \geq n$ for a system of n jobs), the problem $1|r_j, B| \sum w_j C_j$ is NP-complete. Previous NP-hardness results for this problem either restrict that B to be bounded and/or have other constraints on the jobs such as sizes [6] and deadlines [1].

In another direction, we are able to solve two special cases in polynomial time: i.e., the case the number of the processing times is a constant and the case all jobs have a constant number of arrival times. In relation to the former, Chandru, et al., designed an algorithm to minimize total completion time for a system of jobs both with a constant number of processing times and with the same arrival time zero [2,3]. For the latter, our approach requires a novel idea of switching to a scheduling problem for a different job set to solve a major step in the immediate dynamical programming formulation that extends the result [1] of Brucker, et al., for the case of one single arrival time for all jobs. In comparison, Lee and Uzsoy had a pseudo-polynomial time algorithm for the case of jobs with two arrival times [5].

We present the NP-hardness result in Section 2. Then, in Section 3, we discuss the polynomial time algorithm for the case when the number of job processing times is bounded by a constant. In Section 4, our polynomial time algorithm for jobs with a constant number of arrival times is presented. All our results hold for the weighted total completion time (and thus the average response time) and improve previous results. Even though our studies focus on the case when B is unbounded, we expect some ideas in establishing our results can be extended to the case B is bounded.

2 NP-Completeness Proof

The case of $B = +\infty$ can be solved in polynomial time if all jobs arrive at time zero, by a dynamic programming algorithm of Brucker et al [1]. In this section, we show that it becomes NP-complete when jobs arrive at different times.

Theorem 1. *The problem* $1|r_j, B = +\infty| \sum w_j C_j$ *is NP-hard.*

Proof. We will prove it by reduction from the well known PARTITION problem, which is described as the following:

Given a set $G = \{a_1, a_2, \cdots, a_m\}$ of m positive integers, is there a subset X of $\{1, 2, \cdots, m\}$ such that

$$\sum_{j \in X} a_j = A = \frac{1}{2} \sum_{j=1}^{m} a_j$$

Given an instance of PARTITION, we construct an instance of our problem as follows:

There are three types of jobs and a total of $2m + 1$ jobs. For each job $j(j = 1, 2, \cdots, m)$, let $p_j = jm^2 A^4 + a_j$, $r_j = \frac{1}{2}j(j-1)m^2 A^4$, and $w_j = a_j((j+1)m^2 A^4)^{-1}$. For job $m+j(j = 1, 2, \cdots, m)$, let $p_j = jm^2 A^4$, $r_j = \frac{1}{2}j(j-1)m^2 A^4$, and $w_j = a_j((j+1)m^2 A^3)^{-1}$. The last job, numbered $2m+1$, has a sufficient large positive weight M and its arrival time is $\frac{1}{2}m^3(m+1)A^4 + A$ and its processing time is ϵ for a sufficiently small $\epsilon > 0$. For simplicity, we may set the processing time of job $2m+1$ to be zero: $p_{2m+1} = 0$ (alternatively, we may set its processing time to be one and multiplying the processing times of other jobs by a sufficiently large constant).

The time it takes to construct the instance is obviously polynomial. We show that the PARTITION problem has a solution if and only if for the optimal solution of the scheduling problem,

$$\sum_{j=1}^{n} w_j C_j \leq \sum_{j=1}^{n} w_j(r_j + p_j) + A + \frac{4}{3m^2 A}. \tag{2.1}$$

Let $s_j = \{1, 2, \cdots, j\}$. Suppose that the PARTITION problem has a solution X, let $Y = \bar{X}$. The desired schedule is defined as following: First, each job $m+j$ is grouped in batch B_j. Without loss of generality, we may assume that X contains job m. If $j \in X$, then job j is grouped in the same batch with $m+j$. If $j \in Y$, then job j is grouped in the next batch, i.e., the same batch as $m+j+1$. The last job with processing time zero is scheduled in a batch of only one job and the starting time is its arrival time. We will prove that (2.1) holds for this schedule.

For the jobs j which grouped with $m + j$ in the same batch, i.e. $j \in X$

$$v_j = C_j - (r_j + p_j) = \sum_{i \in X \cap s_{j-1}} a_i.$$

Therefore,

$$w_j v_j = \frac{a_j \sum\limits_{i \in X \cap s_{j-1}} a_i}{(j+1)m^2 A^4} < \frac{a_j}{(j+1)m^2 A^3}.$$

For job j that is grouped in the same batch as $m + j + 1$ (i.e. $j \in Y$), we have

$$v_j = C_j - (r_j + p_j) = (j+1)m^2 A^4 + \sum_{i \in X \cap s_{j-1}} a_i - a_j$$

and

$$w_j v_j = a_j + \frac{a_j \left(\sum\limits_{i \in X \cap s_{j-1}} a_i - a_j \right)}{(j+1)m^2 A^4}$$

$$< a_j + \frac{a_j}{(j+1)m^2 A^3}$$

For each job with processing time $(j + 1)m^2 A^4$, we have

$$v_j = C_j - (r_j + p_j) = \sum_{i \in X \cap s_j} a_i.$$

It follows that

$$w_j v_j = \frac{a_j \sum\limits_{i \in X \cap s_j} a_i}{(j+1)m^2 A^3} < \frac{a_j}{(j+1)m^2 A^2}.$$

Therefore we have the following inequalities,

$$\sum w_j v_j = \sum_{j \in X} \frac{a_j}{(j+1)m^2 A^3} + \sum_{j \in Y} \left(a_j + \frac{a_j}{(j+1)m^2 A^3} \right) + \sum_{j \in X} \frac{a_j}{(j+1)m^2 A^2}$$

$$\leq \sum_{j \in Y} a_j + 2 \sum_{j \in X} \frac{a_j}{(j+1)m^2 A^2} + \sum_{j \in Y} \frac{a_j}{(j+1)m^2 A^3}$$

$$< \sum_{j \in Y} a_j + \frac{4}{3m^2 A}.$$

Suppose conversely that there is a schedule satisfying (2.1). We denote by X the set of jobs j that is grouped in the same batch with job $m + j$ and $Y = \{1, 2, \cdots, m\} - X$. First, we point out that none of the jobs $m + j$ ($j =$

$1, 2, \cdots, m$) can be delayed to next batch to be scheduled. In fact, suppose to the contrary that there exists a job $m + j$ that is delayed to $(j + 1)$-th batch or later, then its response time is equal to or larger than $(j + 1)m^2 A^4$, and the added amount of weighted cost will be equal to or greater than A. Second, $X \neq \emptyset$. Otherwise, $Y = J$, for each job $j \in Y$, which is delayed to next batch or latter, $v_j \geq (j + 1)m^2 A^4 - O(A)$. Then $w_j v_j \geq a_j - o(m^2 A^3)$, which is a contradiction to (2.1). Third, because of the large weight of the last job, we know that it must be scheduled at its arrival time. Therefore, we must have

$$\sum_{j \in X} a_j \leq A \qquad (2.2)$$

Finally, each job $j \in Y$ will be scheduled later with job $m + j + 1$ in one batch. Then

$$v_j = C_j - (r_j + p_j) \geq (j + 1)m^2 A^4 + \sum_{i \in X \cap s_{j-1}} a_i - a_j$$

and

$$w_j v_j = a_j + \frac{a_j}{(j + 1)m^2 A^4} \left(\sum_{i \in X \cap s_{j-1}} a_i - a_j \right)$$

$$\geq a_j - \frac{a_j^2}{(j + 1)m^2 A^4}$$

$$> a_j - \frac{1}{(j + 1)m^2 A^2}.$$

Thus we have

$$\sum_{j \in Y} w_j v_j \geq \sum_{j \in Y} a_j - \frac{1}{m A^2}.$$

Combining it with (2.1), we establish

$$\sum_{j \in Y} a_j \leq \sum_{j \in Y} w_j v_j + \frac{1}{m A^2}$$

$$< \sum_{j=1}^{n} w_j v_j + \frac{1}{m A^2}$$

$$< A + \frac{4}{3m^2 A} + \frac{1}{m A^2}.$$

By the fact that all a_j are integers we have

$$\sum_{j \in Y} a_j \leq A.$$

Applying the inequality (2.2), and the fact that $\sum_{i \in X \cup Y} a_i = 2A$, it follows that a schedule that satisfies (2.1) is sufficient to solve the partition problem.

3   Job set with a constant number of processing times

First, we present a solution for the case jobs have the same processing time (i.e., job length) P. Then, we show how to extend this to the case when there are a constant number of job lengths. The presentation is for the case when all weights of jobs are the same but the dynamic programming formulation trivially applies to the general weighted case.

When all jobs are of the same length, we scale it to be one $P = 1$. We index all jobs in their order of arrival: $r_1 \leq r_2 \leq \cdots \leq r_n$. Let $\Omega = \{r_j + i | j = 1, 2, \cdots, n; i = 0, 1, 2, \cdots, n\}$. Obviously, $m = |\Omega| \leq n(n + 1)$. We label the numbers t_k in Ω such that $t_1 < t_2 < \cdots < t_m$. Let $X_k = \{J_j | r_j \leq r_k\}, X_{k,j} = X_k \setminus X_j, j < k$.

Observation 1 *For any subset of J, there is an optimal schedule, in which the starting time of every batch is a point in Ω.*

Let $OPT(t_k, X)$ denote the optimal total completion time for a case when the available jobs at time t_k is X and jobs with arrival times later than t_k is the same as those in the job set J. For simplicity, we discuss the case when all weights are the same $w_i = 1, 1 \leq i \leq n$. The general case is similar. First, we have $OPT(t_m, X) = |X|(t_m + 1)$, which will be the base case for our dynamic programming algorithm. In fact, we need only to consider that case when X is of form $X_{k,j}$ (this observation is especially important when the jobs are not of the same weights). At the second last time t_{m-1}, the optimal value $OPT(t_{m-1}, X_{k,j})$ of the schedule is obtained by taking the minimum of the optimal value postponing all jobs to t_m and that corresponds to a solution with some jobs starting at t_{m-1} and the rest starting at $t_{m-1} + 1$ (if it happens to be t_m). In the general case, we apply the following algorithm.

Algorithm 1

Step 1 Compute the optimal values $OPT(t_m, J)$, and $OPT(t_m, J \setminus X_k)$ for all $X_k, k = 1, 2, \cdots, n$, and t_m. Clearly

$$OPT(t_m, J) = n(t_m + 1), OPT(t_m, J \setminus X_i) = (n - |X_i|)(t_m + 1).$$

Step 2 For $t_k \in [r_i, r_{i+1}), 1 \leq i \leq n - 1$, and $i < j$

$$OPT(t_k, J) = \min\{(t_k + 1)|X_i| + OPT(t_k + 1, J \setminus X_i), OPT(t_{k+1}, J)\}$$

$$OPT(t_k, J \setminus X_j) = \min\{(t_k + 1)|X_{j,i}| + OPT(t_k + 1, J \setminus X_j), OPT(t_{k+1}, J \setminus X_j)\}$$

Here $t_k + 1$ is an element of Ω, for which the value of associated optimal schedule has been obtained at the current step. The desired optimal solution value is then equal to $OPT(r_1, J)$. The corresponding optimal schedule can be found by backtracking. At any t_k there are up to n terms and each requires two operations. Since there are $O(n^2)$ elements in Ω, the algorithm terminates in a total of $O(n^3)$ operations.

We comment that the algorithm can be improved to solve the problem where $B < n$ and all job processing times are identical.

Next we discuss the problem with the number of the processing times being a constant. Without loss of generality, the set of processing times is $\{P_1 \leq P_2 \leq \cdots \leq P_K\}$ with K a fixed constant. We index all jobs in the nonincreasing order of their release times, $r_1 \leq r_2 \leq \cdots \leq r_n$.

Consider

$$\Re = \{t_i | t_i = r_j + x_1 P_1 + x_2 P_2 + \cdots + x_k P_K, j = 1, 2, \cdots, n; \sum_{i=1}^{K} x_i \leq n\}$$

Similar to the uniform job length case, it is easy to see that there is an optimal schedule for which each t_i represents a possible starting time of a batch of jobs. Clearly, the number of points in $\Re$ is at most nC_{n+K-1}^{K}.

Index all elements t_is of $\Re$ such that $t_i \leq t_j$, for $i < j$. Let $|\Re| = m$ and $t_1 = r_1, t_m = r_n + nP_K$. Resorting to the similar analysis to observation 1, we have

Observation 2 *For any subset of J, there is an optimal schedule such that the starting time of every batch is a point in $\Re$.*

Another conclusion is about the characteristic of jobs in the same batch in an optimal schedule.

By observation 2, we know that there may be some jobs arriving at time t_k, no job arrives during (t_i, t_{i+1}). Let

$$\mathcal{X} = \{x_1 P_1 + x_2 P_2 + \cdots + x_k P_k : \sum_{i=1}^{K} x_i \leq n\}$$

denote the collection of multi-sets of x_i jobs of length P_i, $i = 1, 2, \cdots, k$. Let $OPT(t, X)$ denote the objective value of the optimal schedule for job set X restrict the possible starting time at t.

Observation 3 *In any optimal schedule, the maximum job length of every batch is smaller than the minimum job length of jobs that are available at the starting time of the batch but not processed in the batch.*

Given a multi-set X, denote by X^j the multi-set obtained from X by removing from X all the jobs of length no more than P_j. Let $Y(\alpha, \beta)$ denote set of jobs with arrival times in $(\alpha, \beta]$. We now are ready to present our algorithm.

Algorithm 2

Step 1 Compute, at time t_m, $OPT(t_m, X)$ for all $X \in \mathcal{X}$.

Step 2 For each $X \in \mathcal{X}$, compute
$$OPT(t_k, X \cup Y(t_{k+1}, t_m)) :=$$
$$\min\{(t_k + P_j)|X - X^j| + OPT(t_k + P_j, X^j \cup Y(t_{k+1}, t_m)),$$
$$OPT(t_{k+1}, X \cup Y(t_{k+1}, t_m)) \}$$

The desired optimal value is $OPT(t_1, J)$. Notice that the number of possible t_k is a polynomial in n when k is a fixed constant. $|\mathcal{X}|$ is also a polynomial in n when k is a fixed constant. Therefore, for any fixed t_k the above requires computation of a polynomial number of terms. Therefore, the total number of operations is a polynomial in n when k is a fixed constant.

When the jobs are weighted, a job of length P_i and weight w_i can be treated as w_i jobs of length P_i (the input size is $O(\log w_i + \log P_i)$). The above dynamic programming approach still works. It also remains a polynomial time algorithm (polynomial in $O(n + \log w_i + \log P_i)$) since the size of the multi-sets can be presented by the sum of weights (coded in binary). In conclusion, we have

Theorem 2. *The problem* $1|r_j, B = +\infty| \sum w_j C_j$ *is polynomially solvable if the number of job lengths is a constant.*

4   Job system with finite number of arrival times

In this section, we consider the special case where there are a constant number of job arrival times.

Theorem 3. *The problem* $1|r_j, B = +\infty| \sum w_j C_j$ *is polynomially solvable if the number of job arrival times is a constant.*

When the release times of all jobs are the same (say, at zero), a dynamic programming algorithm was proposed by Brucker, et al., in [1]. We denote by $OPT_0(J)$ the optimal total completion time for a job set J all the jobs of which arrive at time zero.

Consider the case of two arrival times for the jobs, let P be the set of jobs arriving at time zero, Q the set of jobs arriving at time r. The job system is denoted by $((P, 0); (Q, r))$ and the optimal total completion time is denoted by $OPT((P, 0); (Q, r))$.

Denote by P^k the set of the first k shortest jobs in P. In the optimal schedule for our problem, by Observation 3, there must exist some k such that, jobs in P^k are scheduled to start before time r, and jobs in $P - P^k$ are postponed to start on or after time r. Let S^* denote the optimal schedule. Let $TC(S^*, A)$ denote the sum of completion times of jobs in A according to the optimal schedule S^* and $MS(S^*, A)$ denote the makespan of jobs in A according to the optimal schedule S^*. For this particular k, we consider two possibilities. First, $MS(S^*, A) < r$. For this simple case, $OPT((P, 0); (Q, r)) = OPT_0(P^k) + (n - k)r + OPT_0(Q \cup (P - P^k))$; otherwise, $OPT((P, 0); (Q, r)) = TC(S^*, P^k) + (n - k)MS(S^*, P^k) + OPT_0(Q \cup (P - P^k))$ (recall that $OPT_0(J)$ is the optimal total completion time for a job set J all the jobs of which arrive at time zero). The first case can be solved in polynomial time by applying twice the algorithm in [1], one to the job set P^k and another to $Q \cup (P - P^k)$. Our problem can be reduced to the second case, that is, to find a schedule S^* that achieving the following minimum:

$$OPT((P, 0); (Q, r)) = \min_{\substack{all\ k,\ all\ S}} \{TC(S, P^k) + (n-k)MS(S, P^k) + OPT_0(Q \cup (P - P^k))\}.$$

In this formula, we can calculate the value of the last term for every k independent of the schedule S. Then, we need to find a schedule S^* that minimize the sum of the first two terms. Our solution is a construction that reduces the determination of the schedule S^* for P^k to an optimal scheduling problem of a job system with the same arrival time.

For each k, take $n - k$ jobs with sufficient large processing time M, say $M = P_1 + P_2 + \cdots + P_n$. Let their arrival time to be all at time zero. We denote this subset of jobs by M_{n-k}. Consider the scheduling problem of the job set $P^k \cup M_{n-k}$, arriving at time zero. Jobs in M_{n-k} must be scheduled at the end because of their large processing time. Then the optimal solution $OPT_0(P^k \cup M_{n-k})$ for this fixed k can be written as the following:

$$\min_{\text{all } T} TC(T, P^k) + (n - k)MS(T, P^k) + OPT_0(P^k \cup M_{n-k}).$$

By the algorithm in [1], the last term of this formula can be calculated independent of the schedule T. To minimize the above term, we seek a schedule T^* which minimizes the sum of the first two terms. This is exactly the same sub-task for the above two stage problem for the same fixed k.

Algorithm 3

Step 1 Schedule optimally the jobs in P^k by the dynamic programming algorithm in [1] to get a schedule S_k with makespan M^k, $1 \leq k \leq |P|$.

Step 2 If $M^{|P|} \leq r$, then optimally schedule Q as in [1]. The schedule combining these two schedules is the optimal one for J.

Step 3 Otherwise, for each k, find the optimal schedule T^k for the job set $P^k \cup M_{n-k}$ all with arrival time zero.

Step 4 For each $k = 1, 2, \cdots, |P|$ with $TC(S_k, P^k) < r$, calculate
$$C_k := \min\{ TC(S_k, P^k) + (n - k)r + OPT_0(Q \cup (P - P^k)),$$
$$TC(T^k, P^k) + (n - k)MS(T^k, P^k) + OPT_0(Q \cup (P - P^k)) \},$$
otherwise set $C_k := TC(T^k, P^k) + (n - k)MS(T^k, P^k) + OPT_0(Q \cup (P - P^k))$.

Step 5 Solve
$$OPT((P, 0); (Q, r)) = \min_{\text{all } k}\{C_k\}.$$

The correctness is shown as above and the time complexity of the algorithm is $O(n^2 \ln n)$ since it takes $O(n \log n)$ time to find the optimal solution to the scheduling problem with n jobs arriving at the same time.

When there are fixed number K of distinct arrival times $r_1 < r_2 < \cdots < r_K$, again define P^{k_i} to be the jobs that arrives at time r_i, $1 \leq i \leq K - 1$ and their processing time started before the next arrival time r_K in the optimal schedule S^*. Again, dependent on the values of $\alpha = MS(S^*; \{(P^{k_i}, r_i) : i = 1, 2, \cdots, K - 1\})$, there are two cases.

If $\alpha \leq r_K$, then it reduces two problems: one optimally scheduling the jobs in $\{(P^{k_i}, r_i) : i = 1, 2, \cdots, K - 1\}$ (one of $K - 1$ arrival times); one optimally solving the scheduling problems of the rest jobs, all arrive at the same time r_K.

Otherwise, $\alpha > r_K$. Similar to the case with two arrival times, we may move all other jobs to arrive at time r_{K-1} and change their processing times to a sufficiently large number to determine the optimal schedule for jobs in $\{(P^{k_i}, r_i) : i = 1, 2, \cdots, K - 1\}$, to solve a problem with $K - 1$ arrival times.

The correct solution depends on our correctly determining the set of jobs $\{(P^{k_i}, r_i) : i = 1, 2, \cdots, K-1\}$, which can be done by brute force in n^K guesses. Each requires solving two problems of $K-1$ arrival times. Therefore, the total time is $O(2^K n^{K!} n \log n)$ which is polynomial in n for fixed constant K.

5 Discussion and remarks

In this paper we study the batch machine scheduling problem. Our focus is for the case the batch processing machine can handle an unbounded number of jobs. This is a valid assumption in many cases. More importantly, the requirement poses a challenge very different from ordinary scheduling problems. With a deep understanding for this case, we hope to shed light on more usual cases with $B : 1 < B < n$.

The problem we solve here has been considered as very difficult by Brucker, et al. [1]. We have made significant progress by showing that the general problem to be NP-hard and presenting dynamic programming algorithms with novel ideas for two special cases.

References

1. P.Brucker, A. Gladky, C.N.Potts,H.Hoogeveen, M.Y.Kovalyov, C.N.Potts, T autenhahn and S.L. Van De Velde. Scheduling a batching machine, *Journal of Scheduling*, **1**, 31-54,(1998).
2. V.Chandru,C.Y.Lee and R.Uzsoy, Minimizing total completion time on batch processing machine,*International Journal of Production Research* **31**,2097-2121(1993).
3. V.Chandru,C.Y.Lee and R.Uzsoy, Minimizing total completion time on a batch processng machine with job family,*Operations Research Letters* **13**,61-65(1993).
4. R.L.,Graham, Lawler, J.K.Lenstra and Kan.Rinnooy, Optimization and approximation in deterministic sequencing and scheduling. *Annal of Discrete Math.* 5(1979).
5. C.Y.Lee and Uzsoy,R., Minimizing makespan on single batch processing machine with dynamic job arrivals, Research Report, Dept. Industrial and System Engineering, Univ. Florida, Jan, 1996.
6. R. Uzsoy, Scheduling a single batch processing machine with non-idential job sizes, *International J. Production Research* **23**, 1615-1635 (1994).

Approximation Algorithms for Bounded Facility Location*

Piotr Krysta** and Roberto Solis-Oba

Max-Planck-Institut für Informatik, Im Stadtwald, D-66123 Saarbrücken, Germany.
krysta, solis@mpi-sb.mpg.de

Abstract. The *bounded k-median problem* is to select in an undirected graph $G = (V, E)$ a set S of k vertices such that the maximum distance from a vertex $v \in V$ to S is at most a given bound d and the average distance from vertices V to S is minimized. We present randomized algorithms for several versions of this problem. We also study the bounded version of the uncapacitated facility location problem. For this latter problem we present extensions of known deterministic algorithms for the unbounded version, and we prove some inapproximability results.

1 Introduction

The *bounded k-median* problem is to select in an undirected graph $G = (V, E)$, a set S of k vertices (called *centers*) such that the maximum distance from a vertex $v \in V \setminus S$ to S is at most a given bound d and the average distance from vertices in $V \setminus S$ to S is minimized. This is a generalization of the k-median problem [1]. As an application, suppose that in some city it is desired to locate a set of k fire departments so that the average distance that a fire crew must travel to get to any building in the city is minimized. Moreover, for safety reasons, it is also required that the maximum time needed to move a crew to any point in the city does not exceed a certain bound. (Otherwise, when the fire crew finally reaches a fire site, the fire might have already destroyed everything there.)

Let $G = (V, E)$ be a graph with minimum dominating set of size k. If all edge lengths are 1 and $d = 1$, then the bounded k-median problem is equivalent to the minimum dominating set problem. The bounded k-median problem is also a generalization of the k-center problem, since in the bounded k-median problem it is desired to minimize the number of vertices at distance 2 from the centers.

Another related problem is the *bounded uncapacitated facility location problem*. Here, given a graph $G = (V, E)$, there is a set $F \subseteq V$ of possible locations for service facilities. Each site $i \in F$ has a cost f_i for opening a facility there. The cost of servicing a request from a vertex $v \in V \setminus F$ is the distance from v to the nearest facility. The placement of facilities must be such that the maximum cost for servicing a request is at most a given bound d. The goal is to find locations

* This work was supported in part by EU ESPRIT LTR No. 20244 (ALCOM-IT).
** Supported by DFG Graduiertenkolleg Informatik, Universität des Saarlandes.

for the facilities so that the total cost of servicing the vertices $v \in V \setminus F$ plus the total cost for opening facilities is minimized.

Although the bounded k-median problem has been studied before, most of the previous work centers on exact algorithms for solving the problem in exponential time [2,6]. The k-median problem is a prototypical clustering problem with a large set of applications (see e.g. [14]). Lin and Vitter [16] showed that the problem does not admit a polynomial time approximation scheme unless $P = NP$. They also designed an algorithm that for any value $\epsilon > 0$, finds a solution of value $(1 + \epsilon)$ times the optimum, but using $(1 + 1/\epsilon)(\ln n + 1)k$ centers [16].

For metric spaces Charikar et al. [3] presented an algorithm for the problem with a performance ratio of $O(\log k \log \log k)$, while Lin and Vitter [17] gave an algorithm that finds a solution of value $2(1 + \epsilon)$ times the optimum using $(1 + 1/\epsilon)k$ centers. Recently Arora et al. [1] designed a randomized algorithm for the problem when the set of vertices lay on the plane. For any value $\epsilon > 0$ the algorithm finds with high probability a solution of value at most $1 + \epsilon$ times the optimum in $O(n^{O(1+\frac{1}{\epsilon})})$ time. Very recently, Charikar et al. [4] designed the first constant-factor approximation algorithm for the metric k-median problem.

The uncapacitated facility location problem is a classical problem of Operations Research [8]. Guha and Khuller [11] have shown that the metric version of the problem is Max SNP-hard, and that it cannot be approximated with a factor better than 1.463 unless $NP \subseteq DTIME(n^{\text{poly} \log n})$. Shmoys et al. [18] presented an algorithm for the problem in metric spaces with performance ratio of 3.16. Algorithms with better ratios of 2.41 and 1.74 were later given in [11] and [7]. When the vertices of the graph lay on the plane Arora et al. [1] designed a randomized algorithm with performance ratio $1 + \epsilon$ and running time $O(n^{O(1+\frac{1}{\epsilon})})$, for any $\epsilon > 0$.

Our Results. The *service cost* of a vertex is defined as the distance from the vertex to its closest center. We prove that the bounded k-median problem is Max SNP-hard even when all edge lengths are 1 and the bound on the service cost is $d = 2$. Moreover, by extending ideas of [11], we prove that the solution of the problem cannot be approximated in polynomial time within a factor smaller than 1.367 unless $NP \subseteq DTIME(n^{O(\log \log n)})$. Even if we allow the use of at most $1.2784\,k$ centers, we show that the problem is still inapproximable within 1.2784, unless $NP \subseteq DTIME(n^{O(\log \log n)})$. These results are described in Section 2.

We present a technique that allows us to design randomized approximation algorithms for several versions of the bounded k-median problem when all edge lengths are 1 (in all these algorithms the stated performance ratio and number of centers are expected values): (i) an 1.4212-approximation algorithm for the bounded k-median problem that uses at most $2k$ centers and has maximum service cost $2d$, (ii) an 1.7768-approximation algorithm for the bounded k-median problem that uses at most k centers, but has maximum service cost $3d$, (iii) an 1.9321-approximation algorithm for the bounded k-median problem with maximum service cost $3d$ when the vertices have weights $\{1, +\infty\}$. For the bounded k-median problem we also give a deterministic 1.5-approximation algorithm that produces a solution with at most $2k$ centers and service cost at most $2d$.

We give approximation algorithms for a fault tolerant version of the bounded k-median problem: the bounded p-neighbor k-median problem. These algorithms improve on average the approximation ratios of the algorithms [13,15] for the unbounded k-center and p-neighbor k-center problems in the case of unit edge lengths. These results appear in Section 3.

For the case of arbitrary edge lengths, we extend algorithms of Guha and Khuller [11], Chudak [7], Shmoys et al. [18] and of Lin and Vitter [17] for the k-median, and the capacitated and uncapacitated facility location problems. These algorithms have the same performance guarantees as the original ones, and additionally they bound the maximum service cost of every vertex. These results are presented in Section 4.

2   Inapproximability Results

We can prove, by giving a reduction from the Max 3-Sat problem, that even when all edge lengths are equal to 1, the bounded k-median problem cannot be approximated within a constant factor smaller than $1 + 4.3 \cdot 10^{-5}$ unless $P = NP$.

Theorem 1. *The bounded k-median problem is Max SNP-hard even when all edge lengths are 1 and all vertices must be serviced at distance at most $d = 2$.*

We have also proven that if there exists a polynomial time α-approximation algorithm for the bounded k-median problem with ratio $\alpha < 1.367$, then $NP \subseteq DTIME(n^{O(\log \log n)})$. This result holds even if all edge lengths are 1 and $d = 1$. Our proof is based on ideas of Guha and Khuller [11].

Let X be a finite set of elements and $S \subseteq X$. We say that set S *covers* all its elements. Given a family of sets $S = \{S_1, \ldots, S_l\}$ such that $\cup_{i=1}^{l} S_i = X$, the minimum set covering problem asks for the smallest number of sets in S that cover X. Let $G = (V, E)$ be an undirected graph with unit length edges. Given two vertices $u, v \in V$ we say that u *dominates* v if either $u = v$ or u and v are adjacent. Let the minimum size of a dominating set in G be k.

Lemma 1. *If there is a polynomial time algorithm A_{Dom} that for any constant β picks βk vertices of G dominating $c'n$ vertices of G, where $c' > 1 - \frac{1}{e^\beta}$, then $NP \subseteq DTIME(n^{O(\log \log n)})$.*

This lemma is proven by showing how to use algorithm A_{Dom} to build an algorithm for the minimum set covering problem with performance ratio smaller than $\ln |X|$, which by a result of Feige [10] implies that $NP \subseteq DTIME(n^{O(\log \log n)})$.

Theorem 2. *Let $d = 1$ and $\alpha < 1.367$. If there exists an α-approximation algorithm for the bounded k-median problem that keeps the service cost of all vertices to at most 2, then $NP \subseteq DTIME(n^{O(\log \log n)})$.*

Proof. Let $G = (V, E)$ be a graph with minimum dominating set of size k. Assume that there is an α-approximation algorithm for the bounded k-median problem that keeps all service costs to at most $2d$. Run the algorithm on G and let

$S \subseteq V$ be the set of centers selected. Let $V_1 \subseteq V$ be the set of vertices at distance 1 from S. Let the number of vertices dominated by S be cn, i.e. $|S \cup V_1| = cn$, where $c < 1$. Since G has a dominating set of size k, there exists a solution to the bounded k-median problem of cost $n - k$. The above α-approximation algorithm finds a solution of cost at most $\alpha(n - k)$. In this solution there are $cn - k$ vertices at distance 1 from S, and $n - cn$ vertices at distance 2 from S. Thus the cost of this solution is $cn - k + 2(n - cn) \leq \alpha(n - k)$, and hence $\alpha \geq \frac{2 - c - \frac{k}{n}}{1 - \frac{k}{n}}$. By Lemma 1, $c \leq 1 - \frac{1}{e}$, and $\alpha \geq \frac{1 + \frac{1}{e} - \frac{k}{n}}{1 - \frac{k}{n}} > 1.367$. $\qquad\square$

Even if more than k centers are allowed, the problem is hard to approximate.

Lemma 2. *Let $\delta \geq 2$ and $0 \leq \epsilon \leq 0.2784$ be fixed constants. The bounded k-median problem in which it is allowed that a solution contains up to $(1 + \epsilon)k$ centers and the service cost for the vertices must be at most δd, cannot be approximated within a factor smaller than $1 + \frac{1}{e^{1+\epsilon}}$ unless $NP \subseteq DTIME(n^{O(\log \log n)})$.*

3   Uniform Cost Bounded k-Median Problem

The *uniform cost bounded k-median* problem is the bounded k-median problem restricted to graphs with unit length edges and $d = 2$. Let $G = (V, E)$ be an undirected graph with unit length edges. The *k-center* problem is to find the smallest distance d' such that there is a set S of k vertices for which the distance $d(v, S)$ from any vertex $v \in V$ to S is at most d'. Assume that the smallest size of a dominating set of G is k. Then the optimum solution for the k-center problem on G has value 1.

Hochbaum and Shmoys [13] designed a simple 2-approximation algorithm A_{HS} for the k-center problem. The algorithm finds the smallest edge length ℓ_e such that the following procedure selects a set S of size at most k. Set $S = \emptyset$ and then repeatedly do the following until $V = \emptyset$. Select some vertex $v \in V$ and add it to S. Remove from V all vertices within distance $2\ell_e$ from v. For a proof of the correctness of the algorithm see [13]. If we run A_{HS} on G, it selects a set S of k centers and places each vertex $v \in V$ at distance at most 2 from S.

Lemma 3. *Algorithm A_{HS} is a $(2 - \frac{k}{n-k})$-approximation algorithm for the uniform cost bounded k-median problem.*

3.1   A Randomized Algorithm

We can formulate the uniform cost bounded k-median problem with maximum service cost d as the following integer linear program, IP.

$$\text{Min } \sum_{i,j \in V} c_{ij} x_{ij} \tag{1}$$

$$\text{s.t. } \sum_{i \in N_j} x_{ij} = 1 \;\; \forall_{j \in V} \tag{2}$$

$$x_{ij} \leq y_i \qquad \forall_{j \in V}, \forall_{i \in N_j} \tag{3}$$

$$\sum_{i \in V} y_i \leq k \tag{4}$$

$$x_{ij}, y_i \in \{0, 1\} \; \forall_{i,j \in V} \tag{5}$$

In this program c_{ij} is the distance from vertex i to j, and N_j is the set of vertices at distance at most d from j. Set N_j includes vertex j. The meaning of the variables is as follows: $y_i = 1$ if and only if vertex i is chosen as a center, and $x_{ij} = 1$ if and only if i is a center, $i \in N_j$, and i is a closest center to j. Let LP be the linear program obtained by relaxing constraints (5) to $0 \leq x_{ij}, y_i \leq 1$.

Let $\lambda \in [0,1)$ be a fixed constant whose value will be specified later. Our algorithm is as follows. If $k > \frac{\lambda}{1+\lambda}n$, then we use algorithm A_{HS}. Otherwise, we solve LP and then round the solution using a rounding procedure based on ideas from Shmoys et al. [18] and Chudak [7]. Let (x, y) be an optimal solution to LP. By results of [7] we can assume that (x, y) is *complete*, i.e. : if $x_{ij} > 0$, then $x_{ij} = y_i$, for every $i, j \in V$. Our algorithm chooses independently with probability y_i each vertex i as a center. Let $S_1 = \{i \mid y_i > 0 \text{ and } i \text{ is a center}\}$. For each $j \in V$, if $j \in S_1$ then set $x_{jj} = 1$. Otherwise, if one of the vertices $i' \in N(j)$ is a center then set $x_{i'j} = 1$. We say that a vertex j is *serviced* by a center i if $x_{ij} = 1$. We run algorithm A_{HS}, and let S_2 be the set of centers that it chooses. Each unserviced vertex is assigned to its closest center in $S_1 \cup S_2$. Note that all vertices are serviced at distance at most $2d$.

For each vertex $j \in V$, let $C_j = \sum_{i \in V} c_{ij} x_{ij}$ denote the fractional cost of servicing j. The expected cost $E(cost(j))$ of servicing j in this rounded solution is bounded in the following Lemma.

Lemma 4. *For each vertex $j \in V$, $C_j + x_{jj} = 1$, and $E(cost(j)) \leq C_j + q(C_j + x_{jj})$, where q is a value no larger than $\frac{1}{e}$.*

Theorem 3. *There is a randomized 1.4212-approximation algorithm for the uniform cost bounded k-median problem, that produces a solution with expected number of centers $2k$. The algorithm services each vertex at distance at most 2.*

Proof. We first argue for the bound on the expected number of centers. In the first phase the algorithm chooses independently each $i \in V$ as a center with probability y_i, so by constraint (4), $E(|S_1|) = \sum_{i \in V} y_i \leq k$. In the second phase, A_{HS} selects at most k centers.

We now prove the bound on the approximation ratio. We consider two cases, if $k \leq \frac{\lambda}{1+\lambda}n$ then by Lemma 4, the expected cost of the solution is $E(cost) \leq \sum_{j \in V} C_j + q(C_j + x_{jj})$. Let L be the value of an optimal solution of LP, then $E(cost) \leq (1+q)L + q\sum_{j \in V} x_{jj} = (1 + q + \lambda q)L + q\sum_{j \in V}(x_{jj} - \lambda C_j)$.

Let $r = \sum_{j \in V}(x_{jj} - \lambda C_j)$. By Lemma 4, $C_j = 1 - x_{jj}$, so $r = \sum_{j \in V}((1 + \lambda)x_{jj} - \lambda)$. From constraint (3), $x_{jj} \leq y_j$ for each vertex j, thus $r \leq \sum_{j \in V}((1 + \lambda)y_j - \lambda)$. Now by constraint (4) and letting $n = |V|$, we get $r \leq (1 + \lambda)k - \lambda n$. Since $k \leq \frac{\lambda}{1+\lambda}n$, then $r \leq 0$. Thus $E(cost) \leq (1 + q + \lambda q)L$, and the expected performance ratio is $r_2 = 1 + q + \lambda q \leq 1 + (1 + \lambda)\frac{1}{e}$.

On the other hand, if $k > \frac{\lambda}{1+\lambda}n$ we use algorithm A_{HS}, but we select $2k$ centers. The value of the solution that the algorithm finds is $2n - 6k$, and so the performance ratio is $r_1 = \frac{2n - 6k}{n - k} < 2 - 4\lambda$. By choosing λ so that $r_1 = r_2$, the performance ratio of the overall algorithm is approximately 1.4212.   □

3.2   A Greedy Algorithm

Now we describe a simple greedy 1.5-approximation algorithm for the uniform cost bounded k-median problem that uses at most $2k$ centers. Let S be a set of vertices and $N(S)$ be the neighbors of S. The algorithm sets $S \leftarrow \emptyset$ and then it greedily adds to S the vertex with largest number of neighbors not in $N(S)$. The process is repeated until there are k vertices in S.

Lemma 5. *Let $G = (V, E)$ be a graph with a dominating set S^* of size k. The above algorithm finds a set S such that $|N(S)| \geq \frac{n-k}{2}$.*

Proof. Partition the vertices $V \setminus S^*$ into k disjoint groups N_ℓ, $\ell \in S^*$, such that every vertex in N_ℓ is adjacent to vertex ℓ. Index the groups in non-increasing order of size. Let S_i, $0 \leq i \leq k$ be the set of the i first vertices chosen by the algorithm. We show by induction on i that $|N(S_i)| \geq \frac{1}{2}|\bigcup_{\ell \leq i} N_\ell|$ for all i.

The claim clearly holds for $i = 1$. Assume that it also holds for all $j < i$. If $|N(S_{i-1}) \cap (\bigcup_{\ell \leq i} N_\ell)| \geq \frac{1}{2}|N(S_i^*)|$ the claim follows. Otherwise, there must be a vertex $\ell \in S^*$, $\ell \leq i$, such that $|N(S_{i-1}) \cap N_\ell| \leq \frac{1}{2}|N_\ell|$. In the i-th iteration the algorithm selects a vertex u with $|N(\{u\}) \cap N(S_{i-1})| \geq |N_\ell| \geq |N_i|$. By induction hypothesis $|N(S_{i-1})| \geq \frac{1}{2}|\bigcup_{\ell < i} N_\ell|$, hence $|N(S_i)| \geq \frac{1}{2}|\bigcup_{\ell \leq i} N_\ell|$. $\square$

Theorem 4. *There is a deterministic 1.5-approximation algorithm for the uniform cost bounded k-median problem, that produces a solution with at most $2k$ centers, and services each vertex at distance at most 2.*

Proof. To the set S found by the above algorithm add the set of centers selected by algorithm A_{HS} for the k-center problem. The new set has at most $2k$ vertices and every vertex $v \in V \setminus S$ is at distance at most 2 from their nearest vertex in S. Moreover at least $(n-k)/2$ vertices are neighbors of centers, and so the value of the solution is at most $\frac{3(n-k)}{2}$. Since the optimum solution has value at least $n - k$ the claim follows. $\square$

3.3   A Randomized Algorithm for $d = 3$

In this section we relax a little the constraint on the distance bound d, by making $d = 3$, and get an algorithm with performance ratio 1.7768. This algorithm uses the clustering technique of Shmoys et al. [18]. The idea is to ensure for each vertex that there is always some center within distance 3 from it. The algorithm is as follows. If $k > \frac{\lambda}{\lambda+1} n$ then use algorithm A_{HS}. Otherwise, find an optimal complete solution (x, y) to LP. For each vertex $j \in V$, let $N(j) = \{i \in V \mid x_{ij} > 0\}$. Partition the set of vertices in the following manner. Choose any vertex $i \in V$ and create a cluster Q_i that includes all vertices $j \in V$ such that $N(i) \cap N(j) \neq \emptyset$. Remove the vertices in Q_i from V, and repeat this process until V is empty. Let C be the set of vertices that serve as indices for the clusters Q_i.

Divide the vertices into two sets: $A = \bigcup_{j \in C} N(j)$ and $B = V \setminus A$. For each cluster Q_i, the algorithm selects exactly one vertex from $N(i)$ as a center in such a way that the probability that vertex $j \in N(i)$ is chosen as the center is $x_{ij} / \sum_{k \in N(i)} x_{ik}$. Then each vertex $i \in B$ is selected as a center independently with probability y_i.

Consider a vertex $j \in Q_i$. If vertex $k \in N(j)$ is a center, then j is serviced at distance 1. But if none of the neighbors $N(j)$ is a center then j is serviced at distance at most 3. To see why, let $i' \in Q_i$ be a center. Since i' is a neighbor of i, and j is at distance at most 2 from i the claim follows. Let B_j be a random variable denoting the distance between j and the center from $N(i)$. Let D_j denote the event that none of the vertices in $N(j)$ are centers. Then we can prove the following result.

Lemma 6. *For each $j \in V$, $E(cost(j)) \leq (1 - \frac{q}{C_j})C_j + 3q$, where $q \leq \frac{1}{e}$.*

Using Lemmas 6 and 4 we obtain the following results.

Lemma 7. *In this algorithm, for each $j \in V$, $E(cost(j)) \leq C_j + 2q(C_j + x_{jj})$.*

Theorem 5. *There is a randomized 1.7768-approximation algorithm for the uniform cost bounded k-median problem, that produces a solution with expected number of centers k and in which every vertex is serviced within distance 3.*

Proof. From the completeness of (x, y) it follows that for each $i \in V$ the probability that i is a center is y_i. Thus $E(\#centers) = \sum_{i \in V} y_i \leq k$, by constraint (4). The rest of the proof is analogous to the proof of Theorem 3.   □

3.4   Weights on the Vertices

Consider a graph $G = (V, E)$ with unit length edges and weights w_i on the vertices. The weights can be either 1 or ∞. The weighted bounded k-median problem is to find a set S of vertices such that $\sum_{i \in S} w_i \leq k$, $d(v, S) \leq d$ for each $v \in V$, and $\sum_{v \in V} d(v, S)$ is minimized. This problem is interesting because it allows us to exclude some vertices as possible centers. Our results improve on average the 3-approximation algorithm of Hochbaum and Shmoys [13] for weighted k-center problem in the special case of weights $\{1, +\infty\}$ on vertices.

We extend the techniques of Section 3.3 to design a randomized 1.9321-approximation algorithm for the weighted bounded k-median problem. Let A_{HS}^W be the 3-approximation algorithm from [13] for the weighted k-center problem. It is easy to see that A_{HS}^W is a 3-approximation algorithm for the weighted bounded k-median problem. Proceeding as in Lemma 3 we obtain the following result.

Lemma 8. *Algorithm A_{HS}^W is a $(3 - \frac{k}{n-k})$-approximation algorithm for the weighted bounded k-median problem, that services each vertex at distance at most 3.*

We can write the weighted bounded k-median problem as an integer linear program, that we call IP$_2$. This is the same as IP, with constraint (4) replaced by $\sum_{i \in V} w_i y_i \leq k$. Let LP$_2$ be the linear program relaxation of IP$_2$.

The algorithm for weighted bounded k-median is like the algorithm of Section 3.3. If $k > \frac{\lambda}{\lambda+1}n$ then use algorithm A_{HS}^W. Otherwise, solve LP_2 and find a complete solution (x, y). Then we define clusters like before, but this time we create clusters only around vertices i with $y_i > 0$. Notice that $y_i = 0$ if $w_i = +\infty$. Using similar arguments as those in Section 3.3 we can prove the following result.

Theorem 6. *There is a randomized* 1.9321-*approximation algorithm for the weighted bounded k-median problem, when the vertex weights are either* 1 *or* ∞, *that finds a set of centers of expected total cost at most k. In this solution each vertex is serviced at distance at most* 3. *Moreover, for each vertex v, the probability that v is serviced at distance* 3 *is at most* $\frac{1}{e}$.

3.5 Fault Tolerant Problem

The bounded p-neighbor k-median problem is given a value d, find a set S of k centers such that for each vertex $v \in V \setminus S$ there are at least p centers in S within distance d of v, and $\sum_{v \in S} d(v, S)$ is minimized. We can also handle the case when $d(v, S)$ is the sum of the distances from v to its p closest centers. In this section we are interested in the case when all edge lengths are 1 and $d = 2$.

The p-neighbor k-center problem consists in finding the smallest distance d for which there is a set S of k vertices such that every vertex $v \in V \setminus S$ is at distance at most d from p vertices from S. Khuller et al. [15] have designed a 2-approximation algorithm for the problem, and when $p = 2$ we can modify the solution that this algorithm finds so that the number of vertices within distance d from their closest centers is at least $k/2$. The idea is that if the set S_d of vertices within distance d from S has size smaller than $k/2$ then some of these vertices can be exchanged from vertices in S to get a new solution in which the set S_d is larger. Since the algorithm involves analyzing several cases, and its description is slightly complicated we omit it. We call this algorithm A_N.

The bounded p-neighbor k-median problem can be modeled as the integer program IP of Section 3.1, with the constraint (2) replaced by $\sum_{i \in N_j} x_{ij} = p, \forall_{j \in V}$. We can solve the linear program relaxation of this integer program and then round the solution as in Section 3.1 but using algorithm A_N instead of A_{HS}. We can show that if $k \leq \frac{2\lambda}{1+\lambda}n$, then for each vertex j, $E(cost(j)) \leq C_j + \frac{1}{2}q(C_j + x_{jj})$, where $q \leq \frac{1}{e^2}$; $E(\#centers) \leq 2k$; and $E(cost) \leq (1 + \frac{q}{2} + \lambda\frac{q}{2})L$, where L is the value of an optimum solution for the linear program. The performance ratio of the algorithm in this case is $r_2 = 1 + (1 + \lambda)\frac{1}{2e^2}$.

If $k > \frac{2\lambda}{1+\lambda}n$ we use algorithm A_N to get a solution of value $2n - 4.5k$. The performance ratio of the algorithm in this case is $r_1 = \frac{2n-4.5k}{n-k} < 2 - \frac{5\lambda}{1-\lambda}$. Selecting λ so that $r_1 = r_2$, we get $r_1 \approx 1.0782$.

Theorem 7. *There is a randomized* 1.0782-*approximation algorithm for the uniform cost bounded 2-neighbor k-median problem that produces a solution with expected number of centers at most $2k$ and in which every vertex is at distance at most* 2 *from* 2 *of the centers.*

We have also designed an algorithm for arbitrary value of p with performance ratio $\frac{1+\lambda}{pe^p}$.

4   Facility Location Problems with Arbitrary Lengths

In this section we consider the bounded uncapacitated facility location and the bounded k-median problems with arbitrary edge lengths.

Given a graph $G = (V, E)$, let f_i be the cost of selecting vertex i as a center. The bounded uncapacitated facility location problem can be stated as the following integer program, that we call IP_3.

$$\text{Min } \sum_{i,j \in V} c_{ij} x_{ij} + \sum_{i \in V} f_i y_i$$
$$\text{s.t.} \quad \sum_{i \in N^d(j)} x_{ij} = 1 \qquad \forall_{j \in V} \tag{6}$$
$$x_{ij} \leq y_i \qquad \forall_{i,j \in V}$$
$$x_{ij}, y_i \in \{0, 1\} \qquad \forall_{i,j \in V}$$

where $N^d(j) = \{i \in V \mid c_{ij} \leq d\}$ and c_{ij} is the distance from i to j. We solve the linear programming relaxation of IP_3, and if $f_i = 1$ for all vertices i, we can use ideas from Chudak [7] and Guha and Khuller [11] to round this solution to get an algorithm with the following performance ratio. The details will appear in the full version of the paper.

Theorem 8. *If $f_i = 1$ for each vertex $i \in V$, then there is a deterministic 2.104-approximation algorithm for the bounded uncapacitated facility location problem that services each vertex at a distance at most $2d$.*

If we relax the constraint on the maximum distance from a vertex to its closest center, we can use ideas of Section 3.3 to get an algorithm with better performance ratio. Build the dual for the linear program relaxation of IP_3 and solve it optimally. Let v_j be a dual variable corresponding to the primal constraint (6). We solve the linear program relaxation of IP_3 and perform the clustering step of Section 3.3 but we create clusters around vertices i with smallest value $v_i + C_i$. The rest of the algorithm is as in Section 3.3.

Theorem 9. *There is a deterministic 1.736-approximation algorithm for the bounded uncapacitated facility location problem that services each vertex at distance at most $3d$.*

In the capacitated facility location problem each center can be assigned a maximum fixed demand u (at most u vertices can be assigned to each center). Shmoys et al. [18] designed a 5.69-approximation algorithm for the metric version of the problem that assigns to each center a demand of at most $4.24 \cdot u$. We can extend their algorithm to get the following result.

Theorem 10. *There is a deterministic 5.69-approximation algorithm for the metric bounded capacitated facility location problem that assigns to each center a demand of at most $4.24 \cdot u$, and services each vertex at distance at most $3d$.*

We have also designed an approximation algorithm for the bounded k-median problem with arbitrary edge lengths. This algorithm extends the $2(1 + 1/\varepsilon)$-approximation algorithm of [17] for the metric k-median problem, that produces a solution with at most $(1 + \varepsilon)k$ centers (for any $\varepsilon > 0$).

Theorem 11. *Given any $\varepsilon > 0$, there is a deterministic $2(1+\frac{1}{\varepsilon})$-approximation algorithm for the metric bounded k-median problem, that produces a solution with at most $(1+\varepsilon)k$ centers. Each vertex is serviced at distance at most $2d$.*

References

1. S. Arora, P. Raghavan and S. Rao, *Approximation schemes for Euclidean k-medians and related problems*, Proceedings of the 30th Annual ACM Symposium on Theory of Computing, 106–113, 1998.
2. O. Berman and E.K. Yang, *Medi-center location problems*, Journal of the Operational Research Society, **42**, 313–322, 1991.
3. M. Charikar, Ch. Chekuri, A. Goel and S. Guha, *Rounding via trees: deterministic approximation algorithms for group steiner trees and k-median*, Proceedings of the 30th Annual ACM Symposium on Theory of Computing, 114–123, 1998.
4. M. Charikar, S. Guha, É. Tardos, and D.B. Shmoys, *A constant-factor approximation algorithm for the k-median problem*, to appear in the Proceedings of the 31st ACM Symposium on Theory of Computing, 1999.
5. S. Chaudhuri, N. Garg and R. Ravi, *The p-neighbor k-center problem*, Information Processing Letters, **65**, 131–134, 1998.
6. I.C. Choi and S.S. Chaudhry, *The p-median problem with maximum distance constraints: a direct approach*, Location Science, **1**, 235–243, 1993.
7. F. Chudak, *Improved approximation algorithms for uncapacitated facility location*, in Integer Programming and Combinatorial Optimization, volume 1412 of Lecture Notes in Computer Science, 180–194, 1998.
8. G. Cornuejols, G.L. Nemhauser, and L.A. Wolsey, *The uncapacitated facility location problem*, in Discrete Location Theory (Mirchandani and Francis, editors), Wiley, New York, pp. 119–171, 1990.
9. Z. Drezner, editor, *Facility location. A survey of applications and methods*, Springer-Verlag, New York, 1995.
10. U. Feige, *A threshold of $\ln n$ for approximating set cover*, Proceedings of the 28th Annual ACM Symposium on Theory of Computing, 314–318, 1996.
11. S. Guha and S. Khuller, *Greedy strikes back: improved facility location algorithms*, Proceedings of the 9th Annual ACM-SIAM Symposium on Discrete Algorithms, 649–657, 1998.
12. D.S. Hochbaum, *Various notions of approximations: good, better, best, and more*, in *Approximation Algorithms for NP-hard Problems* (D. Hochbaum, editor), PWS Company, Boston, 1997.
13. D.S. Hochbaum and D.B. Shmoys, *A unified approach to approximation algorithms for bottleneck problems*, Journal of the ACM, **33**, 533–550, 1986.
14. A.K. Jain and R.C. Dubes, *Algorithms for clustering data*, Prentice Hall, 1981.
15. S. Khuller, R. Pless and Y. J. Sussmann, *Fault tolerant k-center problems*, to appear in Theoretical Computer Science, 1999.
16. J.H. Lin and J.S. Vitter, *ϵ-approximations with minimum packing constraint violation*, Proceedings 24th ACM Symposium on Theory of Computing, 771-782.
17. J.H. Lin and J.S. Vitter, *Approximation algorithms for geometric median problems*, Information Processing Letters, **44**, 245–249, 1992.
18. D.B. Shmoys, É. Tardos, and K. Aardal, *Approximation algorithms for facility location problems*, Proceedings of the 29th ACM Symposium on Theory of Computing, 265–274, 1997.

Scheduling Trees onto Hypercubes and Grids Is $\mathcal{NP}$-complete

Satoshi Tayu

School of Information Science
Japan Advanced Institute of Science and Technology
1-1 Asahidai, Tatsunokuchi-machi, Ishikawa 923-1292, Japan
tayu@jaist.ac.jp

Abstract. In the last two or three decades, the task scheduling onto parallel processing systems have been extensively studied. The structure of the parallel processing systems of the scheduling problem which many researchers have studied is restricted to be fully connected. Further, in the most cases, they have studied without communication delay. However, realistic models of parallel processors, such as hypercubes, grids and so forth, are not fully connected and the communication delay have much effect on the completion time of tasks. In this paper, we consider and formulate the scheduling onto non-fully connected parallel processing systems taking communication delay into consideration. We also show that the problems to schedule unit-size-tree-tasks onto hypercubes and grids are $\mathcal{NP}$-complete.

1 Introduction

The problem to execute parallel algorithms onto parallel processors is formulated as graph scheduling problems. The scheduling problem is widely investigated in connection with the parallel computing. The instance of the problem generally consists of two parts; the one is a set of resources which corresponds to a parallel processor and the other is a set of tasks with data flows and some weight functions. A parallel processor consists of processing elements (PEs) with connection wires each of which connects two of them. A parallel processor is said to be *complete* or *fully connected* if any two PEs are connected by a connection wire.

In the general scheduling problem, the parallel processor is assumed to be fully connected. Moreover, in those studies, any data transmitted between tasks is assumed to have unit size or sometimes the researchers did not consider the communication delay. In the later case, a parallel algorithm is often regarded as a partial ordered set (poset) $(J, \prec)$, which is a task set J together with a partial order $\prec$ on J, with a weight function $W : J \rightarrow \{0, 1, \cdots\}$. For two tasks s and s', $s \prec s'$ implies that, s must have been finished before s' starts to be operated. The objective function of the scheduling is generally defined as the total completion time. The following is a general task scheduling problem and shown to be $\mathcal{NP}$-complete by Karp in 1972 [5].

T. Asano et al. (Eds.): COCOON'99, LNCS 1627, pp. 251–260, 1999.
© Springer-Verlag Berlin Heidelberg 1999

General Shceduling problem:

Given a set $J = \{v_1, v_2, \cdots, v_n\}$ of n tasks, a partial order $\prec$ on J, weight function $W : J \rightarrow \{0, 1, \cdots\}$, a set V of PEs, and a time limit t, determine whether there exists a total time function τ from J to $\{0, 1, \cdots, t - 1\}$ and a mapping $\phi : J \rightarrow V$ such that

1. if $v_i \prec v_j$, then $\tau(v_i) + W(v_i) \leq \tau(v_j)$,
2. $\tau(v) + W(v) \leq t$ for each $v \in J$, and
3. $\phi(v_i) \neq \phi(v_j)$ if $\tau(v_i) < \tau(v_j) + W(v_j)$ and $\tau(v_j) < \tau(v_i) + W(v_i)$.

ϕ assigns each task to a PE and τ does to the beginning time of the task to be operated. For a task v, $W(v)$ represents the execution time (task weight) which v takes to be executed.

It is known that the problems with some restrictions are still $\mathcal{NP}$-complete. For example, the following scheduling problems obtained from General Scheduling problem by adding some restrictions have been shown to be still $\mathcal{NP}$-complete.

Single-execution-time scheduling [11]:
Restriction: the weight of each task is one, i.e., $W(v_i) = 1$ for $v_i \in J$.

Two-PEs, one- or two-time scheduling. [11]:
Restriction: $|V| = 2$ and $W(v_i) = 1$ or 2.

Two-PEs, interval-order scheduling. [7]:
Restriction: partial order $\prec$ is an interval order and $|V| = 2$.

Single-execution-time, opposing-forest scheduling. [4]:
Restriction: $W(v_i) = 1$, and the partial order $\prec$ is an opposing forest.

There are also some positive results. Scheduling problems can be solve in polynomial time if $\prec$ and W have certain restrictions. The followings are known to be in $\mathcal{P}$:

In- or out-forest unit-size scheduling. [4]:
Restriction: $W(v) = 1 \, \forall v \in J$, $\prec$ forms an in- or out-forest.

Interval-ordered, unit-size scheduling. [7]:
Restriction: $W(v) = 1 \, \forall v \in J$, $\prec$ forms an interval graph.

2-PEs, unit-execution-time scheduling [2,3]:
Restriction: $|V| = 2$, $W(v) = 1 \, \forall v \in J$.

Moreover, for some above $\mathcal{NP}$-complete problems, there are approximation or heuristic algorithms for a certain class of task sets [2,3,6,10].

As mentioned above, many researchers have not taken the communication delay into account. Some scheduling problems with communication delay have been investigated [7,8]. However, in most cases, the structure of parallel processors are restricted to be fully connected and, sometimes, the number of PEs is 2 or 3. There are few results for non-fully connected parallel processor and only a few studies consider the data confliction on the routings of distincts data when the researchers construct heuristic algorithms.

We will investigate scheduling problems onto parallel processor which is non-fully connected. Therefore, we have to distinguish every data since each data may be transmitted along a route from one PE to another. So, we employ the notion of the directed acyclic graph (dag) representation for the poset one. In this paper, we formulate such task scheduling problems. Further, we show that the schedulings of algorithms onto hypercubes and grids are $\mathcal{NP}$-complete even if the task graph is restricted to be a rooted tree whose task weight and data size both are unity.

2 Definition

An undirected graph G is a pair of a vertex set, denoted by $V(G)$ and an edge set, denoted by $E(G)$, where $E(G)$ is the set of unordered pairs of vertices. An edge $e = \{u, v\}$ is said to connect between u and v. u and v are called *endvertices* of e. Similarly, a directed graph D is a pair of a vertex set $V(D)$ and an arc set $A(D)$. $A(D)$ is the set of ordered pairs of vertices. An arc $a = (u, v)$ is said to connect from u to v. $t(a) = u$ and $h(a) = v$ are called the *tail* and *head* of a, respectively. The *in-degree* of v is the number of vertices from each of which an arc connects to v, i.e., $\#\{u|(u, v) \in A(D)\}$, and the *out-degree* of v is the number of vertices to each of which an arc connects from v, i.e., $\#\{u|(v, u) \in A(D)\}$.

We denote by $[v_0, v_1, \cdots, v_p]$ an undirected path a path consisting of the vertex set $\{v_0, v_1, \cdots, v_p\}$ and the edge set $\{\{v_i, v_{i+1}\}|0 \le i \le p-1\}$. Also, we denote by $(v_0, v_1, \cdots, v_p)$ a directed path consisting of the vertex set $\{v_0, v_1, \cdots, v_p\}$ and the arc set $\{(v_i, v_{i+1})|1 \le i \le p - 1\}$. By definition, $[v_0, v_1, \cdots, v_p] = [v_p, v_{p-1}, \cdots, v_0]$. The *length* of an undirected (resp. a directed) path P, denoted by $|P|$, is the number of edges (resp. arcs) of P. Note that (u, v) may represent an arc or, sometimes, a directed path with length one. In the latter case, we will specify if we need to distinguish from arcs.

For an undirected graph G and $u, v \in V(G)$, the *distance* between u and v, denoted by $d_G(u, v)$, is the minimum length of a path connecting between u and v. Similarly, for a directed graph D and $u.v \in V(D)$, the distance from u to v, also denoted by $d_D(u, v)$, is the minimum length of a directed path connecting from u to v. A directed tree is called a *in-tree* (resp. *out-tree*) if the out-degree (resp. in-degree) of any vertex except the root is 1. Then, the out-degree (resp. in-degree) of the root is 0 and there exist a path from any vertex to the root (resp. from the root to any vertex). We call both trees 'rooted trees'. For an arc (u, v) of an out-tree, u is called the *parent* of v and denoted by par(v).

An *embedding* $\langle \varphi, \varrho \rangle$ of an undirected graph G into an undirected graph H is defined by a one-to-one mapping $\varphi : V(G) \to V(H)$, together with a mapping ϱ that maps an edge $\{u, v\} \in E(G)$ onto an undirected path of H connecting between $\varphi(u)$ and $\varphi(v)$. The *dilation* of $e \in E(G)$ under $\langle \varphi, \varrho \rangle$ is the length of $\varrho(\{u, v\})$. The dilation of $\langle \varphi, \varrho \rangle$ is the maximum dilation over all edges of G. Note that there exist an embedding of G into H with dilation 1 if and only if G is a subgraph of H.

The structure of a parallel algorithm can be represented by a dag T. The vertex set $V(T)$ corresponds to the task set and an arc of $A(T)$ corresponds to a flow of data. For $\nu, \nu' \in V(T)$, the existence of an arc (ν, ν') implies that operation ν' needs the result of operation ν. Let $W_V : V(T) \rightarrow \{0, 1, \cdots\}$ and $W_A : A(T) \rightarrow \{0, 1, \cdots\}$ be weight functions of $V(T)$ and $A(T)$, respectively. Define $W(v) = W_V(v)$ for $v \in V(T)$ and $W(a) = W_A(a)$ for $a \in A(T)$. A task graph $T = (T, W)$ of a parallel algorithm is a dag T together with the weight function W. If W return 1 for any vertex or arc, we simply write (T, W) by T.

A parallel processor can be represented by an undirected graph G. Each vertex of G corresponds to a PE of the parallel processor and each edge of G corresponds to a connection wire connecting between two PE's to which the end-vertices of the edge corresponds. To show the $\mathcal{NP}$-completeness, it is sufficient to consider the case that $W_V(v) = 1 \ \forall v \in V(T)$ and $W_A(a) = 1 \ \forall v \in A(T)$. So, a parallel algorithm $T = (T, W)$ is written by a dag T.

Now, we will prepare for the definition of a scheduling of T onto G. For a task $\nu \in V(T)$, $\phi(\nu)$ represents the processor which operates ν and $\tau^\phi(\nu)$ represents the starting time of the operation of ν. τ^ϕ is a mapping of $V(G)$ onto non-negative integers such that $\tau^\phi(\nu) < \tau^\phi(\nu')$ if an arc (ν, ν') of T exists, and $\tau^\phi(\nu) + W(\nu) = \tau^\phi(\nu) + 1$ corresponds to the completion time of a process ν. Let ρ be a mapping $(\nu, \nu') \in A(T)$ to paths of H connecting between $\phi(\nu)$ and $\phi(\nu')$. ρ is called a *routing function* of T onto G under ϕ. For any $a = (\nu, \nu') \in A(T)$ and the path $\rho(a) = [u_0, u_1, \cdots, u_q]$ of G with $u_0 = \phi(t(a))$, let $\tau_{\rho(a)}$ be a mapping from $E(\rho(a))$ onto positive integers such that $\tau_{\rho(a)}(\{u_i, u_{i+1}\}) < \tau_{\rho(a)}(\{u_{i+1}, u_{i+2}\})$ for $0 \leq i \leq q - 2$. Let $\tau^\rho = \{\tau_{\rho(a)} | \ a \in A(T)\}$ and τ be the pair of τ^ϕ and τ^ρ.

The tupple $\langle \phi, \rho, \tau \rangle$ is called a *scheduling* of T onto G if following 4 time constraint conditions hold: (After stating those conditions, we will explain what they mean.)

Condition 1. Let a be any arc of T. Put $p = |\rho(a)|$, $[w^0, w^1, \cdots, w^p] = \rho(a)$, and $e_i = \{w^i, w^{i+1}\}$ $(0 \leq i \leq p)$ so that $\phi(t(a)) = w^0$. Then, if $p \geq 1$,
$$\tau^\phi(t(a)) + 1 \leq \tau_{\rho(a)}(e_0),$$
$$\tau_{\rho(a)}(e_i) + 1 \leq \tau_{\rho(a)}(e_{i+1}) \ (0 \leq i \leq p - 2), \text{ and}$$
$$\tau_{\rho(a)}(e_{p-1}) + 1 \leq \tau^\phi(h(a)).$$

Condition 2. For any positive integer s and any edge $e \in E(G)$, there is at most one arc $a \in A(T)$ such that $\rho(a)$ includes e and $\tau_{\rho(a)}(e) = s$.

Condition 3. For any distinct tasks $\nu, \nu' \in V(T)$, $\tau^\phi(\nu) \neq \tau^\phi(\nu')$ if $\phi(\nu) = \phi(\nu')$.

Condition 4. For any $a \in A(T)$, $\tau^\phi(h(a)) \geq \tau^\phi(t(a)) + 1$.

Recall that $W_V(v) = 1$. The completion time of $\langle \phi, \rho, \tau \rangle$ is defined as

$$\max_{\nu \in V(T)} \tau^\phi(\nu) - \min_{\nu \in V(T)} \tau^\phi(\nu) + 1.$$

Condition 1 represents the times for data through links in the routing along which the data are transmitted. Condition 2 implies that no two data are transmitted through the same connection wire at the same time. Condition 3 implies that no two distinct tasks can operated on the same PE at the same time. Condition 4 implies that if a task ν' needs the result of another task ν, ν' operated after ν. Condition 2 should be changed as follows: If $e \in E(\rho(a)) \cap E(\rho(a'))$ and

$\tau_{\rho(a)}(e) = \tau_{\rho(a')}(e)$ then both a and a' correspond to the same data. But, for the reason of simplicity, we adopt the previous one.

By Condition 1, we have the following, directly:

Lemma 1. *For any* $(\nu, \nu') \in A(T)$, *if* $\phi(\nu) \neq \phi(\nu')$ *then* $\tau^\phi(\nu') \geq \tau^\phi(\nu) + 2$. $\square$

Lemma 2. *For any* $(\nu, \nu') \in A(T)$, *if* $\tau^\phi(\nu') = \tau^\phi(\nu) + 2$ *and* $\phi(\nu) \neq \phi(\nu')$ *then* $\{\phi(\nu), \phi(\nu')\} \in E(G)$. $\square$

The scheduling problem can be defined as follows:

Unit-size Scheduling Problem (USP):
Given a task graph (dag) T, an undirected graph G, and an integer t, determine whether there exists a scheduling of T onto G with completion time at most t.

Also, the problem scheduling onto hypercubes as follows.

Unit-size Scheduling Problem onto hypercubes (USP$_{hc}$):
Given a dag T and integers n and t, determine whether there exists a scheduling of T onto Q_n with completion time at most t.

3   Complexity of Scheduling Problems

The scheduling problem onto complete graphs is $\mathcal{NP}$-complete. Though, the problem can be solved in polynomial time if the instance is restricted to be an out-forest (resp. in-forest), where an out-forest (resp. in-forest) is a dag each of whose components is an out-tree (resp. in-tree). In our model, the communication delay has a great effect on the completion time since parallel processors are non-fully connected. Taking it into account, the problem become more complicated.

In this section, we show that the scheduling problems onto hypercubes and grids are $\mathcal{NP}$-complete even if the instance is restricted to be a rooted tree with unit-size task weights and unit-size data.

Unit-size tree scheduling problem onto hypercubes (SP$_{hc}^{tr}$):
This problem is USP$_{hc}$ with restriction of T to be a rooted tree.

The following problem is shown to be $\mathcal{NP}$-complete in [12].

Tree-cube Embedding Problem (EP$_{hc}^{tr}$):
Given a rooted tree T and an integer n, determine whether there exists a dilation one embedding of T into Q_n or not.

For two decision problems Π_1 and Π_2, we denote by $\Pi_1 \propto \Pi_2$ the fact that Π_1 can be reduce to Π_2 in polynomial time.

We show $\mathcal{NP}$-completeness of USP$_{hc}^{tr}$ by reducing EP$_{hc}^{tr}$ to it. It is easily to be seen that SP$_{hc}$ is in $\mathcal{NP}$. Since SP$_{hc}^{tr} \propto$ SP$_{hc}$, we only need to show the following lemma.

Lemma 3. EP$_{hc}^{tr} \propto$ USP$_{hc}^{tr}$.

Proof. Let T be an instance of $\mathrm{EP}_{\mathrm{hc}}^{\mathrm{tr}}$. Choose a vertex r of T. For any $v \in V(T)$, let the *level* of v, denoted by $\mathrm{lev}(v)$, be the distance between r and v. Let η be the maximum level of a vertex of T. Then, the out-tree T' with root r which is an instance of $\mathrm{USP}_{\mathrm{hc}}^{\mathrm{tr}}$ can be constructed as follows:

1. Replace each edge $e = \{u, v\}$ by an arc (u, v), where $d_T(r, v) = d_T(r, u) + 1$. Let $A^r(T)$ be the set of such arcs, i.e., $\{(\mathrm{par}(\nu), \nu) | \nu \in V(T) \setminus \{r\}\}$.
2. For each vertex v connect a new directed path $P(v) = (v_0, v_1, \cdots, v_{2\eta - 2\mathrm{lev}(v)})$ by marging v and v_0.

It is sufficient to show that there exists an embedding of T into Q_n with dilation 1 if and only if there exists a scheduling of T' onto Q_n with completion time $2\eta + 1$.

There are one-to-one correspondence between $E(T)$ and $A^r(T)$. By definition, for $v = v_0 \in V(T)$,

$$|E(P(v))| = 2\eta - 2\mathrm{lev}(v). \tag{1}$$

Let $A_{\overline{T}}(T')$ be the set of arcs of T' not included in $A^r(T)$. That is, $A_{\overline{T}}(T') = \bigcup_{v \in V(T)} A(P(v))$. We give examples of T in Fig. 3 (a) and T' in Fig. 3 (b).

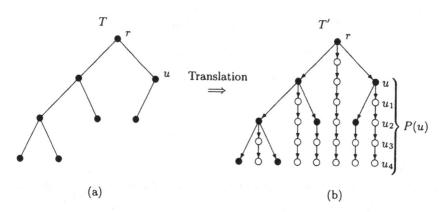

Fig. 1. Translation of an instance T.

For $\forall v \in V(T')$, let the *source* of v, denoted by $\mathrm{src}(v)$, be the vertex such that $v \in V(P(\mathrm{src}(v)))$. Then, v is one of vertices of the path added at $\mathrm{src}(v) \in V(T)$. In Fig. 3 (b), $\mathrm{src}(u_i) = u$ for $0 \leq i \leq \eta - 2\mathrm{lev}(u)$. By definition,

$$a \in A^r(T) \quad \Leftrightarrow \quad \mathrm{src}(t(a)) \neq \mathrm{src}(h(a)). \tag{2}$$

Moreover, if (ν, ν') is an arc of T' and $\nu' \in V(T)$ then $\nu \in V(T)$ and so we have the following:

Proposition 1. *If* $(\nu, \nu') \in A(T')$ *then* $\{\nu, \nu'\} \in E(T)$ *or* $\mathrm{src}(\nu) = \mathrm{src}(\nu')$.   □

We first show the necessity of the lemma.

Suppose that there exists an embedding $\langle \varphi, \varrho \rangle$ of T into Q_n with dilation 1. We will construct a tupple $\langle \phi, \rho, \tau \rangle$ and prove that it is a scheduling of T' onto Q_n with completion time $2\eta + 1$.

Let $\phi : V(T') \to V(Q_n)$ be the mapping such that $\phi(\nu) = \varphi(\mathrm{src}(\nu))$. Since φ is one-to-one,

$$\phi(\nu) = \phi(\nu') \Leftrightarrow \mathrm{src}(\nu) = \mathrm{src}(\nu') \tag{3}$$

For any arc $(\nu, \nu') \in A(T')$, let

$$\rho((\nu, \nu')) = \begin{cases} [\phi(\nu)] & \text{if } \phi(\nu) = \phi(\nu') \\ [\phi(\nu), \phi(\nu')] & \text{if otherwise.} \end{cases}$$

Note that the path $[v]$ is the trivial graph consisting of a vertex v.

Proposition 2. ρ *is a routing function of* T' *onto* Q_n *under* ϕ.

Proof (of Proposition 2). Let $a = (\nu, \nu')$ be any arc of T'.

If $\{\nu, \nu'\} \in E(T)$, by the definition of ρ and Proposition 1, $\rho((\nu, \nu'))$ has exactly one edge $\{\phi(\nu), \phi(\nu')\}$. Moreover, $\{\phi(\nu), \phi(\nu')\} = \{\varphi(\nu), \varphi(\nu')\} \in E(Q_n)$ since $\langle \varphi, \varrho \rangle$ is a dilation 1 embedding of T into Q_n. So, $\rho((\nu, \nu'))$ is a path of Q_n connecting between $\phi(\nu)$ and $\phi(\nu')$.

Otherwise, $\phi(\nu) = \phi(\nu') = \varphi(\mathrm{src}(\nu))$ and so $\rho(a) = [\varphi(\mathrm{src}(\nu))]$ is a path of Q_n with length 0. Thus, we have the proposition.   □

Let $\tau^\phi : V(T') \to \{0, 1, \cdots\}$ be the mapping such that

$$\tau^\phi(\nu) = 2d_{T'}(r, \mathrm{src}(\nu)) + d_{T'}(\mathrm{src}(\nu), \nu). \tag{4}$$

As mentioned above $\mathrm{src}(\nu) = \nu$ for any $\nu \in V(T)$. Then,

$$\begin{aligned} \tau^\phi(\nu) &= 2d_{T'}(r, \mathrm{src}(\nu)) \\ &= 2d_T(r, \nu) \text{ if } \nu \in V(T). \end{aligned} \tag{5}$$

Note that, if $a \in A^r(T)$ then $\rho(a)$ includes exactly one edge and otherwise no edge. So, when we discuss $\tau_{\rho(a)}(e)$ for $e \in E(\rho(a))$, we only need to consider the case $a \in A^r(T)$. For any $a = (\nu, \nu') \in A^r(T)$, we define $\tau_{\rho(a)}$ for edges $\{\nu, \nu'\}$ of $\rho(a)$ onto positive integers as follows:

$$\tau_{\rho((\nu, \nu'))}(\{\nu, \nu'\}) = 2d_T(r, \nu) + 1. \tag{6}$$

Now, we show that $\langle \phi, \rho, \tau \rangle$ satisfies Conditions from 1 to 4 and that its completion time is $2\eta + 1$.

For any arc $a \in A^r(T)$,

$$\tau^\phi(t(a)) = 2\mathrm{lev}(t(a)), \tag{7}$$

$$\tau_{\rho(a)}(e) = 2\mathrm{lev}(t(a)) + 1, \text{ and} \tag{8}$$

$$\tau^\phi(h(a)) = 2\mathrm{lev}(h(a)) \tag{9}$$

$$= 2\mathrm{lev}(t(a)) + 2.$$

Equations (7) and (9) follow from (5). Equation (8) follows from (6). So, $\tau^\phi(t(a)) + 1 \leq \tau_{\rho(a)}(e)$ and $\tau_{\rho(a)}(e) + 1 \leq \tau^\phi(h(a))$. Thus, we have Condition 1.

We show Condition 2 by contradiction. Assume that there exist two distinct arcs a and a' such that both $\rho(a)$ and $\rho(a')$ contain an edge $e \in E(Q_n)$ and $\tau_{\rho(a)}(e) = \tau_{\rho(a')}(e)$. By the definition of ρ, $\phi(t(a)) \neq \phi(h(a))$ since $\rho(a)$ includes an edge. Then from (2) and (3), $a \in A^r(T)$ and then $\{h(a), t(a)\} \in E(T)$. So, we have $\{\varphi(h(a)), \varphi(t(a))\} = \{\phi(h(a)), \phi(t(a))\} = e$. Similarly, we have $\{\varphi(h(a')), \varphi(t(a'))\} = e$. Thus, we have $\{h(a), t(a)\} = \{h(a'), t(a')\}$. since φ is one-to-one. This contradicts to T' is a (directed) tree since if otherwise, T' has parallel or anti-parallel arcs a and a'. Therefore, $\langle \phi, \rho, \tau \rangle$ satisfies Condition 2.

Let ν and ν' be any arcs of T'. If $\phi(\nu) = \phi(\nu')$, then using (3) ν and ν' have the same source s. Since path $P(s)$ includes ν and ν', $d_{T'}(s, \nu) \neq d_{T'}(s, \nu')$. Thus, from (4) $\tau^\phi(\nu) \neq \tau^\phi(\nu')$ and so, we have Condition 3.

Let a be any arc of T'. Then, $d_{T'}(r, h(a)) = d_{T'}(r, t(a)) + 1$. If $|E(\rho(a))| = 1$, $a \in A^r(T)$ and thus, from (5) we have Condition 4. Otherwise, $a \in A(P(v))$ for some $v \in V(T)$, and then from (3) and (4) we have $\tau^\phi(h(a)) = \tau^\phi(t(a)) + 1$. Therefore, we also have Condition 4.

So, $\langle \phi, \rho, \tau \rangle$ satisfies Conditions from 1 to 4. Thus, by Proposition 2, $\langle \phi, \rho, \tau \rangle$ is a scheduling of T' onto Q_n.

The rest of the proof of necessity is to show that the completion time is at most $2\eta + 1$. From (4), $\min_{\nu \in V(T)} \tau^\phi(\nu) = \tau^\phi(r) = 0$. Let ν be any vertex in $V(T')$ and $v = \text{src}(\nu)$. Note that $\text{dis}_{T'}(r, v) = \text{lev}(v)$. Then, from (1) and (4)

$$\tau^\phi(\nu) \leq 2\text{lev}(v) + |E(P(v))|$$
$$= 2\eta.$$

So, $\tau^\phi(v) + 1 - \min_{\nu \in V(T)} \tau^\phi(\nu) \leq 2\eta + 1 - 0$. Thus, we have the necessity. Next, we show the sufficiency of the lemma.

Suppose that there exists a scheduling $\langle \phi, \rho, \tau \rangle$ of T' onto Q_n with completion time $2\eta + 1$. We show that there exists an embedding of T into Q_n with dilation 1.

Without loss of generality, we suppose that $\tau^\phi(r) = 0$. Then, for $\nu \in V(T')$

$$\tau^\phi(\nu) \leq 2\eta \tag{10}$$

since the completion time is $\max_v \tau^\phi(v) + 1 - \tau^\phi(r) = 2\eta + 1$. Recall that $P(v) = (v_0, v_1, \cdots, v_{2\eta - 2\text{lev}(v)})$. Define that $q(v) = |P(v)| = 2\eta - 2\text{lev}(v)$. By Condition 4,

$$\tau^\phi(v_j) \geq \tau^\phi(v_{j-1}) + 1 \ (1 \leq j \leq q(v)) \tag{11}$$

and so, we have

$$\tau^\phi(v_i) \geq \tau^\phi(v) + i, \text{ and} \tag{12}$$
$$\tau^\phi(v_i) \leq \tau^\phi(v_{q(v)}) - (q(v) - i). \tag{13}$$

From (10), $\tau^\phi(v_q) \leq 2\eta$. Then, from (13), we have

$$\tau^\phi(v_i) \leq 2\eta - (q(v) - i)$$
$$\leq 2\text{lev}(v) + i. \tag{14}$$

Proposition 3. *For any $v \in V(T)$, $\phi(v) = \phi(v_{q(v)})$, $\tau^{\phi}(v_{q(v)}) = 2\eta$, and $\tau^{\phi}(v) = 2\mathrm{lev}(v)$. Further, $\{\phi(\mathrm{par}(v)), \phi(v)\} \in E(Q_n)$ if $v \neq r$.*

Proof (of Proposition 3). We show the proposition by induction on the level of v.

Note that r is the only vertex with level 0. From (12) and (14), the proposition holds for r since $\tau^{\phi}(r) = 2\mathrm{lev}(r) = 0$.

Suppose that the proposition holds for any vertex with level at most k. Let v be any vertex with level $k + 1$ and u be the parent of v. Then, $\mathrm{lev}(u) = k$. By induction hypothesis, $\tau^{\phi}(u) = 2k$. Then, from (12) and (14), for $2k \leq i \leq 2\eta$,

$$\tau^{\phi}(u_i) = 2k + i. \tag{15}$$

By Condition 4, $\tau^{\phi}(v) \geq 2k + 1$. Further, from (14), $\tau^{\phi}(v) \leq 2k + 2$. So, we have

$$2k + 1 \leq \tau^{\phi}(v) \leq 2k + 2. \tag{16}$$

Then, applying $i = 1$ and 2 to (15), we have $\phi(u) \neq \phi(v)$. So, from (16) and Lemma 1, we have $\tau^{\phi}(v) = 2k + 2 = 2\mathrm{lev}(v)$. Then, from (12), we have $\tau^{\phi}(v_{q(v)}) = 2\eta$. Moreover, by Lemma 2, $\{\phi(u), \phi(v)\} \in E(Q_n)$ since $\tau^{\phi}(v) = \tau^{\phi}(u) + 2$. □

By $\tau^{\phi}(v_{q(v)}) = 2\eta$ in Proposition 3, for any distinct vertices u and v in $V(Q_n)$,

$$\begin{aligned} \phi(u) &= \phi(u_{q(u)}) \\ &\neq \phi(v_{q(v)}) \\ &= \phi(v). \end{aligned} \tag{17}$$

Equation (17) follows from (3). So, the partial mapping φ of ϕ on $V(T)$ is one-to-one. Then, defining $\rho(\{u, v\})$ to be $[\varphi(u), \varphi(v)]$, $\langle \varphi, \varrho \rangle$ is, by Proposition 3, an embedding of T into Q_n with dilation 1. This completes the proof of sufficiency.

Thus, we have the proposition. □

By Lemma 3, we can obtain the following, directly:

Theorem 1. *USP$_{hc}^{tr}$ and SP$_{hc}^{tr}$ are $\mathcal{NP}$-complete.* □

We have just discussed scheduling problems for hypercubes. The similar arguments also hold for grids. The problem scheduling unit-size tree onto grid (USP$_{gr}^{tr}$) is defined as USP^{tr} by adding the restriction of G to be $(m \times n)$-grid for given integers m and n. The $\mathcal{NP}$-completeness of the following problem is obtained directly from [1].

Tree-grid Embedding Problem (EP$_{gr}^{tr}$):
Given a rooted tree T, integers m and n, determine whether there exists a dilation one embedding of T into $(m \times n)$-grid or not.

Notice that that the proof of Lemma 3 did not use the characteristics of hypercubes. So, appling the similar arguments of the proof of Lemma 3 we have EP$_{gr}^{tr} \propto$ USP$_{gr}^{tr}$ and then we have the following:

Theorem 2. *USP$_{gr}^{tr}$ and SP$_{gr}^{tr}$ are $\mathcal{NP}$-complete.* □

4   Conclusion

In this paper, we formulate the problem scheduling tasks onto parallel processors. In the formulation, we take the communication delay into account which have not been considered in most of old studies. Our model is more complicated than that of old ones but it can reflect the realistic problems more precisely.

Further, we have investigated the complexity of the problem which the parallel processor is restricted to be hypercubes (or grids) and concluded that those scheduling problems are $\mathcal{NP}$-complete even if the structure of the instance parallel algorithm is restricted to be a rooted tree. The arguments of Theorems also hold for the algorithms restricted to be special class of trees but we omit the details.

Acknowledgement

The author is grateful to Professor Y. Kajitani and Associate professor M. Kaneko for his encouragements. The author would like to thank Professor S. Ueno for helpful discussions. The research is supported by the CAD21 Project Group of TIT.

References

1. Bhatt, S., Cosmadakis, S.: The Complexity of Minimizing Wire Lengths in VLSI Layouts, Inform. Process. Lett., **25** (1987) 263–267.
2. Fujii, M., Kasami, T., Ninomiya, K.: Optimal sequencing of two equivalent processors, SIAM J. Appl. Math. **17** (1969)
3. Gabow, H.: An almost linear algorithm for two-processor scheduling, J. ACM, **29** (1982) 766–780
4. Hu, T.: Parallel sequencing and assembly line problems, Operations Research, **9** (1961) 841–848
5. Karp, R.: Reducibility among combinatorial problems, Technical Rep., Department of Computer Science, University of California at Berkeley (1972)
6. Kaufman, M.: An almost-optimal algorithm for the assembly line scheduling problem, IEEE Trans. Comput., **c-23** (1974) 1169–1174
7. Papadimitoriou, C., Yannakakis, M.: Scheduling interval-ordered tasks, SIAM J. Comput., **8** (1979) 405–409
8. Rewini, H., Lewis, T.: Scheduling parallel Program Task onto Arbitrary Target Machines, J. Parallel and Distributed Computing, **9** (1990) 138–153
9. Rewini, H., Lewis, T., Ali, H.: Task scheduling in parallel and Distributed systems, Prentice Hall, (1994)
10. Sethi, R.: Scheduling graphs on two processors, SIAM J. Comput., vol. 5 no. 1, pp. 73–82, 1976.
11. Ulman, J.: NP-complete scheduling problems, J. Comput. System Sci. **10** (1975) 384–393
12. Wagner, A., Corneil, D.: Embedding trees in a hypercube is NP-complete, SIAM J. Comput., **19** (1990) 570–590

Approximations of Weighted Independent Set and Hereditary Subset Problems

Magnús M. Halldórsson[1,2]

[1] Science Institute, University of Iceland, Reykjavik, Iceland. mmh@hi.is
[2] Department of Informatics, University of Bergen, Norway.

Abstract. The focus of this study is to clarify the approximability of the important versions of the maximum independent set problem, and to apply, where possible, the technique to related hereditary subgraph and subset problem. We report improved performance ratios for the Independent Set problem in weighted general graphs, weighted bounded-degree graphs, and in sparse graphs. Other problems with better than previously reported ratios include Weighted Set Packing, Longest Subsequence, Maximum Independent Sequence, and Independent Set in hypergraphs.

1 Introduction

An *independent set*, or a stable set, in a graph is a set of mutually non-adjacent vertices. The problem of finding a maximum independent set in a graph, INDSET, is one of most fundamental combinatorial NP-hard problem. It serves also a the primary representative for the family of subgraph problems that are hereditary under vertex deletions. We are interested in finding approximation algorithms that yield good performance ratios, or guarantees on the quality of the solution they find vis-a-vis the optimal solution.

The focus of this paper is to present improved performance ratios for three major versions of the independent set problem: in weighted graphs, bounded-degree graphs and sparse graphs. We also apply some of the methods to a number of related (or not-so related) problems that obey certain hereditariness property, most of which had not been approximated before.

INDSET has been shown to be hard to approximate on general graphs through advances in the study of interactive proof systems. In particular, Håstad [9] showed it hard to approximate within $n^{1-\epsilon}$, for any $\epsilon > 0$, unless NP-hard problems have randomized polynomial algorithms. The best performance ratio known is $O(n/\log^2 n)$, due to Boppana and Halldórsson [3].

For bounded-degree graphs, Halldórsson and Radhakrishnan [13] gave the first asymptotic improvement over maximal solutions, obtaining a ratio of $O(\Delta/\log\log\Delta)$. Schiermeyer [18] recently improved this considerably; we show in the full version of this paper that a modified version of his algorithm leads to a $O(\Delta/\sqrt{\log\Delta})$ approximation. For small values of Δ, an algorithm of Berman

T. Asano et al. (Eds.): COCOON'99, LNCS 1627, pp. 261–270, 1999.

and Fujito [2] attains the best bound known of $(\Delta + 3)/5$. For sparse graphs, the best bound claimed in the literature is $(2\overline{d} + 3)/5$ [12], for a greedy algorithm complemented by a preprocessing technique. See the survey [10] for a more complete description of earlier results.

Our main results are as follows. We study in Section 2 an elementary general partitioning technique that yields non-trivial performance ratio for a large class of problems satisfying a property that we call *semi-heredity*. All results holds for for weighted versions of the problems. We obtain a $O(n/\log n)$ approximation for INDEPENDENT SET IN HYPERGRAPHS, LONGEST COMMON SUBSEQUENCE, MAX SATISFYING LINEAR SUBSYSTEM, and MAX INDEPENDENT SEQUENCE. We strengthen the ratio for problems that do not contain a forbidden clique, obtaining a $O(n(\log\log n/\log n)^2)$ performance ratio for INDSET and MAX HEREDITARY SUBGRAPH. (All problems are defined in their respective sections.)

In Section 3, we consider another elementary strategy, partitioning the vertices into weight classes. It easily yields that weighted versions on any class of graphs are approximable within $O(\log n)$ of the respective unweighted case, but this overhead factor reduces to a constant in the case of ratios in the currently achievable range, giving a $O(n/\log^2 n)$ ratio for WIS. We also match the best ratio known for the weighted set packing problem of $O(\sqrt{m})$, where m is the number of base elements.

In Section 4, we consider approximations based on semi-definite programming (SDP) relaxations and the theta function of Lovász. Vishwanathan [19] observed that an improved performance ratio for the independent set problem in bounded-degree graphs of $O(\Delta \log\log \Delta/\Delta)$ could be obtained from results on SDP. We generalize this approximation in one direction to weighted graphs, and in another direction to the same function of the average degree.

Notation: Let $G = (V, E)$ be a graph, let n denote its number of vertices and let Δ ($\overline{d}$) denote its maximum (average) degree. WIS takes as input instance (G, w), where G is a graph and $w : V \mapsto \mathbf{R}$ is a vector of vertex weights, and asks for a set of independent vertices whose sum of weights is maximized. The maximum weight of an independent set in instance (G, w) denoted by $\alpha(G, w)$, or $\alpha(G)$ on unweighted graphs. Let $|S|$ denote the cardinality of a set S.

We say that a problem is *approximable within* $f(n)$, if there is a polynomial time algorithm which on on any instance with n distinguished elements returns a feasible solution within a $f(n)$ factor from optimal.

2   Partitioning into easy subproblems

We consider a collection of problems that involve finding a feasible subset of the input of maximum weight. The input contains a collection of n *distinguished elements*, each carrying an associated non-negative rational weight. Each set of distinguished elements uniquely induces a candidate for a *solution*, which we assume is efficiently computable from the set. The weight of a solution is the sum of the weights of the distinguished elements in the solution.

A property is said to be *hereditary* if whenever a set S of distinguished elements corresponds to a feasible solution, any subset of S also corresponds to a feasible solution. A property is *semi-hereditary* if under the same circumstances, any subset S' of S uniquely induces a feasible solution, possibly corresponding to a superset of S'.

Hereditary graph properties are special cases of these definitions. A property of graphs is *hereditary* if whenever it holds for a graph it also holds for its induced subgraphs. For a hereditary graph property, the associated subgraph problem is that of finding a subgraph of maximum vertex-weight satisfying the property. Here, the vertices form the distinguished elements.

Our key tool is a simple partitioning idea, that has been used in various contexts before.

Proposition 1. *Let Π be a semi-hereditary subset property. Suppose that given an instance I, we can produce t instances $I_1, I_2, \ldots, I_t$ that cover the set of distinguished elements (i.e. each distinguished element is contained in at least one I_i). Further, suppose we can solve exactly the maximum Π-subset problem on each I_i. Then, the largest of these t solutions yields an approximation of the maximum Π-subset of I within t.*

In the remainder we describe applications of this approach to a number of particular problems.

Partition into small subsets

Proposition 2. *Let Π be a semi-hereditary property for which feasibility can be decided in time at most polynomial in the size of the input and at most simply exponential in the number of distinguished elements. Then, the maximum weighted Π-subgraph can be approximated within $n/\log n$.*

We achieve this by arbitrarily partitioning the set of distinguished elements into $n/\log n$ sets each with $\log n$ elements. For each subset of each set, obtain the candidate solution for this subset and determine feasibility. By our assumptions, each step can be done in polynomial time, and in total at most $2^{\log n} \cdot n/\log n = n^2/\log n$ sets are generated and tested. By this procedure, we find optimal solutions within each of the $n/\log n$ sets. Since the optimal solution of the whole is divided among these sets, the performance ratio is at most $n/\log n$.

We apply Proposition 2 to several problems featured in the compendium on optimization problems [4]:

Weighted Independent Sets in Hypergraphs: Given a hypergraph, or a set system, $(S, \mathcal{C})$ where S is a set of weighted base elements (vertices) and $\mathcal{C} = \{C_1, C_2, \ldots, C_n\}$ is a collection of subsets of S, find a maximum weight subset S' of vertices such that no subset C_i is fully contained in S'.

Hofmeister and Lefmann [14] analyzed a Ramsey-theoretic algorithm generalizing that of [3], and showed its performance ratio to be $O(n/(\log^{(r-1)} n))$ for the case of r-uniform hypergraphs. It is straightforward to verify the heredity thus a $O(n/\log n)$ performance ratio holds by Proposition 1.

Longest Common Subsequence: Given a finite set R of strings from a finite alphabet Σ, find a longest possible string w that is a subsequence of each string x in R. The problem is clearly hereditary, and feasibility can be tested for each string x in R separately via dynamic programming. Hence, by applying Proposition 2, partitioning the smallest string in the input, we obtain a performance ratio of $O(m/\log m)$, where m is the size of the smallest string.

Max Satisfying Linear Subsystem: Given a system $Ax = b$ of linear equations, with A an integer $m \times n$ matrix and b an integer m vector, find a rational vector $x \in \mathcal{Q}^n$ that satisfies the maximum number of equations.

This problem is clearly hereditary, since any subset of a feasible collection of equations is also feasible. Feasibility of a given system can be solved in polynomial time via linear programming. Hence, $O(m/\log m)$ approximation follows from Proposition 2. This holds equally if the variables are restricted to take on binary values or if equality is replaced by inequalities $(>, \geq)$. It also holds if a particular set of constraints/equations are required to be satisfied by a solution.

Max Independent Sequence: Given a graph, find a maximum length sequence $v_1, v_2, \ldots, v_m$ of independent vertices such that, for all $i < m$, a vertex v_i' exists which is adjacent to v_{i+1} but is not adjacent to any v_j for $j \leq i$. This problem was introduced by Blundo (see [4]).

First observe that solutions to the problem are hereditary: if $v_1, v_2, \ldots, v_m$ is an independent sequence, then so is any subsequence $v_{a_1}, v_{a_2}, \ldots, v_{a_x}$. This is because, for all $i < x$, there exists a node v_i' that is adjacent to $v_{a_{i+1}}$ but not adjacent to any v_j for $j < a_{i+1}$ and hence not to any v_{a_j} for $j \leq i$. Feasibility of a solution can be tested in time polynomial in the size of the input. Independence is easily tested by testing all pairs in the proposed solution. A valid set can be turned into a valid sequence by inductively finding the element adjacent to a vertex outside the set that is adjacent to no other unselected vertex.

Thus, we obtain an $O(n/\log n)$ approximation via Proposition 2. We can also argue strong approximation hardness bounds.

Proposition 3. MAX INDEPENDENT SEQUENCE *is no easier than* INDSET, *within 2. Thus, it is hard to approximate within* $n^{1-\epsilon}$, *for any* $\epsilon > 0$, *unless* $NP = ZPP$.

Theorem 1. *Weighted versions of* INDSET IN HYPERGRAPHS, MAX HEREDITARY SUBGRAPH *and* MAX INDEPENDENT SEQUENCE *can be approximated within* $O(n/\log n)$.

Weighted Independent Sets and Other Hereditary Graph Properties

A theorem of Erdős and Szekeres [6] on Ramsey numbers yields an efficient algorithm [3] for finding either cliques or independent sets of non-trivial size.

Fact 1 (Erdős, Szekeres) *Any graph on n vertices contains a clique on s vertices and independent set on t vertices such that* $\binom{s+t-2}{s-1} \geq n$.

Theorem 1 implies that we can obtain either an independent set of size at least $\log^2 n$, or a clique of size at least $\log n/2 \log\log n$. We now apply this theorem repeatedly, and form a partition into $O(n(\log\log n/\log n)^2)$ classes, each of which is either an independent set or a (not necessarily disjoint) union of $\log n/\log\log n$ different cliques. WIS can be solved on the latter type by exhaustively checking all $(\log n/\log\log n)^{\log n/\log\log n} = O(n)$ possible combinations of selecting up to one vertex from each clique.

This argument can be extended to a large class of hereditary subgraph problems.

Theorem 2. MAX WEIGHTED HEREDITARY SUBGRAPH *can be approximated within $O(n(\log\log n/\log n)^2)$, for properties that fail for some cliques or some independent set.*

Proof. Without loss of generality assume the property fails for cliques of size s. Once we have applied the above partitioning, we can still compute an optimal solution on each part by exhaustively checking all combinations of selecting at most $s - 1$ vertices from each clique. That number is still at most $(\log n/\log\log n)^{s\,\log n/\log\log n}$, which is $poly(n)$ for fixed s. In the case that the property fails for some independent set, we exchange the roles of independent sets and cliques in our partioning routine with no change in the results. $\square$

3 Partitioning into weight classes

We now consider a simple general strategy to obtain approximations to weighted subgraph problems, that comes within a $\log n$ factor from the unweighted case and often within less.

Theorem 3. *Let Π be a hereditary subgraph problem. Suppose Π can be approximated within ρ on unweighted graphs (or on a subclass thereof). Then, the vertex-weighted version can be approximated within $O(\rho \cdot \log n)$.*

Proof. Consider the following strategy. Let W be the maximum vertex weight. Delete all vertices of weight at most W/n. Let V_i be the set of vertices whose weight lies in $(W/2^i, W/2^{i-1}]$, for $i = 1, 2, \ldots, \lg n$. Run the ρ-approximate algorithm on the V_i, ignoring the weights. Output the solution HEU found of maximum weight.

We claim that the performance ratio of this method is at most $2\rho \lg n + 1$. First, note that the set of vertices of small weight add up to at most W, or less than that of HEU. Second, the optimal solution on the remaining vertices is at most the sum of the optimal solutions of the V_i, each of which is at most 2ρ times the weight of the heuristic solution of the respective subgraph. The additional factor of 2 coming from the rounding of the weights. $\square$

We note that the logarithmic loss in approximation is caused by logarithmic decrease in subgraph sizes. However, when the performance function is close to linear, as is the case today, decrease in subgraph size affects performance only slightly. We illustrate this with WIS, matching the known approximation for unweighted graphs.

Theorem 4. *If a hereditary subgraph problem can be approximated within* $g(n) = n^{1-\Omega(1/\log\log n)}$, *then its weighted version can also be approximated within* $O(g(n))$. *In particular,* WIS *can be approximated within* $O(n/\log^2 n)$.

Proof. Let G be a graph partitioned into subgraphs $V_1, \ldots, V_{\log n}$ as in Theorem 3, let OPT be an optimal solution and HEU the heuristic solution found. Observe that the function g satisfies $g(N) = O(g(n) \cdot N/n)$ when $N \geq n/\lg^2 n$, and $g(N) = O(g(n)/\log^2 n)$ when $N \leq n/\lg^2 n$,

Let L be the set of indices i that satisfy

$$w(V_i \cap OPT) \geq w(OPT)/2\lg n, \tag{1}$$

and note that since $|L| \leq \lg n$, $\sum_{i \notin L} w(V_i \cap OPT) \leq w(OPT)/2$.

The performance ratio ρ of our algorithm is

$$\rho = \sum_i \frac{w(V_i \cap OPT)}{w(HEU)} \leq 2\frac{\sum_{i \in L} w(V_i \cap OPT)}{w(HEU)}. \tag{2}$$

Suppose for some $\ell \in L$, $|V_\ell| < n/\lg^2 n$. By (1), $w(V_i \cap OPT) \leq (2\lg n)w(V_\ell \cap OPT)$, for all i. Thus,

$$\rho \leq 4\lg n \frac{\sum_{j \in L} w(V_\ell \cap OPT)}{w(HEU)} \leq 8\lg^2 n \cdot g(|V_\ell|) = O(g(n)).$$

Otherwise, $g(|V_j|) = O(g(n) \cdot |V_j|/n)$ for all $j \in L$, and from (2),

$$\rho \leq 4\sum_i g(|V_i|) = \frac{g(n)}{n}\sum_{i \in L} O(|V_i|) = O(g(n)).$$

The result for WIS follows from [3].   □

Weighted Set Packing: The WSP problem is as follows. Given a set S of m base elements, and a collection $\mathcal{C} = \{C_1, C_2, \ldots, C_n\}$ of weighted subsets of S, find a subcollection $\mathcal{C}' \subseteq \mathcal{C}$ of disjoint sets of maximum total weight $\sum_{C_i' \in \mathcal{C}'} w(C_i')$.

By forming the intersection graph of the given hypergraph, a weighted set packing instance can be transformed to a weighted indepedndent set instance on n vertices. Hence, approximations of WIS — as a function of n — carry over to WSP.

For approximations of unweighted set packing *as a function of* m $(= |S|)$, Halldórsson, Kratochvíl, and Telle [11] gave a simple $\sqrt{m}$-approximate greedy algorithm, and noted that $m^{1/2-\epsilon}$-approximation is hard via [9]. We observe that the positive results hold also for the weighted case, by a simple variant of the greedy method.

Theorem 5. WSP *can be approximated within* $\sqrt{m}$ *in time proportional to the time it takes to sort the weights.*

Proof. The algorithm initially removes all sets of cardinality $\sqrt{m}$ or more. It then greedily selects sets of maximum weight that are disjoint with the previously selected sets.

For any iteration i, let X_i be the set selected by the algorithm and Z_i be the sets eliminated in the iteration. Observe that each set in Z_i is of weight at most that of X_i. Also, the optimal solution contains at most $\sqrt{m}$ sets from Z_i, since each set in Z_i has a private element in common with X_i and X_i contains at most $\sqrt{m}$ elements. Hence, in every iteration, the contribution added to the algorithm's solution is at least $\sqrt{m}$-th fraction of what the optimal solution could get.

Also, the optimal solution contains fewer than $\sqrt{m}$ sets of cardinality greater than $\sqrt{m}$. Since the algorithm contains at least the weight of the maximum weight set, this is at most $\sqrt{m}$ times the algorithm's solution. Combined, the optimal solution is of weight at most $2\sqrt{m}$ times the algorithm's solution. □

4 Applying semi-definite programming

A fascinating polynomial-time computable function $\vartheta(G)$ introduced by Lovász [16] has the remarkable "sandwiching" property that it always lies between two NP-hard functions, $\alpha(G) \leq \vartheta(G) \leq \overline{\chi}(G)$. This property suggests that it may be particularly suited for obtaining good approximations to either function. While some of those hopes have been dashed [7], a number of fruitful applications have been found and it remains the most promising candidate for obtaining improved approximations [8].

Karger, Motwani and Sudan [15] gave improved approximations for k-colorable graphs via the theta function, followed by Alon and Kahale [1] that obtained improved approximations for INDSET in the case of linear-sized independent sets. Mahajan and Ramesh [17] showed how these and related algorithms can be derandomized. Vishwanathan [19] observed that an improved performance ratio for the independent set problem of $O(\Delta \log \log \Delta / \log \Delta)$ could be obtained by combining together theorems of [15.1].

We illustrate here how the improved $O(\Delta \log \log \Delta / \log \Delta)$ also applies to weighted independent sets. For this purpose, we give straightforward generalizations of the results of [15] and [1].

An *orthonormal representation* of a graph $G = (V, E)$ is an assignment of a unit vector b_v in Euclidean space to each vertex v of G, such that $b_u \cdot b_v = 0$ if $u \neq v$ and $(u, v) \notin E$. The (weighted) theta function $\vartheta(G, w)$ equals the minimum over all unit vectors d and all orthonormal labelings b_v of

$$\max_{v \in V} \frac{w(v)}{(d \cdot b_v)^2}.$$

An equivalent dual characterization is to define it as the maximum over all unit vectors d and all orthonormal representations b_v of the complement graph $\overline{G}$ of $\sum_{v \in V} (d \cdot b_v)^2 w(v)$. The (unweighted) Lovász number $\vartheta(G)$ is $\vartheta(G, \mathbf{1})$, the theta function on the unit-weighted graph.

We give a weighted version of a result of [15]. For our purposes, it suffices to use the simpler method of "rounding by hyperplanes", as the constant in the exponent is not important.

Proposition 4. *Let G be a weighted graph and $w(G)$ be the sum of the vertex weights. If $\vartheta(\overline{G}) \leq k$, then an independent set in G of weight $\Omega(w(G)/\Delta^{1-1/2k})$ can be constructed with high probability in polynomial time.*

Proof. The bound on $\vartheta(\overline{G})$ implies that one can construct in polynomial time a representation of the vertices of G as vectors in Euclidean space, such that for adjacent vertices i and j the corresponding vectors v_i and v_j satisfy $(v_i \cdot v_j) \leq -\frac{1}{k}$.

Given such a representation, the algorithm selects r hyperplanes at random, dividing $\mathbf{R}^n$ into 2^r partitions. The algorithm examines each of the partitions, examines the set of vertices not adjacent to any other vertex in the same partition, and outputs the set of maximum weight. We give a lower bound on the expected weight of this set.

Let $q = \frac{1}{2} + \frac{1}{\pi k}$ and let $r = 2 + \lceil \log_{1/(1-q)} \Delta \rceil$. Note that $(1-q)^r \leq 1/4\Delta$. Also, it can be shown that $1/\lg 1/(1-q) \leq 1 - 1/2k$, using that $1/(1-q) = 2(1 + 1/(\pi k/2 - 1))$ and that $\ln(1+x) \geq x/(1+x)$. That means that

$$2^r \leq 8\Delta^{1/\lg 1/(1-q)} = O(\Delta^{1-1/2k}).$$

The probability that a random hyperplane separates the vectors associated with two vertices is ϕ/π, where ϕ is the angle between the vectors. When the vertices are adjacent, this probability is

$$\frac{arccos(-1/k)}{\pi} \leq \frac{1}{2} + \frac{1}{\pi k} = q,$$

where the inequality is obtained from the Taylor expansion of $arccos(x)$. Hence, the probability that at least one of the r hyperplanes cuts a given edge is at least

$$p = 1 - (1-q)^r \geq 1 - \frac{1}{4\Delta}.$$

The probability, for a given vertex v, that none of v's neighbors remain in the same partition is then at least $p^\Delta \geq 1 - 1/4$. Hence, if $w(P)$ denotes the weight of a given partition P, the expected weight of the independent set formed by vertices of zero degree within the partition is at least $w(P)(1 - 1/4)$. Averaging over the 2^r partitions, the expected weight of the set output is at least

$$\frac{w(G)(1 - 1/4)}{2^r} = O(w(G)/\Delta^{1-1/2k}). \qquad \square$$

Proposition 5. *If $\vartheta(G, w) \geq 2w(G)/k$ (e.g. if $\alpha(G, w) \geq 2w(G)/k$), then we can find an induced subgraph K in G such that $\vartheta(\overline{K}) \leq k$ and $w(K) \geq w(G)/k$.*

Proof. We emulate [1]. Let d be a unit vector and b_v a representation such that $\vartheta(G, w) = \sum_{v \in V}(d \cdot b_v)^2 w(v)$. For each v, let a_v denote $(d \cdot b_v)^2$. We can then split the sum for $\vartheta(G, w)$ into two parts: those where a_v is small or at most $1/k$ for some breakpoint k, and those where a_v is large. Thus,

$$\vartheta(G, w) = \sum_{a_v \leq 1/k} a_v w(v) + \sum_{a_v \geq 1/k} a_v w(v) \leq w(G)/k + \sum_{a_v \geq 1/k} a_v w(v).$$

Let K be the subgraph induced by vertices v with $a_v \geq 1/k$. If $\vartheta(G, w) \geq 2w(G)/k$, we have that since $a_v \leq 1$ for each vertex v,

$$w(K) = \sum_{v \in V(K)} w(v) \geq \sum_{a_v \geq 1/k} a_v w(v) \geq \vartheta(G, w) - w(G)/k \geq w(G)/k.$$

Also,

$$\max_{v \in K} \frac{1}{(d \cdot b_v)^2} \leq k,$$

hence the Lovász number of K is at most k by its definition. □

Theorem 6. WIS *can be approximated within* $O(\Delta \log \log \Delta / \log \Delta)$.

Proof. Let (G, w) be an instance with $\alpha(G, w) = 2w(G)/k$, for some k. We find via Proposition 5 a subgraph K_k with $\vartheta(\overline{K_k}) \leq k$ and $w(K_k) \geq w(G)/k$. We then find via Proposition 4 an independent set in $K_k \subset G$ of weight at least $\frac{w(G)/k}{\Delta^{1-1/2k}}$. The approximation ratio is then at most $2\Delta^{1-1/2k}$. Alternatively, we can always find an independent set of weight $w(G)/(\Delta + 1)$ by a greedy coloring, for an approximation ratio of $2(\Delta + 1)/k$. Observe that first ratio is increasing with k and the latter decreasing, with breakpoint achieved when $k = \frac{1}{2} \log \Delta / \log \log \Delta$. In this case, both ratios are $O(\Delta \log \log \Delta / \log \Delta)$. □

Sparse graphs

We can also apply the Lovász number to obtain improved approximations for sparse graphs.

Theorem 7. IS *can be approximated within* $O(\overline{d} \log \log \overline{d} / \log \overline{d})$.

Proof. For a graph G of average degree $\overline{d}$, let $k = n/\alpha(G)$, i.e. $\alpha(G) = n/k$. Consider the subgraph K induced by vertices of degree at most $2k\overline{d}$. Then, $\Delta(K) \leq 2k\overline{d}(G)$. At least $k\overline{d}(n - |V(K)|)$ edges are removed, while G contained only $\frac{1}{2}\overline{d}n$ edges. Hence, at most $\frac{1}{2k}n$ vertices are removed, and thus $\alpha(K) \geq \alpha(G)/2$.

Apply this argument to solutions obtained by the theta function. The approximation ratio is by Proposition 4 at most

$$2 \cdot (2k \cdot \overline{d}(G))^{1-1/2k} = O(\overline{d}(G)^{1-1/2k} \cdot k).$$

Recall that a minimum-degree greedy algorithm attains the Turán bound of $n/(\overline{d} + 1)$ [5], and for an $O(\overline{d}/k)$ approximation [12]. The two functions cross when $k = \frac{1}{4} \log \overline{d} / \log \log \overline{d}$, for the desired ratio. □

References

1. N. Alon and N. Kahale. Approximating the independence number via the θ function. *Math. Programming*. To appear.
2. P. Berman and T. Fujito. On the approximation properties of independent set problem in degree 3 graphs. *WADS '95*, LNCS #955, 449–460.
3. R. B. Boppana and M. M. Halldórsson. Approximating maximum independent sets by excluding subgraphs. *BIT*, 32(2):180–196, June 1992.
4. P. Crescenzi and V. Kann. A compendium of NP optimization problems. On-line survey at http://www.nada.kth.se.
5. P. Erdős. On the graph theorem of Turán (in Hungarian). *Mat. Lapok*, 21:249–251, 1970.
6. P. Erdős and G. Szekeres. A combinatorial problem in geometry. *Compositio Math.*, 2:463–470, 1935.
7. U. Feige. Randomized graph products, chromatic numbers, and the Lovász ϑ-function. *Combinatorica*, 17 (1), 79–90, 1997.
8. U. Feige and J. Kilian. Heuristics for finding large independent sets, with applications to coloring semi-random graphs. *FOCS '98*.
9. J. Håstad. Clique is hard to approximate within $n^{1-\epsilon}$. *FOCS '96*, 627–636.
10. M. M. Halldórsson. A survey on independent set approximations. *APPROX '98*, LNCS # 1444, 1–14
11. M. M. Halldórsson, J. Kratochvíl, and J. A. Telle. Independent sets with domination constraints. *ICALP '98*, LNCS # 1443.
12. M. M. Halldórsson and J. Radhakrishnan. Greed is good: Approximating independent sets in sparse and bounded-degree graphs. *Algorithmica*, 18:145–163, 1997.
13. M. M. Halldórsson and J. Radhakrishnan. Improved approximations of independent sets in bounded-degree graphs via subgraph removal. *Nordic J. Computing*, 1(4):475–492, 1994.
14. T. Hofmeister and H. Lefmann. Approximating maximum independent sets in uniform hypergraphs. *MFCS '98*.
15. D. Karger, R. Motwani, and M. Sudan. Approximate graph coloring by semidefinite programming. *J. ACM*, 45, 2, 246–265, 1998.
16. L. Lovász. On the Shannon capacity of a graph. *IEEE Trans. Inform. Theory*, IT-25(1):1–7, Jan. 1979.
17. S. Mahajan and H. Ramesh. Derandomizing semidefinite programming based approximation algorithms. *FOCS '95*, 162–169.
18. I. Schiermeyer. Approximating independent sets in k-clique-free graphs. *APPROX '98*.
19. S. Vishwanathan. Personal communication via J. Radhakrishnan, 1996.

Multi-coloring Trees

Magnús M. Halldórsson[1,6], Guy Kortsarz[2], Andrzej Proskurowski[3],
Ravit Salman[4], Hadas Shachnai[5], and Jan Arne Telle[6]

[1] Science Institute, University of Iceland, Reykjavik, Iceland. mmh@hi.is.
[2] Dept. of Computer Science, Open University, Ramat Aviv, Israel.
guyk@tavor.openu.ac.il.
[3] Dept. of Computer Science, University of Oregon, Eugene, Oregon.
andrzej@cs.uoregon.edu.
[4] Dept. of Mathematics, Technion, Haifa, Israel. maravit@tx.technion.ac.il.
[5] Dept. of Computer Science, Technion, Haifa, Israel. hadas@cs.technion.ac.il.
[6] Dept. of Informatics, University of Bergen, Bergen, Norway. telle@ii.uib.no.

Abstract. Scheduling jobs with pairwise conflicts is modeled by the
graph *multi-coloring* problem. It occurs in two versions: preemptive and
non-preemptive. We study these problems on trees under the sum-of-
completion-times objective. In particular, we give a quadratic algorithm
for the non-preemptive case, and a faster algorithm in the case that all
job lengths are short, while we present a polynomial-time approximation
scheme for the preemptive case.

1 Introduction

In many real-life situations, *non-sharable* resources need to be shared among
users with *conflicting* requirements. This includes traffic intersection control [B92],
frequency assignment to mobile phone users [DO85,Y73], and session management in local area networks [CCO93]. Each user can be identified with a *job*,
the execution of which involves an exclusive use of some resource, in a given period
of time. Indeed, scheduling such jobs, with pairwise conflicts, is a fundamental
problem - in the above areas, as well as in distributed computing [L81,SP88].

The problem of scheduling dependent jobs is modeled as a graph *coloring*
problem, when all jobs have the same (unit) execution times, and as graph
multi-coloring for arbitrary execution times. The vertices of the graph represent
the jobs and an edge in the graph between two vertices represents a dependency
between the two corresponding jobs, which forbids scheduling these jobs at the
same time. More formally, for a weighted undirected simple graph $G = (V, E)$
with n vertices, let the *length* of a vertex v be a positive integer denoted by $x(v)$
and called the *color requirement* of v. A multi-coloring of the vertices of G is
a mapping into the power set of the positive integers, $\Psi : V \mapsto 2^N$, such that
$|\Psi(v)| = x(v)$ and adjacent vertices receive non-intersecting sets of colors.

The traditional optimization goal is to minimize the total number of colors
assigned to G. In the setting of a job system, this is equivalent to finding a
schedule, in which the time when *all* the jobs have been completed is minimized.

T. Asano et al. (Eds.): COCOON'99, LNCS 1627, pp. 271–280, 1999.
© Springer-Verlag Berlin Heidelberg 1999

However, from the point of view of the jobs themselves, an important goal is to minimize the average completion time of the jobs (or equivalently, the sum of the completion times). This optimization goal is the concern of this paper. Formally, in the *sum multi-coloring* (SMC) problem, we look for a multi-coloring Ψ that minimizes $\sum_{v \in V} f_\Psi(v)$, where $f_\Psi(v)$ is the largest color assigned to v by Ψ. This reduces to the *sum coloring* problem in the case of unit color requirements.

There are two variants of the sum multi-coloring problem. In the *preemptive* (p-SMC) problem, each vertex may get any set of colors, while in the *non-preemptive* (np-SMC) problem, the set of colors assigned to each vertex has to be contiguous. The preemptive version corresponds to the scheduling approach commonly used in modern operating systems [SG98], where jobs may be interrupted during their execution and resumed at a later time. The non-preemptive version captures the execution model adopted in real-time systems, where scheduled jobs must run to completion.

In the current paper we study the sum multi-coloring problems on trees. Given the hardness of these problems on general graphs (see below), it is natural to seek out classes of graphs where effective solutions can be obtained efficiently. Trees constitute the boundary of what we know to be efficiently solvable, and represent perhaps the most frequently naturally occurring class of graphs.

A natural application, in which the resulting conflict graph is a tree, is packet routing on a tree network topology: each node can conflict over its neighboring links, either with its parent or children in the tree. Thus, the conflict graph is induced by the network topology. Conflicts among processes running on a single-user machine (e.g. PCs) are typically for shared data. In many operating systems, the creation of a new process is done by 'splitting' an existing process, via a 'fork' system call [SG98]. Thus, the set of processes form a tree where each process is a node. Conflicts over shared data typically occur between a process and its immediate descendents/ancestor in that tree, as these processes will share parts of their codes. Thus, the conflict graph is also a tree.

Related work

The sum multi-coloring problem was introduced by Bar-Noy et al. [BKH+98]. They presented a comprehensive study of the approximability of both the p-SMC and the np-SMC problems, on general and special classes of graphs. We are not aware of other published work on this problem.

The sum coloring problem was introduced by Kubicka [K89], who gave a polynomial algorithm for trees. Jansen [J97] extended the dynamic programming strategy to graphs of bounded treewidth. Also, the sum-coloring problem on general graphs was shown to be hard to approximate within $n^{1-\epsilon}$, for any $\epsilon > 0$ unless $NP = ZPP$ [FK96,BBH+98], as well as hard to approximate within some factor $c > 1$ on bipartite graphs [BK98]. These hardness results also carry over to the multi-coloring generalizations.

Our results

For the np-SMC problem, we give in Section 3 two exact algorithms, with incomparable complexity: the first one is quadratic, i.e., $O(n^2)$ where $|V| = n$, while the second is more effective if the maximum color requirement p is small, running in time $O(np)$. In both cases, non-trivial optimizations have been made to reduce the time complexity. The first algorithm is still more efficient for the special case of paths, running in time $O(n \cdot \log p / \log \log p)$.

Previous dynamic programming algorithms [K89] can be seen to generalize to multi-coloring, leading to algorithms that are polynomial in n and p, e.g. $O(p^2 n \log n)$. They are, however, not polynomial for large p.

For the case of p-SMC, the solvability for trees appears to be a hard question. We present in Section 4 a polynomial time approximation scheme, along with an exact algorithm for a limited special case. Finally, we discuss in Section 5 several generalizations of the problem, to which our algorithms continue to apply, and mention open problems for further study.

2 Definitions and notation

An instance of a multi-coloring problem is a pair (G, x) where $G = (V, E)$ is a graph and $x : V \to \mathbf{N}$ is a vector of *color requirements* (or *lengths*) of the vertices. We denote by $p = \max_{v \in V} x(v)$ the maximum color requirement.

A *multi-coloring* of G is an assignment $\Psi : V \to 2^N$, such that each vertex v is assigned $x(v)$ distinct colors and adjacent vertices receive non-intersecting sets of colors. The *start time* (*finish time*) of a vertex v under Ψ is the smallest (largest) color assigned to v, denoted by $s_\Psi(v) = \min\{i \in \Psi(v)\}$ ($f_\Psi(v) = \max\{i \in \Psi(v)\}$). A multi-coloring Ψ is *contiguous*, or non-preemptive, if for any v, $f_\Psi(v) = s_\Psi(v) + (x(v) - 1)$. The *sum* of a multi-coloring Ψ of an instance (G, x) is the sum of the finish times of the vertices $\sum_{v \in V} f_\Psi(v)$. The minimum sum of a preemptive (non-preemptive) multi-coloring of G is denoted by pSMC(G) (npSMC).

We denote by n the number of vertices of the input instance. For a vertex v, $deg(v)$ is the degree and $N(v)$ the set of neighbors of v. In the context of rooted tree T, let T_v denote the subtree rooted at v. $kids(v)$ denote the set of children of v, and $par(v)$ its parent. Finally, let $[x, y]$ denote the interval of natural numbers $\{x, x+1, \ldots, y\}$.

We shall be needing the following bound on the number of colors in an optimal sum multi-coloring, whose proof is omitted.

Claim 1 *Optimal p-SMC and np-SMC colorings of a tree use at most* $O(p \log n)$ *colors.*

3 Non-preemptive multicoloring

We say that vertex v is *grounded* in a multi-coloring Ψ, if the smallest of v's colors is 1, $s_\Psi(v) = 1$. We say that v is *flanked* in Ψ by a neighbor u, if the smallest color of v is one larger than the largest color of u, that is, $s_\Psi(v) = 1 + f_\Psi(u)$. We call a sequence of vertices $v_0, v_1, \ldots, v_m$ a *grounding sequence* of v_m, if v_0 is grounded and, for all $0 \le i < m$, v_{i+1} is flanked by v_i.

The following observation is called for.

Observation 1 (Flanking property): *In an optimum* npSMC *coloring of a graph, each vertex v is either grounded or has a flanking neighbor.*

A grounding sequence $v_0, v_1, \ldots, v_m$ of a vertex v_m completely determines the coloring of v_m. In fact, $s_\psi(v_m)$ equals the sum of color requirements of $v_0, \ldots, v_{m-1}$ plus 1.

In our search for an optimum npSMC coloring on trees, we examine possible grounding sequences. We note that there are at most n^2 paths (and grounding sequences) in a tree, which easily leads to a polynomial algorithm for trees. We shall introduce additional ideas to reduce the complexity to $O(n \min(n, p))$. Our general approach is based on dynamic programming, where we find a partial solution (projection of an optimal npSMC scheme on a subset of vertices) assuming a given grounding sequence and using previously computed partial solutions to related subproblems.

An $O(n^2)$ algorithm for np-SMC of trees

Assume that the tree T is arbitrarily rooted in vertex r. We give a dynamic programming algorithm that computes bottom-up a matrix A, where $A[u, v]$ contains the minimum cost of a coloring of a subtree T_u under the constraint that u is grounded in v. The desired solution is then given by $\min_u A[r, u]$.

Let $f_v(u)$ denote the finishing time of u when grounded in v. Namely, $f_v(u)$ is the total length of the grounding sequence of u on v, or the sum of the lengths of the vertices on the unique path from u to v.

A is computed as follows. $A[u, v] = f_v(u)$ for a trivial T_u (when u is a leaf of T), while in general, we compute $A[u, v]$ based on all values of $A[w, *]$ for every child w of u in T. This value could be $A[w, v]$, grounding w in the same vertex as it's parent, or the minimum value of $A[w, z]$ over all those descendants z of w that result in a grounding of w that is compatible with the grounding of its parent u in v. This compatibility must satisfy two properties. First, if u is grounded in a vertex v that belongs to the subtree T_w of a child w of u, then w must also be grounded in v. Second, we must ensure that color ranges assigned to u and w do not overlap, $[f_v(u) - x(u) + 1, f_v(u)] \cap [f_z(w) - x(w) + 1, f_z(w)] = \emptyset$. Formally,

$$A[u, v] = f_v(u) + \sum_{w \in kids(u)} \min(A[w, v], \min_{x \in T_w} \{A[w, x] : f_v(u), f_x(w) \text{ compat.}\}) \quad (1)$$

The major inefficiency in computing the values of $A[u, v]$ according to (1) is the need for searching for compatible grounding in the subtrees T_w, resulting in a cubic algorithm. Instead, we ensure that optimal compatible solutions can be found quickly. We first compute, for each vertex in the tree, a sorted list of the n finishing times of the n different ways of grounding the vertex. This can be done in $O(n^2)$ time as follows: 1) in a bottom-up phase compute for each vertex u the sorted list of all lengths of paths from u to vertices in the subtree T_u rooted at u, 2) In a top-down phase the information in these lists is passed down and merged with the bottom-up list at the children to give the desired lists. We omit details in this extended abstract. This computation of sorted lists of finishing times forms the first stage of our algorithm. We next process T in a bottom-up manner. Let the parent of a vertex u $(u \neq r)$ have the sorted list of finishing times $f_{v_1}(par(u)) \leq \ldots \leq f_{v_n}(par(u))$.

The processing of vertex u involves computing the list of optimum costs $c^{v_1}(u), \ldots, c^{v_n}(u)$ where $c^{v_i}(u)$ is the optimum npSMC of the subtree T_u when the grounding of u is compatible with grounding its parent $par(u)$ in v_i. Assume the children of u have computed similar lists of optimum costs for their subtrees and let the sorted list of finishing times for u be $f_{z_1}(u) \leq \ldots \leq f_{z_n}(u)$. We can now simplify Formula 1 computing the minimum cost npSMC of T_u when grounding u in z_i:

$$A[u, z_i] = f_{z_i}(u) + \sum_{w \in kids(u)} c^{z_i}(w) \tag{2}$$

Note that this is correct even if u is a leaf. For each value of i between 1 and n we also need to compute the "prefix minima" $P[u, i] = \min_{1 \leq j \leq i}\{A[u, z_j] : z_j \in T_u\}$ and "suffix minima" $S[u, i] = \min_{i \leq j \leq n}\{A[u, z_j] : z_j \in T_u\}$ over the entries in this array that reflect grounding in a descendant.

Consider the sorted list $f_{v_1}(par(u)) \leq \ldots \leq f_{v_n}(par(u))$ of finishing times for the parent of u. For each $f_{v_i}(par(u))$, the compatible finishing times in the list $f_{z_1}(u) \leq \ldots \leq f_{z_n}(u)$ of u can be found by a single scan through the two lists, producing for each i a left and right pointer l_i, r_i indicating that $f_{v_i}(par(u))$ is compatible with all finishing times from $f_{z_1}(u)$ up to $f_{z_{l_i}}(u)$ and all those from $f_{z_{r_i}}(u)$ up to $f_{z_n}(u)$. To find the minimum cost associated with these compatible finishing times, we use the l_i-th prefix and r_i-th suffix computed earlier:

$$c^{v_i}(u) = \begin{cases} A[u, v_i] & \text{if } v_i \in T_u, \\ \min(P[u, l_i], S[u, r_i], A[u, v_i]) & \text{otherwise.} \end{cases}$$

This gives us the desired list of optimum costs $c^{v_1}(u), \ldots, c^{v_n}(u)$ with $c^{v_i}(u)$ the optimum npSMC of the subtree T_u where the grounding of u is compatible with grounding its parent in v_i. The value of the optimum cost npSMC for T overall is computed by the root vertex r and given by $P[r, n]$. The processing time for each vertex is linear.

Theorem 2. *The np-SMC problem can be solved for a tree in $O(n^2)$ time.* $\square$

In the case of paths, we can improve the complexity by observing that grounding sequences must be short.

Lemma 1. *The maximum number d of vertices in a grounding sequence $v_1, \ldots, v_d$ in a path is $O(\log p / \log \log p)$.*

Proof. Suppose the vertices $v_0, v_1, \ldots, v_d$ $(d > 2)$ form a grounding sequence in an optimum npSMC coloring Ψ^* of a path. Then, we claim that, for each i, $2 \leq i < d$, $x(v_i) \geq (d - i) \sum_{0 \leq j < i} x(v_j)$. It then follows that

$$x(v_{d-1}) \geq (d - 2)! \sum_{1 \leq j < 2} x(v_j) \geq (d - 2)!.$$

Since $p \geq x(v_{d-1}) = d^{\Omega(d)}$, we have the desired bound.

To prove the claim, consider the coloring obtained from Ψ^* by grounding the sequence $v_i, \ldots, v_{d-1}$. This may necessitate flanking v_{i-1} by v_i. The former decreases the cost (with respect to SMC(G, Ψ)) by $(d-i) \sum_{0 \leq j < i} x(v_j)$, while the latter increases the cost by at most $x(v_i)$. The claim then follows from the assumed optimality of Ψ^*.   □

Corollary 1. *The np-SMC problem can be solved for a path in $O(n \log p / \log \log p)$ time.*

An $O(n \cdot p)$ algorithm for np-SMC on trees

We now give an algorithm whose running time is linear in n, whenever $p \geq 1$ is a small constant. Let

$$B(v) = x(v) + \sum_{u \in N(v)} (x(u) + x(v) - 1) .$$

Note, that the finish time of v under any "reasonable" coloring is at most $B(v)$, since each neighbor u of v can delay the completion of v by at most $x(u)$ steps, from its own length, plus $x(v) - 1$, from leaving a "gap" in the set of available colors for v.

We call a finishing time for vertex v an *optimal*, if there exists an optimal coloring of T_v where v has this finishing time. We say that a finishing time $f(v)$ *interferes* with a child u of v, if the respective coloring of v intersects with the coloring which corresponds to u's optimal finish time in T_u.

The algorithm Tree-color proceeds by arbitrarily rooting the tree from some vertex r and then coloring it bottom-up. The coloring of v involves two tasks: (a) evaluating the cost of the possible finish times of v and selecting the optimal one, from which to derive the corresponding minimum multi-coloring sum of T_v, and (b) preparing a set of at most $x(v) + x(par(v)) - 1 < 2p$ alternative finish times for v, in the event that $par(v)$ chooses a finish time that interferes with v. The data required for these computations will be kept in the following integer arrays:

- $inc_v[B(v)]$, in which the ith entry gives the increase in the cost of the optimal coloring of T_v, when the finish time of v is i. The increase is computed relative to initial coloring of T_v, in which the algorithm assumes that $x(v) = 0$.
- $alt_v[2p]$, of alternative finish times for v: $alt_v[j]$ is the optimal finish time for v under the constraint that $par(v)$ has finish time j.

Each vertex v fills the arrays in four phases.

Phase 1: In the initial phase, v fills the array inc_v with values appropriate for the case that no collisions occur with the optimal colors of its children. That is, $inc_v[i] \leftarrow i$, for $x(v) \leq i \leq B(v)$.

Phase 2: In this phase, v updates the array inc_v. For each child $u \in kids(v)$, v adds in at most $x(u) + x(v) - 1$ entries the increase in the cost of the optimal coloring of T_u. Specifically, for any finish time i of v, which interferes with u, v updates the ith entry of inc_v, using the ith entry of the array alt_u, i.e.

$inc_v[i] \leftarrow inc_v[i] + inc_u[alt_u[i]] - inc_u[f(u)]$. The optimal finish time, $f(v)$, is the value i that minimizes $inc_v[i]$.

Phase 3: We next compute two help vectors, containing prefix and suffix minimas of inc_v. That is, for $i = x(v), \ldots, B(v)$, $p[i]$ is the index j that minimizes $inc_v[j]$ under the constraint $j \leq i$. It is computed inductively in linear time by: $p[i]$ is i, if $inc_v[i] \leq inc_v[p[i-1]]$, and $p[i-1]$ otherwise. Similarly, $s[i]$ is the index j that minimizes $inc_v[j]$ under the constraint $j \geq i$, computed by: $s[i]$ is i if $inc_v[i] \leq inc_v[s[i+1]]$ and $s[i+1]$ otherwise. Here, $p[x(v)-1] = s[B(v)+1] = \infty$.

Phase 4: Finally, alternative finish times are computed. For each possible value j of a finish time of $par(j)$, $alt_v[j]$ should be the index minimizing inv_v. The constraint implies that either v is scheduled before $par(v)$, finishing no later than $j - x(par(v))$, or it is scheduled after $par(v)$, finishing no earlier than $j + x(v)$. The index minimizing inc_v in the former case is then given by $p[j - x(par(v))]$, while in the latter case it is given by $s[j + x(v)]$. Thus, we assign $alt_v[j]$ the better of the two possibilities.

Theorem 3. Tree-color *solves np-SMC on trees in $O(np)$ steps.*

Proof. The optimality of the algorithm follows from the fact that for any subtree T_v, we find a finishing time for v that minimizes the cost of coloring T_v. Formally, for any finish time $x(v) \leq i \leq B(v)$, the SMC of T_v is given by

$$SMC(T_v, i) = i + \sum_{u \in kids(v)} SMC(T_u, n_u(i))$$

where $n_u(i)$ is the finish time of child u in a minimum-cost coloring of T_u, when the finish time of v is i. Thus, the SMC of T_v is

$$SMC(T_v) = \min_{x(v) \leq i \leq B(v)} SMC(T_v, i) \ .$$

The algorithm Tree-color follows the above dynamic programming scheme.

For the complexity of the algorithm, consider separately the phases performed by a vertex v. The first phase takes $O(B(v))$ steps. In the second phase, for each child u of v, $x(u) + x(v) - 1$ entries are updated in inc_v, for a combined complexity at most $O(B(v))$. The vectors p and s are computed in time $O(B(v))$, and the $O(p)$ entries of alt_v computed in constant time each. Observe, that $\sum_v B(v) = O(np)$, thus, summing up the complexity over all the vertices yields the theorem. $\qquad\square$

4 Preemptive case

We turn our attention in this section to the preemptive version of the multi-coloring problem. Here, we do not have a polynomial algorithm for trees, nor a proof of NP-hardness. Instead, we give the next best possible: a polynomial-time approximation schema. We also mention an exact algorithm for the case of small color requirements.

PTAS for p-SMC of trees

The algorithm is a standard dynamic programming algorithm, but one that attempts to find a restricted type of a solution. These solutions have the property that there are at most $(1/\epsilon)^{O(\log p)}$ possible colorings of each vertex. Given such a property, a straightforward dynamic programming algorithm will examine the vertices bottom-up, trying each possible coloring of a vertex, and storing the cost of the subtree for each such choice. The main part of the argument is to show the existence of a restricted solution whose sum is within $1 + \epsilon$ of optimal.

For simplicity, we allow multicolorings where *at least* $x(v)$ colors are assigned to each vertex v; clearly, this does not make the problem any easier.

Lemma 2. *Let (G, x) be a bipartite instance, and let $\epsilon > 0$. Let $q = \max_{uv \in E}(x(u) + x(v))$ and let $s_i = \lfloor \epsilon iq \rfloor$, for $i = 0, \ldots \lceil 1/\epsilon \rceil$. Then, there is a contiguous coloring Ψ' of (G, x) using $\lfloor (1+\epsilon)q \rfloor$ colors, such that for each vertex v there are integers j, j' such that Ψ' assigns to v the interval $[s_j, \ldots, s_{j'} - 1]$.*

Proof. Let R, B be a bipartition of G, and let $r = \lfloor (1 + \epsilon)q \rfloor$. Consider the contiguous coloring Ψ_0 where

$$\Psi_0(v) = \begin{cases} [1, x(v)], & \text{when } v \in R \\ [r - x(v) + 1, r], & \text{when } v \in B. \end{cases}$$

Observe, that there are at least $r - q = \lfloor \epsilon q \rfloor$ values that separate the colors assigned to any pair of adjacent vertices. Hence, this coloring can be extended to a coloring Ψ', given by

$$\Psi'(v) = \bigcup_j \{ [s_j, s_{j+1} - 1] : [s_j, s_{j+1} - 1] \cap \Psi_0(v) \neq \emptyset \}. \qquad \square$$

Let $\chi_\Psi = \max_v f_\Psi(v)$ be the makespan (maximum color used) of a multi-coloring Ψ. We now show how a given multi-coloring can be massaged into one satisfying several properties. The idea is to partition the range of possible colors into "layers" of geometrically increasing sizes. We apply Lemma 2 to schedule the colors of all vertices inside each layer, and to provide us with the desired restrictions on the possible colorings. The completion times of the vertices may increase for two reasons: the expansion factors of each level, and because of changes in the highest level that a vertex is colored in, but we can bound both factors by $1 + \epsilon$.

Theorem 4. *Let (G, x) be a bipartite instance, and $\epsilon > 0$. Then, for any multi-coloring Ψ of G, there is multi-coloring Ψ', such that for each vertex v,*

1. *$f_{\Psi'}(v) \leq (1 + \epsilon) f_\Psi(v)$,*
2. *$\Psi'(v)$ is the union of at most $O(\log_{1+\epsilon} \chi_\Psi)$ contiguous segments, and*
3. *There are $O(1/\epsilon)$ choices for the beginning and the end of each segment.*

Proof. Let $\epsilon_0 = \sqrt{1 + \epsilon} - 1$. For $1 \leq i \leq \lfloor \log_{1+\epsilon} \chi_\Psi \rfloor$, let $q_i = \lceil (1 + \epsilon_0)^i \rceil$ and $I_i = [q_{i-1}, q_i - 1]$. Define the instances (G, x_i), where $x_i(v) = |\Psi(v) \cap I_i|$.

Apply Lemma 2 to obtain colorings Ψ_i' on (G, x_i). Form Ψ' by concatenation:

$$\Psi'(v) = \bigcup_i \{x + \sum_{j=0}^{i-1} \lfloor (1 + \epsilon_0)q_j \rfloor : x \in \Psi_i'(v)\}.$$

If the highest color of $\Psi(v)$ was in interval I_i, then $f(v) > q_i$, while

$$f_{\Psi'}(v) \leq \lfloor (1 + \epsilon_0)q_i \rfloor \leq (1 + \epsilon_0)^2 q_{i-1} \leq (1 + \epsilon)f_\Psi(v),$$

establishing part 1 of the theorem. Parts 2 and 3 also follow from properties of the Ψ_i' colorings of Lemma 2. Specifically, start and end points within each interval I_i are of the form $q_{i-1} + j \cdot \epsilon \cdot (q_i - q_{i-1})$ where $0 \leq j \leq \lfloor 1/\epsilon \rfloor$.   □

Theorem 5. *For each $\epsilon > 0$, the p-SMC problem on trees can be approximated within $1 + \epsilon$ factor in time $(p \cdot \log n)^{O(1/\epsilon \cdot \log(1/\epsilon))}$.*

Proof. Let Ψ be an optimal pSMC solution, and recall the properties of the solution Ψ' that Theorem 4 has shown to exist. We now argue that we can find a solution with such properties. For each vertex v and for each layer i, we try the $\lfloor 1/\epsilon \rfloor$ possibilities for the number of $\epsilon(q_i - q_{i-1})$-sized "nibbles" that $\Psi'(v)$ contains in layer i. Thus, we need to maintain a table of size $(1/\epsilon)^{O(\log_{1+\epsilon} \chi_\Psi)}$. Since $\chi_\Psi = O(p \cdot \log n)$ by Claim 1, and $\ln(1 + \epsilon) \leq \epsilon$, the theorem follows.   □

When p is not polynomial in n, it is possible to show using a (rather non-trivial) scaling argument, that we can transform the instance to a new one with $p = O(n \cdot \log n)$, incurring only negligible loss in the approximation ratio (details omitted). This gives the result.

Exact algorithm for small lengths

Solving the p-SMC problem on trees by an exact algorithm seems to be a challenging task (even on paths). We only exhibit a very modest claim.

Claim 2 *The p-SMC problem on trees admits a polynomial solution when $p = O(\log n/\log \log n)$.*

5   Discussion

The exact algorithms that we have given apply to several generalizations of the np-SMC problem on trees. We mention here two such generalizations.

The Optimum Chromatic Cost Problem (see [J97]) generalizes the Sum Coloring problem, in that the color classes come equipped with a cost function $c : Z^+ \to Z^+$, and the objective is to minimize the value of $\sum_{v \in V} c(f(v))$. We can generalize this to multi-colorings, in which case it is reasonable to assume that the color costs are non-decreasing. Our $O(n^2)$ and $O(np)$ algorithms hold then here as well.

The $d(v)$-**coloring** problem comes with edge lengths $d : E \to Z^+$ and asks for an ordinary coloring, where the colors of adjacent vertices are further constrained to satisfy $|f(v) - f(w)| \geq d(vw)$. A non-preemptive multi-coloring instance corresponds roughly to the case where $d(vw) = (x(v) + x(w))/2$. Our algorithms handle this extension equally well, and can both handle the sum objective as well as minimizing the number of colors.

The argument for paths can be revised to hold for the generalized problems, in which case we can argue an $O(\log p)$ bound on the length of a grounding sequence.

Finally, our study leaves a few open problems. Is the p-SMC problem hard on trees? On paths? More generally, for which non-trivial, interesting classes of graphs, is the p-SMC problem solvable in polynomial time? Can we generalize the polynomial algorithms for np-SMC on trees to a larger class of graphs, e.g. outer-planar graphs? Our current arguments rely on a polynomial bound on the number of paths, which only holds for highly restricted extensions of trees.

References

B92.   M. Bell. Future directions in traffic signal control. *Transportation Research Part A*, 26:303–313, 1992.
BBH$^+$98.   A. Bar-Noy, M. Bellare, M. M. Halldórsson, H. Shachnai, and T. Tamir. On chromatic sums and distributed resource allocation. *Information and Computation*, 140:183–202, 1998.
BKH$^+$98.   A. Bar-Noy, M. M. Halldórsson, G. Kortsarz, H. Shachnai, and R. Salman. Sum Multi-Coloring of Graphs. *Proceedings of the Fifth Annual European Symposium on Algorithms*, Prague, July 1999.
BK98.   A. Bar-Noy and G. Kortsarz. The minimum color-sum of bipartite graphs. *Journal of Algorithms*, 28:339–365, 1998.
CCO93.   J. Chen, I. Cidon and Y. Ofek. A local fairness algorithm for gigabit LANs/MANs with spatial reuse. *IEEE Journal on Selected Areas in Communications*, 11:1183–1192, 1993.
DO85.   K. Daikoku and H. Ohdate. Optimal Design for Cellular Mobile Systems. *IEEE Trans. on Veh. Technology*, VT-34:3–12, 1985.
FK96.   U. Feige and J. Kilian. Zero Knowledge and the Chromatic number. *Journal of Computer and System Sciences*, 57(2):187-199, October 1998.
J97.   K. Jansen. The Optimum Cost Chromatic Partition Problem. *Proc. of the Third Italian Conference on Algorithms and Complexity (CIAC '97)*. LNCS 1203, 1997.
K89.   E. Kubicka. The Chromatic Sum of a Graph. PhD thesis, Western Michigan University, 1989.
L81.   N. Lynch. Upper Bounds for Static Resource Allocation in a Distributed System. *J. of Computer and System Sciences*, 23:254–278, 1981.
SP88.   E. Steyer and G. Peterson. Improved Algorithms for Distributed Resource Allocation. *Proceedings of the Seventh Annual Symposium on Principles of Distributed Computing*, pp. 105–116, 1988.
SG98.   A. Silberschatz and P. Galvin. Operating System Concepts. Addison-Wesley, 5th Edition, 1998.
Y73.   W. R. Young. Advanced Mobile Phone Service, Introduction, Background and Objectives. *Bell Systems Technical Report*, 58:1–14, 1973.

On the Complexity of Approximating
Colored-Graph Problems
Extended Abstract

Andrea E.F. Clementi[1], Pierluigi Crescenzi[2], and Gianluca Rossi[2]

[1] Dipartimento di Matematica, Università degli Studi di Roma "Tor Vergata"
Via della Ricerca Scientifica, 00133 Roma, Italy
clementi@axp.mat.uniroma2.it
[2] Dipartimento di Sistemi e Informatica, Università degli Studi di Firenze
Via C. Lombroso 6/17, 50134 Firenze, Italy
piluc@dsi.unifi.it, rossi@dsi.unifi.it

Abstract. In this paper we prove explicit lower bounds on the approximability of some graph problems restricted to instances which are already colored with a constant number of colors. As far as we know, this is the first time these problems are explicitily defined and analyzed. This allows us to drastically improve the previously known inapproximability results which were mainly a consequence of the analysis of bounded-degree graph problems. Moreover, we apply one of these results to obtain new lower bounds on the approximabiluty of the minimum delay schedule problem on store-and-forward networks of bounded diameter. Finally, we propose a generalization of our analysis of the complexity of approximating colored-graph problems to the complexity of approximating approximated optimization problems.

1 Introduction

An *independent set* in a graph $G = (V, E)$ is a set of pairwise non-adjacent nodes. The MAXIMUM INDEPENDENT SET problem consists of finding an independent set of the largest cardinality. This problem is known to be NP-hard [9], to be approximable within factor $O\left(|V|/(\log|V|)^2\right)$ [5], and to be not approximable within factor $|V|^{1-\epsilon}$ for any $\epsilon > 0$ [12] (unless coRP = NP). A *vertex cover* in a graph $G = (V, E)$ is a set of nodes such that each edge has at least one endpoint in the cover. The MINIMUM VERTEX COVER problem consists of finding a cover of the smallest cardinality. This problem is known to be NP-hard [9], to be approximable within factor 2 [19], and to be not approximable within factor 1.1666 [13] (unless P = NP). Finally, a *coloring* of a graph $G = (V, E)$ is a partition of the set of nodes into a collection of pairwise disjoint independent sets, called *colors*. The MINIMUM GRAPH COLORING problem consists of finding a coloring with the smallest number of colors. This problem is known to be NP-hard [9], to be approximable within factor $O\left(|V|(\log\log|V|)^2/(\log|V|)^3\right)$ [11], and to be not approximable within factor $|V|^{1-\epsilon}$ for any $\epsilon > 0$ [7] (unless coRP = NP).

T. Asano et al. (Eds.): COCOON'99, LNCS 1627, pp. 281–290, 1999.

This paper studies *explicit* approximation thresholds for the above three problems when restricted to graphs that are already colored with a constant number of colors. This is of interest since, for some classes of graphs, one can efficiently find a coloring that uses a constant number of colors. For example, if a graph G is a connected graph whose maximum degree is k with $k \geq 3$ and G is not K_{k+1} (that is, the clique of $k + 1$ nodes), then G can be colored in polynomial time with k colors [18][1]. The main reason to study these colored restrictions is that the approximability properties of MAXIMUM INDEPENDENT SET, MINIMUM VERTEX COVER and MINIMUM GRAPH COLORING drastically improve in the case of colored graphs. Indeed, MAXIMUM INDEPENDENT SET and MINIMUM VERTEX COVER restricted to k-colored graphs are approximable within factor $k/2$ and $2 - 2/k$, respectively [14] (as a consequence of this result and of the algorithm of [18], it thus follows, for example, that MINIMUM VERTEX COVER restricted to graphs whose maximum degree is 5 can be approximated within factor 1.6, which is also the best known performance ratio for this kind of graphs). In the case of MINIMUM GRAPH COLORING, a k-coloring of the input graph is clearly a $k/3$-approximate solution (since we can always assume that at least three colors have to be used) and, as far as we know, this is also the best known approximation algorithm.

It is hence natural to ask how well the above graph problems can be approximated when restricted to colored instances or if they admit a *polynomial-time approximation scheme* (in short, PTAS). Given a graph $G = (V, E)$ and a partition of V into k independent sets, the MAXIMUM k-PARTITE GRAPH INDEPENDENT SET problem consists of finding an independent set I of G whose cardinality is maximal. Analogously, we can define MINIMUM k-PARTITE GRAPH VERTEX COVER and MINIMUM k-PARTITE GRAPH COLORING. In [4] the authors give explicit lower bounds on the approximability of MAXIMUM INDEPENDENT SET and MINIMUM VERTEX COVER restricted to graphs whose maximum degree is bounded by a constant $k \geq 3$ (the bounds are 1.0071 and 1.0069, respectively). Since these graphs can be k-colored in polynomial time, these bounds also hold for MAXIMUM k-PARTITE GRAPH INDEPENDENT SET and MINIMUM k-PARTITE GRAPH VERTEX COVER, respectively (hence, these two problems do not admit a PTAS).

The main novelty of our approach is to *explicitly consider the colored restriction of* MAXIMUM INDEPENDENT SET *and* MINIMUM VERTEX COVER. This will allow us to obtain significantly tighter inapproximability results than the previously known ones (see [4]). In particular, our first two results can be summarized as follows:

- If P $\neq$ NP, *then* MAXIMUM 3-PARTITE GRAPH INDEPENDENT SET *is not approximable within a factor smaller than 26/25 unless.*
- If P $\neq$ NP, *then* MINIMUM 3-PARTITE GRAPH VERTEX COVER *is not approximable within a factor smaller than 34/33.*

---

[1] A more popular example are planar graphs that can be 4-colored in polynomial time [2,3].

Clearly, these results also apply to MAXIMUM k-PARTITE GRAPH INDEPENDENT SET and MINIMUM k-PARTITE GRAPH VERTEX COVER, respectively, for any $k \geq 3$. However, since the best known approximation algorithms for the two problems have performance ratio bounded by $k/2$ and $2 - 2/k$, respectively, it would be interesting to find an explicit relation between the value of k and the approximability threshold (similar to the relation proved in [1] where it is shown that MAXIMUM INDEPENDENT SET restricted to graph of degree at most k is not approximable within factor k^ϵ for some $\epsilon > 0$).

Besides being interesting by itself, studying colored-graph problems can also be useful to obtain better inapproximability results for other problems. Indeed, as a consequence of the fact that finding a 3-coloring of a planar graph of degree bounded by 4 is NP-hard [8], we have that MINIMUM 4-PARTITE GRAPH COLORING is not approximable within factor 4/3 (this bound clearly applies to MINIMUM k-PARTITE GRAPH COLORING, for any $k \geq 4$). The third result of this paper will use this bound on the approximability of the coloring colored-graph problem to obtain a new negative result for the problem of finding a minimum delay schedule on *store-and-forward* networks (also known as *packet-switching* networks) [17]. In these networks, each packet is passed from node to node and can cross each edge depending on the available capacity of the edge and of the destination node. We will consider the *oblivious* case in which the routing paths have been precomputed by a global controller. The *minimum delay schedule problem* then consists of finding an assignment of the network resources (i.e., edges and nodes) to the packets in order to minimize the maximum delay among all packets. In [6] it has been shown that this problem is not approximable within factor $n^{1-\epsilon}$ for any $\epsilon > 0$ (unless P = NP) where n denotes the number of packets. However, this result apply only to networks of unbounded diameter. As an application of the inapproximability of MINIMUM 4-PARTITE GRAPH COLORING, we will prove an *explicit approximability threshold for the minimum delay schedule problem on store-and-forward networks of bounded diameter.*

Finally, we will propose a generalization of our analysis of the complexity of approximating colored-graph problems to the complexity of *approximating approximated optimization problems*. We will also see that in some interesting (but rare) cases "free" approximate solutions do not help at all for finding better solutions.

The paper is structured as follows. The rest of this section will introduce basic approximation theory definitions that will be used in the following sections. Sections 2 and 2.1 contain the proof of the inapproximability results for MAXIMUM INDEPENDENT SET and MINIMUM VERTEX COVER, respectively. In Section 3 we use the inapproximability result for MINIMUM k-PARTITE GRAPH COLORING in order to obtain the bound on the approximability of the minimum delay schedule problem on networks of bounded diameter. Finally, in Section 4 we analyze the complexity of approximating approximated optimization problems.

Approximation theory preliminaries Let A be an optimization problem. Given an instance x and a feasible solution y of x, we define the *performance ratio of*

y *with respect to* x *as* $R(x, y) = \max\left\{\frac{m_\mathbf{A}(x,y)}{m_\mathbf{A}^*(x)}, \frac{m_\mathbf{A}^*(x)}{m_\mathbf{A}(x,y)}\right\}$, where $m_\mathbf{A}(x, y)$ denotes the measure of y with respect to x and $m_\mathbf{A}^*(x)$ denotes the optimum measure (whenever the problem the two functions refer to is clear from the parameter types, we will avoid to specify the subscript). The performance ratio is always a number greater than or equal to 1 and is as close to 1 as y is close to an optimum solution.

Let $\mathbf{A}$ be an optimization problem and let T be a polynomial-time algorithm that, for any instance x of $\mathbf{A}$, returns a feasible solution $T(x)$ of x. We say that T is an *r-approximate algorithm for* $\mathbf{A}$ if there exists $r > 1$ such that, for any instance x, the performance ratio of the feasible solution $T(x)$ with respect to x is at most r. Equivalently, we say that $\mathbf{A}$ is *approximable within factor* r.

The basic tool we use to prove the inapproximability results of MAXIMUM 3-PARTITE GRAPH INDEPENDENT SET and MINIMUM 3-PARTITE GRAPH VERTEX COVER is the L-reducibility introduced in [20]. Given two optimization problems $\mathbf{A}_1$ and $\mathbf{A}_2$, we say that $\mathbf{A}_1$ is *L-reducible* to $\mathbf{A}_2$ if two polynomial-time computable functions f and g and two positive constants α and β exist such that:

1. For any instance x of $\mathbf{A}_1$, $f(x)$ is an instance of $\mathbf{A}_2$.
2. For any instance x of $\mathbf{A}_1$ and for any solution y of $f(x)$, $g(x, y)$ is a solution of x.
3. For any instance x of $\mathbf{A}_1$, $m_{\mathbf{A}_2}^*(f(x)) \leq \alpha m_{\mathbf{A}_1}^*(x)$.
4. For any instance x of $\mathbf{A}_1$ and for any solution y of $f(x)$,

$$|m_{\mathbf{A}_1}^*(x) - m_{\mathbf{A}_1}(x, g(x, y))| \leq \beta |m_{\mathbf{A}_2}^*(f(x)) - m_{\mathbf{A}_2}(f(x), y)|.$$

Fact 1 *If* $\mathbf{A}_1$ *is L-reducible to* $\mathbf{A}_2$ *with constants* α *and* β *and* $\mathbf{A}_1$ *is not approximable within a factor smaller than* t, *then* $\mathbf{A}_2$ *is not approximable within a factor smaller than* $t' = \alpha\beta t/(\alpha\beta t - 1)$.

Finally, we will refer to the maximum one-in-three satisfiability problem which is defined as follows. An instance Φ of MAXIMUM ONE-IN-THREE SAT consists of a set $X = \{x_1, \ldots, x_n\}$ of variables and a collection $\varphi = \{C_1, \ldots, C_m\}$ of (one-in-three) clauses with exactly three literals (a literal is a variable or a negated variable in X). Each of these clauses is a Boolean formula which is true if exactly one of its inputs is true. More formally,

$$\texttt{ONE-IN-THREE}(x, y, z) = x\neg y\neg z \vee \neg xy\neg z \vee \neg x\neg yz.$$

The goal of MAXIMUM ONE-IN-THREE SAT is to find a a truth-assignment that makes true the maximum number of clauses. The following result gives a lower bound on the approximability of this problem.

Fact 2 *[22]* MAXIMUM ONE-IN-THREE SAT *is not approximable within factor* $2 - \epsilon$ *for any* $\epsilon > 0$ *(unless* P=NP*).*

2   The colored-graph independent set problem

In this section we will consider the maximum independent set whose instances include a k-partition of the nodes. The main result is that MAXIMUM 3-PARTITE GRAPH INDEPENDENT SET cannot be approximated within a factor less than 1.04 (unless P=NP). To this aim we show an L-reduction from MAXIMUM ONE-IN-THREE SAT to MAXIMUM 3-PARTITE GRAPH INDEPENDENT SET with $\alpha = 13$ and $\beta = 1$. The hardness result for MAXIMUM 3-PARTITE GRAPH INDEPENDENT SET will then follow from Facts 1 and 2.

Let $\Phi = (X, \varphi)$ be an instance of MAXIMUM ONE-IN-THREE SAT. We now define a graph $G(V, E)$ as follows. For each clause $C_j = \texttt{ONE-IN-THREE}(u_1, u_2, u_3)$ of φ, we add to G the following nodes and edges (see Figure 1):

- Three *clause-nodes* c_1^j, c_2^j and c_3^j.
- Six *variable-nodes* u_i^j, $\bar{u}_i^j$: u_i^j is said to be the *companion* of $\bar{u}_i^j$, and vice versa.
- For $i = 1, 2, 3$, $(u_i^j, \bar{u}_i^j) \in E$.
- For $i = 1, 2, 3$, $(c_i^j, \bar{u}_i^j) \in E$ and, for each $h \neq i$ with $1 \leq h \leq 3$, $(c_i^j, u_h^j) \in E$.

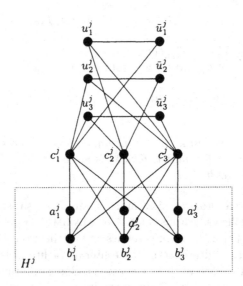

Fig. 1. The subgraph corresponding to the clause $C_j = \texttt{ONE-IN-THREE}(x_1, \bar{x}_2, x_3)$.

Successively, the clause-nodes are interfaced with the gadget H^j shown in the dashed rectangle of Fig. 1: clearly, this gadget satisfies the following property.

Property 1. The graph induced by H^j and the three clause-nodes is 3-partite and contains three maximum independent sets of size 4, each containing *exactly* one of the three clause-nodes.

Finally, in order to ensure the consistency of the truth-value of the literals we let the variable-nodes of all clauses corresponding to the same literal have the same neighborhood: this implies that we must add edges between the two subgraphs corresponding to two different clauses.

It is possible to show that the graph G defined above admits a partition of its nodes into three independent sets. The instance of MAXIMUM 3-PARTITE GRAPH INDEPENDENT SET is then formed by G and this partition.

Let I be an independent set of G. Without loss of generality, we may assume that I does not contain a variable-node corresponding to a variable x and, at the same time, a variable-node corresponding to $\neg x$. Indeed, if this is not true, we can then delete one of the two nodes from I and add its companion to I. Since all variable-nodes corresponding to the same literal have the same neighborhood, this operation will not create any conflict.

Moreover, we may assume that, for each clause C_j, I contains at most one of the clause-nodes corresponding to C_j. Indeed, if I contains two such nodes, then I contains at most one variable-node u_i^j corresponding to C_j. We can then delete from I the clause-nodes and add to it two of the four variable-nodes of C_j which are available. This operation can create a conflict with a clause-node of another clause C_k: however, we can simply delete this clause-node and add the variable-node u_h^k which corresponds to the same literal of u_i^j. The case in which I contains three clause-nodes of C_j can be dealt in a similar way.

Finally, we assume that, for each clause, I contains exactly three variable-nodes and three nodes of the gadget distinct from the clause-nodes. This is due to the fact that at most one clause-node is in I and to Property 1.

In summary, the above assumptions imply that $\mathrm{m}(G, I) = 6m + N_1$, where N_1 denotes the number of clause-nodes in I.

We are now ready to describe how the independent set I is transformed into a truth-assignment τ for φ. For each clause C_j such that I contains one clause-node c_i^j, we assign the value true to the ith literal of C_i. We then assign the value false to all other variables. We have that $\mathrm{m}(G, I) = 6m + N_1 = 6m + \mathrm{m}(\Phi, \tau)$, where the last equality is due to the fact that the number of clause-nodes in I is exactly the number of clauses satisfied by τ. Hence, $\mathrm{opt}(G) = 6m + \mathrm{opt}(\Phi) \leq 13\mathrm{opt}(\Phi)$, where the inequality is due to the fact that, for any instance Φ of MAXIMUM ONE-IN-THREE SAT, $\mathrm{opt}(\Phi) \geq 1/2m$ [22]. Moreover, $\mathrm{opt}(\Phi) - \mathrm{m}(\Phi, \tau) = \mathrm{opt}(G) - \mathrm{m}(G, I)$. That is, our reduction is an L-reduction with $\alpha = 13$ and $\beta = 1$. From this reduction and from Facts 1 and 2, the next result thus follows.

Theorem 3. MAXIMUM 3-PARTITE GRAPH INDEPENDENT SET *is not approximable within a factor smaller than 1.04 unless* P $=$ NP.

2.1   The colored-graph vertex cover problem

It is easy to see that the reduction of the proof of the previous theorem also allows us to prove the next result.

Theorem 4. MINIMUM 3-PARTITE GRAPH VERTEX COVER *is not approximable within a factor smaller than 34/33 unless* P $=$ NP.

3   The colored-graph coloring problem and its application

It is well-known that any planar graph can be colored with 4 colors in polynomial-time [2,3]. Moreover, in [8] it has been proved that 3-coloring a planar graph with maximum degree bounded by 4 is NP-hard. As a consequence of these two results, we have the following theorem.

Theorem 5 ([2,3,8]). *For any $k \geq 4$,* MINIMUM k-PARTITE GRAPH COLOR-ING *is not approximable within a factor smaller than 4/3 unless* P = NP.

We will now apply this result to obtain a new inapproximability result for the minimum delay schedule on store-and-forward networks.

A *Store-and-Forward network* can be represented as a graph, where nodes stand for sites containing resources to be shared (i.e. *buffers*) and edges for communication links (a popular example of store-and-forward network is the butterfly network [17]). Messages flowing in a Store-and-Forward network are represented as *tokens* transmitted by edges of the corresponding graph. Each token fills exactly one buffer contained in the node. Message transmission from one site to an adjacent one is modeled as moving the corresponding token.

Each token in the network follows a route between two fixed nodes: the *source node* (which generates it) and the *sink node* (which eliminates it from the network). We will consider the *oblivious-routing* case only in which the source node selects the entire path before sending the token [17]. Such path cannot be changed after the token has left the source. In this case, each token must carry the whole routing information.

A *schedule* is a strategy to assign free buffers to tokens which are waiting for them. Observe that in the case of oblivious routing a schedule must only solve the conflicts among the tokens which are requiring a buffer in the same node. Given a network and a set of tokens, the MINIMUM DELAY SCHEDULE problem consists of finding a schedule which minimizes the total end-to-end delay, i.e., the difference between the delivery time and the delivery time reachable if the network had unbounded node and channel capacity.

In this section, we will consider a particular network topology, that is, layered networks [17]. *Layered networks* are directed graphs defined as follows: nodes are partitioned in $L + 1 \geq 2$ sets V^i, with $0 \leq i \leq L$ and edge (h, k) can exist only if $h \in V^i$ and $k \in V^{i+1}$ for some $0 \leq i \leq L - 1$. Nodes of V^i are said to be of *level i*.

In [6] it has been proved that MINIMUM DELAY SCHEDULE on layered networks is not approximable within $n^{1-\epsilon}$ for any $\epsilon > 0$, where n denotes the number of tokens. However, the network resulting from the reduction has unbounded diameter: in particular, $L = \Theta(n)$. This case is clearly unrealistic since in most applications the diameter of the network is significantly less than linear (typically, it is logarithmic) [17]. It is then natural to analyze the approximability properties of the MINIMUM DELAY SCHEDULE on networks whose diameter is bounded by a sublinear function. The main result of this section is that even on networks of constant diameter the problem is not approximable within a specified constant factor: the proof (which is here omitted) is based on an L-reduction from MINIMUM k-PARTITE GRAPH COLORING to MINIMUM DELAY SCHEDULE with $\alpha = \beta = 1$ such that the diameter of the resulting layered network is bounded by $2kd$ where d is the maximum degree of the original graph.

Theorem 6. *For any $\delta \geq 32$,* MINIMUM DELAY SCHEDULE *on networks whose diameter is bounded by δ is not approximable within factor $4/3 - \epsilon$ for any $\epsilon > 0$.*

4   Approximating approximated optimization problems

The approximability properties of several optimization problems for which a "promise" is given along with the instance has been recently studied in several papers. For example, in [15] the special case of the minimum coloring problem where the input graph is guaranteed to be k-colorable, for some small k, is considered and the best known (randomized) approximation algorithm in terms of the number of nodes is presented. More recently, in [21] it is shown that whenever an instance of the maximum satisfiability problem is guaranteed to be satisfiable a better approximation algorithm can be obtained with respect to the standard problem. This latter result has been extended to other kinds of constraint satisfaction problems in [22]. Along with these positive results, negative ones have also been obtained: for instance, in [16] it is shown that it is not possible to color a 3-colorable graph with 4 colors in polynomial time (unless P = NP).

Of course, all the above promises are NP-hard statements (otherwise, they do not add any interesting information). Moreover, they basically provide the value of an optimum solution: for instance, in the case of the maximum satisfiability problem the promise states that the optimum value is equal to the number of clauses. However, they do not provide the optimum solution since otherwise they would make the problem a non-sense one.

In the previous sections, we have basically proposed a new kind of promises to be used in order to approximate specific problems: on the one hand, we have relaxed the optimality of the promise (in our case, the cardinality of the partition of the nodes), on the other we have pretended that the promise itself is "constructive" (in our case, the partition is explicitily given). This leads us to the following definition. Given an optimization problem **A** we denote with $\mathbf{A}_{[r=\rho]}$, where ρ is any constant greater than 1, the problem whose instances contains both an instance x of **A** and a solution y of x whose performance ratio is guaranteed to be at most ρ. Clearly, $\mathbf{A}_{[r=\rho]}$ is ρ-approximable (just return the promise y). However, the natural question is whether y can be somehow used in order to obtain a better solution? In other words, does there exist an $\epsilon > 0$ such that $\mathbf{A}_{[r=\rho]}$ is $(\rho - \epsilon)$-approximable?

Observe that, in order to make the promise useful, approximating **A** within ρ must be NP-hard (otherwise, the promise does not add any power to our algorithm). On the other hand, the promise can be so useful to let **A** be solvable in polynomial time or, at least, admit a PTAS.

The results of the previous section can then be reformulated as follows:

- If P $\neq$ NP, then MAXIMUM INDEPENDENT SET$_{[r=1.16]}$ is not approximable within factor $26/25 - \epsilon$ for any $\epsilon > 0$.
- If P $\neq$ NP, then MINIMUM VERTEX COVER$_{[r=1.29]}$ is not approximable within factor $34/33 - \epsilon$ for any $\epsilon > 0$.
- If P $\neq$ NP, then MINIMUM GRAPH COLORING$_{[r=4/3]}$ is not approximable within factor $4/3 - \epsilon$ for any $\epsilon > 0$.

The next result shows that by simple amplification techniques it is possible sometimes to prove that approximate solutions do not help at all (as in the case of MINIMUM GRAPH COLORING).

Theorem 7. *For any $\delta > 0$, MAXIMUM 3-SATISFIABILITY$_{[r=8/7-\delta]}$ is not approximable within factor $r - \epsilon$ for any $\epsilon > 0$ (unless P =NP).*

Proof (Sketch). Let x be a Boolean formula with m_1 for which it is NP-hard to decide whether all clauses are satisfiable or at most $7/8$ of the clauses are satisfiable (the existence of such a formula has been proved in [13]). We then add to x a set of m_2 new clauses based on new m_2 variables which are all satisfiable: more precisely, each of these new clauses is a unitary clause containing the only occurrence of a new variable. The $8/7 - \delta$-approximate solution associated with this new formula is given by a truth assignment that sets to true all the new variables. The value of the original variables is computed by means of the $8/7$-approximation algorithm of [10]. If $m_2 = km_1$, then this solution is at most $1+1/(7+8k)$-approximate: by appropriately choosing k this factor can be smaller than $8/7-\delta$. To obtain a better solution we have to find an approximate solution for x whose performance ratio is smaller than $8/7$: this task is NP-hard. $\square$

The proof of the above theorem is based on the fact that a tight lower bound on the approximability of MAXIMUM 3-SATISFIABILITY is known. Unfortunately, this is a very rare case. For this reason, we believe it should be worth studying the usefulness of approximate solution in order to better solve approximation problems. From a structural complexity point of view, this is the main question left open by this paper.

Acknowledgements We thank Luca Trevisan for several remarks on a preliminary version of this paper. Luca also suggested the proof of Theorem 7.

References

1. ALON, N., FEIGE, U., WIGDERSON, A., AND ZUCKERMAN, D. Derandomized graph products. *Computational Complexity 5* (1995), 60–75.
2. APPEL, K., AND HAKEN, W. Every planar map is four colorable. part I: discharging. *Illinois J. Math. 21* (1977), 429–490.
3. APPEL, K., AND HAKEN, W. Every planar map is four colorable. part II: reducibility. *Illinois J. Math. 21* (1977), 491–567.

4. BERMAN, P., AND KARPINSKI, M. On some tighter inapproximability results. Further improvements. Tech. Rep. TR98-065, ECCC, 1998.

5. BOPPANA, R., AND HALLDÓRSSON, M. M. Approximating maximum independent sets by excluding subgraphs. *Bit 32* (1992), 180–196.

6. CLEMENTI, A., AND DI IANNI, M. On the hardness of approximating optimum schedule problems in store and forward networks. *IEEE/ACM Transaction on Networking 4* (1996), 272–280.

7. FEIGE, U., AND KILIAN, J. Zero knowledge and the chromatic number. In *Proc. Eleventh Ann. IEEE Conf. on Comp. Complexity* (1996), IEEE Computer Society, pp. 278–287.

8. GAREY, M., JOHNSON, D., AND STOCKMEYER, L. Some simplified NP-complete graph problems. *Theoretical Comput. Sci. 1* (1976), 237–267.

9. GAREY, M. R., AND JOHNSON, D. S. *Computers and Intractability: a guide to the theory of NP-completeness.* W. H. Freeman and Company, San Francisco, 1979.

10. GOEMANS, M. X., AND WILLIAMSON, D. P. Improved approximation algorithms for maximum cut and satisfiability problems using semidefinite programming. *J. ACM 42* (1995b), 1115–1145.

11. HALLDÓRSSON, M. M. A still better performance guarantee for approximate graph coloring. *Inform. Process. Lett. 45* (1993a), 19–23.

12. HÅSTAD, J. Clique is hard to approximate within $n^{1-\epsilon}$. In *Proc. 37th Ann. IEEE Symp. on Foundations of Comput. Sci.* (1996), IEEE Computer Society, pp. 627–636.

13. HÅSTAD, J. Some optimal inapproximability results. In *Proc. 29th Ann. ACM Symp. on Theory of Comp.* (1997), ACM, pp. 1–10.

14. HOCHBAUM, D. S. Efficient bounds for the stable set, vertex cover and set packing problems. *Disc. Appl. Math. 6* (1983), 243–254.

15. KARGER, D., MOTWANI, R., AND SUDAN, M. Approximate graph coloring by semidefinite programming. *J. ACM 45* (1998), 246–265.

16. KHANNA, S., LINIAL, N., AND SAFRA, S. On the hardness of approximating the chromatic number. In *Proc. 1st Israel Symp. on Theory of Computing and Systems* (1992), IEEE Computer Society, pp. 250–260.

17. LEIGHTON, F. *Introduction to parallel algorithms and architectures: arrays, trees, hypercubes.* Morgan Kaufmann, San Mateo, CA, 1992.

18. LOVASZ, L. Three short proofs in graph theory. *J. Combin. Theory 19* (1975), 269–271.

19. MONIEN, B., AND SPECKENMEYER, E. Ramsey numbers and an approximation algorithm for the vertex cover problem. *Acta Inf. 22* (1985), 115–123.

20. PAPADIMITRIOU, C. H., AND YANNAKAKIS, M. Optimization, approximation, and complexity classes. *J. Comput. System Sci. 43* (1991), 425–440.

21. TREVISAN, L. Approximating satisfiable satisfiability problems. In *Proc. 5th Ann. European Symp. on Algorithms* (1997), Springer-Verlag, pp. 472–485.

22. ZWICK, U. Approximation algorithms for constraint satisfaction problems involving at most three variables per constraint. In *Proc. 9th Ann. ACM-SIAM Symp. on Discrete Algorithms* (1998), ACM-SIAM, p. 201.

On the Average Sensitivity of Testing Square-Free Numbers

Anna Bernasconi[1], Carsten Damm[2,*], and Igor E. Shparlinski[3,**]

[1] Institut für Informatik, Technische Universität München
D-80290 München, Germany
bernasco@informatik.tu-muenchen.de
[2] Fachbereich für Informatik, Universität Trier
D-54286 Trier, Germany
damm@uni-trier.de
[3] School of MPCE, Macquarie University
Sydney, NSW 2109, Australia
igor@mpce.mq.edu.au

Abstract. We study combinatorial complexity characteristics of a Boolean function related to a natural number theoretic problem. In particular we obtain a linear lower bound on the average sensitivity of the Boolean function deciding whether a given integer is square-free. This result allows us to derive a quadratic lower bound for the formula size complexity of testing square-free numbers and a linear lower bound on the average decision tree depth. We also obtain lower bounds on the degrees of exact and approximative polynomial representations of this function.

1 Introduction

In view of the many applications in modern cryptology Boolean functions related to number theoretic problems are a natural object to study from the complexity viewpoint. Recently for such functions several complexity lower bounds have been obtained for representations like unbounded fan-in Boolean circuits, decision trees, and real polynomials (see [1,3,4,8,17,18]). The two main ingredients of these papers are harmonic analysis and estimates based on number theoretic considerations.

In this paper we focus to a purely combinatorial complexity characteristic: the *average sensitivity* of Boolean functions. The sensitivity of a function f on input w is defined as the number of bits such that flipping one of them will change the value of the function; the sensitivity of f is the maximum of the sensitivity of f on input w over all strings w of a given length; finally, the average sensitivity of f is the average (taken with respect to the uniform distribution) of the sensitivity of f on input w over all w of a given length. These definitions are made precise

* Partially supported by DFG grant Me 1077/14-1
** Partially supported by ARC grant A69700294

T. Asano et al. (Eds.): COCOON'99, LNCS 1627, pp. 291–299, 1999.
© Springer-Verlag Berlin Heidelberg 1999

below. The sensitivity is of interest because it can be used to obtain lower bounds for the CREW PRAM complexity of Boolean functions (see [9,10,16,19]), that is the complexity on a *parallel random access machine* with an unlimited number of all-powerful processors, such that simultaneous reads of a single memory cell by several processors are permitted, but simultaneous writes are not. The average sensitivity is a finer characteristic of Boolean functions which has been studied in a number of papers, see [2,5,14].

Our main result consists in a linear lower bound on the average sensitivity of testing square-free numbers. We recall that an integer x is called **square-free** if there is no prime p such that $p^2|x$. More precisely, we consider the function g which decides whether a given $(n + 1)$-bit odd integer is square-free, that is the function for which

$$g(x_1, \ldots, x_n) = \begin{cases} 1, & \text{if } 2x + 1 \text{ is square-free,} \\ 0, & \text{if } 2x + 1 \text{ otherwise,} \end{cases} \qquad (1)$$

where $x = x_1 \ldots x_n$ is the bit representation of x, $0 \leq x \leq 2^n - 1$ (if necessary we add several leading zeros), and prove that for g a linear lower bounds on the average sensitivity holds. This lower bound is derived by studying the distribution of odd square-free numbers with a fixed binary digit.

We then apply this estimate to derive new lower bounds on the formula size, on the average depth of a decision tree and on the degree of certain polynomial representations for g.

The linear bound on the average sensitivity of g also provides an alternative proof for the statement proved in [3,4] that g does not belong to the class $\mathbf{AC^0}$. On the other hand, even a stronger result has recently been obtained in [1].

2   Basic Definitions

Let $\mathfrak{B}_n = \{0, 1\}^n$ denote the n dimensional Boolean cube.

For a binary vector $a \in \mathfrak{B}_n$ we denote by $a^{(i)}$ the vector obtained from a by flipping its ith coordinate. Now we introduce the main combinatorial parameters of Boolean functions $f : \mathfrak{B}_n \to \{0, 1\}$ considered in this paper.

The **sensitivity** of f at input $a \in \mathfrak{B}_n$ is the number

$$\sigma_a(f) = \sum_{i=1}^{n} \left| f(a) - f(a^{(i)}) \right|.$$

The **sensitivity** of f is defined as

$$\sigma(f) = \max_{x \in \mathfrak{B}_n} \sum_{i=1}^{n} \left| f(x) - f(x^{(i)}) \right|,$$

and the **average sensitivity** of f is

$$s(f) = 2^{-n} \sum_{x \in \mathfrak{B}_n} \sum_{i=1}^{n} \left| f(x) - f(x^{(i)}) \right|.$$

Clearly, $s(f) \leq \sigma(f) \leq n$ for any f.

The average sensitivity of a function f can be equivalently defined as the sum of the **influences** of all variables on f, where the influence of x_i on f, denoted $I_i(f)$, is the probability that flipping the i-th variable of a random Boolean input will flip the output. In other words, $I_i(f)$ is a measure of how influential is the variable x_i in determining the outcome of f. Precisely we have

$$I_i(f) = 2^{-n} \sum_{x \in \mathfrak{B}_n} \left| f(x) - f(x^{(i)}) \right|,$$

which immediately implies

$$s(f) = \sum_{i=1}^{n} I_i(f). \tag{2}$$

Formulae are defined in the following recursive way: the variables x_1, x_2, $\ldots, x_n$ and their negations $\neg x_1, \neg x_2, \ldots, \neg x_n$ are formulae; if F_1, F_2 are formulae, so are $F_1 \wedge F_2$ and $F_1 \vee F_2$. The **size** of a formula F is the number of occurrences of variables in it. Notice that a formula can be equivalently defined as a Boolean circuit whose fan-out of gates is bounded by one.

A **decision tree** with input variables $x_1, \ldots, x_n$ is a rooted binary tree in which each inner node is labeled with a variable and the edges leaving the node are labeled 0 and 1, respectively. Further each leaf v of the tree is labeled with some value $\lambda(v) \in \{0, 1\}$.

A decision tree T in a natural way defines a Boolean function f_T. For an input assignment a the computation proceeds as follows: starting from the root, at each visited inner node a certain variable x_i is tested. The computation proceeds along the edge labeled a_i. Define $f_T(a) = \lambda(v)$, where v is the eventually reached leaf.

For a decision tree T we say input $a \in \mathfrak{B}_n$ exits in depth i (denoted $D_a(T) = i$), if during the computation of $f_T(a)$ exactly i edges are passed.

The **depth** of the tree is $D(T) = \max\{D_a(T) \mid a \in \mathfrak{B}_n\}$ and its **average depth** is

$$\overline{D}(T) = 2^{-n} \sum_{a \in \mathfrak{B}_n} D_a(T).$$

For a Boolean function f let $D(f)$ and $\overline{D}(f)$, respectively, denote the minimal depth and the minimal average depth, respectively, of any decision tree T with $f_T = f$. Clearly, $\overline{D}(f) \leq D(f) \leq n$ for any f.

Further we mention the following definitions from [15]: for a Boolean function $f : \mathfrak{B}_n \to \{0, 1\}$ let the **real degree** of f, denoted by $\Delta(f)$, be the degree of the unique multilinear real polynomial $P(X_1, \ldots, X_n)$ for which

$$f(x_1, \ldots, x_n) = P(x_1, \ldots, x_n)$$

holds for every $(x_1, \ldots, x_n) \in \mathfrak{B}_n$. Here multilinearity means, that each variable appears with degree at most 1.

We also define the **real approximate degree** of f, denoted by $\delta(f)$, as the degree of a multilinear real polynomial $P(X_1, \ldots, X_n)$ for which

$$|f(x_1, \ldots, x_n) - P(x_1, \ldots, x_n)| \leq 1/3$$

holds for every $(x_1, \ldots, x_n) \in \mathfrak{B}_n$. Certainly, in all our results $1/3$ can be replaced by any constant $\gamma < 1/2$.

Clearly, $\delta(f) \leq \Delta(f) \leq n$ for any f.

Throughout the paper we identify integers and their bit representations. In particular, we write $f(x)$ and $x^{(i)}$ for n-bit integers x, assuming that these apply to their bit representations.

3 Relations between Formula Size, Decision Tree Depth, Polynomial Degree, and Average Sensitivity

Let f be a Boolean function on n variables, and $L(f)$ denote the number of occurrences of variables in the minimal-size formula that computes f. Following [13], it is possible to restate Khrapchenko's Theorem, which gives lower bounds on the size of Boolean formulae, as follows. For $A \subseteq f^{-1}(0)$ and $B \subseteq f^{-1}(1)$, we define the $|B| \times |A|$ matrix Q, with $q_{uv} = 1$ if the binary strings $u \in B$ and $v \in A$ differ in exactly one component; otherwise $q_{uv} = 0$. Then

$$L(f) \geq \frac{1}{|A||B|} \left(\sum_{u \in A, \ v \in B} q_{uv} \right)^2.$$

From this version of Khrapchenko's Theorem, we can easily derive a lower bound on the formula size in terms of the average sensitivity.

Lemma 1. *Let f be a Boolean function depending on n variables and let p denote the probability that f takes the value 1. Then*

$$L(f) \geq \frac{1}{4p(1-p)} s(f)^2.$$

Proof. Let $A = f^{-1}(0)$ and $B = f^{-1}(1)$. We obtain the desired lower bound by observing that

$$\sum_{u \in A, \ v \in B} q_{uv} = 2^{n-1} s(f),$$

$|A| = 2^n(1-p)$ and $|B| = 2^n p$. $\square$

Notice that this bound on the formula size in terms of average sensitivity was essentially mentioned also in [2,5].

Lemma 2. *Let f be a Boolean function. Then*

$$\overline{D}(f) \geq s(f).$$

Proof. Let a be an input assignment. The inequality $D_a(T) \geq \sigma_a(f)$ holds since otherwise some untested variable still could decide about the function value. Hence, $\mathbf{E}[D_a(T)] \geq \mathbf{E}[\sigma_a(f)] = s(f)$, where $\mathbf{E}$ denotes the expectation with respect to a uniformly distributed random input a. $\square$

Essentially the same inequality has been proved in [7] using harmonic analysis of Boolean functions.

Finally we mention the following inequalities which are a combination of the identity (2) with Corollary 2.5 of [15] and a weaker version of Lemma 3.8 of [15], respectively.

Lemma 3. *Let f be a Boolean function. Then*

$$\Delta(f) \geq s(f) \qquad and \qquad \delta(f) \geq (s(f)/6)^{1/2}.$$

4 Distribution of Square-Free Numbers

First of all we need a result about the uniformity of distribution of odd square-free numbers with a fixed binary digit.

Let i be an integer, $1 \leq i \leq n$ and let $\mathcal{N}_i$ denote the set of integers x, $0 \leq x \leq 2^n - 1$ such that $2x + 1 \equiv 0 \pmod 9$ and $2x^{(i)} + 1$ is square-free. Let M_i be the number of elements in $\mathcal{N}_i$.

Lemma 4. *For any i, $1 \leq i \leq n$, the bound*

$$M_i = \frac{1}{\pi^2} 2^n + O\left(2^{3n/4}\right)$$

holds.

Proof. It is easy to see that $2x + 1 \equiv 0 \pmod 9$ is equivalent to the condition $x \equiv 4 \pmod 9$. Let $T_i(d)$ be the number of integers x, $0 \leq x \leq 2^n - 1$, such that

$$x \equiv 4 \pmod 9 \qquad and \qquad 2x^{(i)} + 1 \equiv 0 \pmod{d^2}. \tag{3}$$

By applying the inclusion-exclusion principle we derive that

$$M_i = \sum_{\substack{1 \leq d \leq 2^{(n+1)/2} \\ d \equiv 1 \ (\mathrm{mod}\ 2)}} \mu(d) T_i(d),$$

where $\mu(d)$ is the Möbius function. We recall that $\mu(1) = 1$, $\mu(d) = 0$ if d is not square-free and $\mu(d) = (-1)^{\nu(d)}$ otherwise, where $\nu(d)$ is the number of prime divisors of $d \geq 2$.

It is easy to see that $2x + 1$ and $2x^{(i)} + 1$ are relatively prime because they differ by a power of 2. Therefore $T_i(d) = 0$ if $3|d$.

Let us now estimate $T_i(d)$ for d with $\gcd(6, d) = 1$.

The flipping position splits the bits of x into two parts. Let us denote by $t = \max\{i - 1, n - i\}$ the length of the longest part. Then it is obvious that for any fixing of the ith binary digit and all digits of the shortest part we obtain that the system of congruences (3) can be replaced by 2^{n-t} systems of congruences, each of them of the form

$$2^s z + a \equiv 0 \pmod 9 \qquad and \qquad 2^s z + b \equiv 0 \pmod{d^2}$$

with $0 \leq z \leq 2^t - 1$, for some integers s, a and b. Since $\gcd(d, 3) = 1$, applying the Chinese Remainder Theorem we see that these congruences define z uniquely modulo $9d^2$. Thus there are $2^t/9d^2 + O(1)$ such values of z in the interval $0 \leq z \leq 2^t - 1$.

Putting everything together we obtain

$$T_i(d) = 2^{n-t}\left(\frac{2^t}{9d^2} + O(1)\right) = \frac{2^n}{9d^2} + O(2^{n-t}).$$

From the inequality $t \geq (n - 1)/2$ we conclude that

$$T_i(d) = \frac{2^n}{9d^2} + O(2^{n/2}). \tag{4}$$

It is also clear that

$$T_i(d) \leq 2^n/d^2. \tag{5}$$

Let $K \geq 1$ be an integer. Using (4) for $d \leq K$ and (5) for $d > K$, and taking into account that

$$\sum_{K < d \leq 2^{(n+1)/2}} d^{-2} < \sum_{d=K+1}^{\infty} \frac{1}{d(d-1)} = \sum_{d=K+1}^{\infty}\left(\frac{1}{d-1} - \frac{1}{d}\right) = \frac{1}{K},$$

we obtain

$$M_i = 2^n \sum_{\substack{1 \leq d \leq K \\ \gcd(6,d)=1}} \frac{\mu(d)}{9d^2} + O\left(2^{n/2} \sum_{1 \leq d \leq K} 1 + 2^n \sum_{K < d \leq 2^{(n+1)/2}} d^{-2}\right)$$

$$= 2^n \sum_{\substack{1 \leq d \leq K \\ \gcd(6,d)=1}} \frac{\mu(d)}{9d^2} + O\left(2^{n/2}K + 2^n K^{-1}\right).$$

Extending the summation range in the first sum to all integers d introduces an additional error of order $2^n K^{-1}$. Therefore

$$M_i = 2^n \sum_{\gcd(6,d)=1} \frac{\mu(d)}{9d^2} + O\left(2^{n/2}K + 2^n K^{-1}\right) \tag{6}$$

for any integer K, $1 \leq K \leq 2^{(n+1)/2}$.

It is easy to verify that

$$\sum_{\gcd(6,d)=1} \frac{\mu(d)}{d^2} = \left(1 - \frac{1}{4}\right)^{-1}\left(1 - \frac{1}{9}\right)^{-1} \sum_{d=1}^{\infty} \frac{\mu(d)}{d^2} = \frac{3}{2}\sum_{d=1}^{\infty} \frac{\mu(d)}{d^2}.$$

From Theorem 287 of [11] we derive

$$\sum_{\gcd(6,d)=1} \frac{\mu(d)}{d^2} = \frac{9}{\pi^2}.$$

Selecting $K = \lfloor 2^{n/4} \rfloor$ in (6), we obtain the desired result.   □

5   Average Sensitivity of Testing Square-Free Numbers

At this point we are able to derive our main result, namely a linear lower bound on the average sensitivity of testing square-free numbers.

Theorem 1. *For the Boolean function g given by (1) the bound*

$$s(g) \geq \frac{1}{\pi^2} n + o(n)$$

holds.

Proof. It is easy to see that, for any i, $1 \leq i \leq n$

$$I_i(f) \geq 2^{-n} M_i .$$

Since the average sensitivity is defined by the sum of the influences of all variables, applying Lemma 4 we obtain the desired estimate.   □

We can now apply this estimate to derive non-trivial lower bounds for both the formula size and the average decision tree depth of the function (1).

Theorem 2. *For the Boolean function g given by (1) the bound*

$$L(g) \geq \frac{1}{32(\pi^2 - 8)} n^2 + o(n^2) ,$$

holds.

Proof. Let p be the probability that g takes the value 1. By applying the same elementary considerations we used in the proof of Lemma 4, it is easy to show that $p = 8\pi^{-2} + o(1)$. Combining Theorem 1 with Lemma 1 we then derive the desired statement.   □

Using Lemma 2 one proves:

Theorem 3. *For the Boolean function g given by (1) the bound*

$$\overline{D}(g) \geq \frac{1}{\pi^2} n + o(n)$$

holds.

Finally, from Lemma 3 we see:

Theorem 4. *For the Boolean function g given by (1) the bounds*

$$\Delta(g) \geq \frac{1}{\pi^2} n + o(n) \quad \text{and} \quad \delta(g) \geq \frac{1}{6^{1/2}\pi} n^{1/2} + o(n^{1/2})$$

hold.

It is interesting to note that for representations of g as a polynomial over the field $GF(2)$ instead of over the reals the best known lower bound on the degree is of order $\Omega(\log n)$ (see [17]).

Finally, it is worth mentioning that since the average sensitivity of Boolean functions of the class $\mathbf{AC}^0$ does not exceed $(\log n)^{O(1)}$ (as it is shown in [14]), Theorem 1 provides an alternative proof for the statement proved in [3,4] that g does not belong to $\mathbf{AC}^0$. This result has recently been improved in [1], where it is shown that for any prime p, testing square-free numbers as well as primality testing and testing co-primality of two given integers cannot be computed by $\mathbf{AC}^0[p]$ circuits, that is, $\mathbf{AC}^0$ circuits enhanced by MOD_p gates.

Apparently the result of Lemma 4 can be improved by means of some more sophisticated sieve methods (see for instance [12]). However this will not improve our main results. On the other hand, we believe that one can prove the following asymptotic formula for the average sensitivity of testing square-free numbers.

Question 1. Prove that

$$s(g) = \frac{16}{\pi^2}\left(1 - \frac{8}{\pi^2}\right)n + o(n)$$

for the function g given by (1).

References

1. E. Allender, M. Saks and I. E. Shparlinski, 'A lower bound for primality', *Proc. 14 IEEE Conf. on Comp. Compl.*, IEEE, 1999 (to appear).
2. A. Bernasconi, B. Codenotti and J. Simon, 'On the Fourier analysis of Boolean functions', *Preprint* (1996), 1-24.
3. A. Bernasconi, C. Damm and I. E. Shparlinski, 'Circuit and decision tree complexity of some number theoretic problems', *Tech. Report 98-21*, Dept. of Math. and Comp. Sci., Univ. of Trier, 1998, 1–17.
4. A. Bernasconi and I. E. Shparlinski, 'Circuit complexity of testing square-free numbers', *Lect. Notes in Comp. Sci.*, Springer-Verlag, Berlin, **1563** (1999), 47–56.
5. R. B. Boppana, 'The average sensitivity of bounded-depth circuits', *Inform. Proc. Letters*, **63** (1997), 257–261.
6. R. B. Boppana and M. Sipser, 'The complexity of finite functions', *Handbook of Theoretical Comp. Sci., Vol. A*, Elsevier, Amsterdam (1990), 757–804.
7. Y. Brandman, A. Orlitsky and J. Hennessy, 'A spectral lower bound technique for the size of decision trees and two-level AND/OR circuits', *IEEE Transactions on Computers*, **39** (1990), 282–287.
8. D. Coppersmith and I. E. Shparlinski, 'On polynomial approximation of the discrete logarithm and the Diffie–Hellman mapping', *J. Cryptology* (to appear).
9. M. Dietzfelbinger, M. Kutyłowski and R. Reischuk, 'Feasible time-optimal algorithms for Boolean functions on exclusive-write parallel random access machine', *SIAM J. Comp.*, **25** (1996), 1196–1230.
10. F. E. Fich, 'The complexity of computation on the parallel random access machine', *Synthesis of parallel algorithms*, Morgan Kaufmann Publ., San Mateo, CA, 1993, 843–899.
11. G. H. Hardy and E. M. Wright, *An introduction to the number theory*, Oxford Univ. Press, Oxford, 1965.

12. D. R. Heath-Brown, 'The least square-free number in an arithmetic progression', *J. Reine Angew. Math.*, **332** (1982), 204–220.
13. E. Koutsoupias, 'Improvements on Khrapchenko's theorem', *Theor. Comp. Sci.*, **116** (1993), 399–403.
14. N. Linial, Y. Mansour and N. Nisan, 'Constant depth circuits, Fourier transform, and learnability', *Journal of the ACM*, **40** (1993), 607-620.
15. N. Nisan and M. Szegedy, 'On the degree of Boolean functions as real polynomials', *Comp. Compl.*, **4** (1994), 301–313.
16. I. Parberry and P. Yuan Yan, 'Improved upper and lower time bounds for parallel random access machines without simultaneous writes', *SIAM J. Comp.*, **20** (1991), 88–99.
17. I. E. Shparlinski, 'On polynomial representations of Boolean functions related to some number theoretic problems', *Electronic Colloq. on Comp. Compl.*, `http://www.eccc.uni-trier.de/eccc/`, TR98-054, 1998, 1–13.
18. I. E. Shparlinski, *Number theoretic methods in cryptography: Complexity lower bounds*, Birkhäuser, 1999.
19. I. Wegener, *The complexity of Boolean functions*, Wiley-Teubner Series in Comp. Sci., Stuttgart (1987).

Binary Enumerability of Real Numbers
(Extended Abstract)

Xizhong Zheng

Theoretische Informatik, FernUniversität Hagen
58084 Hagen, Germany
xizhong.zheng@fernuni-hagen.de

Abstract. A real number x is called *binary enumerable*, if there is an effective way to enumerate all "1"-positions in the binary expansion of x. If at most k corrections for any position are allowed in the above enumerations, then x is called *binary k-enumerable*. Furthermore, if the number of the corrections is bounded by some computable function, then x is called *binary ω-enumerable*. This paper discusses some basic properties of binary enumerable real numbers. Especially, we show that there are two binary enumerable real numbers x and y such that their difference $x - y$ is not binary ω-enumerable (in fact we have shown that it is even of no "ω-r.e. Turing degree").

1 Introduction

There are many different ways to represent a real number by rational numbers. One of them is the binary expansion. In this way, any real number $x \in [0; 2]$ corresponds naturally to a subset $A \subseteq \mathbb{N}$ in the sense that $x = \sum_{n \in A} 2^{-n}$, i.e., A consists of all "1"-positions in the binary expansion of x. If we choose the finite set A to correspond to the rational number x, then such correspondence is one-to-one. We will call such set A a *binary set* of x and the real number x (which is usually denoted by x_A) is called a *binary real number* of A. It is well known that x_A is a computable real number, if and only if A is a recursive set. If a set A is recursively enumerable (r.e.), then its binary real number x_A is a limit of an increasing computable sequence of rational numbers, i.e., it is left computable (see [13]). As it is observed by Jockusch (see [9]), the converse is not true. That is, there is a left computable real number such that its binary set is not r.e.

It is well known that, a real number x is computable, if and only if its binary expansion is computable, namely, we can effectively write either "0" or "1" one bit after another. A correction in this procedure is not allowed. That is, if some bit "0" or "1" is written at some stage, then it cannot be changed later any more.

In the practice, the situation may be different. If we want to determine some real number x by giving its binary expansion, we will begin with an empty string (i.e., all bits are assumed to be "0" at the beginning) and write longer and longer

T. Asano et al. (Eds.): COCOON'99, LNCS 1627, pp. 300–309, 1999.

strings of "0" and "1" to approximate x. In some stages, we may have to change some bits from "1" back to "0" and vice versa. According to the numbers of changes which are allowed in the this procedure, we get different type of the real numbers. Since additional 0's at the end of our binary expansions do not change its value, we can assume that, before a "1" is written first time to some position, there is already a "0" been written there. Then the first change at any position, if any, can only be from "0" to "1". If at most one change at any position is allowed, then the binary set of corresponding real number is r.e. and such real numbers are called *binary enumerable*. Therefore, x is binary enumerable, if and only if there is an r.e. set $A \subseteq \mathbb{N}$ such that $x = x_A$. Generally, we get a bigger class of real numbers which has less "effectivity", if more changes are allowed. For example, if k changes are allowed for any position, then the corresponding real numbers are called *binary k-enumerable*. More generally, if the number of such changes is bounded by some recursive function (respect to the positions) instead of some fixed number, then the corresponding real numbers are called *binary ω-enumerable*.

It is worth noting that the binary k-enumerability defined above is not symmetric with respect to "0" and "1" positions. Our definition corresponds to the $(k$-)enumerability of binary set A of the real number x_A. That is, x_A is binary k-enumerable iff A is k-r.e. (See Definition 1). Thus, the binary enumerability is not closed under the operation of "$-$" which similar to that the k-r.e.ness of subsets of $\mathbb{N}$ is not closed under the set operations of complement. In fact, our main result of this paper shows that there are binary enumerable real numbers x and y such that $x - y$ is not binary ω-enumerable.

W.l.o.g., we consider only the real numbers in the interval $[0; 2]$ in this paper. For other real numbers, say y, outside this interval, there is a natural number n and an $x \in [0; 1]$ such that $y = n + x$. Intuitively x and y have completely same type of computability in any reasonable sense.

We conclude this section by introducing some notations. Let $\Sigma := \{0, 1\}$ be an alphabet and $\mathbb{N}$ be the set of natural numbers. Denote by Σ^* and Σ^ω the set of binary strings and the set of infinite sequences over Σ, respectively. Strings are denoted by lower case letters u, v, w. The concatenation of two strings x and y is denoted by xy; $|u|$ denotes the length of the string u; λ is the empty string which has the length 0; $<$ is the length-lexicographical ordering on Σ^*. The ith bit of the string u is denoted by $u(i)$, so $u := u(0)u(1)\cdots u(|x|-1)$. For any set $A \subseteq \mathbb{N}$, its characteristic sequence is also denoted by A. For $A \subseteq \mathbb{N}$ and $n \in \mathbb{N}$, let $A \upharpoonright n$ denote the finite initial segment of A below n, i.e. $A \upharpoonright n := \{i : i < n \ \& \ i \in A\}$. We identify this initial segment whith its characteristic string, i.e. $A \upharpoonright n = A(0)A(1)\cdots A(n-1) \in \Sigma^*$. Similarly, we define the initial segment of $u \in \Sigma^*$ below $n \in \mathbb{N}$ by $u \upharpoonright n := u(0)u(1)\cdots u(n-1)$ if $n \leq |u|$ and $u \upharpoonright n := u$ otherwise. If $u = A \upharpoonright n$ for some $n \in \mathbb{N}$, then we denote that $u \sqsubset A$. Thus the length-lexicographical order on Σ^* can be extended to $\Sigma^* \cup \Sigma^\omega$ by $u < A \Longleftrightarrow \exists v(v \sqsubset A \ \& \ u < v)$ and $A < u \Longleftrightarrow \exists v(v \sqsubset A \ \& \ v < u)$ for any $x \in \Sigma^*$ and $A \in \Sigma^\omega$. For $A \subseteq \mathbb{N}$, we define A_L to be the set of all finite binary strings which are "left" the infinite binary sequence A, i.e. $A_L := \{u \in \Sigma^* : u < A\}$.

2 Ershov Hierarchy and Weak Computable Real Numbers

In this section we will recall at first the definition of Ershov Hierarchy on subsets of natural numbers and discuss some relationships between this hierarchy and weak computability of real numbers.

Definition 1 (Putnam, [7] Gold [4] and Ershov [3]).

1. Let $h : \mathbb{N} \to \mathbb{N}$ be any function. A set $A \subseteq \mathbb{N}$ is called h-r.e., if there is a computable sequence $(A_s)_{s \in \mathbb{N}}$ of finite subsets of $\mathbb{N}$ such that
   a) $A_0 = \emptyset$;
   b) $A = \lim\limits_{s \to \infty} A_s := \bigcup\limits_{n=0}^{\infty} \bigcap\limits_{s=n}^{\infty} A_s$; and
   c) $\forall n \in \mathbb{N}(|\{s \in \mathbb{N} : n \in A_{s+1} \triangle A_s\}| \leq h(n))$.
   where $\triangle$ is defined by $B \triangle C := (B \backslash C) \cup (C \backslash B)$. The sequence $(A_s)_{s \in \mathbb{N}}$ is called an *effective h-enumeration* of A.
2. Set $A \subseteq \mathbb{N}$ is called ω-r.e., if A is h-r.e. for some recursive function h. In this case, the sequence $(A_s)_{s \in \mathbb{N}}$ is called an *effective ω-enumeration* of A and h is a *bounding function* of this enumeration.
3. If h is a constant function with $h(x) := k$, then the h-r.e. set $A \subseteq \mathbb{N}$ is called k-r.e. The sequence $(A_s)_{s \in \mathbb{N}}$ is called an *effective k-enumeration* of set A.

So, in particular, the empty set is the unique 0-r.e. set and the 1-r.e. sets are the r.e. sets. The 2-r.e. are usually called *d-r.e.* as they are the differences of the r.e. sets, i.e., A is d-r.e., iff $A = B \backslash C$ for some r.e. sets B and C. Similarly, A is $(k+1)$-r.e., iff there are r.e. set B and k-r.e. set C such that $B \backslash C$. A Turing degree is called *(k-) r.e.* if it contains at least one $(k-)$ r.e. set.

Definition 2. For any $\alpha \in \mathbb{N}^+$ or $\alpha = \omega$, a real number x is called *binary α-enumerable*, if there is an integer n and an α-r.e. set A such that $x = n + x_A$. The class of all binary α-enumerable real numbers is denoted by $\mathbb{B}_\alpha$

Weihrauch and Zheng [13] call a real number x *left (right), computable* if there is an increasing (decreasing) sequence of rational numbers which converges to x. Left and right computable real numbers are called *semi-computable*. The class of semi-computable real numbers is denoted by $\mathbb{C}_1$.

Theorem 1 (Weihrauch and Zheng [13]). *For any $k \in \mathbb{N}$.*

1. *There is a real number $x \in \mathbb{B}_{k+1} \backslash \mathbb{B}_k$ such that x is left computable.*
2. *There is a real number $x \in \mathbb{B}_\omega$ with $x \notin \mathbb{B}_k$ for any $k \in \mathbb{N}$ such that x is left computable.*
3. *There is real number $x \in \mathbb{B}_2$ such that x is not left computable.*

By Theorem 1, a left computable real number can have a $(k+1)$-r.e. binary set which is not k-r.e. or even have an ω-r.e. binary set which is not k-r.e. for any $k \in \mathbb{N}$. But next theorem shows that they must *tt-equivalent* to some r.e. set.

Theorem 2 (Ambos-Spies [1], Jockusch [5]). *For any set $A \subseteq \mathbb{N}$, x_A is left computable, if and only if A_L is r.e. Therefore, if x_A is semi-computable, then A has an r.e. tt-degree.*

The converse of Theorem 2 is not true because it is shown in [13] (Theorem 5) that there are r.e. sets $B, C \subseteq \mathbb{N}$ such that $B \oplus \overline{C}$, which has an r.e. tt-degree, has a non-semi-computable binary real number.

Theorem 3. *A real number x_A is semi-computable, if and only if A_L is k-r.e. for some $k \in \mathbb{N}$.*

Proof. The part of "$\Rightarrow$" is obvious. We prove here the part of "$\Leftarrow$" and only for the case of $k = 2$. For other $k > 2$, the idea is similar.

Let $A_L = B \backslash C$ and $B, C \subseteq \Sigma^*$ are r.e. sets. If C has a least element u_0, then $A_L = B \cap (\Sigma^* \lceil u_0) := \{v \in \Sigma^* : v < u_0 \& v \in B\}$ is an r.e. set. This implies that x_A is left computable.

Suppose that C has no least element. We can define a computable sequence $\{u_s\}_{s \in \mathbb{N}}$ of finite strings such that $u_s \in C$ and $u_{s+1} < u_s$ for any $s \in \mathbb{N}$. That is, $\{u_s\}_{s \in \mathbb{N}}$ is a decreasing sequence which has the limit, say, $x_{A'} =: \lim_{s \to \infty} u_s$. If $A = A'$, then A is right computable. Suppose that $A \neq A'$, hence that $x_A < x_{A'}$. There is an $n \in \mathbb{N}$ such that $x_{A'} - x_A > 2^{-n}$.

If $A'(m) = 0$ for almost all $m \in \mathbb{N}$, hence $(\forall n > m_0)(A'(m) = 0)$ for some $m_0 \in \mathbb{N}$, then $u \geq (A' \lceil m_0)$ for all $u \in C$. This implies that A_L is r.e. again.

Suppose that there are infinite many m such that $A'(m) = 1$. Choose an $m > n$ such that $A'(m) = 1$, let $v =: (A' \lceil m)0$. Then we have $x_A < v < x_{A'}$ since $(x_{A'} - v) \leq 2^{-n}$. It follows that $A_L = B \cap (\Sigma^* \lceil v)$ is an r.e. set again. Thus x_A is left-computable.

As an natural extension of semi-computable real numbers, Weihrauch and Zheng [13] introduced the class of weakly computable real numbers.

Definition 3. A real number x is called *weakly computable* (w.c. for short), if there are two semi-computable real numbers y, z such that $x = y + z$. The set of all weakly computable real numbers is denoted by $\mathbb{C}_2$.

For example, if $A := B \backslash C$ is a d-r.e. set with r.e. sets B and C, then $B \cup C$ is also r.e. set. So x_A is weakly computable because $x_A = x_{B \cup C} - x_C$. More generally, we can show by an easy induction on k, that if A is a k-r.e set, $k \geq 1$ then x_A is a weakly computable real number. Furthermore, The class $\mathbb{C}_2$ is a closed field generated by $\mathbb{C}_1$ as shown in [13]. The class of weakly computable reals has another characterization as follows.

Theorem 4 (Weihrauch and Zheng [13]). *A real number x is weakly computable, iff there is a computable sequence $(x_n)_{n \in \mathbb{N}}$ of rational numbers which converges to x weakly effectively, i.e., the sum of its jumps $\sum_{n=0}^{\infty} |x_{n+1} - x_n|$ is bounded.*

It follows immediately from Theorem 1 that the class of w.c. real numbers extends the class of semi computable real numbers properly, since x_A is w.c. if A is d-r.e. This implies also that there are two semi-computable real numbers such that their sum is not semi-computable.

Now we will prove a result about Turing degrees. By Definition, the recursive Turing degree **0** consists of only recursive sets. We will show that any nonrecursive r.e. degree contains a set whose binary real number is w.c. but not semi-computable. We will also show that the binary sets of weakly computable real numbers do not exhaust all ω-r.e. sets. These results follow immediately flowing observation of Ambos-Spies [1].

Theorem 5 (Ambos-Spies [1]).

1. If B, C are r.e. sets such that $B|_T C$, then $x_{B \oplus \overline{C}}$ is not semi-computable.
2. Assume that $x_{A \oplus \emptyset}$ is weakly computable. Then A is f-r.e. for $f(n) = 2^{3n}$.

Theorem 6. For any nonrecursive r.e. Turing degree **a**, there is a set $A \in$ **a** such that x_A is weakly computable but not semi-computable.

Proof. Let **a** be a nonrecursive r.e. Turing degree. By Sacks' Splitting Theorem [8] there exist two incomparable r.e. degrees $\mathbf{b_0}, \mathbf{b_1}$ such that $\mathbf{a} = \mathbf{b_0} \vee \mathbf{b_1}$. Choose r.e. sets $B_0 \in \mathbf{b_0}$ and $B_1 \in \mathbf{b_1}$ and define set $A =: B_0 \oplus \overline{B_1}$. Obviously, A is a d-r.e set. So x_A is weakly computable. On the other hand, it is not semi-computable by (1) of Theorem 5, since $B_0|_T B_1$. ∎

Theorem 7. There exists a binary ω-enumerable real number x such that x is not weakly computable, i.e., $\mathbb{B}_\omega \subsetneq \mathbb{C}_2$.

Proof. Let $f(n) =: 2^{3n}$ and $g(n) =: 2^{4n}$. Then $g(n) > f(n)$ for all $n \in \mathbb{N}$. By a result of Ershov [3], there is a g-r.e. set A' which is not f-r.e. Let $A =: A' \oplus \emptyset$. By (2) of the Theorem 5, x_A is not weakly computable. But A', hence A is ω-r.e. since g is a recursive function. ∎

3   More Complicated Weakly Computable Real Numbers

In this paper, we will construct two binary enumerable real numbers x and y such that their difference $z := x - y$ is not binary ω-enumerable. In fact we prove even more that the binary set of z has no ω-r.e. Turing degree. Our proof here is a sophisticated finite injury priority construction. In the proof, we will use the following three kinds of the restriction of a subset of $\mathbb{N}$:

$$A \upharpoonright n := \{x \in A : x < n\};$$
$$A \downharpoonright n := \{x \in A : x > n\};$$
$$A \upharpoonright (n; m) := \{x \in A : n < x < m\}.$$

We show at first a technical lemma whose proof is straightforward.

Lemma 1. *Let $A, B, C \subset \mathbb{N}$ be finite sets such that $x_A = x_B - x_C$ and n, m and y be any natural numbers. Then the following hold.*

1. $\max A \leq \max(B \cup C)$;
2. *If $B \downarrow n = C \downarrow n$, then $\max A \leq n$ and $n \in A \iff n \in B \backslash C$,*
3. *If $n, m \in B \backslash C$, $n < y < m$ and $(B \upharpoonright (n; m)) \backslash \{y\} = (C \upharpoonright (n; m)) \backslash \{y\}$, then $n \notin A \iff y \in C \backslash B$;*
4. *If $x_{A_1} = x_{B_1} - x_{C_1}$, $(B \cup C) \upharpoonright (n + 1) = (B_1 \cup C_1) \upharpoonright (n + 1)$ and $n \in B \backslash C$, then $A \upharpoonright n = A_1 \upharpoonright n$.*
5. *Suppose that $m < y < n$, $n \in A$.*
   (5.a) *If $m \in A$, $x_{A'} = x_A - 2^{-y}$ and $A \upharpoonright (m; n) = \emptyset$, then $A' := (A \backslash \{m\}) \cup \{m + 1, m + 1, \ldots, y\}$.*
   (5.b) *If $m \notin A$, $x_{A'} = x_A + 2^{-y}$ and $A \upharpoonright (m; n) = \{m + 1, m + 2, \ldots, y\}$, then $A' := (A \cup \{m\}) \backslash \{m + 1, \ldots, y\}$.*

For the proof of our next result, we recall some standard recursion-theoretical notations. (More detail explanations of these notations can be found in, e.g., [10,6].) Let $(M_e^A : e \in \mathbb{N})$ be an effective enumeration of all (type one) Turing machines (with oracle A) and $(\varphi_e^A : e \in \mathbb{N})$ be the corresponding effective enumeration of all A-computable functions from $\mathbb{N}$ to $\mathbb{N}$ such that φ_e^A is computed by the e-th Turing machine M_e^A with oracle A. $\varphi_{e,s}^A$ is the s-th approximation of φ_e^A. The use-function $u_{e,s}^A(x)$ of the computation $M_e^A(x)$ is defined to be the length of the initial segment of the oracle which is really used in the computation $u_{e,s}^A(x)$. It is important to note that, if $u_{e,s}^A(x) = t$, and $A \upharpoonright t = B \upharpoonright t$, then the computations of $M_e^A(x)$ and $M_e^B(x)$ are completely the same. As usual Turing machine model, we assume always that $u_{e,s}^A(x) \leq s$. This is a very useful estimation because we need only to preserve simply the initial segment $A \upharpoonright s$ so that the computation $M_{e,s}^A(x)$ will not be destroyed. Thus, any change of the membership of the elements $a \geq s$ to A do not destroy this computation. As usual, we always assume that the use-function is nondecreasing about s. As the partial recursive functionals, $M_e^A(x)$ is often denoted by uppercase Greek letters Γ, Δ, Λ, etc. and the corresponding lowercase Greek letters γ, δ, λ are their use-functions, respectively. To simplify the notation, instead of pointing out the subscript s, we will often use the expression like $\Gamma^W(x) \upharpoonright \gamma(x)[s]$ to denote the current value of this expression $\Gamma^W(x) \upharpoonright \gamma(x)$ at the stage s. Then we have, e.g., $A_s = A[s]$, $B_s(x_s) = B(x)[s]$, etc.

As usual, superscript A is always omitted if $A = \emptyset$. We define also $W_e := \text{dom}(\varphi_e)$ and $W_{e,s} := \text{dom}(\varphi_{e,s})$. Then $(W_e : e \in \mathbb{N})$ enumerates all r.e. subsets of $\mathbb{N}$ and $(W_{e,s} : s \in \mathbb{N})$ is a recursive enumeration of W_e.

We usually do not distinguish between a subset of $\mathbb{N}$ and its characteristic function. That is, we have always that $x \in A \iff A(x) = 1$ and $x \notin A \iff A(x) = 0$.

Theorem 8. *There are r.e. sets B and C such that the set A which satisfies $x_A = x_B - x_C$ is not of ω-r.e. Turing degree.*

Proof. (Sketch) We will construct effectively the computable sequences $(A_s : s \in \mathbb{N})$, $(B_s : s \in \mathbb{N})$ and $(C_s : s \in \mathbb{N})$ of finite subsets of $\mathbb{N}$ which satisfy the following conditions:

1. $x_{A_n} = x_{B_n} - x_{C_n}$ for all $n \in \mathbb{N}$;
2. The limits $A := \lim_{n \to \infty} A_n$, $B := \lim_{n \to \infty} B_n$ and $C := \lim_{n \to \infty} C_n$ exist and $x_A = x_B - x_C$ holds;
3. $(B_s : s \in \mathbb{N})$ and $(C_s : s \in \mathbb{N})$ are the recursive enumerations of sets B and C, respectively. Hence B and C are r.e. sets and x_A is weakly computable.
4. The degree $\deg_T(A)$ is not ω-r.e.

The first condition is easy to satisfy. At any stage $s + 1$, we will define directly the finite sets B_{s+1} and C_{s+1} in such a way that $B_s \subseteq B_{s+1}$ and $C_s \subseteq C_{s+1}$ and let A_{s+1} simply to be the unique finite set satisfying $x_{A_{s+1}} = x_{B_{s+1}} - x_{C_{s+1}}$. This satisfy automatically the second and third conditions too.

Now we can discuss the fourth condition. The degree $\deg_T(A)$ is not ω-r.e. means that A is not Turing equivalent to any ω-r.e. set. Then, for meeting the fourth condition, it suffices to make sure that, for all ω-r.e. set V (with the effective ω-enumeration $(V_s : s \in \mathbb{N})$ and corresponding recursive bounding function h), partial recursive functionals Γ and Δ, the following requirements are satisfied:

$$R_{V,\Gamma,\Delta} : A \neq \Gamma^V \text{ or } V \neq \Delta^A.$$

Since the classes of all ω-r.e. sets and all partial recursive functionals are effectively enumerable (see [2,10]), respectively. So these requirements can be effectively enumerated too. We fix $(R_e : e \in \mathbb{N})$ as one of such kind of effective enumeration. Thus all requirements can be given different priorities according to this enumeration, i.e., R_i is of the higher priority than R_j if and only if $i < j$.

The strategy for satisfying a single requirement $R_{V,\Gamma,\Delta}$ is as follows:

First phase: At first we choose arbitrarily a witness $x \in \mathbb{N}$. Then wait for the stage s such that:

$$A_s(x) = \Gamma^V(x)[s] \text{ \& } V \upharpoonright \gamma(x)[s] = \Delta^A \upharpoonright \gamma(x)[s]. \tag{1}$$

(if this never happens, then x is a witness to the success of $R_{V,\Gamma,\Delta}$.) Define $B_{s+1} := B_s \cup \{x\}$, $C_{s+1} := C_s$. It can be shown that $A_{s+1} = A_s \cup \{x\}$. In a later stage, we might hope to remove x from A when it is necessary. Since B should be a r.e. set, we can not simply remove it by taking out x from B. Instead, we achieve that by putting some bigger element to C. Besides, we hope to preserve the computations of $\Delta^A \upharpoonright \gamma(x)[s]$, so we define here a supplementary element y big enough so that $y > \max\{\delta\gamma(x)[s], x\}$. Then into next phase.

Second phase: Wait for some new stage $s' > s$ at which the following hold:

$$A_{s'}(x) = \Gamma^{V \upharpoonright \gamma(x)}(x)[s'] \text{ \& } V \upharpoonright \gamma(x)[s'] = \Delta^{A \upharpoonright \delta\gamma(x)} \upharpoonright \gamma(x)[s']. \tag{2}$$

(if this never happens, then x is also a witness to the success of $R_{V,\Gamma,\Delta}$.) In this case, we hope to remove x from A to force the initial segment $V \upharpoonright \gamma(x)[s']$ to be changed whenever this condition is satisfied later again. This can be achieved by putting y into C, i.e., define $B_{s'+1} := B_{s'}$, $C_{s'+1} := C_{s'} \cup \{y\}$. It can be shown that $x \notin A_{s'+1}$. Then into next phase.

Third phase: Wait for some new stage $s'' > s'$ at which the following hold:

$$A_{s''}(x) = \Gamma^{V \restriction \gamma(x)}(x)[s''] \;\&\; V \restriction \gamma(x)[s''] = \Delta^{A \restriction \delta\gamma(x)} \restriction \gamma(x)[s'']. \qquad (3)$$

(if this never happens, then x is a witness to the success of $R_{V,\Gamma,\Delta}$ again.) Define $B_{s''+1} := B_{s''} \cup \{y\}$, $C_{s''+1} := C_{s''}$ and let $A_{s''+1}$ similar as before. It can be shown that $x \in A_{s''+1}$ again. Now the supplementary element y is used two times and can not be used again. Then choose a new one $y := y + 1$ and go to the second phase.

Because the action in any of above phases changes the membership of x to A, whenever we go from phase X to another phase Y, the value of $A(x)$, hence also the initial segment $V \restriction \gamma(x)$ must be changed. But if we go from phase Y back to phase X later, $A \restriction \delta\gamma(x)$ is recovered to that of last appearance of phase X. So, the initial segment $\Delta^{A \restriction \delta\gamma(x)} \restriction \gamma(x)$, hence also the initial segment $V \restriction \gamma(x)$, is recovered. Because of the ω-r.e.ness of the set V, this can happen at most finitely often. Therefore, after some stages, (2) or (3) will never hold again. Thus the requirement $R_{V,\Gamma,\Delta}$ is finally satisfied by the witness x.

To satisfy all the requirements, we apply the finite injury priority construction. A witness x_e and a supplementary element y_e ($> x_e$) are appointed to the requirement R_e at any stage. Instead of trying to satisfy all requirements in the given order we will attack those requirements first to which the above strategy can be applied. This action should be preserved from injury by lower priority requirements. Therefore whenever the actions of any above phases are made for the requirement R_e, all requirements R_i with $i > e$ will be *initialized*. That is, we redefine the witnesses x_i and supplementary elements y_i of R_i (for $i > e$) bigger than all elements enumerated into B or C so far and also bigger than the $\delta\gamma(x_e)$ to preserve the computations of $\Delta^{A \restriction \delta\gamma(x)} \restriction \gamma(x_e)$ from the injury by lower priority requirements. Still it is not enough, because the witness $x_e[s]$ may be removed from A or put into A not only by putting y_e into C or B as we hoped, but also by putting y_i in to C or B through the actions for R_i with some $i > e$. Then the condition (2) or (3) can be satisfied again without any changes of $V \restriction \gamma(x_e)$, and hence in such a way R_e can requires attentions infinitely often. To avoid this bad situation, we will build a "firewall" between the all supplementary elements y_e's of R_e and the x_i's, y_i's for R_i with $i > e$. Namely, we put another element z_e, which satisfies that $\forall i > e \, (y_e < z_e < x_i, y_i)$, into $B \backslash C$ as well. By (4) of the Lemma 1, it succeeds. Therefore, at any stage s, we will define a witness x_e and two supplementary elements y_e, z_e such that $x_e < y_e < z_e$ and $\forall i > e \, (z_e < x_i)$.

Of course, a requirement could be injured by some higher priority requirements. Since finite many attacks suffices for any given requirement as we saw in the above strategy, every requirement R_e can be injured at most finitely often because there are only finite many requirements which have the higher priority than R_e. So every requirement have enough chances to be satisfied by the above strategy.

We give the exact construction as follows:

Stage 0: Define $A_0 = B_0 = C_0 := \emptyset$. For any $e \in \mathbb{N}$, we define $x_e[0] := 3e + 1$, $y_e[0] := 3e + 2$ and $z_e[0] := 3e + 3$. All requirements R_e are said to be *initialized* at this stage.

Stage $s+1$: Given $A_s, B_s, C_s, x_e[s], y_e[s]$ and $z_e[s]$ for all $e \in \mathbb{N}$. A requirement $R_e(= R_{V,\Gamma,\Delta})$, for some ω-r.e. set V, and partial recursive functionals Γ and Δ, *requires attention* if the following condition holds:

$$A(x_e)[s] = \Gamma^{V \restriction \gamma(x_e)}(x_e)[s] \,\&\, V \restriction \gamma(x_e)[s] = \Delta^{A \restriction \delta \gamma(x_e)} \restriction \gamma(x_e)[s]. \qquad (4)$$

Choose the minimal $e \leq s$, if any, such that the requirement $R_e(= R_{V,\Gamma,\Delta})$ requires attention at this stage. Then do the following:

In all the following cases we will always define the set A_{s+1} as the unique finite subset of $\mathbb{N}$ which satisfies the condition that $x_{A_{s+1}} = x_{B_{s+1}} - x_{C_{s+1}}$. We define also $x_e[s + 1] := x_e[s]$.

Case 1. $x_e[s] \notin B_s$. Let $g(x_e)[s] := \sum_{i=0}^{\gamma(x_e)[s]} h(i)$. Then we define in this case all the following:

$$
\begin{cases}
y_e[s + 1] := \max\{\delta\gamma(x_e)[s], \max(B_s \cup C_s)\} + 1; \\
z_e[s + 1] := y_e[s + 1] + g(x_e)[s] + 1; \\
x_i[s + 1] := z_e[s + 1] + 3i + 1, \text{ for all } i > e; \\
y_i[s + 1] := z_e[s + 1] + 3i + 2, \text{ for all } i > e; \\
z_i[s + 1] := z_e[s + 1] + 3i + 3, \text{ for all } i > e; \\
B_{s+1} := B_s \cup \{x_e[s], z_e[s + 1]\} \cup (C_s \restriction x_e[s]); \\
C_{s+1} := C_s \cup (B_s \restriction x_e[s]).
\end{cases}
$$

Notice that we have that $x_e[s], z_e[s + 1] \in B_{s+1}$, $x_e[s] \in A_{s+1}$, by (2) of Lemma 1, and that

$$x_e[s] = x_e[s + 1] < \max\{\delta\gamma(x_e)[s], \max(B_s \cup C_s)\} < y_e[s + 1]$$
$$< y_e[s + 1] + g(x_e)[s] < z_e[s + 1] < x_i[s + 1] < y_i[s + 1] < z_i[s + 1],$$

for all $i > e$.

Case 2. $x_e[s] \in B_s$. There are still four possibilities as follows:

Case 2.1. $x_e[s] \in A_s \,\&\, y_e[s] \notin C_s$. This means that the supplementary element $y_e[s]$ is not yet used and hence not in $(B \cup C)[s]$. Then define

$$
\begin{cases}
B_{s+1} := B_s \,\&\, C_{s+1} := C_s \cup \{y_e[s]\}; \\
y_e[s + 1] := y_e[s] \,\&\, z_e[s + 1] := z_e[s].
\end{cases}
$$

Case 2.2. $x_e[s] \in A_s \,\&\, y_e[s] \in C_s$. Here $x_e[s] \in A_s$ means that if a supplementary element y_e is put into C, then it must be put into B too. In this situation, a new supplementary element is defined which is still not in C (see case 2.3.) So this case can not happen in fact.

Case 2.3. $x_e[s] \notin A_s \,\&\, y_e[s] \in C_s$. In this case, $y_e[s]$ is already put into C but not yet enumerated into B. We need only put $y_e[s]$ into B and define a new y_e. That is, we define

$$
\begin{cases}
B_{s+1} := B_s \cup \{y_e[s]\} \,\&\, C_{s+1} := C_s; \\
y_e[s + 1] := y_e[s] + 1 \,\&\, z_e[s + 1] := z_e[s].
\end{cases}
$$

Case 2.4. $x_e[s] \notin A_s$ & $y_e[s] \notin C_s$. This can also never happen in fact.

In all of these cases, we define $x_i[s+1] := x_i[s]$; $z_i[s+1] := z_i[s]$ and $y_i[s+1] := y_i[s]$ for all $i < e$. Then *initialize* all lower priority requirements R_i (for $i > e$) by defining

$$x_i[s+1] := 3w[s] + 1; y_i[s+1] := 3w[s] + 2 \text{ and } z_i[s+1] := 3w[s] + 3 \quad (5)$$

where $w[s] := \max(B_{s+1} \cup C_{s+1})$. If the requirements R_i $(i > e)$ has received attention before stage $s + 1$ and was not yet been initialized thereafter, then it is called that R_i is *injured* at this stage by the requirement R_e. We say that the requirement R_e *receives attention* at this stage.

If no requirements require attention at this stage, then every thing remains unchanged and go directly to the next stage.

This ends the construction. We can show that the sets $A := \lim_{s \to \infty} A_s$, $B := \lim_{s \to \infty} B_s$ and $C := \lim_{s \to \infty} C_s$ satisfy the theorem.

References

1. K. Ambos-Spies. A note on recursively approximatable real numbers, *Forschungsberichte Mathematische Logik*, Mathematisches Institut, Universität Heidelberg. Nr. 38, 1998.

2. M. M. Arslanov. Degree structures in local degree theory, in *Complexity, Logic, and Recursion Theory*, Andrea Sorbi (ed.), *Lecture Notes in Pure and Applied Mathematics*, Vol. 187, Marcel Dekker, New York, Basel, Hong Kong, 1997, pp. 49–74.

3. Y. Ershov. On a hierarchy of sets I; II and III, *Algebra i Logika*, 7(1968), no. 1, 47 – 73; 7(1968), no. 4, 15 – 47 and 9(1970), no. 1, 34 – 51.

4. E. M. Gold. Limiting recursion *J. of Symbolic Logic*, 30(1965), 28–48.

5. C. G. Jockusch Semirecursive sets and positive reducibility. Trans. Amer. Math. Soc. 131(1968), 420–436.

6. P. Odifreddi. *Classical Recursion Theory*, vol. 129 of Studies in Logic and the Foundations of Mathematics, North-Holland, Amsterdam, 1989.

7. H. Putnam. Trial and error predicates and solution to a problem of Mostowski, *J. of Symbolic Logic*, 30(1965), 49–57.

8. G. E. Sacks On the degrees less than $0'$, *Ann. of Math.* 77(1963), 211–231.

9. R. Soare Recursion theory and Dedekind cuts, *Trans, Amer. Math. Soc.* 140(1969), 271–294.

10. R. Soare. *Recursively Enumerable Sets and Degrees*, Springer-Verlag, Berlin, Heidelberg, 1987.

11. K. Weihrauch. *Computability*. EATCS Monographs on Theoretical Computer Science Vol. 9, Springer-Verlag, Berlin, Heidelberg, 1987.

12. K. Weihrauch. *An Introduction to Computable Analysis*. (In Preparation.)

13. K. Weihrauch and X. Zheng. A finite hierarchy of recursively enumerable real numbers, MFCS'98, Brno, Czech Republic, August 24 – 28, 1998.

GCD of Many Integers

(Extended Abstract)

Gene Cooperman[1,*], Sandra Feisel[2], Joachim von zur Gathen[2],
and George Havas[3,**]

[1] College of Computer Science, Northeastern University
Boston, MA 02115, USA
gene@ccs.neu.edu
[2] FB Mathematik-Informatik, Universität-GH Paderborn
33095 Paderborn, Germany
{feisel,gathen}@uni-paderborn.de
[3] Centre for Discrete Mathematics and Computing
Department of Computer Science and Electrical Engineering
The University of Queensland, Queensland 4072, Australia
havas@csee.uq.edu.au

Abstract. A probabilistic algorithm is exhibited that calculates the gcd of many integers using gcds of pairs of integers; the expected number of pairwise gcds required is less than two.

1 Introduction

In many algorithms for polynomials in $\mathbb{Z}[x]$, e.g., computation of gcds or factorization, one is interested in making a polynomial primitive, i.e., computing and removing its content. The primitive polynomial remainder sequence needs such gcd computations but is not used in practice: Ho and Yap [7] state that "the primitive PRS has the smallest possible coefficients, but it is not efficient because content computation is relatively expensive". Geddes, Czapor and Labahn [4, Chapter 7], write about the primitive polynomial remainder sequence: "The problem with this method, however, is that each step requires a significant number of GCD operations in the coefficient domain. ... The extra cost is prohibitive when working over coefficient domains of multivariate polynomials."

We remedy this sorry situation for $\mathbb{Z}[x]$ by computing the gcd of random linear combinations of the inputs. More precisely, we solve the following problem.

Given m positive integers $a_1, \ldots, a_m$ with $m > 1$, compute $\gcd(a_1, \ldots, a_m)$ with a small number of pairwise gcds, i.e., gcds of two integers each.

MAIN THEOREM. *This problem can be solved by taking an expected number of less than two pairwise gcds of random linear combinations of the input.*

* Supported in part by NSF Grant CCR-9509783
** Supported in part by the Australian Research Council

T. Asano et al. (Eds.): COCOON'99, LNCS 1627, pp. 310–317, 1999.
© Springer-Verlag Berlin Heidelberg 1999

We stress that we are not doing an *average-case analysis*, as is done in [11, Note 8.2], where the inputs $a_1, \ldots, a_m$ are randomly chosen according to some distribution, but rather a *worst-case analysis*: we are looking for probabilistic algorithms for which we can prove a good bound on the expected number of gcds that is true no matter what the inputs are. A different approach to solving this problem is implicit in [5, Proof of Lemma 1.2] while a brief and less precise treatment appears in [1]. We provide a refined algorithm together with a rigorous analysis and some experimental results.

The use of random linear combinations is readily suggested by the success of that method for polynomials but we do not know how to prove that the naive implementation of this idea works. Our main algorithmic contribution is a clever choice of the range for the random coefficients. There is a natural (heuristic) upper bound on the success probability of any algorithm; with our choice of the range, we can prove a lower bound that is reasonably close to that upper bound. In fact, the lower bound can be moved arbitrarily close to the upper bound, but at the expense of requiring very large coefficients, which limits the practical usefulness.

However, we believe that our method, free of any heuristic or distributional assumption, makes content computation for integer polynomials eminently practical.

2   Iterative gcd computation

By the associativity of the gcd

$$\gcd(a_1, \ldots, a_m) = \gcd(\gcd(a_1, a_2), a_3, \ldots, a_m)$$
$$= \gcd(\ldots (\gcd(\gcd(a_1, a_2), a_3), \ldots, a_m),$$

the most obvious way to solve our problem is to compute the pairwise gcds successively and to stop whenever a result is 1. This iterative computation of pairwise gcds will work well for random inputs, but there are "nasty" inputs on which this method will need $m - 1$ pairwise gcd computations.

If the inputs are randomly chosen integers this naive algorithm usually does not need all the $m - 1$ steps:

Fact 1. *Let two integers a and b be chosen uniformly at random from the positive integers up to N. Then for large N*

$$\text{prob}(\gcd(a, b) = 1) \to \zeta(2)^{-1} = \frac{6}{\pi^2} \sim 0.60793.$$

For a proof, see [8], Section 4.5.2.

In this case, the probability that a_1 and a_2 are already coprime is about 0.6, and for randomly chosen $a_1, \ldots, a_m$, we can therefore expect that the naive algorithm will already stop after about two computations of a pairwise gcd.

Heuristic reasoning says that a prime p divides a random integer a with probability p^{-1}, and m random integers with probability p^{-m}. Assuming the independence of these events, we have

$$\text{prob}(\gcd(a_1, \ldots, a_m) = 1) = \prod_p (1 - p^{-m}) = \zeta(m)^{-1}.$$

This can be made into a rigorous argument showing that the probability that m random integers, as in Fact 1, have trivial gcd tends (with $N \to \infty$) to $\zeta(m)^{-1}$. The first few values are:

m	$\zeta(m)^{-1}$
2	0.608
3	0.832
4	0.924
5	0.964
6	0.983
7	0.992
8	0.996
9	0.998
10	0.999

Although this strategy is quite successful for randomly chosen inputs, there exist "nasty" sequences, in which the $m - 1$ steps in the above algorithm are really necessary.

Example 1. Let $p_1, \ldots, p_m$ be the first m primes, $A = p_1 \cdots p_m$, and $a_i = A/p_i$ for each i. Then $\gcd(a_1, \ldots, a_m) = 1$ but for any proper subset S of $\{1, \ldots, m\}$, $\gcd(\{a_i : i \in S\})$ is non-trivial. So the successive computation of pairwise gcds does not give the right output until $m - 1$ steps in the above algorithm have been performed.

We do not address the question of representing the gcd as a linear combination of the inputs. This problem is considered in [9,6].

3   Random linear combinations

In the case of polynomials over a field F, the use of random linear combinations is known to be successful.

Fact 2. *Let F be a field, $a_1, \ldots, a_m \in F[x]$ be nonzero polynomials of degree at most d, $h = \gcd(a_1, \ldots, a_m)$, $A \subseteq F$ finite, $x_3, \ldots, x_m \in A$ be randomly chosen elements, and $g = a_2 + \sum_{3 \le i \le m} x_i a_i \in F[x]$. Then h divides $\gcd(a_1, g)$, and*

$$\text{prob}(h = \gcd(a_1, g)) \ge 1 - d/\#A.$$

This is based, of course, on resultant theory ([2], [3]), and also works for multivariate polynomials. So for the polynomial case, we expect that with only about one calculation of a pairwise gcd of two polynomials we find the true gcd of many polynomials, provided that A is chosen large enough.

We try to solve the corresponding integer problem by using a similar approach, and consider therefore the following algorithm.

Algorithm 1. *Probabilistic gcd of many integers.*

Input: m positive integers $a_1, \ldots, a_m$ with $a_i \leq N$ for all i, and a positive integer M.
Output: Probably $\gcd(a_1, \ldots, a_m)$.

1. Pick $2m$ uniformly distributed random integers $x_1, \ldots, x_m$ and $y_1, \ldots, y_m$ in $\{1, \ldots, M\}$, and compute $x = \sum_{1 \leq i \leq m} x_i a_i$ and $y = \sum_{1 \leq i \leq m} y_i a_i$.
2. Return $\gcd(x, y)$.

Then $g = \gcd(a_1, \ldots, a_m)$ divides $\gcd(x, y)$, and we can easily check equality by trial divisions by $\gcd(x, y)$. This is a Monte-Carlo algorithm and we want to prove a lower bound for its success probability, i.e., for the probability that $\gcd(x, y)$ equals g.

Due to Fact 1 we cannot expect this probability to be as high as in the polynomial case, but tests show that this strategy seems to have roughly the expected success probability of $6/\pi^2$. For two lists we have tested (with 5000 independent tests each) whether two random linear combinations of the list elements have the same gcd as the list elements. The list "nasty" has 100 elements as described in Example 1, whose largest entry has 220 decimal digits, and "random" is a random list with 100 elements of up to 220 decimal digits.

Table 1. Using random linear combinations

M	nasty list success rate	random list success rate
2	0.7542	0.6066
10	0.6522	0.6016
100	0.6146	0.6078
1000	0.6020	0.6098
30030	0.5952	0.6038

Table 1 shows that one gcd computation of two random linear combinations of the a_i gives the right answer in about 60% of the cases, i.e., in most of the cases, and this seems to be somewhat independent of the size of M. (The higher success rate for the nasty list using low M is readily explained.)

Our task is to *prove* that the probability that $\gcd(x, y) = g$ is high.

Remembering Fact 1, the solution of our problem would be easy if x and y were random numbers, but although the results in Table 1 show that they behave approximatively as if they were uniformly distributed, this is not literally true. So we have to find another way to bound the probability that $g \neq \gcd(x, y)$.

4   A probabilistic estimate

It would, of course, be sufficient to show that our random linear combination x behaves like a random integer in its range. Then the bound $\zeta(s)^{-1}$ would hold for the gcd of s random linear combinations. In any case, we cannot reasonably hope to have a better success probability than this bound. However, we want a (small) bound on M, possibly in terms of the inputs. Then, if $a_1, \ldots, a_m \leq N$, we have

$$x = \sum_{1 \leq i \leq m} x_i a_i \leq mMN = B,$$

but x is clearly not uniformly distributed in $\{1, \ldots, B\}$. Even if it were, the probability that a prime p divides x would not be the exactly desired $1/p$, but only close to it, since B is not necessarily a multiple of p. We circumvent this obstacle by choosing M to be the product of the first r primes. For $r = 6$, we have $M = 2 \cdot 3 \cdot 5 \cdot 7 \cdot 11 \cdot 13 = 30030$. (In fact, for our purposes a multiple of this number suffices.) Then for all primes p up to the rth prime p_r, p divides a random linear combination with probability exactly $1/p$. For our probability estimate, the main contribution becomes

$$\prod_{p \leq p_r} (1 - p^{-s}),$$

just as for the unavoidable $\zeta(s)^{-1} = \prod_p (1 - p^{-s})$. We only use this idea for $s = 2$, and abbreviate

$$\eta_r = \prod_{p \leq p_r} (1 - p^{-2}).$$

We fix the following notation. For $m \geq 2$ and $r \geq 1$, let $a_1, \ldots, a_m$ be positive integers, all at most N; let M be a multiple of $\Pi_r = p_1 \cdots p_r$. Let $x_1, \ldots, x_m$, $y_1, \ldots, y_m$ be uniformly distributed random integers between 1 and M; and let $x = \sum_{1 \leq i \leq m} x_i a_i$ and $y = \sum_{1 \leq i \leq m} y_i a_i$.

For the probability estimate we need the following lemma:

Lemma 1. *If* $\gcd(a_1, \ldots, a_m) = 1$, *then the events that* p_i *divides* x *for* $1 \leq i \leq r$ *are independent.*

PROOF.   Let $1 \leq k \leq r$. Since $\gcd(a_1, \ldots, a_m) = 1$, there exists an index j such that $p_k \nmid a_j$. Then for any $x_1, \ldots x_{j-1}, x_{j+1}, \ldots, x_m \in \mathbb{Z}$ the congruence

$$x_j a_j \equiv -\sum_{i \neq j} x_i a_i \bmod p_k$$

has exactly one solution x_j modulo p_k. Hence, there are p_k^{m-1} solutions with $1 \le x_1, \ldots, x_m \le p_k$ and

$$\sum_{1 \le i \le m} x_i a_i \equiv 0 \bmod p_k.$$

Now let $I \subseteq \{1, \ldots, r\}$ and $q = \prod_{k \in I} p_k$. By the Chinese Remainder Theorem, we have $\prod_{k \in I} p_k^{m-1} = q^{m-1}$ solutions modulo q giving $x \equiv 0 \bmod q$. Since $q \mid M$, this congruence has $q^{m-1} \cdot (M/q)^m$ solutions $1 \le x_1, \ldots, x_m \le M$. Hence,

$$\text{prob}\{q \mid x\} = \frac{q^{m-1} \cdot \left(\frac{M}{q}\right)^m}{M^m} = \frac{1}{q}. \tag{1}$$

In particular, $\text{prob}\{p_i \mid x\} = 1/p_i$ for each $i \le r$, and since (1) holds for each subset of $\{1, \ldots, r\}$, the events are independent. $\qquad\Box$

Theorem 1.
With the above notation, let P be the probability that $\gcd(a_1, \ldots, a_m) = \gcd(x, y)$. Then

$$P > \eta_r - \frac{2.04}{p_r \ln p_r} - \frac{2}{M}\left(\ln\ln M + \frac{1}{\ln^2 M} + \frac{1}{2\ln^2 p_r} - \ln\ln p_r\right)$$
$$- \frac{mN}{M(\ln(mMN) - 3/2)} + \frac{r}{M^2}.$$

The proof of this theorem is by detailed analysis of various cases and uses several bounds from [10].

Corollary 1. *For any $\epsilon > 0$ one can choose r and M such that $P > \zeta(2)^{-1} - \epsilon$.*

Corollary 2. *Let $r \ge 6$. Then*

$$P > 0.5466 - \frac{mN}{M(\ln(mMN) - 3/2)}.$$

PROOF. We have $\eta_r > 1/\zeta(2)$ for all $r \ge 1$, and

$$\frac{2}{M}\left(\ln\ln M + \frac{1}{\ln^2 M} + \frac{1}{2\ln^2 p_r} - \ln\ln p_r\right) < 0.0000983402$$

for $r = 6$. The left hand side expression decreases strictly for increasing r. Hence,

$$P > 0.6079271016 - \frac{2.04}{13 \ln 13} - 0.0000983402 - \frac{mN}{M(\ln(mMN) - 3/2)}$$
$$> 0.5466489662 - \frac{mN}{M(\ln(mMN) - 3/2)}. \qquad \Box$$

Corollary 3.
Let $r \geq 6$, and $M \geq 3mN$ be a multiple of $p_1 \cdots p_r$. Then $P > 51\%$.

PROOF. Let $M \geq 3mN$. Indeed $M \geq 30030$ and we can also assume $m \geq 3$ and $N \geq 2$ (otherwise everything is trivial). Then

$$\ln(mMN) - 3/2 \geq \ln(3 \times 30030 \times 2) - 3/2 > 10,$$

so

$$
\begin{aligned}
P &> 0.5466489662 - \frac{mN}{M(\ln(mMN) - 3/2)} \\
&> 0.5466489662 - \frac{mN}{10M} \\
&> 0.5466489662 - 1/30 \\
&> 0.51. \quad \square
\end{aligned}
$$

So, if M is chosen large enough, Algorithm 1 has a success probability greater than 0.51. The results in Table 1 and other experiments suggest that P is even somewhat larger and rather independent both of the size of M and of its choice as a multiple of the small primes.

Thus we have a Monte-Carlo algorithm for which the result is correct in more than 50% of the cases. Hence after two independent executions of the algorithm we expect that the smaller value of g equals $\gcd(a_1, \ldots, a_m)$.

Acknowledgements

We are grateful to Mark Giesbrecht, Boaz Patt-Shamir and Igor Shparlinski for helpful discussions.

References

1. GENE COOPERMAN AND GEORGE HAVAS, Elementary Algebra Revisited: Randomized Algorithms. In *Randomization Methods in Algorithm Design*, DIMACS Series in Discrete Mathematics and Theoretical Computer Science **43** (1999) 37–44.
2. ANGEL DÍAZ AND ERICH KALTOFEN, On computing greatest common divisors with polynomials given by black boxes for their evaluations. In *ISSAC'95* (Proc. Internat. Sympos. Symbolic and Algebraic Computation), ACM Press (1995) 232–239.
3. JOACHIM VON ZUR GATHEN, MAREK KARPINSKI AND IGOR SHPARLINSKI, Counting curves and their projections. *Computational complexity* **6** (1996), 64–99. (Extended Abstract in Proc. 25th ACM Sympos. Theory of Computing.)
4. K. O. GEDDES, S. R. CZAPOR AND G. LABAHN, *Algorithms for Computer Algebra*. Kluwer Academic Publishers, 1992.
5. MARK GIESBRECHT, Fast computation of the Smith normal form of an integer matrix, In *ISSAC'95* (Proc. Internat. Sympos. Symbolic and Algebraic Computation) ACM Press (1995) 110–118.

6. GEORGE HAVAS, BOHDAN S. MAJEWSKI AND KEITH R. MATTHEWS, Extended gcd and Hermite normal form algorithms via lattice basis reduction. *Experimental Mathematics* **7** (1998) 125–136.

7. CHUNG-YEN HO AND CHEE KENG YAP, The Habicht approach to subresultants. *Journal of Symbolic Computation* **21** (1996). 1–14.

8. DONALD E. KNUTH, *The Art of Computer Programming, Vol.2, Seminumerical Algorithms*. Addison-Wesley, Reading MA, 3rd edition, 1997.

9. BOHDAN S. MAJEWSKI AND GEORGE HAVAS, A solution to the extended gcd problem. In *ISSAC'95* (Proc. Internat. Sympos. Symbolic and Algebraic Computation) ACM Press (1995) 248–253.

10. J. BARKLEY ROSSER AND LOWELL SCHOENFELD, Approximate formulas for some functions of prime numbers. *Illinois J. Math.* **6** (1962), 64–94.

11. RICHARD ZIPPEL, *Effective polynomial computation*. Kluwer Academic Publishers, 1993.

Multi-party Finite Computations [*]

Tomasz Jurdziński[1], Mirosław Kutyłowski[1,2], and Krzysztof Loryś[1]

[1] Institute of Computer Science, Wrocław University
[2] Department of Mathematics and Computer Science, Poznań University

Abstract. We consider systems consisting of a finite number of finite automata which communicate by sending messages. We consider number of messages necessary to recognize a language as a complexity measure of the language. We feel that these considerations give a new insight into computational complexity of problems computed by read-only devices in multiprocessor systems. Our considerations are related to multi-party communication complexity, but we make a realistic assumption that each party has a limited memory.

We show a number of hierarchy results for this complexity measure: for each constant k there is a language, which may be recognized with $k+1$ messages and cannot be recognized with $k-1$ messages. We give an example of a language that requires $\Theta(\log \log n)$ messages and claim that $\Omega(\log \log(n))$ messages are necessary, if a language requires more than a constant number of messages. We present a language that requires $\Theta(n)$ messages. For a large family of functions f, $f(n) = \omega(\log \log n)$, $f(n) = o(n)$, we prove that there is a language which requires $\Theta(f(n))$ messages. Finally, we present functions that require $\omega(n)$ messages.

1 Introduction

Today, communication capacity is one of the most important bottlenecks of computer systems. For this reason, there is a growing interest in communication complexity of problems. From a practical point of view, corresponding complexity measures are often more important than traditional measures such as time and space complexities.

Communication complexity has been introduced as *two-party* communication complexity. For this measure, an input is divided between two processing units that compute together the output [8]. Communication complexity is defined as the number of message bits exchanged between the processing units. Despite many efforts, the number of results obtained for this measure is relatively low, proving that the field belongs to difficult parts of theoretical computer science. The results that have been obtained so far are often very involved (see for instance [8]).

Development of distributed computing systems increased interest in multi-party communication complexity. Also, it became clear that interprocessor communication is a crucial limitation for parallel systems and that good parallel algorithms must be communication efficient. Two-party communication complexity

---

[*] partially supported by Komitet Badań Naukowych, grant 8 T11C 032 15

T. Asano et al. (Eds.): COCOON'99, LNCS 1627, pp. 318–329, 1999.

has been generalized to some extent to multi-party communication complexity. However, these results often assume specific architecture (see for instance [4]).

The approach discussed above is based on the assumption that during a two-party computation communication is the most crucial resource and the size limitation of the internal memory can be neglected in order to keep the picture clear. However, often this is not a good description of a real situation. For instance, on a multiprocessor system, computational capacity of each processing unit might be severely limited.

In this paper we consider a situation when each processing unit has a finite memory, that is, it is a finite automaton. (Our results can be easily extended to the situation, when each of the processing units has some *small* internal memory.) The input is not divided between the processing units. It is stored in a read-only memory and is available for each processing unit. We consider the processing units to be automata with read heads, but their sequential access to input data has no direct influence on the amount of communication between the processing units. Thus, our considerations are relevant for large problems with input data stored in a read-only shared memory, computed by systems of many processing units of limited internal memory communicating by messages.

Research on coordination means of systems of finite automata have been initiated in [3,6]. Somewhat related are also the papers on the trade-off between computation time and space of multihead automata [5,7]. However, these results are based on the fact that in order to examine distant parts of the input the heads must spend quite a long time to get there. So they are not relevant for the size of communication between automata.

A significant step in understanding communication complexity was to apply ideas of Borodin, Cook [2] and Yesha [11] to two-party computation. A trade-off for communication time and space has been shown for problems such as matrix multiplication [1,9]. Here, communication time relates to the number of bits exchanged and space denotes logarithm of the number of internal states. In the strongest version, the computing model assumed is very general (i.e. branching programs). The results mentioned have some impact on what we can say about systems of finite automata, since their computations may be simulated by two "agents" communicating through messages.

Model We consider systems consisting of a finite number of finite automata working independently, except that they may communicate by broadcasting messages. (Since we consider systems of finitely many automata, this is the same, up to a constant factor, as messages delivered in a point-to-point fashion.) Since each automaton has only finitely many internal states, the number of different messages is finite, as well. Transition function of a single automaton depends on its internal state, an input symbol seen and all messages received from the other automata at this moment. The messages are delivered instantly. The automata work synchronously. More formally, we define such systems as follows:

Definition 1. $M = (A_1, A_2, \ldots, A_l)$ *is an l-automata system, if the following conditions hold. There is a finite set Q such that every internal state of any A_i*

is in Q. Each automaton has a single read-only head. Let automaton A_i be in state q and let a be the input symbol seen by A_i. Then the next step is executed by A_i as follows:

(1) A_i broadcasts a message $m = \mu_i(q, a)$ to other automata. If $\mu_i(q, a) = $ nil, then no message is sent by A_i.

(2) A_i changes its internal state to q' and moves its read head r positions to the right ($r \in \{0, 1, -1\}$), where q' and r are given by transition function δ_i:

$$(q', r) = \delta_i(q, a, m_1, \ldots, m_{i-1}, m_{i+1}, \ldots, m_l)$$

where m_j denotes the message sent by automaton A_j during the first phase of the step (if no message is sent by A_j, then $m_j = $ nil).

*   We say that an l-automata system accepts a language L, when a fixed automaton (for instance the first one) enters an accepting state on input x if and only if $x \in L$.*

Complexity classes We define $\text{MESSAGE}_l(f(n))$ as the class of all languages that may be recognized by an l-automata system where at most $f(n)$ messages are sent for every input word of length n. By $\text{MESSAGE}(f(n))$ we denote $\bigcup_{l \in \mathbb{N}} \text{MESSAGE}_l(f(n))$. Note that even $\text{MESSAGE}_2(1)$ contains non-regular languages. Indeed, $\{a^m b^m | m \in \mathbb{N}\} \in \text{MESSAGE}_2(1)$. Similarly, $\{a^m b^m c^m | m \in \mathbb{N}\} \in \text{MESSAGE}_3(1)$ but is not context-free.

New results We prove a number of hierarchy results for MESSAGE complexity measure. The first one is the hierarchy result for a constant number of messages:

Theorem 1. $\text{MESSAGE}_l(k - 1) \subsetneq \text{MESSAGE}_l(k + 1)$, for all $k \geq 1$ and $l > 1$.

We present an example of a language that requires a double-logarithmic number of messages:

Theorem 2. *For $a, b \in \mathbb{N}$, let*

$$L_{a.b} = \{1^{f_1} \# 1^{f_2} \# \ldots \# 1^{f_k} : f_1 = a, f_2 = b, (f_{i-1} \cdot f_{i-2}) | (f_i - 1), f_i > 1 \text{ for } i = 3, \ldots, k\}.$$

If $a, b > 1$ are prime, then $L_{a.b} \in \text{MESSAGE}_2(O(\log \log n))$ and $L_{a.b} \notin \text{MESSAGE}_l(f(n))$ for any $f(n) = o(\log \log n)$ and $l \in \mathbb{N}$.

The following example shows that there are languages requiring $\Omega(n)$ messages:

Theorem 3. *Let $L_{xx} = \{x \# x : x \in \{0, 1\}^*\}$. Then $L_{xx} \notin \text{MESSAGE}_l((f(n)))$, for any $l \in \mathbb{N}$ and $f(n) = o(n)$. (While $L_{xx} \in \text{MESSAGE}_2(n)$.)*

We already know that there are languages requiring $\Theta(\log \log(n))$ and $\Theta(n)$ messages, respectively. It is interesting to see that there is a dense hierarchy between these bounds.

Theorem 4. *There exists a constant $c > 0$ such that for any Turing machine time-constructible function $f(m)$, $f(m) = o(2^{2^{c^m}})$, there exists a language L which can be recognized by a 2-automata system M which uses $O(f^2(\log \log n))$ messages and cannot be recognized by a multi-automata system (with an arbitrary number of automata) which uses $o(f^2(\log \log n))$ messages.*

In particular, for simple functions f such as n^c for $c < 1$, and $\log^c n$ for $c \in \mathbb{N}$ we may obtain a hierarchy result. Finally, we give examples when a superlinear number of messages is necessary:

Theorem 5. *Computing product over $\mathbb{F}_2$ of square matrices with n elements (giving the output through a one-way output tape) requires $\Omega(n^{3/2}/\beta(n))$ messages, for any number of automata in the system and every $\beta(n) = \omega(\log^2(n))$. (By a standard algorithm $O(n^{3/2})$ messages are sufficient.)*

Theorem 5 can be derived from the results on two-party communication complexity [1] through simulating a system of finite automata in a two-party communication model (see the forthcoming full version of the paper). Here we give a relatively simple and direct proof based on Kolmogorov complexity. Also, we may derive some hierarchy results for superlinear number of messages using results known for the class LOGSPACE. However, the last technique does not provide a tight hierarchy. A variant of Theorem 5 gives explicit examples of functions for bounds between n and $n^{3/2}$.

We assume that the reader is familiar with *Kolmogorov complexity* concept ([10]), which is the most important tool in crucial moments of our proofs.

Organization of the paper In this extended abstract we present sketches of the proofs of Theorems 1 (Section 2), 2 and 4 (Section 3), and 5 (Section 4). The details and the omitted proofs will be included in the full version of the paper.

2 Hierarchy for a Constant Number of Messages

We prove Theorem 1 by showing that the following languages separate the classes $\text{MESSAGE}_l(k-1)$ and $\text{MESSAGE}_l(k+1)$:

$$L_k^l = \{x \# y_1 \# y_2 \# \ldots \# y_k \mid \forall i \ D(x, y_i)\}$$

where $D(x, y_i)$ holds if the characters at positions $|x|, 2|x|, \ldots, (l-1)|x|$ in y_i are ones.

First let us see that an l-automata system may recognize L_k^l using $k+1$ messages. First, all automata work in a cycle of length l. During this cycle automaton A_i makes exactly i moves to the right. This is continued until A_1 reaches the endmarker terminating x. Till this moment, A_i has moved $(i-1) \cdot |x|$ positions in y_1. Then the automata check if there are ones on positions $|x|, 2|x|, \ldots, (l-1)|x|$ in y_1. In order to check the form of y_2 all automata move their heads at the same speed to the right, and A_1 sends a message once it reaches the symbol # marking the beginning of y_2. The computation continues this way until all y_i are checked. Finally, $A_2, \ldots, A_l$ send messages saying if all tests were positive. This is organized so that only one message is sent: at step i of the final phase, automaton A_{i+1} sends its message if anything was wrong and the whole system halts in a rejecting state. If the characters seen by A_{i+1} witness that the input is in L_k^l, then A_{i+1} sends no message.

We argue that $L_k^l \notin \mathrm{MESSAGE}_l(k-1)$. Intuitively any two-way deterministic finite automaton cannot "remember" the symbol of y_i on position $|x|$ at configuration in which the position of the automaton on y_i is far from position $|x|$. Moreover, we show that for appropriate inputs the messages are sent only in configurations in which one of the automata is "near" symbol # or an end-marker. It follows that using l automata and sending $k-1$ messages we can check only $(k-1)(l-1)$ positions. But there are $k(l-1)$ positions to be checked. Below we sketch a proof based on this intuition.

Lemma 1. *Let A be a 2-way finite automaton and let q be the number of states of A. Then there is a word $b = b_L b_R$ of length at most $2q^2$ such that:*

- *if the head of A enters b from the left, then it either*
  - *leaves b on the left side without reaching b_R, or*
  - *loops inside b_L, or*
  - *leaves b on the right side and for no $x \in \{0,1\}^*$, the head of A ever leaves b to the left side while working inside bx.*
- *if the head enters b from the right, then analogous properties hold.*

Proof. A simple induction. We commence by considering the state $q_1 \in Q$. If there is any word v that causes A to loop inside v while entered v from the left side in state q_1, then we put $b^{(1)} = v$. Otherwise, if there is any word w that causes A to leave w to the left, when entered w from the left in state q_1, then we put $b^{(1)} = w$ for some such a word w. Otherwise $b^{(1)}$ is an empty word. Then we consider $q_2 \in Q$. If there is a word v' such that A loops inside $b^{(0)}v'$ if entered it from the left in state q_2, then we put $b^{(2)} = b^{(1)}v'$. Otherwise, if there is w' such that A leaves $b^{(1)}w'$ to the left, after some computation within $b^{(1)}w'$ started in state q_2 at the left side, then we put $b^{(2)} = b^{(1)}w'$. Otherwise $b^{(2)} = b^{(1)}$. We proceed this way until we consider all states of A. We put $b_L = b^{(q)}$ (where $q = |Q|$). By reversing directions we construct b_R. ∎

Let b be a word given by Lemma 1 for an automaton A. We consider inputs of a special form consisting of bits separated by word b: if $z = z_1 z_2 \ldots z_n$, $z_i \in \{0,1\}$ for $i \leq n$, then $b_z = b z_1 b z_2 b z_3 \ldots b z_n b$. The purpose of these words b is to enforce a quasi one-way work of A on z. By Lemma 1, we see that if the head of automaton A enters a word b_z from the left (right) and reaches z_1 (z_n), then it leaves b_z to the right (left) after some number of steps. Indeed, since the head comes through b, it cannot loop inside any word starting with b or go back through b in the opposite direction as started.

Our goal now is to show that a finite automaton cannot recognize what symbol stands in the middle of a word, even if we inspect the computation time. In fact, this is not true at some specific situations. We prove our claim for inputs b_z, where z is random in the sense of Kolomogorov complexity. By *clock characterization* of word z we mean the set of tuples (q_1, p, q_2, t) such that if the head of A enters b_z in the state q_1 on side $p \in \{\text{Left, Right}\}$ of b_z, and reaches z_1 for $p = \text{Left}$ or z_n for $p = \text{Right}$, then the head leaves b_z in the state q_2 on the other side of b_z after t steps. Thus, clock characterization gives all information provided by the outcome of the computation on b_z including such "side channel" information as computation time.

Lemma 2. *(Confusion Lemma) Let A be a two-way finite automaton and b be the word given by Lemma 1. Let n be a sufficiently large odd number, $z = z_1 z_2 \ldots z_n$, $z_i \in \{0,1\}$ for $i \leq n$, $K(z) \geq n - c \log n$ for some constant c, and $z_{n/2} = 1$. Then there is a word $z' \neq z$ such that $z'_{n/2} = 0$ and the clock characterizations of z and z' are the same.*

Proof. (sketch) Assume for simplicity that the clock characterization of z consists of a single element $(q', \text{Left}, q'', t)$. Consider a computation of A starting in the state q' at the leftmost symbol of b_z. For $i \leq n$, let q_i be the state of A at the moment when the head reaches z_i for the first time. (So, by definition of b, the head cannot return to z_{i-1} anymore after this moment). We say that subsequence $q_j, q_{j+1}, \ldots, q_{j+h}$ is a cycle if $q_j = q_{j+h}$ and the states $q_j, q_{j+1}, \ldots, q_{j+h-1}$ are different. Obviously, every subsequence of $q_1, \ldots, q_n$ of length q contains a cycle. Let f be the cycle which occurs most frequently in $q_{n/2}, \ldots, q_n$. So, it occurs $\Omega(n)$ times. There exists at least one occurrence of the first state of f in $q_1, q_2, \ldots, q_{n/4}$, say at position j. Indeed, otherwise z would be compressible, i.e. $K(z) < n - c \log n$. Let Z_i be the word obtained from z by moving the subwords of z corresponding to the first i cycles f from the second half of z to position j. In this way, we obtain $\Omega(n)$ words Z_i with the same clock characterization as z.

If the number of bits moved left to obtain Z_i from z is smaller than $n/4$, then the middle bit in Z_i is some bit from the first half of z. Note that this situation occurs for $\Omega(n)$ cases of i. The lemma holds if one of those bits is a zero. Otherwise, z contains ones on positions $n/2 - i|f|$ for $i = 1, 2, \ldots, s$, where $s = \Omega(n)$. This means, however, that we could compress z by $\Omega(n)$ bits, a contradiction.

The proof easily generalizes to clock characterizations containing more, say r, elements. It suffices to consider sequences of r-tuples of states. ∎

Lemma 3. *(Border Lemma) Let M be a multi-automata system that uses at most $c = O(1)$ messages for every input of the form $\{0, 1, \#\}^*$. Then there is $b \in \{0,1\}^*$ and a constant $d \in \mathbb{N}$ such that*

(a) Word b satisfies the claim of Lemma 1 for each automaton of M.
(b) Assume that an automaton of M sends a message while working on an input word $b_{z_1} \# b_{z_2} \# \ldots \# b_{z_m}$, $m \in \mathbb{N}$, $z_i \in \{0,1\}^*$ for $i \leq m$. Then at this moment some of the heads of M occurs at distance at most $d \cdot c \cdot |b|$ from a symbol $\#$ or an endmarker.

Proof. First we generalize the proof of Lemma 1. The first change is that we consider all automata of M in the inductive construction of b_L and b_R. The old construction applies provided that each state $q \in Q$ is replaced by copies $q^{(1)}, \ldots, q^{(l)}$, where $q^{(i)}$ corresponds to automaton A_i of M. At the construction step for $q_j^{(i)}$, we consider automaton A_i. This establishes part (a).

For part (b) we incorporate one more condition into definition of b. Suppose that we have to consider state q corresponding to automaton A_j and that we have already constructed a word b'. We first check if there is a word u such that A_j sends a message while working on $b'u$, if entered from the left in state q and no message is received in the meantime from the other automata of M. If such a word u exist, then we replace b' through $b'u$. Intuitively speaking, if there is a trigger causing A_j to send a message, then such a trigger is built in b.

Now consider a computation of M on an input of the form $b_{z_1} \# b_{z_2} \# \ldots \# b_{z_m}$. We distinguish periods, called later *internal periods*, when no automaton sees an endmarker or an $\#$. Consider an internal period τ. Let A_p denote an automaton that has left an endmarker or an $\#$ at the beginning of period τ. Let ρ denote this endmarker or $\#$. We show that a message can be sent only at the beginning of period τ, when head A_p is still close to ρ. If any automaton sends a message during period τ, then by properties of b, it happens when it scans a single b, i.e. within $O(|b|)$ steps. At this time, the head of A_p makes $O(|b|)$ moves on the input. After receiving the first message we enter a new phase of the internal period, when a second message can be sent. Notice that again it must happen within $O(|b|)$ steps, if it happens at all. Again the distance of A_p from ρ may increase by at most $O(|b|)$. Since totally at most c messages are sent during the whole computation, it follows that the distance of A_p from ρ is $O(c \cdot |b|)$ at the moments when a message is sent during internal period τ.   ∎

At this point we are ready to prove Theorem 1. Assume that M is an l-automata system that recognizes L_k^l with at most $k-1$ messages. Now we consider an input $x \# y_1 \# \ldots \# y_k$ defined as follows. For each i, word y_i has the form b_{z_i}, where b is given by Border Lemma, $x = (1b)^n$, for large enough n, $|y_i| > l \cdot |x|$, and $K(z_i) \geq |z_i| - c \cdot \log |z_i|$, for $i \leq k$. Let the symbols on positions $|x|, 2|x|, \ldots, (l-1)|x|$ in $y_1, y_2, \ldots, y_k$ be called *important* symbols. We assume that all important symbols are ones. Let $y_{i,j}$ be the subword of y_i consisting of symbols on positions $j \cdot |x| - \lfloor \sqrt{n} \rfloor, \ldots, j \cdot |x| + \lfloor \sqrt{n} \rfloor$.

Without loss of generality we assume that M accepts at the moment when all heads stand at the first symbol of the input. By Border Lemma, if a message is sent, at least one head is at distance $O(|b|)$ from an $\#$ or an endmarker. So at most $l-1$ automata have their heads near the important symbols. Hence there are i, j such that no message is sent if a head is inside $y_{i,j}$. By Confusion Lemma, the substring $y_{i,j}$ may be replaced by a string $y'_{i,j}$ having the same clock characterization, but a zero in the middle. Since no message is sent when a head is inside $y_{i,j}$ ($y'_{i,j}$ respectively), the same clock characterizations guarantee the same result at the end of the computation (note that the heads are outside $y_{i,j}$ ($y'_{i,j}$) at this moment). A contradiction, since the modified word is not in L_k^l.

∎ Theorem 1.

3   Hierarchy Between $\log \log n$ and n

In this section we prove Theorems 2 and 4. The lower bound arguments are based on Lemma 4. It says that $\Omega(m)$ messages are needed in order to check whether the lengths of consecutive blocks of ones in a word of the form $\#1^{d_1}\#1^{d_2}\ldots\#1^{d_m}\#$ satisfy equality $d_i = \alpha(d_{i-1}, d_{i-2})$ for $i = 3, \ldots, m$ and a function $\alpha : \mathbb{N}^2 \to \mathbb{N}$ defined as below.

Lemma 4. *Let $\alpha : \mathbb{N}^2 \to \mathbb{N}$ be a function such that for any $a_1 > a_2$ and b it holds that $\alpha(a_1, b) \geq a_1$ and $\alpha(a_1, b) > \alpha(a_2, b)$. Let $f : \mathbb{N} \to \mathbb{N}$. Let $L_{\alpha, f}$ be an infinite subset of such that if a word $z = x \star 1^{d_1}\#\ldots\#1^{d_m} \star y$ belongs to $L_{\alpha, f}$,*

$$\{x \star w \star y \mid x, w, y \in (\Sigma \backslash \{\star\})^*, w = 1^{d_1}\#\ldots\#1^{d_m}, m \in \mathbb{N}, d_i = \alpha(d_{i-1}, d_{i-2}) \text{ for } i = 3\ldots m\}$$

then $m \geq f(n)$, where $n = |z|$. If $\alpha(x, y) = x$ for any $y > 0$, then we assume additionally that $d_1 \geq g(m)$ and $d_2 \geq g(m)$ for some nondecreasing function $g : \mathbb{N} \to \mathbb{N}$ that is not bounded by a constant.
Then $L_{\alpha, f} \notin \text{MESSAGE}(h(n))$, for any $h(n) = o(f(n))$.

Proof. Assume that there exists a multi-automata system S that recognizes language $L_{\alpha, f}$ and uses $h(n) = o(f(n)) = o(m)$ messages on w for every input $x \star w \star y \in L_{\alpha, f}$.

First we shall show that no message is sent on a "long" sequence of "long" blocks of ones in the middle part of an input $z \in L_{\alpha, f}$. Then we shall find two blocks $\#1^{d_i}\#$ and $\#1^{d_j}\#$ such that the system does not distinguish between z and a word z' obtained from z by extending the block $\#1^{d_i}\#$ and shortening the block $\#1^{d_j}\#$ by some number of ones.

Let Q_S be the set of states of the automata of S. We focus our attention on a set $W_m \subseteq L_{\alpha, f}$ containing all words $x \star 1^{d_1}\#\ldots\#1^{d_m} \star y \in L_{\alpha, f}$ such that $|d_i| \geq 2|Q_S|!$ for $i > m/2$. We consider messages sent during computation of S on a word $x \star w \star y \in W_m$. Let $w' = \#1^{d_u}\#\ldots\#1^{d_{u+a-1}}\#$ be a subword of w such that no message was sent while any automaton was reading w'. Since $h(n) = o(n)$, we can choose w' so that $u > m/2$ and a is a large constant (to be determined later), if m is large enough.

We inspect more closely behavior of S on w'. For each block of ones in w', we define a *trace* of computation on w': for $u \leq i \leq u + a - 1$, trace T_i is the set of tuples

$$(A_j, q, s, k, q_f, s_f, q_t, s_t)$$

where A_j is an automaton of S, $q, q_f, q_t \in Q$, $s, s_f, s_t \in \{Left, Right\}$, satisfying the following properties: during a computation of A_j on w' started at side s of w' in state q, block i of 1's is entered for the kth time on side s_f in state q_f and is left on side s_t in state q_t. Observe that $k \leq |Q_S|$, so the number of possible traces is bounded by some function of $|Q_S|$. But $|Q_S|$ is a constant, so we may take a number a bigger than the number of possible traces. Then there are two blocks of 1's in w', say the ith and the jth, with the same traces.

Let us look closer at computation of automata of S on the ith and the jth blocks in w'. Each automaton entering such a block either leaves it on the same side, or goes to the opposite end. In the first case, the automaton cannot reach more than $|Q_S|$ symbols inside the block, since otherwise it would loop and go to the opposite end. In the second case, after at most $|Q_S|$ steps the automaton enters a loop. The number of symbols read during the loop, say r, is at most $|Q_S|$. Observe that we may extend the block of ones by a multiple of r and the automaton reaches the opposite end in the same state. We may also shorten it by a multiple of r provided that we leave place for initial steps before going into the loop. In our construction we shorten or extend the blocks of ones by $|Q_S|!$ ones. The trick is that this number is a multiple of any r considered above. The second point is that the blocks have length at least $2|Q_S|!$, so we leave place for automata to get into the loop or return to the side of arrival within the remaining part of the block.

In order to fool S, we replace 1^{d_i} by $1^{d_i + |Q_S|!}$ and 1^{d_j} by $1^{d_j - |Q_S|!}$. As already noticed, this leaves the traces unchanged. It follows that blocks i and j are traversed in the same way and the same number of times as before. The only difference is that the runs through block i are longer than before, and the runs through block j are shorter than before. However, the additional time lost on block i is equal to the time saved on block j. This follows from the fact that each automaton goes through the same loops on both blocks and that we have extended and shortened them by the same number of ones. Since each automaton spends the same time on w' as before, S cannot detect any change. However, the changed word does not satisfy the conditions from the definition of $L_{\alpha, f}$. ∎

Now we prove the lower bound of Theorem 2. By Lemma 4, it suffices to show that the number of blocks of ones is $\Omega(\log \log n)$ for infinitely many words $x \in L_{a,b}$. If a and b are prime, $x = 1^a \# 1^b \# 1^{f_3} \# \ldots \# 1^{f_k} \in L_{a,b}$, then $f_i \geq f_{i-1} \cdot f_{i-2} + 1$ for $i \leq k$ by Chinese Reminder Theorem. For the shortest word $x \in L_{a,b}$ with k blocks of ones we have even $f_i = f_{i-1} \cdot f_{i-2} + 1$. Then it is easy to check that $f_k = \Omega(a^{F_{k-3}} b^{F_{k-2}})$ and $f_k = O(2^k a^{F_{k-3}} b^{F_{k-2}})$ where F_i is the ith Fibonacci number. Furthermore $f_k \geq k + \sum_{i=1}^{k-1} f_i$ for large enough k. So, $f_k \leq n \leq 2 f_k$ and therefore $k = \Theta(\log \log n)$.

A two-automata system may recognize $L_{a,b}$ as follows:

1. The first automaton checks that the lengths of the first two blocks of ones are equal to a and b.
2. For $i = 3, \ldots, k$ the system checks that $f_i = 1 \mod f_{i-2}$ and $f_i = 1 \mod f_{i-1}$ (if not, then it rejects)
3. Accept

Checking that $f_i = 1 \mod f_{i-1}$ is performed as follows: automaton A_1 goes from the left to the right over block i while automaton A_2 loops between the "ends" of block $i - 1$. They finish when the automaton A_1 reaches symbol $\#$ and sends a message. Since A_2 may remember in states if it has left the symbol $\#$ in the previous step, the result may be easily established.

Now we sketch the proof of Theorem 4. Let M be a 2-tape Turing machine which makes exactly $f(n)$ steps on input 1^n. We construct a language L_M which requires $\Theta(f^2(\log \log n))$ messages. L_M consists of words of the form

$z_k = b_k \& c_k \& h_k \& l_k \star e_k\star$, where subwords b_k, c_k, h_k are responsible for "computing" $f(k)$, and l_k, e_k guarantee that $f^2(k)$ messages are necessary. Let b_k be the word in $L_{2,3}$ consisting of k blocks of ones and $c_k = 1^k$. In this way we fix the input 1^k for M. (Word b_k guarantees that the input length is double-exponential in the length of c_k on which we simulate machine M.) Then $h_k = k_1 \# k_2 \# \ldots \# k_{f(k)}$, where k_i is the contents of the tapes of M (in consecutive configurations) with head positions marked with symbols denoting the state of M. Finally, $l_k = (1^{f(k)-1}\#)^{f(k)}$ and $e_k = (1^{f^2(k)}\#)^{f^2(k)}$.

It is easy to check that if $|z_k| = n$, then $k = \Omega(\log \log n)$. The bound $\Omega(f^2(\log \log n))$ on the number of messages necessary to recognize L_M follows from Lemma 4, since we have to check that $f^2(k)$ blocks of ones in e_k have the same length. It is tedious but relatively easy to show that $O(f^2(k))$ messages suffice to check the whole structure of z_k.   ∎ Theorem 4

4   Superlinear Size of Communication

In this section we sketch the proof of Theorem 5. We assume that the first input matrix is listed row by row, the second matrix is listed column by column (rows and columns are separated by # symbols). Moreover, the output matrix is printed out column by column. An implementation of the standard algorithm for multiplying matrices requires $O(n^{3/2})$ messages. So the non-trivial part of this theorem is to show the lower bound claimed.

Consider an l-automata system M computing product of matrices over $\mathbb{F}_2$. We count the write operations on the output tape as messages (this does not change the number of messages significantly). Consider a computation of M on an input $x\#y$ describing two $N \times N$ matrices X and Y, $n = 2N^2 + 2N$, such that $K(x\#y) \geq 2N^2 - c\log 2N^2$. We divide computation into phases during which the automata generate $s(N)$ bits of the output (number $s(N)$ will be chosen in a moment). We prove the following key property:

Claim A. *Let functions s and k satisfy $s(N) = \omega(\log N)$, $s(N) = o(N/k(N))$. Then during computation on $x\#y$ at least $k(N)$ messages are sent during each phase. So totally, $N^2 \cdot k(N)/s(N)$ messages must be sent.*

W.l.o.g. we may assume that $s(N)$ divides N. Then during each phase the output bits generated belong to the same column. Consider a phase $\mathcal{E}$ on input $x\#y$ and its output bits $c_{p,j}, \ldots, c_{p+s(N)-1,j}$ of column c_j. These bits are the scalar products of rows $w_p, \ldots, w_{p+s-1}$ of X by column y_j of Y. Assume that during phase $\mathcal{E}$ less than $k(N)$ messages are sent. We show that this contradicts the assumption that $K(x\#y) \geq 2N^2 - c\log N^2$.

By our assumption, there is an interval B in y_j of length $b = N/(l \cdot k(N))$ such that no message is sent while any automaton scans B during phase $\mathcal{E}$. Computation of M on B may be described by a *clock characterization*. For each automaton it contains tuples describing state of automaton while entering and leaving B and the time spent on B. Since there are constant number of states at which an automaton may enter and leave B and $O(\log n)$ bits suffice to describe computation time, each clock characterization has $O(\log n)$ bits. Obviously, the output bits generated during phase $\mathcal{E}$ are uniquely determined by input $x\#y$ without B, the initial configuration of M at the beginning of phase $\mathcal{E}$, and clock characterization of B. All these data may be described by a string of length $2N^2 - b + O(l \cdot \log n)$. The next claim shows that we may uniquely describe B by a relatively short string, given the output of M during phase $\mathcal{E}$.

Claim B . *There are at most $2^{b-s(N)}$ vectors u of length b such that replacing the contents of B by u inside input $x\#y$ does not change the output bits* $c_{p,j}, c_{p+1,j}, \ldots, c_{p+s(N)-1,j}$.

Let Z be the submatrix, consisting of those elements of rows $w_p, \ldots, w_{p+s-1}$ of X, which are multiplied by elements of B while determining the output of phase $\mathcal{E}$ by the standard algorithm. Clearly, Z is an $s(N) \times b$ matrix. We claim that the rows of Z are linearly independent. Otherwise, we could represent a row of Z as a linear combination of its remaining rows, say $Z_k = \sum_{i \neq k} v_i \cdot Z_i$. Then we could encode Z by the vectors $Z_1, \ldots, Z_{k-1}, Z_{k+1}, \ldots, Z_{s(n)}$, number k and the coefficients $v_1, \ldots, v_{s(n)}$. Such a representation would consist of $|Z| - b + s(N) + O(\log N)$ bits. This would lead to compression of the input $x\#y$ into $2N^2 - b + s(N) + O(\log N)$ bits, so below $2N^2 - c \log 2N^2$.

Let e denote the output bits of phase $\mathcal{E}$. Observe that given e, the rows $w_p, \ldots, w_{p+s-1}$, and the contents of y_j except B, the product $Z \cdot B$ is uniquely determined. The number of solutions of equation $Z \cdot B = d$ is $2^{b-s(N)}$, since Z contains $s(N)$ linearly independent rows and the number of (unknown) bits in B equals b. ∎ Claim B

Input $x\#y$ can be encoded by: $x\#y$ without B and $\#$ symbols (together $2N^2 - b$ bits), the configuration of M at the beginning of phase $\mathcal{E}$, the position of B, the clock characterization of B (all of length $O(l \log n)$), and the index of the contents of B in the lexicographic ordering among all strings u that generate output bits $c_{p,j}, \ldots, c_{p+s(N)-1,j}$ (at most $b - s(N)$ bits, by Claim B). The length of this encoding is $(2N^2 - b) + O(l \cdot \log n) + (b - s(N)) = 2N^2 + O(l \cdot \log n) - s(N)$. Since $s(N) = \omega(\log N)$, $\log n \approx 2 \log N$, and l is a constant, this encoding is shorter than $2N^2 - c \log N^2$. Contradiction. ∎ Claim A

Let $\beta(n) = \omega(\log^2 n)$, $\gamma(n) = \beta(n)/\log^2(n)$. We take $s(N) = \gamma^{1/3}(n) \log n$ and $k(N) = N/(\gamma^{2/3}(n) \log N)$. These functions satisfy conditions of Claim A, so we get that $N^3/\beta(n)$ messages are necessary. This concludes the proof of Theorem 5.

References

1. P. Beame, M. Tompa, P. Yan, *Communication-Space Tradeoffs for Unrestricted Protocols*, SICOMP 23 (1994), 652–661.
2. A. Borodin, S. Cook, *A Time-Space Tradeoff for Sorting on a General Sequential Model of Computation*, SICOMP 11 (1982), 287–297.
3. A.O. Buda, *Multiprocessor Automata*, IPL 25 (1987), 157–161.
4. M. Dietzfelbinger, *The Linear-Array Problem in Communication Complexity Resolved*, in Proc. STOC'97, 373–382.

5. P. Duriš, Z. Galil, *A Time-Space Tradeoff for Language Recognition*, MST 17 (1984), 3–12.
6. P. Duriš, T. Jurdziński, M. Kutyłowski, K. Loryś, *Power of Cooperation and Multihead Finite Systems*, in Proc. ICALP'98, 896–907.
7. M. Karchmer, *Two Time-Space Tradeoffs for Element Distinctness*, TCS 47 (1986), 237–246.
8. E. Kushilevitz, N. Nisan, *Communication Complexity*, Cambridge University Press, 1997.
9. T. Lam, P. Tiwari, M. Tompa, *Trade-offs between Communication and Space*, JCSS 45 (1992), 296–315.
10. M. Li, P. Vitanyi, *An Introduction to Kolmogorov Complexity and its Applications*, Springer-Verlag, 1993.
11. Y. Yesha, *Time-Space Tradeoffs for Matrix Multiplication and the Discrete Fourier Transform of Any General Sequential Random-Access Computer*, JCSS 29 (1984) 183–197.

Probabilistic Local Majority Voting for the Agreement Problem on Finite Graphs [*]

Toshio Nakata[1], Hiroshi Imahayashi[2], and Masafumi Yamashita[2]

[1] Department of Information Education, Fukuoka University of Education,
Akama Munakata Fukuoka, 811-4192, JAPAN.
nakata@fukuoka-edu.ac.jp
[2] Department of Computer Science and Communication Engineering,
Kyushu University,
Hakozaki Higashi-ku Fukuoka 812-8581, JAPAN.
{imahayas, mak}@csce.kyushu-u.ac.jp

Abstract. Motivated by the study of the deterministic local majority polling game by Peleg et al., this paper investigates a repetitive probabilistic local majority polling game on a finite connected graph by formulating it as a Markov chain. We mainly investigate the probability that the system reaches a given absorbing state and characterize when the probability attains the maximum (and minimum).

Keywords : distributed computing, local majority voting, graph theory, agreement and consensus problem, Markov chain.

1 Introduction

Taking the opinion that a majority of the members support is the most natural way to reach an agreement. If every member can observe the opinion of each member, (s)he can trivially confirm the majority opinion. What may happen if (s)he can observe only the opinions of her/his neighbors?

A deterministic local majority voting system was studied by a lot of researchers, mainly motivated by agreement problems in distributed systems (see e.g., [1], [4] and [5]). Especially Peleg [5] surveys recent developments concerning the process of local majority voting. These papers basically investigate how many members are necessary and sufficient (for all members) to agree on the opinion they support.

They naturally model the system by a finite connected graph $G = (V, E)$: A subset M of V is called a *monopoly*, if for any $v \in V$, members in M form a majority of the vertices adjacent to v (including v itself). Then how small can M be? Linial et al. [4] discussed the problem as packing and covering problems of graphs. They showed that M is $\Omega(\sqrt{n})$ and gave a graph with M of size $O(\sqrt{n})$, where $n = |V|$. Bermond and Peleg [1] studied some of it modifications, r-monopoly and self-ignoring monopoly.

---

[*] Dedicated to Professor Izumi Kubo on occasion of his 60th birthday

T. Asano et al. (Eds.): COCOON'99, LNCS 1627, pp. 330–338, 1999.

Peleg [6] also discussed a repetitive deterministic majority polling game on a finite graph: intuitively all members synchronously change their opinions to their local majority opinion, and repeat this process. Clearly the finiteness implies that the dynamical system is eventually periodic or stationary, and of course there are graphs and initial configurations such that the game will not end up by all vertices agreeing on a opinion.

We are interested in such a (distributed) algorithm for each vertex, who can observe the opinions of its neighbors, that all vertices obeying the algorithm eventually agree on the opinion that is supported by a majority of the vertices initially. As we saw in above, the deterministic repetitive local majority polling game is not powerful enough for agreeing on the global majority opinion, let alone a one-shot deterministic local majority polling.

In this paper, we study a repetitive probabilistic majority polling game as the basis for designing randomized agreement algorithms, which was suggested in [5, 8.1.5 Other variants]. To this end, we formalize the system using a Markov chain with absorbing states. In our setting, vertices have a value either 0 or 1. Initially 0 is the global majority value; hence we wish that all vertices eventually agree on 0.

We first observe that all vertices eventually agree on a state which is either $\mathbf{0} = (0, \cdots, 0)$ or $\mathbf{1} = (1, \cdots, 1)$ with probability 1 (Theorem 1). Sometimes they agree on incorrect state $\mathbf{1}$. Theorem 2 gives the probability that the system ends up with undesirable decision. By virtue of this theorem, we can easily calculate the probability. It is a natural guess that the state of some vertices contributes to the eventual decision more than others. We also investigate which vertices are important in the above sense.

2   The Model

Let $G(V, E)$ be a finite connected graph with order $|V| = n$. We assign to each vertex $v \in V$, a value $\xi(v) \in \{0, 1\}$. A configuration (with respect to G) is the set of values the vertices have and is denoted by $\xi = (\xi(v_1), \cdots, \xi(v_n)) \in \Xi = \{0, 1\}^V$. Let $\Gamma(v) = \{v\} \cup \{u \in V : \{u, v\} \in E\}$ denotes the neighbors of v, including v itself. Then $\Gamma(v)$ is the set of vertices whose values affect the local majority polling carried by v. The number of vertices having value $\epsilon \in \{0, 1\}$ in ξ is denoted by $N_\xi(v, \epsilon)$, i.e., $N_\xi(v, \epsilon) = |\{w \in \Gamma(v) : \xi(w) = \epsilon\}|$ and

$$N_\xi(v, 0) + N_\xi(v, 1) = |\Gamma(v)| \quad \text{for } v \in V. \tag{1}$$

Now we formalize a natural probabilistic model of local majority voting: For $\xi \in \Xi$ and $v \in V$, the probability $q_\xi(v, \epsilon)$ that v changes its value from $\xi(v)$ to ϵ is defined as

$$q_\xi(v, \epsilon) = \frac{N_\xi(v, \epsilon)}{|\Gamma(v)|} \quad \text{for } \epsilon \in \{0, 1\}.$$

Example 1. Consider a graph with vertex set $V = \{v_1, v_2, v_3\}$ shown in Figure 1. Currently, value 0 is assigned to vertices v_1 and v_2, while v_3 has value 1. This configuration is denoted by $\xi = (\xi(v_1), \xi(v_2), \xi(v_3)) = (0, 0, 1)$. Then

$$q_\xi(v_1, 0) = \frac{2}{2}, \quad q_\xi(v_2, 0) = \frac{2}{3}, \quad q_\xi(v_3, 0) = \frac{1}{2},$$

$$q_\xi(v_1, 1) = \frac{0}{2}, \quad q_\xi(v_2, 1) = \frac{1}{3}, \quad q_\xi(v_3, 1) = \frac{1}{2}.$$

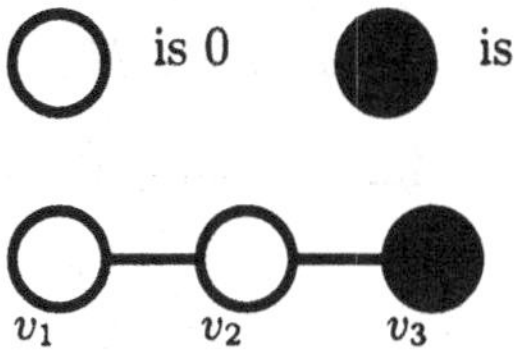

Fig. 1. Illustration of the graph used in Example 1.

We finally assume that the update actions by vertices are independent, so that the transition probability $p(\xi, \eta)$ from a configuration ξ to a configuration η is calculated by

$$p(\xi, \eta) = \prod_{v \in V} q_\xi(v, \eta(v)).$$

In the next section, we investigate this probabilistic model.

3   Results

We first observe that $p(\xi, \eta)$ is a transition probability in the sense of a Markov chain.

Lemma 1.

$$p(\xi, \eta) \geq 0 \quad \text{and} \quad \sum_{\eta \in \Xi} p(\xi, \eta) = 1 \quad \text{for } \xi, \eta \in \Xi.$$

Proof. Since $q_\xi(v, \eta(v)) \geq 0$ for any $v \in V$, $p(\xi, \eta) = \prod_{v \in V} q_\xi(v, \eta(v)) \geq 0$ holds. By (1), we obtain the following:

$$\sum_{\eta \in \Xi} p(\xi, \eta) = \sum_{\eta \in \Xi} \prod_{v \in V} \frac{N_\xi(v, \eta(v))}{|\Gamma(v)|}$$

$$= \sum_{\eta(v_2) \in \{0,1\}} \cdots \sum_{\eta(v_n) \in \{0,1\}} \prod_{v \in V \setminus \{v_1\}} \frac{N_\xi(v, \eta(v))}{|\Gamma(v)|} \sum_{\eta(v_1) \in \{0,1\}} \frac{N_\xi(v_1, \eta(v_1))}{|\Gamma(v_1)|}$$

$$= \sum_{\eta(v_2)} \cdots \sum_{\eta(v_n)} \prod_{v \neq v_1} \frac{N_\xi(v, \eta(v))}{|\Gamma(v)|} = \cdots = \sum_{\eta(v_n)} \frac{N_\xi(v_n, \eta(v_n))}{|\Gamma(v_n)|} = 1. \quad \Box$$

By virtue of Lemma 1, we can introduce a finite Markov chain on Ξ with the transition probability $P = (p(\xi, \eta))_{\xi, \eta \in \Xi}$. The elements of P^k are denoted by $p^{(k)}(\xi, \eta)$ for $k \in \mathbf{N} \cup \{0\}$. Set $\mathbf{0} = (0, \cdots, 0), \mathbf{1} = (1, \cdots, 1) \in \Xi$ and $\hat{\Xi} = \Xi \setminus \{\mathbf{0}, \mathbf{1}\}$. We show the following theorem.

Theorem 1. *For an arbitrary finite connected graph $G(V, E)$, we have the following classification:*

(i) $\mathbf{0} \in \Xi$ *and* $\mathbf{1} \in \Xi$ *are absorbing states, i.e.,* $p(\mathbf{0}, \mathbf{0}) = p(\mathbf{1}, \mathbf{1}) = 1$.
(ii) *For any* $\xi \in \hat{\Xi}$ *and* $\eta \in \Xi$, *there exists* $k \in \mathbf{N} \cup \{0\}$ *such that* $p^{(k)}(\xi, \eta) > 0$.

By a general theory of the finite Markov chain, Theorem 1 implies the following claims:

CLAIM

(i) $\hat{\Xi}$ is transitive, i.e., any $\xi \in \hat{\Xi}$ is absorbed into either $\mathbf{0}$ or $\mathbf{1}$ in a finite number of steps with probability 1.
(ii) For any initial distribution π, there exists a (unique) stationary distribution $(\lim_{k \to \infty} \pi P^k)$ whose form is $(\overset{0}{p_1}, 0, \cdots, 0, \overset{1}{p_2})$, where $p_1 + p_2 = 1$, i.e., the support of its stationary distribution is $\{\mathbf{0}, \mathbf{1}\} \subset \Xi$.

Now we prepare Lemma 2 (iii) to prove Theorem 1. Items (i) and (ii) of the lemma are for proving it. In the lemma, we use the Hamming distance $d(\xi, \eta) = \sum_{v \in V} |\xi(v) - \eta(v)|$ for $\xi, \eta \in \Xi$.

Lemma 2. (i) *For any* $\xi \in \Xi$, $p(\xi, \xi) > 0$.
(ii) *Suppose that* $d(\xi, \eta) \leq 2$ *and there exist* $v \neq w \in V$ *such that*

$$\xi(v) \neq \xi(w), \quad \xi(u) = \eta(u) \quad \text{for } w \in \Gamma(v), \ u \in V \setminus \{v, w\},$$

then $p(\xi, \eta) > 0$.
(iii) *If* $\xi \in \hat{\Xi}$ *and* $\eta \in \Xi$ *satisfy* $d(\xi, \eta) = 1$, *then* ξ *is reachable to* η, *i.e., there exists* $k \in \mathbf{N} \cup \{0\}$ *such that* $p^{(k)}(\xi, \eta) > 0$.

Proof. (i) By definition, $N_\xi(v, \xi(v)) \geq 1$ for any $v \in V$. So $q_\xi(v, \xi(v)) > 0$ for any $v \in V$. Therefore $p(\xi, \xi) = \prod_{v \in V} q_\xi(v, \xi(v)) > 0$ for $\xi \in \Xi$.

(ii) If $d(\xi, \eta) = 0$, then we have already proved in the above. Suppose that $d(\xi, \eta) = 1$. Without loss of generality we can assume that

$$\xi(v) \neq \eta(v) \tag{2}$$

and $\xi(u) = \eta(u)$ for $u \in V \setminus \{v\}$. By the same argument as in Lemma 2 (i), we have that $q_\xi(u, \eta(u)) > 0$ for $u \in V \setminus \{v\}$. By the assumption $\xi(v) \neq \xi(w)$ for $w \in \Gamma(v)$, we have that $\xi(w) = \eta(v)$ by (2). Since $N_\xi(v, \eta(v)) = N_\xi(v, \xi(w)) \geq 1$, we obtain that $q_\xi(v, \eta(v)) > 0$. So we deduce that $p(\xi, \eta) > 0$. We can also prove the case of $d(\xi, \eta) = 2$ in the same method.

(iii) By assumption, suppose that $\xi(v) \neq \eta(v)$, $\xi(v') \neq \eta(v')$ for $v' \in V \setminus \{v\}$. Without loss of generality we can assume that $1 = \xi(v) \neq \eta(v) = 0$. If there exists $w \in \Gamma(v)$ such that $\xi(w) = 0$, then we have already proved in the case of

$d(\xi, \eta) = 1$ of Lemma 2 (ii). So we assume that $\xi(w) = 1$ for any $w \in \Gamma(v)$. Since $\xi \neq 1$, there exists $u \in V \setminus \Gamma(v)$ such that $\xi(u) = 0$. By the connectivity of the graph, there exists a path from v to u. Let $r \in V \setminus \{v\}$ be the intersectant vertex of the path and $\Gamma(v)$. Define the state $\xi' \in \Xi$ which satisfies that $0 = \xi'(r) \neq \xi(r) = 1$, $\xi'(s) = \xi(s)$ for any $s \in V \setminus \{r\}$. Then we obtain that $p(\xi', \eta) > 0$. In fact, since $0 = \eta(v) \neq \xi(v) = \xi'(v) = 1$, $1 = \eta(r) \neq \xi'(r) = 0$, and $\eta(s) = \xi'(s)$ for any $s \in V \setminus \{v, r\}$, we can apply the case of $d(\xi, \eta) = 2$ of Lemma 2 (ii) to it. So it is sufficient to show that ξ is reachable to ξ'. In the path from v to u, let $u' \in V$ be the first vertex which satisfies $\xi(u') = 0$. Put $u' = p_0, p_1, \cdots, p_q = r$ the vertices in the path from u' to r respectively. Then we have that $\xi(p_k) = 1$ for $k = 1, \cdots, q$. Now we define $\zeta^1 \in \Xi$ such that $0 = \zeta^1(p_1) \neq \xi(p_1) = 1$ and $\zeta^1(s) = \xi(s)$ for $s \in V \setminus \{p_1\}$. Moreover for $k = 2, \cdots, q$ define $\zeta^k \in \Xi$ such that $\zeta^k(p_{k-1}) = 1, \zeta^k(p_k) = 0$ and $\zeta^k(s) = \zeta^{k-1}(s)$ for $s \in V \setminus \{p_{k-1}, p_k\}$. Then we have that $\xi' = \zeta^{p_q}$. Using Lemma 2 (ii), we obtain that $p(\xi, \zeta^1) > 0$, $p(\zeta^k, \zeta^{k+1}) > 0$ for $k = 1, \cdots, q - 1$. Therefore we obtain that

$$p^{(q)}(\xi, \xi') \geq p(\xi, \zeta^1) \prod_{k=1}^{q-1} p(\zeta^k, \zeta^{k+1}) > 0.$$

This completes the proof of Lemma 2.   □

Proof of Theorem 1. The first item (i) is trivial, so we concentrate on the second one. Suppose that $d(\xi, \eta) = l$ for $\xi \in \hat{\Xi}$ and $\eta \in \Xi$. The case $l = 0$ is trivial by definition. The case $l = 1$ is by Lemma 2. Finally if $l \geq 2$, then there exists a $\xi' \in \hat{\Xi}$ such that

$$d(\xi, \xi') = 1, \quad d(\xi', \eta) = l - 1. \tag{3}$$

By Lemma 2 (iii), ξ is reachable to ξ'. Using this method inductively, we can prove it by the second case.   □

We next consider the absorbing probability that starting from a given initial configuration $\xi \in \hat{\Xi}$, the system is absorbed into a given absorbing configuration (i.e., $\mathbf{0}$ or $\mathbf{1}$). Now we claim the following lemma by a general theory of the Markov chain.

Lemma 3. *Let p_ξ be an absorbing probability into $\mathbf{0}$ starting at the initial configuration ξ. Then we obtain the following:*

$$P\mathbf{p} = \tilde{\mathbf{p}}, \tag{4}$$

where $P = (p(\xi, \eta))_{\xi, \eta \in \Xi}$, $\mathbf{p} = (p_\eta)_{\eta \in \Xi}$, $\tilde{\mathbf{p}} = (\tilde{p}_\xi)_{\xi \in \Xi}$ and

$$\tilde{p}_\xi = \begin{cases} 1 & \text{if } \xi = \mathbf{0}, \\ 0 & \text{if } \xi = \mathbf{1}, \\ p_\xi & \text{otherwise.} \end{cases}$$

Conversely there exists a unique $\mathbf{p}$ which satisfies (4).

Standard textbooks on the probability theory contain a similar statement, see e.g., Feller [2, P.403 Theorem 2]. We can prove Lemma 3 in the same way.

Theorem 2. *The absorbing probability p_ξ to $\mathbf{0}$ with the initial configuration ξ is the following:*

$$p_\xi = \frac{\sum_{v\in V, \xi(v)=0} |\Gamma(v)|}{\sum_{w\in V} |\Gamma(w)|} \left(= \frac{\sum_{v\in V, \xi(v)=0} |\Gamma(v)|}{2|E| + |V|} \right). \tag{5}$$

Moreover for any initial distribution $\pi = (\pi_\xi)_{\xi\in\Xi}$, the absorbing probability p_π to $\mathbf{0}$ is the mean of (5) with respect to π, that is,

$$p_\pi = \sum_{\xi\in\Xi} \pi_\xi p_\xi, \quad where \ \pi_\xi \geq 0, \ \sum_{\xi\in\Xi} \pi_\xi = 1. \tag{6}$$

Proof. By definition it is clear that

$$\sum_{w\in V} |\Gamma(w)| = 2|E| + |V|.$$

To prove (5), we will check the following equation by Lemma 2:

$$\sum_{\eta\in\Xi} p(\xi, \eta)p_\eta = \tilde{p}_\xi \quad for \ \xi \in \Xi.$$

If ξ is $\mathbf{0}$ or $\mathbf{1}$, then it is trivial. So it is sufficient to prove that

$$\sum_{\eta\in\Xi} \left(\prod_{v\in V} \frac{N_\xi(v, \eta(v))}{|\Gamma(v)|} \right) \left(\sum_{\eta(w)=0} |\Gamma(w)| \right) = \sum_{\xi(v)=0} |\Gamma(v)| \quad for \ \xi \in \hat{\Xi}. \tag{7}$$

Let (LHS) be the left hand side of (7), then by (1) we deduce that

$$\begin{aligned}
(\text{LHS}) &= \sum_{\eta(v_1)\in\{0,1\}} \cdots \sum_{\eta(v_n)\in\{0,1\}} \prod_{v\in V} \frac{N_\xi(v, \eta(v))}{|\Gamma(v)|} \sum_{\eta(w)=0} |\Gamma(w)| \\
&= \frac{N_\xi(v_1, 1)}{|\Gamma(v_1)|} \sum_{\eta(v_2)} \cdots \sum_{\eta(v_n)} \prod_{v\neq v_1} \frac{N_\xi(v, \eta(v))}{|\Gamma(v)|} \sum_{w\neq v_1, \eta(w)=0} |\Gamma(w)| \\
&\quad + \frac{N_\xi(v_1, 0)}{|\Gamma(v_1)|} \sum_{\eta(v_2)} \cdots \sum_{\eta(v_n)} \prod_{v\neq v_1} \frac{N_\xi(v, \eta(v))}{|\Gamma(v)|} \left(\sum_{w\neq v_1, \eta(w)=0} |\Gamma(w)| + |\Gamma(v_1)| \right) \\
&= \sum_{\eta(v_2)\in\{0,1\}} \cdots \sum_{\eta(v_n)\in\{0,1\}} \prod_{v\neq v_1} \frac{N_\xi(v, \eta(v))}{|\Gamma(v)|} \sum_{w\neq v_1, \eta(w)=0} |\Gamma(w)| \\
&\quad + N_\xi(0, v_1) \sum_{\eta(v_2)\in\{0,1\}} \cdots \sum_{\eta(v_n)\in\{0,1\}} \prod_{v\neq v_1} \frac{N_\xi(v, \eta(v))}{|\Gamma(v)|}.
\end{aligned}$$

Using the calculation of Lemma 1, we obtain that

$$\begin{aligned}
(\text{LHS}) &= \sum_{\eta(v_2)\in\{0,1\}} \cdots \sum_{\eta(v_n)\in\{0,1\}} \prod_{v\neq v_1} \frac{N_\xi(v, \eta(v))}{|\Gamma(v)|} \sum_{w\neq v_1, \eta(w)=0} |\Gamma(w)| + N_\xi(0, v_1) \\
&= \cdots = \sum_{v\in V} N_\xi(0, v) = \sum_{\xi(v)=0} |\Gamma(v)|.
\end{aligned}$$

By (5), for the Dirac distribution $\delta_\xi = (0, \cdots, 0, \overset{\xi}{1}, 0, \cdots, 0)$ whose support is only $\{\xi\} \subset \Xi$, there exists a (unique) $\theta_\xi = (\overset{0}{p_\xi}, 0, \cdots, 0, \overset{1}{p'_\xi})$ which satisfies that

$$\lim_{k \to \infty} \delta_\xi P^k = \theta_\xi, \quad \theta_\xi P = \theta_\xi, \quad p_\xi + p'_\xi = 1.$$

For any initial distribution $\pi = (\pi_\xi)_{\xi \in \Xi}$, we have that $\pi = \sum_{\xi \in \Xi} \pi_\xi \delta_\xi$. So we obtain that

$$\lim_{k \to \infty} \pi P^k = \lim_{k \to \infty} \sum_{\xi \in \Xi} \pi_\xi \delta_\xi P^k = \sum_{\xi \in \Xi} \pi_\xi \theta_\xi.$$

This completes the proof of Theorem 2. □

Example 2. By virtue of Theorem 2, we can easily calculate the absorbing probability starting at the configuration of the left graph into **0**, that is,

$$\frac{3+3+2+6}{3+3+2+6+3+3} = \frac{14}{20} = \frac{7}{10}.$$

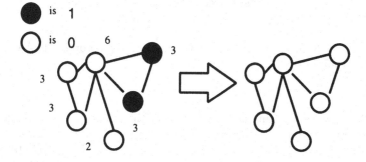

For $V = \{v_1, \cdots, v_n\}$, without loss of generality we can assume that

$$\deg(v_1) \leq \cdots \leq \deg(v_n). \tag{8}$$

For $m = 0, \cdots, n$, set

$$\Xi_m = \{\xi = (\xi(v_1), \cdots, \xi(v_n)) \in \Xi : |\{v \in V : \xi(v) = 1\}| = m\}.$$

Before closing this section, we would like to present three useful corollaries.

Corollary 1. *For $m = 0, \cdots, n$, the set of configurations in Ξ_m that maximize the absorbing probability p_ξ is $\{\xi \in \Xi_m : \xi(v_1) = \cdots = \xi(v_m) = 1, \xi(v_{m+1}) = \cdots = \xi(v_n) = 0\}$.*

Proof. Suppose that there exist $1 \leq i < j \leq n$ such that $\xi(v_i) = 0$, $\xi(v_j) = 1$ and $\deg(v_i) < \deg(v_j)$. Now put $\xi' \in \Xi_m$ such that $\xi'(v_i) = 1$, $\xi'(v_j) = 0$ and $\xi'(w) = \xi(w)$ for $v_i, v_j \neq w \in V$. Then we have that $p_\xi < p_{\xi'}$. So ξ does not attain the maximum. Conversely if $\xi \in \Xi_m$ satisfies that $\xi(v_1) = \cdots = \xi(v_m) = 1$ and $\xi(v_{m+1}) = \cdots = \xi(v_n) = 0$ then clearly ξ attain the maximum by (8). □

Corollary 2. *For any finite connected graph $G(V, E)$ and configuration $\xi \in \Xi_1$, the absorbing probability $p_\xi(G(V, E))$ to $\mathbf{0}$ satisfies*

$$\max_{|V|=n, \xi \in \Xi_1} p_\xi(G(V, E)) = 1 - \frac{2}{n^2 - 2n + 4}, \quad \min_{|V|=n, \xi \in \Xi_1} p_\xi(G(V, E)) = \frac{2n - 2}{3n - 2}.$$

The maximum and minimum are respectively achieved by the following configurations (on graphs).

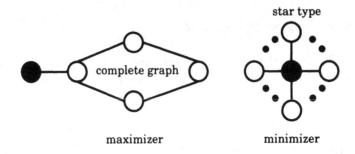

maximizer   minimizer

Fig. 2. The graph and configuration that maximizes the absorbing probability to $\mathbf{0}$ and those for minimizing it.

Proof. We show the maximum case. It is clear that there exists a maximizer by the finiteness of graphs. Fix G and $\xi \in \Xi_1$ with $\xi(v) = 1, \xi(w) = 0$ for $w \in V \backslash \{v\}$ which satisfies that $\deg(v) = k$ for $k = 1, \cdots, n - 1$ and $\sum_{w \in V} |\Gamma(w)| = L$ for $2n \leq L \leq n^2$. Let $G' = (V', E')$ be a graph which satisfies that $V' = V$ and $E \neq E' = E \cup \{w_1, w_2\}$ for $w_1, w_2 \in V \backslash \{v\}$. Then we have that $p_\xi(G) = \frac{L-(k+1)}{L} < \frac{L-(k+1)+2}{L+2} = p_\xi(G')$. Therefore if we would like $p_\xi(G)$ to be large, $G \backslash \{v\}$ should be a complete graph for each $k = \deg(v)$. Now suppose that $G \backslash \{v\}$ is complete. Then we have that $p_\xi(G) = \frac{(n-1)^2+k}{(n-1)^2+2k+1}$. It is clear that $\frac{(n-1)^2+k}{(n-1)^2+2k+1} \leq \frac{(n-1)^2+1}{(n-1)^2+3}$ for $k = 1, \cdots, n - 1$. Hence we obtain the desired result. We can prove the minimum case in the same manner described above. $\square$

For any finite connected graph if π is the uniform distribution on Ξ, then $p_\pi = 1/2$ by Theorem 2.

Corollary 3. *For any finite connected graph $G(V, E)$ with $|V| = n$, if π is the uniform distribution on Ξ_1, i.e.,*

$$\pi = (0, \cdots, 0, \overbrace{\frac{1}{n}, \cdots, \frac{1}{n}}^{\Xi_1}, 0, \cdots, 0)$$

then $p_\pi = 1 - 1/n$.

Proof. It is clear that $|\Xi_1| = n$. By (6), we obtain the following:

$$1 - p_\pi = 1 - \sum_{\xi \in \Xi_1} \frac{1}{n} \pi_\xi = \frac{1}{n} \sum_{\xi \in \Xi_1} (1 - \pi_\xi) = \frac{1}{n} \sum_{\xi \in \Xi_{n-1}} \pi_\xi$$

$$= \frac{1}{n} \sum_{\xi \in \Xi_{n-1}} \frac{\sum_{w \in V, \xi(w)=0} |\Gamma(w)|}{\sum_{v \in V} |\Gamma(v)|} = \frac{1}{n} \frac{\sum_{w \in V} |\Gamma(w)|}{\sum_{v \in V} |\Gamma(v)|} = \frac{1}{n}. \quad \square$$

4   Conclusion

This paper investigated a repetitive probabilistic local majority polling game on a finite connected graph by formulating it as a Markov chain. From the view of all members of a (distributed) system agreeing on the correct opinion that is supported by a majority in the system, the repetitive deterministic local majority polling discussed by Peleg et al. is not powerful enough. This motivated our study. We mainly investigated the probability that the system ends up with the correct agreement reflecting the global majority opinion.

Acknowledgments. The authors would like to thank three anonymous referees for their helpful suggestions. In particular, one of them kindly showed us a similar work independently done by Hassin and Peleg [3]. This work was supported in part by a Scientific Research Grant-in-Aid from the Ministry of Education, Science, and Culture of Japan and Research for the Future Program from Japan Society for the Promotion of Science: Software for Distributed and Parallel Supercomputing.

References

1. J-C. Bermond and D. Peleg, "The power of small coalitions in graphs," *Proc. 2nd Structural Information & Communication Complexity,* Olympia, Greece, Carleton Univ. Press, 173-184, 1995.
2. W. Feller, "An Introduction to Probability Theory and its Applications Vol. I (3rd ed.), " New York: John Wiley & Sons, 1950.
3. Y. Hassin and D. Peleg, "Distributed probabilistic polling and applications to proportionate agreement," *Proc. 26th International Colloquium on Automata, Languages, and Programming,* Prague, Czech Republic, July 1999 (to appear).
4. N. Linial, D. Peleg, Y. Rabinovich and M. Saks, "Sphere packing and local majorities in graphs," *Proc. 2nd Israel Symposium on Theoretical Computer Science,* IEEE Computer Soc. Press, 141-149, 1993.
5. D. Peleg, "Local majority voting, small coalitions and controlling monopolies in graphs: A review," Technical Report, Weizmann Institute, 1996.
   http://www.wisdom.weizmann.ac.il/Papers/trs/CS96-12/abstract.html
6. D. Peleg, "Size bounds for dynamic monopolies," *Discrete Applied Mathematics,* 86, 263-273, 1998.

A Dynamic-Programming Bound for the Quadratic Assignment Problem

Ambros Marzetta[1] and Adrian Brüngger[2]

[1] International Computer Science Institute
Berkeley
U.S.A.
ambros.marzetta@iaeth.ch

[2] Novartis Pharma AG
Basel
Switzerland
adrian.bruengger@pharma.novartis.com

Abstract. The quadratic assignment problem (QAP) is the NP-complete optimization problem of assigning n facilities to n locations while minimizing certain costs. In practice, proving the optimality of a solution is hard even for moderate problem sizes with $n \approx 20$.

We present a new algorithm for solving the QAP. Based on the dynamic-programming paradigm, the algorithm constructs a table of subproblem solutions and uses the solutions of the smaller subproblems for bounding the larger ones. The algorithm can be parallelized, performs well in practice, and has solved the previously unsolved instance NUG25. A comparison between the new dynamic-programming bound (DPB) and the traditionally used Gilmore–Lawler bound (GLB) shows that the DPB is stronger and leads to much smaller search trees than the GLB.

1 The Quadratic Assignment Problem

The quadratic assignment problem (QAP) [9,5] is the problem of assigning n facilities to n locations such that the total cost of the assignment is minimized. The total cost consists of a linear contribution from the assignment of the facilities (e.g., the cost of establishing each facility in the chosen location) and a quadratic contribution from the interaction within each pair of facilities depending on their locations (e.g., communication or transport costs).

We denote the set of facilities by $\mathcal{F}$, the set of locations by $\mathcal{L}$, and the size of both sets by $|\mathcal{F}| = |\mathcal{L}| = n$. A feasible solution corresponds to a permutation π that assigns each facility $i \in \mathcal{F}$ to a location $\pi(i) \in \mathcal{L}$. Let the linear costs be given by an $n \times n$ matrix C with c_{ik} equal to the cost of assigning facility $i \in \mathcal{F}$ to location $k \in \mathcal{L}$. The quadratic costs (in the Koopmans–Beckmann version of the problem) can be broken into a flow component F and a distance component D. For each pair of facilities $i, j \in \mathcal{F}$ a flow $f_{i,j}$ is known, and for each pair of locations $k, l \in \mathcal{L}$ the distance $d_{k,l}$ is known. The cost between facilities i and j is then $f_{i,j} \cdot d_{\pi(i),\pi(j)}$.

T. Asano et al. (Eds.): COCOON'99, LNCS 1627, pp. 339–348, 1999.
© Springer-Verlag Berlin Heidelberg 1999

We can now state the quadratic assignment problem as

$$QAP(F, D, C) = \min_{\pi: \mathcal{F} \to \mathcal{L}} \left(\sum_{i,j \in \mathcal{F}} f_{i,j} \cdot d_{\pi(i), \pi(j)} + \sum_{i \in \mathcal{F}} c_{i, \pi(i)} \right) .$$

The QAP is NP-complete, since the traveling salesman problem and the maximum clique problem are special cases of it [7]. Moreover, it is hard in a practical sense because the solution of instances with $n \geq 20$ usually consumes huge amounts of computing time, whereas traveling salesman problems with thousands of cities have been solved exactly. Although excellent (often optimal) solutions for the QAP are easy to find by heuristic (local search) methods, proving optimality is hard.

Related to the QAP is the linear assignment problem (LAP)

$$LAP(C) = \min_{\pi} \sum_{i \in \mathcal{F}} c_{i, \pi(i)} ,$$

which can be solved in $O(n^3)$ time. It often arises in the computation of lower bounds for the QAP.

Various applications of the QAP, such as circuit board layout or keyboard design, are described in [9]. Often, the distance matrix exhibits a regular structure, the locations being the vertices of a hypercube or of a rectangular grid.

In the last few years, researchers have used the library of QAP instances QAPLIB [4] to measure and compare the performance of their algorithms. The Nugent series is the most popular because its instances cover a range of complexities where progress is easily measurable. Prior to this work, the largest Nugent instances that could be solved were NUG22 [2] and NUG24. In this paper, we present an algorithm that solves NUG25. The algorithm is based on branch-and-bound, but uses a novel bounding function which is evaluated efficiently in a dynamic-programming approach.

2   Branch and Bound for the QAP

Algorithms for the exact solution of the QAP are usually based on the branch-and-bound paradigm. A branch-and-bound algorithm recursively decomposes a problem into subproblems by partitioning the set of feasible solutions into disjoint subsets. These subproblems are the nodes of a search tree. Any feasible solution of a subproblem is a feasible solution of the original problem, and the optimal solution of the original problem is the optimal solution of some subproblem. The two major components of a branch-and-bound algorithm are a *branching rule* which defines the decomposition and a *bounding function* which prunes the search tree by identifying subtrees that do not contain the optimal solution.

There are two main evaluation orders for the nodes of a branch-and-bound tree. *Depth-first* evaluation uses little memory, but evaluates more nodes than necessary unless a good initial solution is known. *Best-first* evaluation stores subproblems in a (potentially exponentially large) priority queue and evaluates the minimum number of subproblems.

We now define the subproblems of the QAP given by the set of facilities $\mathcal{F}$, the set of locations $\mathcal{L}$, the flow matrix F, the distance matrix D, and the linear-cost matrix C. Given any subsets $\mathcal{F}^0 \subseteq \mathcal{F}$ and $\mathcal{L}^0 \subseteq \mathcal{L}$ of equal size and an assignment π^0 of the facilities in $\mathcal{F}^0$ to the locations in $\mathcal{L}^0$, the corresponding subproblem consists of minimizing the objective function over the assignments π with $\pi(i) = \pi^0(i)$ for all $i \in \mathcal{F}^0$. We will show that this subproblem is again a quadratic assignment problem $\mathrm{QAP}(F', D', C')$ (Fig. 1). Defining the set of unassigned facilities $\mathcal{F}' := \mathcal{F} \backslash \mathcal{F}^0$ and the set of unassigned locations $\mathcal{L}' := \mathcal{L} \backslash \mathcal{L}^0$, we can expand the objective function:

$$
\min_{\pi : \mathcal{F} \to \mathcal{L}, \pi | \mathcal{F}^0 = \pi^0} \left(\sum_{i,j \in \mathcal{F}} f_{i,j} \cdot d_{\pi(i),\pi(j)} + \sum_{i \in \mathcal{F}} c_{i,\pi(i)} \right)
$$

$$
= \min_{\pi : \mathcal{F}' \to \mathcal{L}'} \left(\sum_{i,j \in \mathcal{F}'} f_{i,j} \cdot d_{\pi(i),\pi(j)} \right.
$$

$$
+ \sum_{i \in \mathcal{F}', j \in \mathcal{F}^0} f_{i,j} \cdot d_{\pi(i),\pi^0(j)}
$$

$$
+ \sum_{i \in \mathcal{F}^0, j \in \mathcal{F}'} f_{i,j} \cdot d_{\pi^0(i),\pi(j)}
$$

$$
+ \sum_{i,j \in \mathcal{F}^0} f_{i,j} \cdot d_{\pi^0(i),\pi^0(j)}
$$

$$
+ \sum_{i \in \mathcal{F}'} c_{i,\pi(i)}
$$

$$
\left. + \sum_{i \in \mathcal{F}^0} c_{i,\pi^0(i)} \right)
$$

$$
= \mathrm{QAP}(F', D', C') + const
$$

Thus, the subproblem is a smaller-sized QAP over $\mathcal{F}'$ and $\mathcal{L}'$, where F', D' and C' are defined by

$$
\begin{aligned}
f'_{i,j} &= f_{i,j} && \text{for } i, j \in \mathcal{F}' \\
d'_{k,l} &= d_{k,l} && \text{for } k, l \in \mathcal{L}' \\
c'_{i,k} &= c_{i,k} + \sum_{j \in \mathcal{F}^0} f_{i,j} \cdot d_{k,\pi^0(j)} + f_{j,i} \cdot d_{\pi^0(j),k} && \text{for } i \in \mathcal{F}', k \in \mathcal{L}'
\end{aligned}
$$

Notice that the matrices F' and D' depend only on the sets $\mathcal{F}'$ and $\mathcal{L}'$, but not on the fixed assignment π^0. This means that many subproblems in the search tree differ from one another only in C' rather than F' and D', a fact that we will exploit.

Essential for the performance of a branch-and-bound algorithm is the bounding function, which computes a lower bound for the optimal solution by solving a

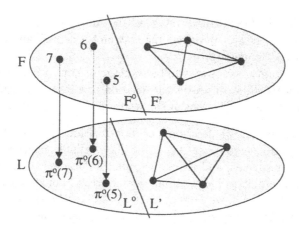

Fig. 1. A subproblem of a QAP. The size of the original QAP is $|\mathcal{F}| = |\mathcal{L}| = n = 7$. When the assignment π^0 of the three facilities in $\mathcal{F}^0 = \{5, 6, 7\}$ to the three locations in $\mathcal{L}^0$ is fixed, a smaller QAP of size $|\mathcal{F}'| = |\mathcal{L}'| = n' = 4$ remains. The new flow matrix F' and the new distance matrix D' depend only on $\mathcal{F}'$ and $\mathcal{L}'$, respectively, whereas the new linear cost matrix C' depends on both of them as well as on the permutation π^0.

relaxed problem. The stronger the bound is, the larger the subtrees pruned during the search. Since the bounding function is called for every node of the search tree, the total computation time depends on the time used to compute a single bound as well as on the size of the search tree. It follows that there is a trade-off between quickly computed but weak bounds and strong but slow bounds. For small instances, the Gilmore–Lawler bound (GLB) leads to the shortest computations. Although it is a weak bound, it is computed efficiently by solving a linear assignment problem. Extrapolation of experiments suggests that other bounds are asymptotically better [6], but their superiority applies only to instances that are too large to be solved today.

Looking at current programs that solve the largest QAP instances (usually by depth-first branch-and-bound and often on large parallel machines), we notice three deficiencies which suggest directions for improvement:

– Current programs take no advantage of the large memory (main memory and disk) available. This situation can be understood as an adaptation to the computers available earlier, when memory was scarce and expensive. The space–time trade-off which is ubiquitous in algorithm design suggests that algorithms using more memory might be faster.
Because the depth of the search tree is at most n, depth-first branch-and-bound search uses little memory. One might think that best-first evaluation profits from the large memory available, but for the QAP, both evaluation orders are equivalent: Because a good approximate solution is guessed easily, best-first search would evaluate the same number of subproblems as depth-first.

- Current programs take no advantage of the similarity of QAP subproblems that arises from F' and D' depending only on the sets $\mathcal{F}'$ and $\mathcal{L}'$ rather than on the fixed assignment π^0.
- Current programs cannot take full advantage of any special structure of the distance matrix. If, for instance, the locations are in a rectangular grid, only basic rotations and reflections of the whole grid are exploited.

In the following section, we describe a novel algorithm based on these observations. The algorithm gains speed by using the available memory, exploits the similarity of subproblems, and is parallelizable. It becomes even more efficient when the distance matrix has a regular structure, although such regularity is unnecessary. In Section 4, we present the results of applying the new algorithm to a large, previously unsolved QAP instance and assess the quality of the new bound.

3   A New Relaxation: The Dynamic-Programming Bound

The main contribution of this article is a new lower bound for QAP subproblems, called the *dynamic-programming bound*, which is the sum of a QAP and a LAP:

$$\mathrm{dpb}(F', D', C') := \mathrm{QAP}(F', D', 0) + \mathrm{LAP}(C')$$

$$= \min_{\pi} \sum_{i,j \in \mathcal{F}'} f_{i,j} \cdot d_{\pi(i),\pi(j)} + \min_{\pi} \sum_{i \in \mathcal{F}'} c_{i,\pi(i)}$$

$$\leq \min_{\pi} \left(\sum_{i,j \in \mathcal{F}'} f_{i,j} \cdot d_{\pi(i),\pi(j)} + \sum_{i \in \mathcal{F}'} c_{i,\pi(i)} \right)$$

$$= \mathrm{QAP}(F', D', C')$$

Whereas the second term, $\mathrm{LAP}(C')$, can be solved in polynomial time, the solution of the first term, $\mathrm{QAP}(F', D', 0)$, is as expensive as the exact solution of $\mathrm{QAP}(F', D', C')$. The bounding function, however, is applied to millions of subproblems, many of which are equivalent in the sense of having the same F' and D'. It therefore makes sense to store the values of $\mathrm{QAP}(F', D', 0)$ in a table T, so that only one QAP per subproblem equivalence class is computed.

In essence, the dynamic-programming algorithm computes a table of subproblem solutions; it progresses from small subproblems to larger ones and finally to the solution of the original problem. For every table entry, it performs a branch-and-bound search, basing the evaluation of lower bounds on values computed in earlier steps.

To minimize both space and time requirements, the table must be as small as possible and the value of every $\mathrm{QAP}(F', D', 0)$ computed must be used for bounding as many subproblems with the same F' and D' as possible. This objective and the dependence of F' and D' on $\mathcal{F}'$ and $\mathcal{L}'$ determine the branching rule to be used. Assuming that $\mathcal{F} = \{1, \ldots, n\}$, we define a branching rule so that all subproblems of size n' have the same facility subset $\mathcal{F}' = \{1, \ldots, n'\}$ and thus the same flow submatrix F'. With this branching rule, the equivalence class of a subproblem is determined by $\mathcal{L}'$ alone, so that the table has at most 2^n entries. This number is small compared to the number of subproblems evaluated when the Gilmore–Lawler bound is used. A distance matrix D containing equivalent submatrices reduces the table size even more.

We define a function s which maps each location set $\mathcal{L}'$ to its equivalence class and allocate a table entry $T[s(\mathcal{L}')]$ for every equivalence class. These definitions enable us to bring the bounding function into its final form by replacing $QAP(F', D', 0)$ by a table lookup:

$$DPB(\mathcal{L}', C') := T[s(\mathcal{L}')] + LAP(C')$$

We distinguish the bounds $DPB()$ and $dpb()$, for we can speed up the computation of the table by noting that the subproblems $QAP(F', D', 0)$ need not be solved exactly by a full branch-and-bound; it suffices to compute a lower bound for them by beginning a best-first branch-and-bound search and stopping after an arbitrary number of steps. The smallest bound of all nodes in the priority queue then is a valid lower bound on the original problem and is stored in the table. This modification of the algorithm saves execution time in exchange for making $DPB()$ a weaker bound than $dpb()$.

One can think of the original problem as if its solution was the final table entry. For this final table entry, however, we want to compute an exact solution rather than only a lower bound, and thus the algorithm performs a complete branch-and-bound search on the original problem rather than a partial one. Whereas it evaluates the subproblems best-first to find lower bounds in the priority queue, it evaluates the original problem depth-first to save memory.

The complete algorithm is shown in Table 1. Its correctness is based on the following invariants for the dynamic-programming table T:

- Every entry of T is assigned a value exactly once.
- After $T[s(\mathcal{L}')]$ has been assigned a value, $T[s(\mathcal{L}')] \leq QAP(F', D', 0)$, and thus $DPB(\mathcal{L}', C') \leq dpb(F', D', C') \leq QAP(F', D', C')$.
- When $T[s(\mathcal{L}')]$ is being computed, all $T[s(\mathcal{L}'')]$ with $|\mathcal{L}''| < |\mathcal{L}'|$ are known.
- A branch-and-bound search generates only subproblems of a smaller size than the original problem and thus reads only entries of T which have been assigned a value.

The inner FOR-loop can be parallelized easily, and the final depth-first branch-and-bound search can be parallelized with the aid of a search library [3,8]. The dynamic-programming algorithm is therefore suitable for massively parallel computation.

4   Computational Results: NUG25 Is Solved

To show the practicality of the dynamic-programming algorithm, we applied it to NUG25 [1]. The amount of time necessary to solve this instance on a sequential

```
PROCEDURE DynProg(n, F, D, C);
BEGIN
   Set all entries of the table T to undefined;
   FOR n' := 1 TO n - 1 DO
      F' := {1,...,n'};
      FOR ALL L' ⊆ L WITH |L'| = |F'| DO
         IF T[s(L')] = undefined THEN
            Find a lower bound for the subproblem QAP(F', D', 0) by (partial)
            best-first branch-and-bound using DPB as bounding function, and
            store it in T[s(L')]
         END
      END
   END;
   Solve the original problem QAP(F, D, C) by depth-first branch-and-bound
   using DPB as bounding function
END DynProg;
```

Table 1. The dynamic-programming algorithm for the QAP.

computer when using branch-and-bound with various lower bounds was recently estimated to be at least 45 years [6].

In NUG25, 25 facilities are to be assigned to 25 locations. The locations are placed on a 5×5 grid, the distance matrix D is defined by the L_1 metric (Manhattan distance), and the cost matrix C is the zero matrix. The regularity of the grid reduces the number of QAPs to be solved from $2^{25} \approx 34$ million to only four million (Fig. 2). A table of four million integers can easily be kept in main memory.

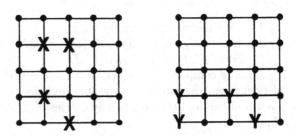

Fig. 2. Using the regularity of a distance matrix. The 25 locations of NUG25 are located in a 5×5 grid. The two four-location subsets marked by X and Y, respectively, give equal distance submatrices D' (up to row and column renumbering), and are mapped to the same equivalence class by s. The 2^{25} subsets are mapped to four million equivalence classes.

Not all four million subproblems had to be solved exactly. To speed up the computation, only the three million subproblems up to size 14 were solved optimally. For the remaining QAPs of sizes 15 to 24, a lower bound was computed by performing a best-first branch-and-bound search for a time proportional to the problem size. The final problem (the original problem of size 25) was, of course, solved exactly.

The program was implemented in parallel and run on between 64 and 128 processors of an Intel Paragon and nine DEC Alpha processors. Building the table up to size 24 took 10 days of computing time. Solving the final problem took another 20 days.

After these long computations, the question arises whether the result can be trusted. There is always the possibility of hardware faults, programming errors, or bad manipulation of data by the operator. As a consistency check, redundant computations were performed. A 10-percent random sample of QAPs of each size was taken. The exact solutions of the QAPs of size less than 15 were verified by an independently developed QAP solver. For the larger QAPs (sizes 15–25), upper bounds were computed by simulated annealing. All table entries tested were smaller than or equal to the solutions found by simulated annealing.

Quality of the Dynamic-Programming Bound

To evaluate the quality of the dynamic-programming bound, we compared it to the Gilmore–Lawler bound (GLB). First, we computed both GLB and DPB along 10000 randomly chosen paths in the search tree of NUG25 (Table 2). In the upper part of the search tree, the DPB is almost 12 percent better than the GLB. The deeper in the tree a node is, the closer both bounds become, but even at depth 13, the DPB is 2.2 percent better than the GLB. For the two million nodes that were inspected, the DPB outperformed the GLB in every single node.

The relevant question now is whether this improvement actually reduces the size of the search tree. Again, we chose 10000 random paths in the search tree of NUG25 and measured their length for both GLB and DPB (Fig. 3). A shorter path means that the path was cut off at a higher level in the tree; shorter paths therefore induce a smaller search tree. For neither of the bounding methods a path shorter than six or longer than 12 was found. The average path length for the DPB was 8.4 and thus much smaller than the average path length for the GLB, which was 10.0.

5   Conclusions

We have presented a new approach to solving quadratic assignment problems based on dynamic programming. The new lower bounding function is stronger than the Gilmore–Lawler bound and makes the solution of larger problems possible. In particular, the previously unsolved instance NUG25 was solved by the algorithm in much less time than estimated to be necessary for other algorithms. As the dynamic-programming algorithm uses a table of size $O(2^n)$, the size of the instances that it can solve is limited by the available memory rather than by computing time.

depth	GLB	DPB	difference [%]
1	2910	3250	11.7
2	2960	3271	10.5
3	3050	3304	8.3
4	3153	3381	7.2
5	3268	3480	6.5
6	3377	3561	5.4
7	3491	3663	4.9
8	3602	3749	4.1
9	3661	3801	3.8
10	3705	3832	3.4
11	3736	3844	2.9
12	3749	3833	2.2
13	3764	3847	2.2

Table 2. Comparison of dynamic-programming bound and Gilmore–Lawler bound. The nodes on 10000 random paths in the search tree were bounded by both the GLB and the DPB. The average bound for each depth is given, a higher number corresponding to a stronger bound. The rightmost column shows the relative quality difference $\frac{DPB - GLB}{GLB}$. The DPB is stronger than the GLB at all depths.

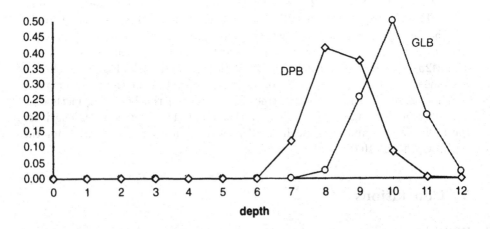

Fig. 3. Effect of the bounding function on search-tree size. To estimate how much the DPB's being stronger than the GLB affects search-tree size, the lengths of 10000 random paths through both search trees were measured. The graphs show the relative number of paths ending at a given depth. It is evident that the DPB generates shorter paths (average length 8.4 rather than 10.0) and therefore a much smaller search tree than the GLB.

Acknowledgments

We thank Jürg Nievergelt for his helpful advice and Jens Clausen for providing the QAP solver used to verify the results.

References

1. Adrian Brüngger. *Solving Hard Combinatorial Optimization Problems in Parallel: Two Case Studies.* PhD thesis, ETH Zürich, 1997.
2. Adrian Brüngger, Ambros Marzetta, Jens Clausen, and Michael Perregaard. Solving large-scale QAP problems in parallel with the search library ZRAM. *Journal of Parallel and Distributed Computing*, 50:157–169, 1998. http://wwwjn.inf.ethz.ch/ambros/zram/jpdc_qap.pdf.
3. Adrian Brüngger, Ambros Marzetta, Komei Fukuda, and Jürg Nievergelt. The parallel search bench ZRAM and its applications. *Annals of Operations Research.* to appear; http://wwwjn.inf.ethz.ch/ambros/aor_zram.ps.gz.
4. Rainer E. Burkard, Stefan E. Karisch, and Franz Rendl. QAPLIB—a quadratic assignment problem library. *Journal of Global Optimization*, 10:391–403, 1997. http://serv1.imm.dtu.dk/~sk/qaplib/.
5. Eranda Çela. *The Quadratic Assignment Problem: Theory and Algorithms.* Kluwer Academic Publishers, 1998.
6. Jens Clausen, Stefan E. Karisch, Michael Perregaard, and Franz Rendl. On the applicability of lower bounds for solving rectilinear quadratic assignment problems in parallel. *Computational Optimization and Applications*, 10:127–147, 1998. http://serv1.imm.dtu.dk/~sk/diku2496.ps.
7. Michael R. Garey and David S. Johnson. *Computers and Intractability: A Guide to the Theory of NP-Completeness.* Freeman, New York, 1979.
8. Ambros Marzetta. *ZRAM: A Library of Parallel Search Algorithms and Its Use in Enumeration and Combinatorial Optimization.* PhD thesis, ETH Zürich, 1998. http://www.inf.ethz.ch/publications/diss.html.
9. P. Pardalos and H. Wolkowicz, editors. *Quadratic Assignment and Related Problems,* volume 16 of *DIMACS Series in Discrete Mathematics and Theoretical Computer Science.* American Mathematical Society, 1993.

A New Approach for Speeding Up Enumeration Algorithms and Its Application for Matroid Bases

Takeaki Uno

Dept. Industrial Engineering and Management,
Tokyo Institute of Technology,
2-12-1 Oh-okayama, Meguro-ku, Tokyo 152, Japan. uno@me.titech.ac.jp

Abstract. We propose a new approach for speeding up enumeration algorithms. The approach does not rely on data structures deeply, instead utilizes analysis of computation time. It speeds enumeration algorithms for directed spanning trees, matroid bases, and some bipartite matching problems. We show one of these improved algorithms: one for enumerating matroid bases. For a given matroid $\mathcal{M}$ with m elements and rank n, an existing algorithm runs in $O(T)$ time per base. We improved the time complexity to $O(T/n)$. or $O(T/m(m-n))$.

1 Introduction

Enumeration algorithms for many graph and geometry objects have been developed, and one of the most important and interesting fields of related to these algorithms devoted to making them faster. Although there are numerous algorithms and numerous ways to improve enumeration algorithms, especially for spanning trees and paths, no general technique or framework for their improvements has been proposed. Most enumeration algorithms have been taking advantage of data structures that speed their iterations. So if we can not speed the iterations, we cannot obtain a fast algorithm in the usual way.

In this paper, we describe a new approach, "trimming and balancing" to making enumeration algorithms faster. It is not based on the use of data structures and it speeds enumeration algorithms without speeding iterations. It can thus be used to improve many enumeration algorithms that have not been improved in the usual way. By applying our approach, we can reduce the time complexity of the enumeration algorithm for perfect matchings from O(m) per matching to O(n), that of the enumeration algorithm for maximum matchings from O(m) per matching to O(n), and that of the enumeration algorithm for directed spanning trees from O(n) per tree to O($\log n$). Here m and n denote the number of edges and vertices of the given graph. In this paper, we also report an enumeration algorithm for matroid bases that has improved by using the trimming and balancing approach. The problem of enumerating matroid bases includes some other enumeration problems, spanning trees and maximal sets of

T. Asano et al. (Eds.): COCOON'99, LNCS 1627, pp. 349–359, 1999.

independent columns of a matrix. There are some studies on enumeration algorithms for these problems, especially spanning trees [2,1]. Although several algorithms achieve optimal time and space complexities, they require the use of data structures that are not simple. Our Trimming and Balancing algorithms for these problems attain smaller, or at least not larger time complexities without using complicated data structures.

Our approach adds two new phases to an enumeration algorithm, which we call a trimming phase and a balancing phase. An iteration of a modified algorithm may take much more computation time than that of the original algorithm, but the total time complexity of the modified algorithm is often considerably smaller than that of the original algorithm. Some our algorithms take much more computation time in an iteration than the original algorithm, but it attains smaller time complexities. In the next section, we describe the framework of trimming and balancing. A trimming and balancing algorithm surely terminate in short time, but to prove it is not so easy by using usual analysis of time complexity. Hence we also describe a technique for analyzing time complexities in the next section.

2   Framework of Trimming and Balancing

To explain the concept of trimming and balancing, we use here a simple binary partition algorithm for enumerating of all the bases in the given matroid $\mathcal{M}$. The structure of the algorithm is quite simple. For a given matroid $\mathcal{M} = (E, \mathcal{I})$, we denote the cardinality of its grand set by m and we denote its rank by n. For a base B of $\mathcal{M}$, we denote the elementary circuit of $e \notin B$ by $Cir(B, e)$. We also denote the elementary cut set of an element $e \in B$ by $Cut(B, e)$. For an element e of $\mathcal{M}$, we denote the matroid obtained by contracting e by $\mathcal{M}/e$, and the matroid obtained by deleting e by $\mathcal{M} \setminus e$. To obtain elementary circuits and elementary cuts, the simple algorithm uses an *independent set oracle algorithm*, or an *elementary circuit oracle algorithm*. The former checks whether the given set is independent or not, and the latter outputs the elementary circuit included in the union of the given base and element. We denote the time complexity of these oracle algorithms by $T_{ind}(m, n)$ and $T_{cir}(m, n)$. To contract and delete an element of the matroid, we use matroid subroutines whose time complexity are denoted by $T_{cnt}(m, n)$ and $T_{del}(m, n)$. Assume that the contraction and the deletion of k elements simultaneously also take only $T_{cnt}(m, n)$ and $T_{del}(m, n)$ time.

By using these algorithms, we can obtain a simple binary partition algorithm. For a given matroid $\mathcal{M} = (E, \mathcal{I})$ and its base B, we find a partitioning element $e \in B$ and an element $f \notin B$ such that $B \setminus e \cup f$ forms a base. If there is another base in $\mathcal{M}$, such a pair of elements always exists. We spend $O((m-n)T_{cir}(m, n))$ or $O((m-n)nT_{ind}(m, n))$ time to find them. By using e, we divide the problem into two subproblems: enumerating all bases including e and enumerating all those not including e. All bases including e are bases of $\mathcal{M}/e$. All bases not including e are also bases of $\mathcal{M} \setminus e$. Hence they can be enumerated by recursive calls: one inputs $\mathcal{M}/e$, and the other inputs $\mathcal{M} \setminus e$. Thus we obtain an enumeration algorithm simply by using algorithms for constructing these two matroids, and it runs in $O((m-n) \min\{T_{cir}(m,n), nT_{ind}(m,n)\} + T_{cnt}(m,n) + T_{del}(m,n))$ time. We denote the computation time by $T(m,n)$.

Our trimming and balancing approach reduces the time complexity of enumeration algorithms like this one by adding a trimming phase and a balancing phase. The trimming phase removes unnecessary parts from the input. In the above algorithm considered here, for example, unnecessary parts are elements included in all bases, or included in no base. These elements can be contracted or removed, hence we can transform the problem into a smaller one. The new problem includes many outputs for its size rather than the original problem. We show in a later section that a matroid $\mathcal{M}$ including no unnecessary element contains $\Omega(n(m-n))$ bases. By the trimming phase, the computation time of an iteration will be not so large for the number of outputs, so the total computation time per output will be small.

The trimming phase does not always decrease the total computation time. Suppose that the algorithm input is a matroid composed of one circuit $\{e_1, ..., e_m\}$. As we can see, no element is unnecessary. Now we suppose that the algorithm chooses the element e_1 as a partitioning element. Since only one base does not include e_1, one of the subproblems terminates immediately. To construct the other subproblem, e_1 is contracted and we obtain a matroid almost identical to the original matroid: one composed of only one circuit with $m - 1$ elements. Hence the total computation time will be $O(T(m, m-1) + T(m-1, m-2) + ... + T(2, 1))$. Under the condition that $T(m, n)$ is polynomial, the sum is $O(mT(m, n))$, which is $O(T(m, n))$ per base.

Why does the worst case running time not decrease? The answer is that the way of partitioning the problem is not good. We therefore reduce the time complexity by adding a balancing phase. It partitions the original problem into "good subproblems" such as subproblems that are not small after trimming by, for example, choosing a good element for partitioning. If we add a balancing phase, the structure of the recursion will be balanced. Hence, if the number of outputs is small for the input size, then the both of the subproblems generated will be small, and the total time complexity will also be small.

The problem of enumerating bases of a circuit can be partitioned into one subproblem which is to enumerate bases including elements $\{e_1, ..., e_{\lfloor m/2 \rfloor}\}$, and another subproblem which is to enumerate bases including $\{e_{\lfloor m/2 \rfloor+1}, ..., e_m\}$. All the bases include $m-1$ elements of $\{e_1, ..., e_m\}$, the problem is surely partitioned. By using this partitioning rule in this case, we can partition the problem into two subproblems of almost equal sizes, thus the total computation time will be reduced to $O((\log m)T(m, m-1))$ or smaller. In fact, this way of trimming can not attain our result. In later section, we describe the other way of trimming which is utilized in our algorithm for enumerating matroid bases.

Next we show our technique for analyzing the time complexity of enumeration algorithms. Before explaining our analysis, we introduce a virtual tree which is called an *enumeration tree*. Enumeration tree is defined for an enumeration algorithm and an input, and captures the structure of the recursive calls occurring in the algorithm. For a given enumeration algorithm and its input, let $\mathcal{V}$ be a vertex set whose elements correspond to all recursive calls occurring in the algorithm. We consider an edge set $\mathcal{E}$ on $\mathcal{V} \times \mathcal{V}$ such that each whose edge connects two vertices if and only if a recursive call corresponding to one of the vertices occurs in the other. Since the structure of a recursive algorithm contains no circulation, the graph $\mathcal{T} = (\mathcal{V}, \mathcal{E})$ forms a tree. This tree is called an enumeration tree of the algorithm. The root vertex of the tree corresponds to the start of the algorithm. To analyze enumeration algorithms, we show some properties of enumeration trees which are satisfied for any input.

To analyze the time complexity of an enumeration algorithm, we consider the following distribution rule on the enumeration tree. Let $D(x)$ be the number of descendants of a vertex x of $\mathcal{T}$, and $T(x)$ be an upper bound of the computation time on x. Suppose that $\hat{T}$ is an upper bound of $\max_{x \in T}\{T(x)/D(x)\}$. Our analysis uses an arbitrary constant number α. Using these upper bounds and this constant, we distribute the computation time. The distribution is done from a parent vertex to its children in the top-down manner. Let $T_p(x)$ be the computation time distributed to x by the parent of x. x has computation time $T_p(x)+T(x)$. We store $\alpha\hat{T}$ of $T_p(x)+T(x)$ on x, and distribute $T_p(x)+T(x)-\alpha\hat{T}$ to the children of x. We have two distribution rules type 1 and 2 as the following. In each case, we distribute the computation time of each child recursively. Type 1 is that each child y receives computation time proportional to the number of the descendants of y. The computation time distributed to y is $(T_p(x)+T(x)-\alpha\hat{T})D(y)/(D(x)-1)$. Type 2 is that each child y receives computation time proportional to the computation time of y. If a child y receives more than $\alpha\hat{T}D(y)$ then we distribute as the same way of type 1. The computation time distributed to y is $(T_p(x) + T(x) - \alpha\hat{T})T(y)/(D(x) - 1)$.

By this distribution rule, some vertices may receive much computation time. Thus we define excess vertices as those satisfying $T_p(x) + T(x) > \alpha\hat{T}D(x)$ and we stop the distribution on the excess vertices. The children of an excess vertex receive no computation time from their parent. Note that the above distribution rule is also applied to the descendants of excess vertices. By this new rule, $T_p(x)$ for any vertex x is bounded by $\alpha\hat{T}D(x)$ since the computation time distributed from a parent to its child is proportional to the number of descendants of the child.

After this distribution, no vertex except excess vertices has more than $O(\hat{T})$ on it. We then distribute the computation time on each excess vertex to all its descendants uniformly. Since the computation time on any excess vertex x is bounded by $(\alpha+1)\hat{T}D(x)$, each descendant receives at most $(\alpha+1)\hat{T}$ time from an excess ancestor. Let X^* be an upper bound of the maximum number of the excess vertices on a path from the root to a leaf. Using X^*, we have an upper bound $O(\hat{T}X^*)$ of the time complexity per iteration. From these, we obtain the following theorem.

Theorem 1. *An enumeration algorithm runs in $O(\hat{T}X^*)$ time per iteration.*

$\square$

Our analysis requires $\hat{T}$ and X^*. To obtain a good upper bound of the computation time, we have to set X^* and $\hat{T}$ to sufficiently good values. As a candidate of $\hat{T}$, we can utilize $\max_{x \in T}\{T(x)/\bar{D}(x)\}$ where $\bar{D}(x)$ is a lower bound of $D(x)$. Although it is hard to identify excess vertices in the enumeration tree, we can obtain an efficient upper bound. Let x and y be excess vertices such that y is an ancestor of x and no other excess vertex is in the path P_{yx} from y to x in the enumeration tree. Note that P_{yx} has at least one internal vertex.

Lemma 2. *If we use type 2 rule on all vertices, a vertex w of $P_{yx} \setminus y$ satisfies the condition that $T(w) > \frac{\alpha-1}{\alpha} \sum_{u \in C(w)} T(u)$ where $C(w)$ is the set of children of w.*

Proof. We show that all vertices w of $P_{yx} \setminus y$ satisfy the condition that $T_p(w) < (\alpha-1)\hat{T}(w)$ under the assumption that $P_{yx} \setminus y$ includes no such vertex. Any child of y satisfies the condition since y is an excess vertex. Suppose that a vertex w of $P_{yx} \setminus y$ satisfies the condition that $T_p(w') \leq (\alpha - 1)T(w')$, where w' is the parent of w. From the assumption, we have

$$
\begin{aligned}
T_p(w) &= ((T_p(w') + T(w') - \alpha\hat{T})T(w)) \; / \; (\textstyle\sum_{u \in C(w')} T(u)) \\
&\leq (\alpha T(w')T(w)) \; / \; (\textstyle\sum_{u \in C(w')} T(u)) \\
&\leq (\alpha T(w)\tfrac{\alpha-1}{\alpha} \textstyle\sum_{u \in C(w)} T(u)) \; / \; (\textstyle\sum_{u \in C(w')} T(u)) \quad = \quad (\alpha - 1)T(w)
\end{aligned}
$$

Since $T(w) \leq \hat{T}D(w)$, we have $T_p(w) \leq (\alpha-1)\hat{T}D(w)$. This implies that $T_p(x) + T(x) \leq \alpha\hat{T}D(x)$, and contradicts that x is an excess vertex. □

From this lemma, we can obtain X^* by estimating an upper bound of the number of vertices satisfying this condition in any path from the root to a leaf. Similarly, we can obtain the following corollary.

Corollary 3. *If we use type 1 rule on all vertices and $\hat{T} = \max_{x \in T}\{T(x)/\bar{D}(x)\}$, a vertex w of $P_{yx} \setminus y$ satisfies that $\bar{D}(w) > \sum_{u \in C(w)} \frac{\alpha}{\alpha+1} \bar{D}(u)$.* □

These conditions can be easily checked, and are often sufficient to analyze. In the next section, we describe a trimming and balancing algorithm for matroid bases. To see trimming and balancing algorithms for directed spanning trees or perfect matchings, refer [3,4].

3 Trimming and Balancing Algorithms for Matroid Bases

In this section, we describe algorithms for trimming phase and balancing phase of our enumeration algorithm for matroid bases. Our trimming algorithm is quite simple. For a given matroid, if an element of the grand set forms a circuit, then it is included in no base. We call the element a *loop*. In the matroid obtained by deleting all loops, the set of bases is preserved. If the elementary cut of an element is composed of only the element, it is included in all bases. These elements can be contracted. The trimming algorithm contracts or deletes these unnecessary elements. These conditions for all elements can be checked by finding the elementary circuits for all elements outside a base. This can be done in $O((m-n)\min\{nT_{ind}(m,n), T_{cir}(m,n)\})$.

We next consider the balancing algorithm. By partitioning the given matroid, we obtain two matroids $\mathcal{M}/e$ and $\mathcal{M}\setminus e$. If one is a sufficiently smaller than the other after trimming, the enumeration tree of the algorithm may be biased. Hence we use a balancing algorithm to avoid it. To obtain our balancing algorithm, we use the following properties of $\mathcal{M}/e$ and $\mathcal{M}\setminus e$ for a base B of $\mathcal{M}$.

Property 1 *Suppose that $e \in B$ and B' is the base of $\mathcal{M}/e$ obtained by $B\setminus e$. The following conditions hold for $\mathcal{M}/e$ and B'. (1) For any element $f \notin B$, the difference between $Cir(B,f)$ and $Cir(B',f)$ is either e or nothing. (2) For any element $f \in B\setminus e$, $Cut(B,f) = Cut(B',f)$.*   □

Property 2 *The following conditions hold for $e \notin B$. (1) For any element $f \notin B$, the elementary circuits of f on B in $\mathcal{M}$ and of f on B in $\mathcal{M}\setminus e$ are equal. (2) For any element $f \in B$, the difference between the elementary cuts $Cut(B,f)$ in $\mathcal{M}$ and $Cut(B,f)$ in $\mathcal{M}\setminus e$ is either nothing or e.*   □

Using these properties, we construct the balancing algorithm. First we examine the case in which some elements of E are loops in $\mathcal{M}/e$. In this case, we partition the original problem in a way that is based on the following properties.

Property 3 *For two elements e and e' of E such that $\{e,e'\}$ is a circuit, B is a base including e if and only if $B\setminus e \cup e'$ is a base.*   □

Since $\mathcal{M}$ is trimmed, any loop $\{e'\}$ of $\mathcal{M}/e$ composes a circuit of $\mathcal{M}$ with e. Thus we can see that all bases of $\mathcal{M}$ including a loop e' of $\mathcal{M}/e$ can be enumerated by enumerating all bases including e. All bases including e' are constructed from the bases including e simply by replacing e with e'.

Suppose that $e_2, ..., e_k$ are loops in $\mathcal{M}/e_1$. We then divide the problem into the subproblems of enumerating bases including each e_i, and of enumerating bases not including any e_i. Since any matroid $\mathcal{M}/e_i$ can be constructed from $\mathcal{M}/e_1$ simply by renaming e_1 to e_i, the time complexity of an iteration does not increase.

Next we examine the case in which some elements of $\mathcal{M}\setminus e$ have elementary cuts consisting of only those elements. The partitioning method we use in this case is based on the following lemma.

Lemma 4. *For a trimmed matroid $\mathcal{M}$, let $e_1, ..., e_k$ be elements satisfying the condition that any base not including e_1 always includes $e_2, ..., e_k$. All bases include at least $k-1$ of these elements.*

Proof. Suppose that there is a base not including e_i and e_j. From the assumption, e_i and e_j are not e_1. Since $\mathcal{M}$ is trimmed, by removing e_1 and adding an element to the base, we obtain a base not including e_1 and e_j, or not including e_1 and e_i. This contradicts the assumption.   □

Lemma 5. *Let $e_1, ..., e_k$ be elements satisfying the condition that any base not including e_1 always includes $e_2, ..., e_k$. For any base not including e_1, there is a base obtained by exchanging e_1 and e_i. Conversely, for any i and any base including e_1, there is a base obtained by exchanging e_1 and e_i.*

Proof. To prove this lemma, we show that any circuit includes all e_i, or includes no e_i. Suppose that a circuit C includes e_i and does not include e_j. By removing e_i from C, we can obtain an independent set. Let B be a base including the independent set. From Lemma 4, B includes e_j. Since the matroid is trimmed, there is a base not including e_j. Hence there is an element e' whose elementary circuit includes e_j. Since $Cir(B, e_i)$ is C, e' is not e_i. By exchanging e' and e_j, we obtain a base including neither e_i nor e_j. This contradicts Lemma 4.

We now have that any circuit includes all e_i or no e_i. Hence for any base not including e_1, we have a base obtained by exchanging e_1 and e_i. For any i and any base including e_1, we also have a base obtained by exchanging e_1 and e_i.   □

Let us consider that $e_2, ..., e_k$ are included in any base of $\mathcal{M} \setminus e_1$. In this case, we partition the problem into the subproblems of enumerating all bases not including each e_i and of enumerating all bases including all e_i. Since any matroid $\mathcal{M} \setminus e_i$ can be constructed from $\mathcal{M} \setminus e_i$ by renaming e_1 to e_i, each additional subproblem is constructed in O(1) time. We now describe the details of the algorithm as follows. Assume that $\mathcal{M}$ is trimmed.

ALGORITHM: T&B_FOR_MATROID_BASES $(\mathcal{M})$
Step 1: Choose an element $e \in B$.
Step 2: Find a base B' not including e.
Step 3: Generate and trim $\mathcal{M}/e$ and $\mathcal{M} \setminus e$.
Step 4: If one of the subproblems loses some elements, partition the problem again by using the balancing algorithm.
Step 5: Solve the subproblems by recursive calls.

This algorithm takes $O(T(m, n))$ time on an iteration. we bound the total time complexity by a quite small order with the use of our analysis in the next section.

4   Bounding the Amortized Time Complexity

To analyze the time complexity, we use the enumeration tree T of our algorithm. For a vertex x of T, we denote the matroid by $\mathcal{M}_x$. which a recursive call corresponding to a vertex x inputs. Since x corresponds to an iteration, leaves correspond to bases one-to-one, and any internal vertex has at least two children. Hence the number of iterations does not exceed twice the number of bases. We denote the size of the grand set of $\mathcal{M}_x$ by m_x and the rank by n_x.

We first establish a lower bound $\bar{D}(x)$. It is given by the lower bound on the number of bases in a trimmed matroid.

Lemma 6. *A trimmed matroid $\mathcal{M}$ includes at least $(m - n)n$ bases.*

Proof. We prove the lemma by induction. Let B be a base of $\mathcal{M}$. For each element e in B, there is an element not in B whose elementary circuit includes e. By exchanging e and the element, we obtain another base of $\mathcal{M}$. The generated bases are distinct, hence we have at least n other bases in the matroid. For each element e' not in B, we have an elementary circuit including at least one element of B. By exchanging e' and an element of the circuit, we can obtain $m-n$ distinct bases. Hence we have at least $m-n$ bases in the matroid. Therefore the condition of the lemma holds if $n = 1$ or $m - n = 1$.

Assume that the condition of the lemma holds if $m - n < k$ and $n \leq k'$ hold, or $m - n \leq k$ and $n < k'$ hold. Under this assumption, we prove that the condition of the lemma holds for the matroid $\mathcal{M}$ with $m - n = k > 1$ and $n = k' > 1$. Let e be an element of the grand set of $\mathcal{M}$. If the matroids $\mathcal{M}/e$ and $\mathcal{M} \setminus e$ lose no element by the trimming algorithm, $\mathcal{M}$ includes at least $(m - n - 1)n + (m - n)(n - 1) = 2(m - n)n - m$ bases. Since $m - n$ and $n > 1$, we have $2(m - n)n - m > (m - n)n$.

Let us consider the other cases. If $\mathcal{M}/e$ loses k ($1 \leq k \leq m - n$) elements by the trimming algorithm, then the balancing algorithm partitions the bases of $\mathcal{M}$ to bases of at least $k + 1$ trimmed matroids with $m - k - 1$ elements and the rank $n - 1$. Since these matroids have at least $(m - k - 1 - (n - 1))(n - 1)$ bases from the assumption, $\mathcal{M}$ has at least $(k + 1)(m - k - n)(n - 1) \geq (m - n)n$ bases.

In the case that $\mathcal{M} \setminus e$ loses k ($1 \leq k \leq n$) elements, the bases of $\mathcal{M}$ are partitioned to bases of $k + 1$ matroids with $m - k - 1$ elements and the rank $n - k$. Since all these matroids have at least $(m - k - 1 - (n - k))(n - k)$ bases by the assumption, $\mathcal{M}$ has at least $(k + 1)(m - n - 1)(n - k) \geq (m - n)n$ bases. $\quad\square$

Next we set X^* to an upper bound on excess vertices. We set α to 30. From Corollary 3, for X^*, we can use an upper bound of the number of vertices x in a path from the root to a leaf satisfying the condition (a) that $\bar{D}(x) < \frac{\alpha}{\alpha+1} \sum_{u \in C(x)} \bar{D}(u)$. We show which vertices of T satisfy the condition. We first consider the case that $m_x - n_x = 1$ or $n_x = 1$. In this case, the balancing algorithm generates $m_x - n_x$ or n_x subproblems with only one element in the grand set. This element may be an excess vertex, although the number of this type of excess vertices in the path from the root to a leaf is at most 1.

Next we consider the case that $m_x - n_x, n_x > 1$. If both generated subproblems of x lose no element by the trimming algorithm, the number of bases included in $\mathcal{M}/e$ or $\mathcal{M} \setminus e$ is at least $(m_x - n_x - 1)n_x + (m_x - n_x)(n_x - 1) = 2(m_x - n_x)n_x - m_x$. For a fixed m_x, $(m_x - n_x)n_x$ takes its minimum value $2m_x - 2$ when $n_x = 2$ and $n_x = m_x - 2$. Since $2m_x - 2 \geq m_x/30$ for any $m_x > 2$, we have $(m_x - n_x)n_x \leq \frac{29}{30}(2(m_x - n_x)n_x - m_x)$. Hence x does not satisfy the condition (a). Suppose that x is an excess vertex. Then a subproblem of x loses k elements by the trimming algorithm. Thus we have $\frac{30}{29}(m_x - n_x)n_x > (k + 1)(m_x - k - n_x)(n_x - 1)$ or $\frac{30}{29}(m_x - n_x)n_x > (k + 1)(m_x - n_x - 1)(n_x - k)$ from the proof of lemma 6. Hence we have $\frac{1.04 n_x}{n_x - 1} > \frac{(k+1)(m_x - k - n_x)}{(m_x - n_x)}$, or $\frac{1.04(m_x - n_x)}{m_x - n_x - 1} > \frac{(k+1)(n_x - k)}{n_x}$.

The former condition holds only for $k > m_x - n_x - 4$, and the latter condition holds only for $k > n_x - 4$. Thus any child y of x satisfies $m - n \leq 4$ or $n \leq 4$. The rank and the cardinality of the grand set decrease strictly in the children, hence there are at most 8 the above excess vertices in a path from the root to a leaf. Therefore we can set X^* to 9.

Since $T(x) = O(T(m_x, n_x))$, we set $\hat{T}$ to $\max_x \{O(T(m_x, n_x)/\bar{D}(x))\} = O(T(m, n)/(m_x - n_x)n_x)$. We obtain the following theorem.

Theorem 7. *We can enumerate all matroid bases in $O(T(m, n))$ preprocessing time and $O(T(m, n)/n(m - n)\})$ time per base, if the time complexities of oracle algorithms are depend on the input size of contracted and deleted matroids.* $\square$

5   On Graphic Matroids

In the following sections, we show some applications of our algorithm for some enumeration problems of matroid bases. Here we see the case of graphic matroids. The grand set of a graphic matroid $\mathcal{M}$ is given by the edge set E of an undirected graph $G = (V, E)$, and its independent set is given by a forest of the graph. If the graph is connected, any base is a spanning tree. Hence the problem is that of enumerating all spanning trees of given undirected graphs. There are numerous studies for this problem, and numerous algorithms have been proposed and improved. The time complexities of the enumeration algorithms have been reduced from $O(|E| + |V|)$ to $O(|V|)$ and to $O(1)$ per spanning tree. And their space complexities have been reduced from $O(|V||E|)$ to $O(|E| + |V|)$. The methods by which these algorithms were improved relied on some advanced data structures, but our algorithm in this subsection requires neither complicated algorithm nor advanced data structure.

Since any spanning tree of G has $|V| - 1$ edges, we have that the number of elements m is $|E|$ and the rank n is $|V| - 1$. A base of a graphic matroid can be found by a graph search algorithm in $O(m + n)$ time. The contraction and the deletion of the matroid are also easy. $\mathcal{M}/e$ is given by the graph obtained by contracting the edge e, and $\mathcal{M} \setminus e$ is given by the graph obtained by deleting e. Since any independent set of the matroid includes no cycle, a circuit of the matroid is a cycle in G. Thus an elementary circuit oracle algorithm with $O(n)$ running time can be constructed.

We thus have $T_{cnt}(m, n) = O(n)$, $T_{del}(m, n) = O(m - n)$, and $T_{cir}(m, n) = O(n)$. The time complexity of an iteration of the trimming and balancing algorithm using these oracle algorithms is bounded by $O((m - n)n)$. Since all loops are detected in $O(m)$ time and all elements with elementary cuts composed of only themselves can be found by the 2-connected component decomposition in $O(m+n)$ time, the time complexity of the trimming algorithm can be reduced to $O(m+n)$. Therefore the trimming and balancing algorithm takes $O(m+n)$ time for preprocessing and $O(\frac{O((m-n)n)}{(m-n)n}) = O(1)$ time per spanning tree. It requires only $O(m + n)$ memory. These complexities are equal to those of the optimal algorithm reported by Shioura et al.[2]. Here we have the following theorem.

Theorem 8. *All bases of a graphic matroid can be enumerated by a trimming and balancing algorithm in $O(m + n + N)$ time and $O(m + n)$ memory where N denotes the number of bases.* $\square$

6   On Linear Matroids

For an m-by-n matrix M with $m > n$, the grand set of the linear matroid is given by the set of column vectors, and an independent set is given by a set of independent column vectors. The independent set family is given by the correction of all the independent sets. If the given matrix is not degenerate, the bases of the linear matroid are all submatrices composed of n independent column vectors. For a given non-singular m-by-n matrix, we consider the problem of enumerating all these n-by-n non-singular submatrices, which are composed of n independent column vectors. For this problem, Y. Matsui [1] proposed an enumeration algorithm with a time complexity of $O(n^2)$ per base.

A base of the linear matroid can be found easily. By some LDL or LU decomposition algorithms, we can obtain a non-singular submatrix B in at most $O(n^2m)$ time. Our algorithm for linear matroids utilizes an elementary circuit oracle algorithm. To obtain a fast elementary circuit oracle algorithm, we consider a linear matroid equivalent to $\mathcal{M}$. The matroid is given by a matrix $U(M, B)$ generated from M as follows. All ith column vectors of B are replaced by the unit vector e_i whose ith element is 1 and other elements are 0. A column vector $x \notin B$ is replaced by the vector y satisfying $By = x$. Since a set of column vectors of M is independent if and only if the set of column vectors corresponding to them is independent in $U(M, B)$, the linear matroid given by $U(M, B)$ is equivalent to $\mathcal{M}$. We enumerate bases of the matroid instead of the bases of $\mathcal{M}$.

As column vectors of B are unit vectors, the elementary circuit of a column vector x can be found easily. The ith column vector of B is in $Cir(B, x)$ if and only if the ith element of x is not 0. In each iteration, we exchange ith column vector of B and the other column vector x and thereby obtain the other base B'. x is generally not a unit vector, thus we have to obtain $U(M, B')$. Since only x is not a unit vector in B', for any column vector z of M, we can obtain a vector y satisfying $B'y = z$ in $O(n^2)$ time. Thus we take $O(n^2(m - n))$ time to obtain $U(M, B')$.

We next show contract and delete operation of an element. The deleting operation is simple, since it involves only the deletion of a column vector. And because in each iteration we delete a column vector outside the base, we do not need to obtain $U(M, B')$. To contract the matroid by the ith column vector x_i, we fix x_i in B and consider only the bases including x_i. Since x_i is included in any base, we can restrict the rest of the column vectors to the linear subspace orthogonal to x_i. Hence we consider the matrix obtained by deleting ith element from all column vectors of M. As the linear matroid given by the matrix is equivalent to $\mathcal{M}/x_i$, we enumerate the bases of the matroid instead of the bases of $\mathcal{M}/x_i$.

Because each of the above operations takes at most $O(n^2(m-n))$ time, for an iteration x inputting an m_x-by-n_x matrix, the time complexity is $O(n_x^2(m_x - n_x))$ Since $n^2(m - n)/n(m - n) = n$, we obtain the following theorem.

Theorem 9. *All bases of the linear matroid given by an m-by-n matrix are enumerated in $O(n^2m)$ preprocessing time and $O(n)$ time per base.*   □

7 On Matching Matroids

Matching matroids are given by bipartite graphs. For a bipartite graph $G = (V_1 \cup V_2, E)$, the grand set of a matching matroid is given by V_1, and an independent set of the matroid is given by a set of vertices of V_1 covered by a matching of G. A base B of a matching matroid is obtained by finding a maximum cardinality matching M of G. For a vertex $v \in V_1 \setminus B$, $u \in B$ is in $Cir(B, v)$ if and only if there is an alternating path from v to u in G. An alternating path is a path of G satisfying that edges in M and edges not in M appear in the path alternatively. By exchanging edges of M and the other edges along the alternating path from v to u, we obtain the other maximum matching M' which covers v and does not cover u. To obtain the elementary circuit of v, we find the vertices of the base to which there are some alternating paths from v. We can find them in $O(m + n)$ time by using a graph search algorithm [3].

To delete a vertex from a matching matroid is easy. It is done by deleting the vertex from G. The contracted matroid is not obtained by reforming G. Instead, we simply mark the vertex to contract, and never output marked vertices when we find elementary circuit. By this modification of the elementary circuit oracle algorithm, we can obtain the contracted matching matroid.

An iteration x takes $O((|E_x| - |V_x|)(|E| + |V|))$ time where E_x and V_x denote the edge set and vertex set of the input graph of x. Note that E is the edge set of the graph of the original problem. From our analysis, we know that the time complexity of this enumeration algorithm is $\frac{O((|E_x| - |V_x|)(|E| + |V|))}{|V_x|(|E_x| - |V_x|)} = O(|E| + |V|)$ per iteration. Therefore we have the following theorem.

Theorem 10. *All bases of a matching matroid given by a bipartite graph $G = (V, E)$ can be enumerated in $O(|E| + |V|)$ time per base.* $\qquad\square$

References

1. Y. Matsui, "An Algorithm for Generating All the Bases of Equality Systems," Research Report, Dept. of Math. Engineering and Info. Physics, University of Tokyo, Tokyo (to appear).
2. A.Shioura, A.Tamura and T.Uno, "An Optimal Algorithm for Scanning All Spanning Trees of Undirected graphs," SIAM J.Comp.**26**, pp.678-692 (1997).
3. T.Uno, "Algorithms for Enumerating All Perfect, Maximum and Maximal Matchings in Bipartite Graphs," LNCS **1350**, Springer-Verlag, pp.92-101 (1997).
4. T.Uno, "A New Approach for Speeding Up Enumeration Algorithms" LNCS **1533**, Springer-Verlag, pp.287-296 (1998)

On Routing in Circulant Graphs

Jin-Yi Cai[1,*], George Havas[2,**], Bernard Mans[3], Ajay Nerurkar[1,***],
Jean-Pierre Seifert[4], and Igor Shparlinski[3,†]

[1] Department of Computer Science and Engineering,
State University of NY at Buffalo, Buffalo, NY 14260-2000, USA.
{cai,apn}@cs.buffalo.edu.
[2] Centre for Discrete Mathematics and Computing,
Department of Computer Science and Electrical Engineering,
The University of Queensland, Qld 4072, Australia.
havas@csee.uq.edu.au.
[3] Department of Computing,
Macquarie University, Sydney, NSW 2109, Australia.
{bmans,igor}@mpce.mq.edu.au
[4] Department of Mathematics and Computer Science,
University of Frankfurt, 60054 Frankfurt on the Main, Germany.
seifert@cs.uni-frankfurt.de

Abstract. We investigate various problems related to circulant graphs – finding the shortest path between two vertices, finding the shortest loop, and computing the diameter. These problems are related to shortest vector problems in a special class of lattices. We give matching upper and lower bounds on the length of the shortest loop. We claim NP-hardness results, and establish a worst-case/average-case connection for the shortest loop problem. A pseudo-polynomial time algorithm for these problems is also given. Our main tools are results and methods from the geometry of numbers.

Keywords: Circulant graphs, Shortest paths, Loops, Diameter, Lattices

1   Introduction

Circulant graphs have a vast number of applications in telecommunication networking, VLSI design and distributed computation [5,21,23]. We study various routing problems in circulant graphs such as finding the shortest path between two vertices, finding the shortest loop and the diameter. We establish relations between these routing problems and the problem of finding the shortest vector in the L_1-norm in a lattice.

We recall that an n-vertex *circulant graph* G is a graph whose adjacency matrix $A = (a_{ij})_{i,j=0}^{n-1}$ is a circulant. That is, the ith row of A is the cyclic shift

* Supported in part by NSF grant CCR-9634665 and a J. S. Guggenheim Fellowship.
** Supported in part by ARC grants A49702337 and A49801415.
*** Supported in part by NSF grant CCR-9634665.
† Supported in part by ARC grant A69700294.

T. Asano et al. (Eds.): COCOON'99, LNCS 1627, pp. 360–369, 1999.
© Springer-Verlag Berlin Heidelberg 1999

of the zeroth row by i, so that $a_{ij} = a_{0,j-i}$ for $i, j = 0, \ldots, n-1$ (adjacency matrix subscripts are taken modulo n). We consider undirected graphs, that is, $a_{ij} = a_{ji}$ for $i, j = 0, \ldots, n-1$ and there are no self-loops ($a_{ii} = 0$ for $i = 0, \ldots, n-1$).

Therefore with every circulant graph one can associate a set S of positive integers which shows which pairs of nodes are connected. Two nodes u and v of the graph are connected if and only if $\pm(u-v) \pmod{n} \in S$. Given S we denote the corresponding n-vertex graph by $\langle S \rangle_n$. We denote the group of residues modulo n by $\mathbb{Z}_n$. Without loss of generality, we always assume that $S \subseteq \mathbb{Z}_n$.

Let $S = \{s_1, \ldots, s_m\} \subseteq \mathbb{Z}_n$ with $1 \leq s_i \leq n/2$ for $i = 1, \ldots, m$. Given two vertices u and v of the graph $\langle S \rangle_n$, a path from u to v can be described by an integer vector $\mathbf{x} = (x_1, \ldots, x_m) \in \mathbb{Z}^m$ such that there are x_i steps of the form $w \to w + s_i$ if $x_i \geq 0$, or there are $|x_i|$ steps of the form $w \to w - s_i$ if $x_i < 0$, for $i = 1, \ldots, m$. It is easy to see that any permutation of a sequence of steps in a path from u to v produces another path from u to v.

Starting from u we arrive at v if and only if $\sum_{i=1}^{m} x_i s_i \equiv v - u \pmod{n}$. It is natural to aim to minimize the L_1-norm $|\mathbf{x}|_{L_1}$ which is the length of the shortest path between u and v. For any given n and $S = \{s_1, \ldots, s_m\}$, this minimal number depends only on $u - v$. Accordingly for $w \in \mathbb{Z}_n$ we denote by $\Omega(w) = \Omega_{n,S}(w)$ the set of solutions $\mathbf{x} = (x_1, \ldots, x_m) \in \mathbb{Z}^m$ of the congruence $\sum_{i=1}^{m} x_i s_i \equiv w \pmod{n}$. Define $D_{n,S}(w) = \min\{|\mathbf{x}|_{L_1} : \mathbf{x} \in \Omega(w), \mathbf{x} \neq 0\}$. The condition $\mathbf{x} \neq 0$ is relevant only when $w = 0$ and in this case $L(n, S) = D_{n,S}(0)$ is the length of the shortest loop. We note that $D(n, S) = \max_{1 \leq w \leq n-1} D_{n,S}(w)$ is the diameter of $\langle S \rangle_n$.

Let $a_1, \ldots, a_n$ be linearly independent vectors in $\mathbb{R}^n$. The set of all integral linear combinations of the a_i forms a (n-dimensional) lattice L, and the a_i are called a basis of L. A lattice can also be abstractly defined as a discrete additive subgroup of $\mathbb{R}^n$. The *determinant* of a lattice L is denoted by $\det(L)$. If $a_1, \ldots, a_n$ is a basis of L, then $\det(L) = |\det(a_1, \ldots, a_n)|$. It is invariant under a change of basis. We remark that $\Omega(0)$ is a lattice and $\Omega(w)$ is a shifted (affine) lattice, and so we can apply existing tools for studying short vectors in lattices and their shifts.

We formulate the following problems.

Shortest-Path: Given a set $S = \{s_1, \ldots, s_m\} \subseteq \mathbb{Z}_n$ and a residue $w \in \mathbb{Z}_n$, $w \neq 0$, find $D_{n,S}(w)$ and a vector $\mathbf{x} \in \Omega(w)$ for which $D_{n,S}(w) = |\mathbf{x}|_{L_1}$.

Shortest-Loop: Given a set $S = \{s_1, \ldots, s_m\} \subseteq \mathbb{Z}_n$, find $L(n, S)$ and a vector $\mathbf{x} \in \Omega(0)$ for which $L(n, S) = |\mathbf{x}|_{L_1}$.

Diameter: Given a set $S = \{s_1, \ldots, s_m\} \subseteq \mathbb{Z}_n$, find $D(n, S)$.

For general graphs these are well known problems. Efficient polynomial time algorithms have been developed for various shortest path problems. However, for the class of circulant graphs, there is an important distinction to be made, and that concerns the natural input size to a problem. For a general graph it is common to consider that the input size is of order n^2, which is the number of elements in the adjacency matrix. However, any circulant graph can be described by only m integers $1 \leq s_1, \ldots, s_m \leq n/2$. In this representation the input size is of order $m \log n$. Thus polynomial time algorithms for general graphs may exhibit exponential complexity in the special case of circulant graphs.

Despite quite active interest in the above problems, motivated by various applications of circulant graphs, very few algorithmic results are known except for some special partial cases such as $m = 2$, see [10,11,19,30]. These papers use quite elementary number theoretic considerations. Thus one of the purposes of this paper is to introduce new techniques in this area, namely the relatively modern theory of geometric lattices [1,2,3,4,6,7,8,13,15,16,17,18,20,22,24,27,28,29] as well as more classical tools from the geometry of numbers [9] and combinatorial number theory [12,14].

First, we give estimates on the shortest loop length in terms of associated lattices. We provide upper and lower bounds which are close, within a factor of approximately two.

Then, we briefly discuss the case when the input n is given in unary. We provide an algorithm specifically tailored for the circulant graphs which solves the *Shortest-Loop* problem in polynomial time for this input measure. In contrast, we show that the *Shortest-Path* problem is NP-hard in the context of the more concise representation.

Finally, we claim that three lattice problems, which are believed to be hard, are reducible to an average-case instance of the *Shortest Vector Problem* (**SVP**) for the homogeneous lattice family defined by circulant graphs, under a certain distribution on such lattices.

2   Estimates of Shortest Vectors

In this section we give estimates for the shortest loop length $L(n, S)$ in terms of the lattice defined by the congruence $\sum_{i=1}^{m} a_i x_i \equiv 0 \pmod{n}$. For simplicity we focus on the case where the modulus n is a prime p. This can be generalized to a composite modulus via the Chinese Remainder Theorem. In fact taking the lattice view, it is natural to consider the more general setting where a set of s congruences is given, $A\mathbf{x} \equiv 0 \pmod{p}$ with $\mathbf{x} \in \mathbb{Z}^m$, where $A = (a_{ij}) \in \mathbb{Z}_p^{s \times m}$, which defines a lattice L of dimension m. The case with the circulant graph is $s = 1$. This lattice is of dimension m, independent of s, since $p \cdot e_i \in L$, for all i, where e_i is the ith canonical basis vector (for $\mathbb{Z}^m$). Note that in general $p \cdot e_i$ do not form a basis for L.

We now compute the determinant of L. The map $\mathbf{x} \mapsto A\mathbf{x} \pmod{p}$ defines a homomorphism from $\mathbb{Z}^m$ to $\mathbb{Z}_p^s$, the kernel of which is our lattice L. Thus the cardinality $|\mathbb{Z}^m/L| \leq p^s$ (equality holds if and only if A has rank m over $\mathbb{Z}_p$). Hence $\det(L) \leq p^s$, with equality if and only if A has rank m over $\mathbb{Z}_p$.

Let $B_m(r) = \{\mathbf{x} \in \mathbb{R}^m : |\mathbf{x}|_{L_1} \leq r\}$ be the L_1-norm ball of radius r in m-dimensional space. Then the volume $\mathrm{vol}(B_m(r)) = r^m \cdot \frac{2^m}{m!}$.

By Minkowski's Theorem, see Theorem 2 of Chapter 3 of [9], if $\mathrm{vol}(B_m(r)) \geq 2^m \det(L)$, then there is a non-zero vector $x \in L \cap B_m(r)$. Because $2^m p^s \geq 2^m \det(L)$, a sufficient condition on r is $r \geq \sqrt[m]{m!} p^{s/m} \approx \frac{m}{e} p^{s/m}$. (The approximation is based on $\sqrt[m]{m!} = \frac{m}{e}(1 + o(1))$ as $m \to \infty$.) Thus we have

Theorem 1. *For any system $Ax \equiv 0 \pmod{p}$, where $A \in \mathbb{Z}_p^{s \times m}$, there is a solution $\mathbf{x} \in \mathbb{Z}^m$, $\mathbf{x} \neq 0$, and $|\mathbf{x}|_{L_1} \leq \sqrt[m]{m!}p^{s/m}$. In particular, for any circulant graph $\langle S \rangle_p$, $L(p, S) \leq \sqrt[m]{m!}p^{1/m}$, where $m = |S|$.*

Let $N_m(r) = |\{\mathbf{x} \in \mathbb{Z}^m : x_i \geq 0,$ and $x_1 + \cdots + x_m \leq r\}|$. It is elementary to show that $N_m(r) = \binom{r+m}{m}$. Thus $\mathbb{Z}^m \cap B_m(r) = \{\mathbf{x} \in \mathbb{Z}^m : |\mathbf{x}|_{L_1} \leq r\}$ has cardinality at most $2^m \binom{r+m}{m}$.

Now to each non-zero $\mathbf{x} \in \mathbb{Z}^m \cap B_m(r)$, there are exactly $p^{m-1} - 1$ non-zero coefficient sequences $(a_1, \ldots, a_m) \in \mathbb{Z}_p^m$ for which $\sum_{i=1}^m a_i x_i \equiv 0 \pmod{p}$. Altogether this can account for at most $(2^m \binom{r+m}{m} - 1)(p^{m-1} - 1)$ non-zero sequences $(a_1, \ldots, a_m) \in \mathbb{Z}_p^m$. Thus if $p^m - 1 > (2^m \binom{r+m}{m} - 1)(p^{m-1} - 1)$, then some $(a_1, \ldots, a_m) \neq 0$ has no solution with norm at most r. The following bound is certainly a sufficient condition for this: $p \geq 2^m \binom{r+m}{m}$. Since $\binom{r+m}{m} \leq ((r+m)e/m)^m$, after some simple calculation, we have the sufficient condition $r < m \left(\frac{1}{2e} p^{1/m} - 1 \right)$.

Theorem 2. *There are circulant graphs $\langle S \rangle_p$ with $L(p, S) \geq m \left(\frac{1}{2e} p^{1/m} - 1 \right)$.*

We note that this lower bound almost matches the upper bound in Theorem 1, within a factor of about 2. Also the analysis here can be given in terms of a matrix $A \in \mathbb{Z}_p^{s \times m}$, in which case, we replace the quantity $p^{1/m}$ by $p^{s/m}$.

Unfortunately the above considerations do not seem to apply to the non-homogeneous case. Nevertheless, some number theoretic results of [12,14] allow us to deal with this case as well. In particular, we obtain an upper bound on $D(n, S)$, which, in some cases, is quite tight.

Theorem 3. *For any circulant graph $\langle S \rangle_p$ with $m \geq 0.5 \lfloor (4p - 7)^{1/2} \rfloor + 1$, $D(p, S) \leq 0.5 \lfloor (4p - 7)^{1/2} + 1 \rfloor \sim p^{1/2}$, where $|S| = m$.*

Proof. For any set $T \subseteq \mathbb{Z}_p^*$ of cardinality $|T| = \lfloor (4p - 7)^{1/2} \rfloor + 1$, it is shown in [12] that any element $w \in \mathbb{Z}_p$ can be represented in the form $w \equiv \sum_{t \in W} t \pmod{p}$ for some subset $W \subseteq T$ of cardinality at most $|T|/2$. It is obvious that the cardinality of the set $R = S \cup -S$ is at least $2m - 1$. Thus selecting an arbitrary subset $T \subseteq R$ of cardinality $|T| = \lfloor (4p - 7)^{1/2} \rfloor + 1 \leq 2m - 1$ we obtain the required result. $\square$

We remark that if $m \sim p^{1/2}$ and $S = \{1, \ldots, m\}$ then, obviously, $D(p, S) \geq p/m \sim p^{1/2}$, thus Theorem 3 is tight for such circulant graphs.

3   Polynomial Time Algorithms for Small n

When n is given as a unary input and the time complexity is measured in terms of n, there are well known polynomial time algorithms to find both the length of the shortest path between any two vertices of a graph with n vertices, as well as a shortest path itself. This applies to circulant graphs as a special case. However, even in this case, due to the symmetry present in any circulant graph, considerable savings can be realized in time complexity, compared to using a general graph algorithm.

We concentrate on the *Shortest-Loop* problem. The definition of a *loop* here is specifically tailored for the circulant graph, and is not merely a specialization of a loop in general graphs. This is because there is a "commutativity" of the underlying group $\mathbb{Z}_n$, and so we exclude certain "trivial loops". For example, suppose there are two distinct step sizes s_1 and $s_2 \in S$. The closed path $\langle 0, s_1, s_1 + s_2, s_2, 0 \rangle$, using steps $s_1, s_2, -s_1, -s_2$ respectively, would be considered a loop in a general graph, but is excluded here.

Assume we are given $S = \{s_1, \ldots, s_m\} \subseteq \mathbb{Z}_n$. The idea of the following algorithm for finding loops is to try to compute all shortest paths connecting 0 and some $is_k \pmod{n}$, using only steps from $S' = \{s_1, \ldots, s_{k-1}\}$. We do this for all i and then for all k, and finally take the minimum.

Fix k, $1 \leq k \leq m$. Let $g_k = \gcd(s_k, n)$. Let $s'_k = s_k/g_k, n_k = n/g_k$. Then $\gcd(s'_k, n_k) = 1$, and we can solve integral linear equations $s'_k x - n_k y = \pm 1$. The general solution has the form $x = \pm[x_0 + tn_k]$ and $y = \pm[y_0 + ts'_k]$, for some x_0 and y_0, and where all $t \in \mathbb{Z}$ are admissible. Choose t so that the absolute value $|x| = |x_0 + tn_k|$ is minimum. This will set $|x| \leq \lfloor n_k/2 \rfloor$ and the corresponding t is essentially unique. We denote this minimizing x by x_k.

Next, for each j, $1 \leq j \leq \lfloor n_k/2 \rfloor$, we compute the shortest path from 0 to $js_k \bmod(n)$, in the graph $\langle S' \rangle_n$ defined by $S' = \{s_1, \ldots, s_{k-1}\}$. The minimum path length in this graph is $D_{n,S'}(js_k)$.

$$\text{Let } f(i,k) = \begin{cases} j + D_{n,S'}(js_k) & \text{if } i \equiv \pm js_k \pmod{n}, 1 \leq j \leq \lfloor n_k/2 \rfloor, \\ \infty & \text{otherwise, that is, } g_k \nmid i, \end{cases} \tag{1}$$

where $1 \leq i \leq n - 1$, and let $f(0, k) = n_k$.

Note that, for $1 \leq i \leq n - 1$, $g_k = \gcd(s_k, n) \mid i$ if and only if there is some j, where $1 \leq j \leq \lfloor n_k/2 \rfloor$, such that $i \equiv \pm js_k \pmod{n}$. Moreover, in this case, such a $\pm j$ is unique, (except in the case where n_k is even and $i = \frac{n_k}{2} s_k$, in which case there are exactly two values $+\frac{n_k}{2}$ and $-\frac{n_k}{2}$.)

To see this, one direction is trivial: if $i \equiv \pm js_k \pmod{n}$ for some j, then $g_k = \gcd(s_k, n) \mid i$. Now suppose $g_k \mid i$. Then i belongs to the subgroup generated by g_k in $\mathbb{Z}_n$, which is also generated by s_k. This subgroup has order n_k. Hence there is some j within the specified range $1 \leq j \leq \lfloor n_k/2 \rfloor$, such that

$$i \equiv \pm js_k \pmod{n} \tag{2}$$

$j \neq 0$ since $i \not\equiv 0 \pmod{n}$. In the general case when $g_k \mid i$, except when n_k is even and $i = \frac{n_k}{2} s_k$, the uniqueness of such a $\pm j$ is quite obvious. For otherwise we have $((\pm j) - (\pm j'))s_k \equiv 0 \pmod{n}$ which implies that $n_k \mid ((\pm j) - (\pm j'))$. And the only possibility for this to hold is $j = j' = n_k/2$ but they took opposite signs in (2), and $i = \frac{n_k}{2} s_k$. Indeed in this case the $\pm j$ in (2) is not unique, but nonetheless the value of $f(i, k)$ is uniquely defined.

This shows that the function $f(i, k)$ is well defined, and given any algorithm for finding shortest paths, this gives us an algorithm to compute $f(i, k)$. In particular this gives a polynomial time algorithm for $f(i, k)$.

Now we claim that the minimum loop size

$$L(n, S) = \min_{0 \leq i \leq n-1, 1 \leq k \leq m} f(i, k). \tag{3}$$

Clearly the minimum value on the right in (3) produces the length of some loop. Since all $s_k \neq 0$, using s_k alone, one can form a loop with exactly $f(0, k) = n_k$ steps (and no less). Thus the minimum is finite. In the case $1 \leq i \leq n - 1$, the value $f(i, k)$ gives the length of some loop, if such a loop exists, that consists of step sizes $s_1, \ldots, s_k$ only, with a non-zero number of step size s_k.

Let $\mathcal{L}$ be a loop achieving $L(n, S)$. We show that our minimum value on the right of (3) is at least as small. This establishes (3).

If $\mathcal{L}$ involves only one step size s_k, then $f(0, k) = n_k$ gives the shortest loop in this case. Let $\mathcal{L}$ involve more than one step size, and let k be the largest ℓ such that $\mathcal{L}$ involves a non-zero number x_ℓ of steps of size s_ℓ. Since $\mathcal{L}$ is minimal, $s_k x_k \not\equiv 0 \pmod{n}$, for otherwise, by deleting the steps involving s_k we still have a non-trivial loop.

Let i be the least modulus of $-s_k x_k \pmod{n}$, where $1 \leq i \leq n - 1$, then $\sum_{\ell=1}^{k-1} s_\ell x_\ell \equiv i \pmod{n}$, and x_ℓ is the number of steps in $\mathcal{L}$ involving s_ℓ. Since the subgroup of $\mathbb{Z}_n$ generated by s_k has order n_k, there is a j, $1 \leq j \leq \lfloor n_k/2 \rfloor$, such that $-s_k x_k \equiv \pm j s_k \pmod{n}$. This implies that $-x_k \equiv \pm j \pmod{n_k}$. Since $|\mathbf{x}|_{L_i}$ is minimal, $x_k = \pm j$.

After taking away the steps involving s_k, what remains is a path from 0 to $i \equiv -s_k x_k \pmod{n}$ using only steps $s_1, \ldots, s_{k-1}$, the minimum length of which has been found in the equation (1) defining $f(i, k)$.

This completes the description of our algorithm to compute $L(n, S)$. Now, we estimate its time complexity. Let $\langle S_l \rangle_n$ be the graph defined by $S_l = \{s_1, \ldots, s_l\}$, where $1 \leq l \leq m$. To compute $f(i, k)$, where $i \equiv \pm j s_k \pmod{n}$, we need to know the value of $D_{n, S_{k-1}}(j s_k)$. We first do a breadth-first search on $\langle S_l \rangle_n$, for $l = 1, \ldots, m$, to compute the length of the shortest path from 0 to every other vertex in each of these graphs. This can be done in a total time of $O(nm^2)$. The rest of the computation takes time $O(nm \log^{O(1)} n)$. Therefore we have the following result

Theorem 4.
The Shortest-Loop Problem is solvable in time $O(nm^2 + nm \log^{O(1)} n)$.

A more careful design of the algorithm can reduce the time to $O(nm)$.

4 NP-hardness for the *Shortest-Path* Problem

In this section we claim that the *Shortest-Path* problem is NP-hard. We can also show that this problem is NP-hard to approximate within a factor $n^{1/\log \log n}$ using the result of [13] and its recent improvement due to Ran Raz [26]. This will be discussed in the full paper, which will also contain omitted proofs. We first state a decision problem version of the *Shortest-Path* problem. This problem is NP-complete, and the NP-hardness of the optimization problem follows easily.

Shortest-Path Decision Problem(SPDP)
Instance: Given in binary an integer n, a set $S = \{s_1, \ldots, s_m\} \subseteq \mathbb{Z}_n$, a residue $w \in \mathbb{Z}_n$, $w \neq 0$, and a bound b.
Question: Is $D_{n,S}(w) \leq b$?
Size: $\text{length}(n) + \sum_{i=1}^{m} \text{length}(s_i) + \text{length}(w) + \text{length}(b)$.

Theorem 5. *The Shortest-Path Decision Problem (SPDP) is NP-complete.*

The proof is by reducing the well-known Exact-3-Cover NP-complete decision problem to the Shortest-Path Decision Problem.

5   The Hardness for Homogeneous Systems

In the last section we claimed that the optimization problem of finding $D_{n,S}(w)$ for a general right hand side w is NP-hard. This corresponds to the SVP in a special class of *affine* lattices. The homogeneous case where $w = 0$ is the SVP in a special class of *homogeneous* lattices, namely those definable by a single congruence of the form $\sum_{i=1}^{m} s_i x_i \equiv 0 \pmod{n}$. In this section, we focus on a worst-case/average-case connection.

There has been a considerable amount of work on the SVP for homogeneous lattices. Van Emde Boas [29] showed that it is NP-hard for the L_∞-norm (see also [20]). The NP-hardness for every other L_p-norm, especially for the L_2-norm, was open until a recent breakthrough by Ajtai [2], who showed that the problem for the L_2-norm is NP-hard under randomized reductions. Moreover Ajtai [2] also showed that it is NP-hard to approximate the shortest non-zero vector in an m-dimensional lattice within a factor of $(1 + 2^{-m^k})$, for a sufficiently large constant k. This was improved by Cai and Nerurkar [8] to a factor of $(1 + \frac{1}{m^\epsilon})$, for any $\epsilon > 0$. They [8] also noted explicitly that the proof works for arbitrary L_p-norms, $1 \leq p < \infty$. Micciancio further improved this factor to $2^{1/p} - \epsilon$ for the L_p-norm, $p \geq 1$ [24]. Approximations to the closest vector in a lattice were considered in [4,13]. However, in all these reductions, the lattices constructed do not fall into the category of lattices defined in terms of circulant graphs, that is, definable by a single congruence. Nevertheless, we claim that the SVP for such lattices is, in fact, NP-hard under randomized reductions. We do so by reducing the general SVP to it. Such a reduction was given previously by Paz and Schnorr [25] but our result uses more elementary ideas. We have the following theorem:

Theorem 6. *The SVP for the class of (homogeneous) lattices defined by circulant graphs is NP-hard under randomized reductions. Moreover, it remains NP-hard to find an approximate shortest vector in this class of lattices within a factor of $2 - \epsilon$, for any constant $\epsilon > 0$.*

Ajtai, in a separate work [1], established a reduction from the worst-case complexity of some problems believed to be hard, to the average-case complexity of the approximate SVP for a certain class of random lattices. This worst-case/average-case connection is the only such provable reduction known for a problem in NP believed to be hard and has generated a lot of interest [3,6,7,15,16,17,18]. We state three problems which are believed to be hard for any constants c_1, c_2 and c_3.

(P1) Find $\lambda_1(L)$, the length of a shortest non-zero vector in an m-dimensional lattice L, up to a polynomial factor m^{c_1}.

(P2) Find the shortest non-zero vector in an m-dimensional lattice, where the shortest vector is unique up to a polynomial factor m^{c_2}.

(P3) Find a basis in an m-dimensional lattice whose maximum length is the smallest possible, up to a polynomial factor m^{c_3}.

Let $\mathbb{Z}_q^{r \times l}$ denote the set of $r \times l$ matrices over $\mathbb{Z}_q$. For every r, l, q, $\Omega_{r,l,q}$ denotes the uniform distribution on $\mathbb{Z}_q^{r \times l}$. For every $X \in \mathbb{Z}_q^{r \times l}$, the set $\Lambda(X) = \{y \in \mathbb{Z}^l \mid Xy \equiv 0 \pmod q\}$ defines a lattice of dimension l. $\Lambda_{r,l,q}$ denotes the probability space of lattices consisting of $\Lambda(X)$ by choosing X according to $\Omega_{r,l,q}$. Let $q = r^c$ be an arbitrary but fixed polynomial of r. By Minkowski's Theorem, see Theorem 2 of Chapter 3 of [9], it can be proved that, $\forall c \, \exists c'$ such that $\forall \Lambda(X) \in \Lambda_{r,c'r,r^c}$, $\lambda_1(\Lambda(X)) \leq r$. (In fact, this bound can be improved to $\Theta(r^{1/2+\epsilon})$.) Let $l = c'r$ where c' depends on c as indicated above. Let $\Lambda = \Lambda_{r,l,q}$. Ajtai [1] showed:

Theorem 7. *Let γ be any constant. If there is a probabilistic polynomial time algorithm $\mathcal{A}$ such that, with non-trivial probability $1/r^{O(1)}$, $\mathcal{A}$ finds a short non-zero vector v, $|v|_{L_2} \leq r^\gamma$, for a uniformly chosen lattice in the class Λ indexed by r, then there is a probabilistic polynomial time algorithm $\mathcal{B}$ that solves the three problems (P1), (P2) and (P3) in the worst case, with high probability, for some constants c_1, c_2 and c_3.*

We claim here that these problems are also reducible to an average-case instance of the SVP for the lattice family defined by circulant graphs, under a certain distribution on such lattices. This is not quite an NP-hardness proof, since these problems (P1), (P2) and (P3) are not known to be NP-hard. In fact there is evidence that they are not. Goldreich and Goldwasser showed that approximating the shortest lattice vector within a factor of $O(\sqrt{m/\log m})$ is not NP-hard assuming the polynomial time hierarchy does not collapse [15]. Cai showed that finding an $m^{1/4}$-unique shortest lattice vector is not NP-hard unless the polynomial time hierarchy collapses [6]. On the other hand, these problems have resisted all attempts at finding a (probabilistic) polynomial time algorithm to date. The best polynomial time algorithm is essentially the L^3 basis reduction algorithm [22] which achieves an approximation factor of $2^{m/2}$ for the SVP. Schnorr improves this factor to $(1+\epsilon)^m$, still exponential in m and with the running time depending badly on ϵ in the exponent [27]. Thus it is reasonable to assume that these problems (P1), (P2) and (P3) are computationally hard.

Our proof establishes that for a certain distribution of circulant graphs, or equivalently for the special class of lattices, the average-case complexity of approximating the shortest vector in a lattice within any polynomial factor is at least as hard as these three problems in the worst-case. We use a construction that reduces the problem of finding a short vector in a lattice $\Lambda(X)$ to the problem of finding a short solution vector to an appropriate congruence.

Theorem 8. *Let α be any number. Given a vector $\mathbf{v} \in \Lambda(X)$, $0 < |\mathbf{v}|_{L_1} \leq \alpha$, a solution vector $\mathbf{x}$ to the congruence $\sum_{i=1}^{l+r} s_i x_i \equiv 0 \pmod{n}$, with $0 < |\mathbf{x}|_{L_1} \leq (r+1)\alpha$, can be computed in polynomial time, and given a solution $\mathbf{x}$ to the congruence, $0 < |\mathbf{x}|_{L_1} \leq \beta$, a vector $\mathbf{v} \in \Lambda(X)$ with $0 < |\mathbf{v}|_{L_1} \leq \beta$, can be computed in linear time.*

We now apply this reduction to show that, to solve a congruence of the form $\sum_{i=1}^{r} s_i x_i \pmod{n}$ is as hard on the average, under a suitable distribution on the s_i and n, as problems **(P1)**, **(P2)** and **(P3)** are in the worst case. We get the following worst-case/average-case connection.

Theorem 9. *Let $\gamma > 2.5$ be any constant in Theorem 7 and let $l = \Theta(r)$ as above. If there is a probabilistic polynomial time algorithm $\mathcal{C}$ such that, with non-trivial probability $1/r^{O(1)}$, $\mathcal{C}$ finds a solution vector x of L_1-norm length at most r^γ for the congruence $\sum_{i=1}^{l+r} s_i x_i \equiv 0 \pmod{n}$, where the s_i and n are chosen according to distribution D_{r,r^γ}, then there is a probabilistic polynomial time algorithm $\mathcal{D}$ that solves the three problems **(P1)**, **(P2)** and **(P3)** in the worst case, with high probability, for some constants c_1, c_2 and c_3.*

References

1. M. Ajtai. Generating hard instances of lattice problems. In *Proc. 28th Annual ACM Symposium on the Theory of Computing* (1996) 99–108.
2. M. Ajtai. The shortest vector problem in L_2 is NP-hard for randomized reductions. In *Proc. 30th Annual ACM Symposium on the Theory of Computing* (1998) 10–19.
3. M. Ajtai and C. Dwork. A public-key cryptosystem with worst-case/average-case equivalence. In *Proc. 29th ACM Symposium on Theory of Computing* (1997) 284–293.
4. S. Arora, L. Babai, J. Stern, and Z. Sweedyk. The hardness of approximate optima in lattices, codes, and systems of linear equations. In *Proc. 34th IEEE Symposium on Foundations of Computer Science (FOCS)*, 724-733, 1993.
5. J.-C. Bermond, F. Comellas and D. F. Hsu. Distributed loop computer networks: A survey. *Journal of Parallel and Distributed Computing* **24** (1995) 2–10.
6. J-Y. Cai. A relation of primal-dual lattices and the complexity of shortest lattice vector problem. *Theoretical Computer Science* **207** (1998) 105–116.
7. J-Y. Cai and A. Nerurkar. An improved worst-case to average-case connection for lattice problems. In *Proc. 38th IEEE Symposium on Foundations of Computer Science* (1997) 468–477.
8. J-Y. Cai and A. Nerurkar. Approximating the SVP to within a factor $\left(1 + \frac{1}{\dim^\epsilon}\right)$ is NP-hard under randomized reductions. In *Proc. 13th Annual IEEE Conference on Computational Complexity* (1998) 46–55.
9. J. W. S. Cassels. *An introduction to the geometry of numbers.* Springer-Verlag, 1959.
10. N. Chalamaiah and B. Ramamurty. Finding shortest paths in distributed loop networks. *Inform. Proc. Letters* **67** (1998) 157–161.
11. Y. Cheng and F. K. Hwang. Diameters of weighted loop networks. *J. of Algorithms* **9** (1988) 401–410.

12. J. A. Dias da Silva and Y. O. Hamidoune. Cyclic spaces for Grassmann derivatives and additive theory *Bull. Lond. Math. Soc.* **26** (1994) 140–146.

13. I. Dinur, G. Kindler and S. Safra. Approximating-CVP to within almost-polynomial factors is NP-hard. 1998.

14. P. Erdös and H. Heilbronn On the addition of residue classes mod *p* *Acta Arithm.* **9** (1964) 149–159.

15. O. Goldreich and S. Goldwasser. On the limits of non-approximability of lattice problems. In *Proc. 30th Annual ACM Symposium on the Theory of Computing* (1998) 1–9.

16. O. Goldreich, S. Goldwasser and S. Halevi. Collision-free hashing from lattice problems. 1996. Available as TR96-042 from *Electronic Colloquium on Computational Complexity* at http://www.eccc.uni-trier.de/eccc/.

17. O. Goldreich, S. Goldwasser and S. Halevi. Eliminating decryption errors in the Ajtai-Dwork cryptosystem. *Advances in Cryptology - CRYPTO '97* (editor B. Kaliski Jr.), Lecture Notes in Computer Science 1294 (Springer Verlag, 1997) 105–111.

18. O. Goldreich, S. Goldwasser and S. Halevi. Public-key cryptosystems from lattice reduction problems. Advances in Cryptology - CRYPTO '97 (editor B. Kaliski Jr.), Lecture Notes in Computer Science 1294 (Springer Verlag, 1997) 112–131.

19. D. J. Guan. Finding shortest paths in distributed loop networks. *Inform. Proc. Letters* **65** (1998) 255–260.

20. J. C. Lagarias. The computational complexity of simultaneous diophantine approximation problems. In *Proc. 23rd IEEE Symposium on Foundations of Computer Science* (1982) 32–39.

21. F. T. Leighton. *Introduction to parallel algorithms and architectures: Arrays, trees, hypercubes.* M. Kaufmann, 1992.

22. A. K. Lenstra, H. W. Lenstra and L. Lovász. Factoring polynomials with rational coefficients. *Mathematische Annalen* **261** (1982) 515–534.

23. B. Mans. Optimal Distributed algorithms in unlabeled tori and chordal rings. *Journal on Parallel and Distributed Computing* **46** (1997) 80–90.

24. D. Micciancio. The shortest vector in a lattice is hard to approximate to within some constant. In *Proc. 39th IEEE Symposium on Foundations of Computer Science* (1998) 92–98.

25. A. Paz and C. P. Schnorr. Approximating integer lattices by lattices with cyclic factor groups. Automata, Languages and Programming, 14th International Colloquium, *Lecture Notes in Computer Science* 267 (Springer-Verlag, 1987) 386-393.

26. R. Raz. Personal communication.

27. C. P. Schnorr. A hierarchy of polynomial time basis reduction algorithms. *Theoretical Computer Science* **53** (1987) 201–224.

28. J-P. Seifert and J. Blömer. On the complexity of computing short linearly independent vectors and short bases in a lattice. To appear in the proceedings of STOC 1999.

29. P. van Emde Boas. Another NP-complete partition problem and the complexity of computing short vectors in lattices. *Technical Report 81-04*, Mathematics Department, University of Amsterdam, 1981.

30. J. Žervonik and T. Pisanski. Computing the diameter in multi-loop networks. *J. of Algorithms* **14** (1993) 226–243.

Minimum Congestion Embedding of Complete Binary Trees into Tori

Akira Matsubayashi and Ryo Takasu

Department of Information Science, Utsunomiya University,
Utsunomiya 321-8585, Japan
mbayashi@is.utsunomiya-u.ac.jp

Abstract. We consider the problem of embedding complete binary trees into 2-dimensional tori with minimum (edge) congestion. It is known that for a positive integer n, a $2^n - 1$-vertex complete binary tree can be embedded in a $(2^{\lceil n/2 \rceil} + 1) \times (2^{\lfloor n/2 \rfloor} + 1)$-grid and a $2^{\lceil n/2 \rceil} \times 2^{\lfloor n/2 \rfloor}$-grid with congestion 1 and 2, respectively. However, it is not known if $2^n - 1$-vertex complete binary tree is embeddable in a $2^{\lceil n/2 \rceil} \times 2^{\lfloor n/2 \rfloor}$-grid with unit congestion. In this paper, we show that a positive answer can be obtained by adding wrap-around edges to grids, i.e., a $2^n - 1$-vertex complete binary tree can be embedded with unit congestion in a $2^{\lceil n/2 \rceil} \times 2^{\lfloor n/2 \rfloor}$-torus. The embedding proposed here achieves the minimum congestion and an almost minimum size of a torus (up to the constant term of 1). In particular, the embedding is optimal for the problem of embedding a $2^n - 1$-vertex complete binary tree with even n into a square torus with unit congestion.

1 Introduction

The problem of efficiently implementing parallel algorithms on parallel machines has been studied as the graph embedding problem, which is to embed the communication graph underlying a parallel algorithm within the processor interconnection graph for a parallel machine with minimal communication overhead. It is well known that the dilation and/or congestion of the embedding are lower bounds on the communication delay, and the problem of embedding a guest graph within a host graph with minimal dilation and/or congestion has been extensively studied. In particular, it was pointed out by Kim and Lai [2] that minimal congestion embeddings are very important for a parallel machine that uses circuit switching for node-to-node communication.

In this paper, we consider minimal congestion embeddings of complete binary trees in tori. Complete binary trees are well known as one of the most fundamental communication graphs for divide-and-conquer algorithms. Also, tori are well known as one of the most popular processor interconnection graphs for parallel machines.

Gordon [1] showed that for a positive integer n, a $2^n - 1$-vertex complete binary tree denoted by $C(n)$ can be embedded into a $(2^{\lceil n/2 \rceil} + 1) \times (2^{\lfloor n/2 \rfloor} + 1)$-grid with unit congestion. Zienicke [4] showed that $C(n)$ can be embedded into

T. Asano et al. (Eds.): COCOON'99, LNCS 1627, pp. 370–378, 1999.

a $2^{\lceil n/2 \rceil} \times 2^{\lfloor n/2 \rfloor}$-grid with congestion 2. The same result under a constraint of row-column routing was shown by Lee and Choi [3].

Although it is an interesting question to ask if $C(n)$ is embeddable in a $2^{\lceil n/2 \rceil} \times 2^{\lfloor n/2 \rfloor}$-grid with unit congestion, we have no answer for the problem. Lee and Choi [3] mentioned that this would be negative.

Since a torus contains the grid of the same side lengths as a subgraph, we can immediately obtain from the results of [1,4,3] that $C(n)$ can be embedded in a $(2^{\lceil n/2 \rceil} + 1) \times (2^{\lfloor n/2 \rfloor} + 1)$-torus and a $2^{\lceil n/2 \rceil} \times 2^{\lfloor n/2 \rfloor}$-torus with congestion 1 and 2, respectively. However, it is not known whether $C(n)$ is embeddable in a $2^{\lceil n/2 \rceil} \times 2^{\lfloor n/2 \rfloor}$-torus with unit congestion. In this paper, we give a positive answer for the question by proving the following theorem:

Theorem 1. *For a positive integer n, $C(n)$ can be embedded into a $2^{\lceil n/2 \rceil} \times 2^{\lfloor n/2 \rfloor}$-torus with unit congestion.*

We construct an embedding satisfying the condition of Theorem 1 by using Gordon's embeddings [1]. The embedding proposed here achieves the minimum congestion and an almost minimum size of a torus (up to the constant term of 1). In particular, the embedding is optimal for the problem of embedding $C(n)$ with even n into a square torus with unit congestion.

The paper is organized as follows: Some definitions are given in Sect. 2. In Sect. 3, we review the Gordon's embeddings. Based on the results, we prove Theorem 1 in Sect. 4.

2   Preliminaries

Let G be a graph and let $V(G)$ and $E(G)$ denote the vertex set and edge set of G, respectively.

The *(two dimensional) $m_1 \times m_2$-grid* denoted by $M(m_1, m_2)$ is the graph with vertex set $\{(i, j) \mid 0 \le i < m_1, 0 \le j < m_2\}$ and edge set $\{((i, j), (i + 1, j)) \mid 0 \le i < m_1 - 1, 0 \le j < m_2\} \cup \{((i, j), (i, j + 1)) \mid 0 \le i < m_1, 0 \le j < m_2 - 1\}$. The *(two dimensional) $m_1 \times m_2$-torus* denoted by $D(m_1, m_2)$ is the graph obtained from $M(m_1, m_2)$ by adding *wrap-around edges* $((i, 0), (i, m_2 - 1))$ $(0 \le i < m_1)$ and $((0, j), (m_1 - 1, j))$ $(0 \le j < m_2)$. We denote $M(m, m)$ and $D(m, m)$ by $M^2(m)$ and $D^2(m)$, respectively.

An *embedding* $\langle \phi, \rho \rangle$ of a graph G into a graph H is defined by a one-to-one mapping $\phi : V(G) \to V(H)$, together with a mapping ρ that maps each edge $(u, v) \in E(G)$ onto a set of edges of H which induces a path connecting $\phi(u)$ and $\phi(v)$. The *dilation* of $\langle \phi, \rho \rangle$ is $\max_{e_G \in E(G)} |\rho(e_G)|$. The *(edge) congestion* of $\langle \phi, \rho \rangle$ is $\max_{e_H \in E(H)} |\{e_G \in E(G) \mid e_H \in \rho(e_G)\}|$.

For an embedding $\varepsilon = \langle \phi, \rho \rangle$ of a graph G into a graph H, let ϕ^ε and ρ^ε denote ϕ and ρ, respectively. For $U \subseteq V(G)$, let $\phi^\varepsilon(U) = \{\phi^\varepsilon(v) \mid v \in U\}$. Moreover, for $S \subseteq E(G)$, let $\rho^\varepsilon(S) = \bigcup_{e \in S} \rho^\varepsilon(e)$.

3 Gordon's Embeddings

In this section, we review the embeddings given in [1] which embed complete binary trees into grids with unit congestion.

Let G_1, G_2, and G_3 be graphs. For an embedding $\varepsilon_1 = \langle \phi_1, \rho_1 \rangle$ of G_1 into G_2 and a dilation-1 embedding $\varepsilon_2 = \langle \phi_2, \rho_2 \rangle$ of G_2 into G_3, we denote by $\varepsilon_2 \circ \varepsilon_1$ the embedding $\langle \phi_3, \rho_3 \rangle$ of G_1 into G_3 defined by $\phi_3 : u \in V(G_1) \mapsto \phi_2(\phi_1(u))$ and $\rho_3 : e \in E(G_1) \mapsto \rho_2(\rho_1(e))$. It should be noted that since $\rho_1(e)$ is a set of edges which induces a path of G_2 and the dilation of ε_2 is one, $\rho_2(\rho_1(e))$ is a set of edges which induces a path of G_3.

For an embedding ε of a graph into $M^2(m)$, we denote $\psi_m \circ \varepsilon$ by $\bar{\varepsilon}$, where ψ_m is the dilation-1 embedding, or the autoisomorphism of $M^2(m)$ which maps $(i,j) \in V(M^2(m))$ $(0 \le i \le m-1, 0 \le j \le m-1)$ to $(m-1-i, m-1-j)$. We define that w_m, x_m, y_m, and z_m are the dilation-1 embeddings of $M^2(m)$ into $M^2(2m-1)$ such that $(i,j) \in V(M^2(m))$ $(0 \le i \le m-1, 0 \le j \le m-1)$ is mapped to vertices (i,j), $(i, j+m-1)$, $(i+m-1, j)$, and $(i+m-1, j+m-1)$, respectively, of $M^2(2m-1)$.

For embeddings ε and ε' of a graph G into $M(m_1, m_2)$, we write $\varepsilon | \varepsilon'$ if ε and ε' satisfy the following conditions:

- $(i, m_2 - 1) \notin \phi^\varepsilon(V(G))$ or $(i, 0) \notin \phi^{\varepsilon'}(V(G))$ for $0 \le i \le m_1 - 1$.
- $((i, m_2 - 1), (i+1, m_2 - 1)) \notin \rho^\varepsilon(E(G))$ or $((i, 0), (i+1, 0)) \notin \rho^{\varepsilon'}(E(G))$ for $0 \le i \le m_1 - 2$.

We write $\varepsilon / \varepsilon'$ if ε and ε' satisfy the following conditions:

- $(m_1 - 1, j) \notin \phi^\varepsilon(V(G))$ or $(0, j) \notin \phi^{\varepsilon'}(V(G))$ for $0 \le j \le m_2 - 1$.
- $((m_1 - 1, j), (m_1 - 1, j+1)) \notin \rho^\varepsilon(E(G))$ or $((0, j), (0, j+1)) \notin \rho^{\varepsilon'}(E(G))$ for $0 \le j \le m_2 - 2$.

Lemma A (Gordon[1]). *For an even integer n, there exists an embedding of $C(n+2)$ into $M^2(2m+1)$ $(m = 2^{n/2})$ with unit congestion if there exist embeddings W, X, Y, and Z satisfying the following condition:*

Condition 1.

(a) W, X, Y, and Z are embeddings of $C(n)$ into $M^2(m+1)$ with unit congestion.
(b) $W|X$, $Z|Y$, $W/\overline{Y}$, $X/\overline{Z}$.
(c) $(m, m) \notin \phi^\varepsilon(V(C(n)))$ for $\varepsilon \in \{W, Z\}$.
(d) $(m, 0) \notin \phi^\varepsilon(V(C(n)))$ for $\varepsilon \in \{X, Y\}$.
(e) $(m, m/2) \notin \phi^\varepsilon(V(C(n)))$ for $\varepsilon \in \{W, X, Y, Z\}$.
(f) $\{((m, j), (m, j+1)) \mid m/2 \le j < m\} \cap \rho^\varepsilon(E(C(n))) = \emptyset$ for $\varepsilon \in \{W, Z\}$.
(g) $\{((m, j), (m, j+1)) \mid 0 \le j < m/2\} \cap \rho^\varepsilon(E(C(n))) = \emptyset$ for $\varepsilon \in \{X, Y\}$.
(h) $\{((i, m/2), (i+1, m/2)) \mid m/2 \le i < m\} \cap \rho^\varepsilon(E(C(n))) = \emptyset$ for $\varepsilon \in \{W, X, Y, Z\}$.
(i) ϕ^ε maps the root of $C(n)$ to $(m/2, m/2)$ for $\varepsilon \in \{W, X, Y, Z\}$.

$\square$

This lemma can be proved by constructing a desired embedding, which is obtained by (i) embedding four $C(n)$'s with $w_{m+1} \circ W$, $x_{m+1} \circ X$, $y_{m+1} \circ \overline{Y}$, and $z_{m+1} \circ \overline{Z}$, (ii) mapping the root r of $C(n+2)$, r's child c_1, and the other child c_2 to (m,m), $(m,m/2)$, and $(m,3m/2)$, respectively, (iii) and connecting r, c_i ($i = 1, 2$), and c_i's children with the shortest paths as shown in Fig. 1. It is easy to see that this is an embedding of $C(n+2)$ into $M^2(2m+1)$ with unit congestion. We denote by $F_n(W, X, Y, Z)$ the embedding of $C(n+2)$ into $M^2(2m+1)$ which is constructed as described above from four embeddings W, X, Y, and Z satisfying Condition 1 for an even integer n and $m = 2^{n/2}$.

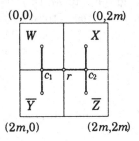

Fig. 1. Embedding of $C(n+2)$ into $M^2(2m+1)$

Theorem B (Gordon[1]). *For an even integer $n \geq 8$, there exist embeddings P_n, Q_n, R_n, S_n, and T_n of $C(n)$ into $M^2(m+1)$ ($m = 2^{n/2}$) with unit congestion such that the following conditions are satisfied:*

Condition 2.

(a) $(0,0) \notin \phi^\varepsilon(V(C(n)))$ for $\varepsilon \in \{P_n, R_n, S_n\}$.
(b) $\{(0,m), (m,0), (m,m/2)\} \cap \phi^\varepsilon(V(C(n))) = \emptyset$ for $\varepsilon \in \{P_n, Q_n, R_n, S_n, T_n\}$.
(c) $(m,m) \notin \phi^\varepsilon(V(C(n)))$ for $\varepsilon \in \{P_n, Q_n, S_n, T_n\}$.

Condition 3.

(a) $\{((0,j), (0, j + 1)) \mid 0 \leq j < m\} \cap \rho^\varepsilon(E(C(n))) = \emptyset$ for $\varepsilon \in \{P_n, Q_n, R_n, S_n\}$.
(b) $\{((m, j), (m, j+1)) \mid 0 \leq j < m\} \cap \rho^\varepsilon(E(C(n))) = \emptyset$ for $\varepsilon \in \{P_n, Q_n, S_n, T_n\}$.
(c) $\{((i,0), (i + 1, 0)) \mid 0 \leq i < m\} \cap \rho^\varepsilon(E(C(n))) = \emptyset$ for $\varepsilon \in \{P_n, R_n, S_n, T_n\}$.
(d) $\{((i,m), (i+1, m)) \mid 0 \leq i < m\} \cap \rho^\varepsilon(E(C(n))) = \emptyset$ for $\varepsilon \in \{P_n, Q_n, R_n, S_n, T_n\}$.
(e) $\{((m, j), (m, j + 1)) \mid 0 \leq j < m/2\} \cap \rho^{R_n}(E(C(n))) = \emptyset$.
(f) $\{((i, m/2), (i + 1, m/2)) \mid m/2 \leq i < m\} \cap \rho^\varepsilon(E(C(n))) = \emptyset$ for $\varepsilon \in \{P_n, Q_n, R_n, S_n, T_n\}$.

Condition 4. $P_n|Q_n$, $P_n|S_n$, $Q_n|Q_n$, $P_n|R_n$, $S_n|R_n$, $T_n|R_n$, $S_n|S_n$, $T_n|Q_n$, $R_n|P_n$, $Q_n|S_n$, $S_n|T_n$.

Condition 5. $P_n/\overline{S_n}$, $Q_n/\overline{P_n}$, $Q_n/\overline{R_n}$ $S_n/\overline{R_n}$, $R_n/\overline{T_n}$, $\overline{S_n}/P_n$, $\overline{P_n}/R_n$, $\overline{R_n}/T_n$, $\overline{S_n}/Q_n$.

Condition 6. ϕ^ε maps the root of $C(n)$ to $(m/2, m/2)$ for $\varepsilon \in \{P_n, Q_n, R_n, S_n, T_n\}$.

□

We describe here outlines of the definitions of Q_n and S_n, which are used to construct our embedding. Q_n and S_n are defined recursively as follows: First, Q_8 and S_8 are defined as shown in Fig. 2. Next, for an even integer $n \geq 8$, we assume that there exist P_n, Q_n, R_n, and S_n. Q_{n+2} and S_{n+2} are defined as $F_n(Q_n, Q_n, R_n, P_n)$ and $F_n(S_n, S_n, R_n, P_n)$, respectively (Fig. 3). See [1] for the complete definitions.

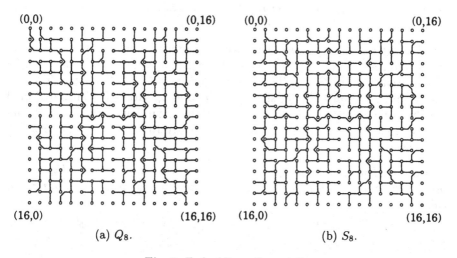

(a) Q_8.   (b) S_8.

Fig. 2. Embeddings Q_8 and S_8

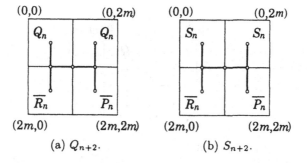

(a) Q_{n+2}.   (b) S_{n+2}.

Fig. 3. Recursive constructions of Q_{n+2} and S_{n+2} ($n \geq 8$)

4   Proof of Theorem 1

In this section, we prove Theorem 1 by a sequence of lemmas.

Lemma 2. *For an even integer $n \geq 8$, $Q_n/\overline{S_n}$.*

Proof. It is easy from Fig. 2 to see that $Q_8/\overline{S_8}$. Thus, we show that $Q_{n+2}/\overline{S_{n+2}}$ for an even integer $n \geq 8$. It follows from Theorem B (Condition 5) that $\overline{P_n}/R_n$ and hence $\overline{R_n}/P_n$. Moreover, it follows from Theorem B ((a) and (b) in Condition 2) that $(0,0) \notin \phi^{R_n}(V(C(n)))$ and $(0,m) \notin \phi^{P_n}(V(C(n)))$. From these facts and the definitions of Q_{n+2} and S_{n+2} (Fig. 3), we have that $Q_{n+2}/\overline{S_{n+2}}$.   □

Lemma 3. *For an even integer $n \geq 8$, Q_n, Q_n, S_n, and S_n satisfy Condition 1 for W, X, Y, and Z, respectively.*

Proof. Since Q_n and S_n satisfy Conditions 2 through 6 by Theorem B, it is not difficult to see that (c), (d), (e), and (i) in Condition 1 are satisfied. Also, (b) in Condition 1 is satisfied by Theorem B (Condition 4) and Lemma 2. Then, (f) and (g) in Condition 1 are satisfied since Q_n and S_n satisfy (b) in Condition 3. Finally, (h) in Condition 1 is satisfied by the definitions of Q_n and S_n and the fact that for $n \geq 10$, P_{n-2} and R_{n-2} satisfy (d) and (c) in Condition 3, respectively.   □

Lemma 4. *For an even integer $n \geq 10$, there exists an embedding U_n of $C(n)$ into $M^2(2^{n/2}+1)$ with unit congestion such that the following condition is satisfied:*

Condition 7. $U_n|U_n, U_n/U_n$, and $\{(2^{n/2},0),(0,2^{n/2}),(2^{n/2},2^{n/2})\} \cap \phi^{U_n}(V(C(n))) = \emptyset$.

Proof. Let $n \geq 10$ be an even integer and $m = 2^{n/2}$. From Theorem B, there exist embeddings Q_{n-2} and S_{n-2} with unit congestion such that Conditions 2 through 6 are satisfied. We define as shown in Fig. 4 that $U_n = F_{n-2}(Q_{n-2}, Q_{n-2}, S_{n-2}, S_{n-2})$, which is an embedding of $C(n)$ into $M^2(m+1)$ with unit congestion by Lemmas A and 3.

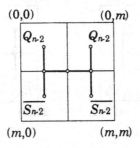

Fig. 4. Embedding U_n

The following claims shows that U_n satisfies Condition 7.

Claim 5. $U_n | U_n$.

Proof. Immediate from the definition of U_n and Lemma 3.

End of proof of Claim 5

Claim 6. U_n / U_n.

Proof. It follows from Theorem B (Condition 5) that $\overline{S_n}/Q_n$. Moreover, it follows from Theorem B ((a) and (b) in Condition 2) that $\{(0, m/2), (0, 0)\} \cap \phi^{Q_{n-2}}(V(C(n-2))) \subseteq \{(0,0)\}$ and $\{(m/2, 0), (m/2, m/2)\} \cap \phi^{\overline{S_{n-2}}}(V(C(n-2))) = \emptyset$. From these facts and the definition of U_n, we have that U_n/U_n.

End of proof of Claim 6

Claim 7. $\{(m, 0), (0, m), (m, m)\} \cap \phi^{U_n}(V(C(n))) = \emptyset$.

Proof. As shown in the proof of Claim 6, it follows that $(0, m/2) \notin \phi^{Q_{n-2}}(V(C(n-2)))$ and $\{(m/2, 0), (m/2, m/2)\} \cap \phi^{\overline{S_{n-2}}}(V(C(n-2))) = \emptyset$. Thus, the claim holds by the definition of U_n.

End of proof of Claim 7

Thus, U_n satisfies Condition 7. Therefore, the proof of Lemma 4 is completed.

□

Lemma 8. *For an even integer $n \geq 10$, there exists an embedding of $C(n)$ into $D^2(2^{n/2})$ with unit congestion.*

Proof. Let $m = 2^{n/2}$, $C = C(n)$, $M = M^2(m+1)$, and $D = D^2(m)$. We define that $\theta_m : (i, j) \in V(M) \mapsto (i \bmod m, j \bmod m) \in V(D)$ and $\lambda_m : (u, v) \in E(M) \mapsto (\theta_m(u), \theta_m(v)) \in E(D)$.

By Lemma 4, there exists an embedding U_n of C into M such that Condition 7 is satisfied. We construct from U_n a desired embedding $\langle \phi, \rho \rangle$ of C into D. Let $\phi : v \in V(C) \mapsto \theta_m(\phi^{U_n}(v))$, and for an edge $(u, v) \in E(C)$, let $\tau((u, v)) = \{\lambda_m(e) \mid e \in \rho^{U_n}((u, v))\}$. Since λ_m maps two adjacent edges of M to two adjacent ones of D by definition, $\tau((u, v))$ induces a connected subgraph of D which contains $\phi(u)$ and $\phi(v)$. Thus, there exists a subset of $\tau((u, v))$ which induces a path connecting $\phi(u)$ and $\phi(v)$[1]. We define that $\rho((u, v))$ is the subset of $\tau((u, v))$.

Since U_n satisfies Condition 7, it follows that ϕ is a one-to-one mapping of $V(C)$ to $V(D)$ and that for distinct edges e and e' of C, $\tau(e)$ and $\tau(e')$ are disjoint. Thus, $\langle \phi, \rho \rangle$ is an embedding of C into D with unit congestion.   □

It is not difficult to see that there exists an embedding of $C(n)$ into $D^2(2^{n/2})$ for even integer $n \leq 8$. Figure 5 shows examples of such embeddings for $n = 6$ and $n = 8$. Thus, we have the following lemma, which proves Theorem 1 for the case that n is even:

Lemma 9. *For an even integer n, there exists an embedding of $C(n)$ into $D^2(2^{n/2})$ with unit congestion.*   □

---

[1] Indeed, $\tau((u, v))$ itself induces a path for U_n constructed in the proof of Lemma 4.

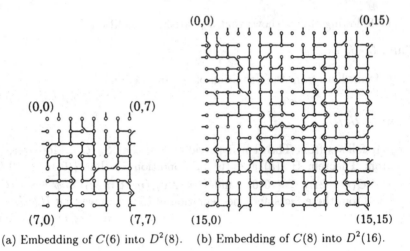

(a) Embedding of $C(6)$ into $D^2(8)$.   (b) Embedding of $C(8)$ into $D^2(16)$.

Fig. 5. Congestion-1 Embeddings of $C(n)$ into $D^2(2^{n/2})$ for $n = 6$ and $n = 8$. Wrap-around edges are represented by half lines

It remains to show that Theorem 1 holds for the case of odd n. For an even integer $n \geq 10$, we can obtain from the definition of U_n constructed in the proof of Lemma 4 an embedding U'_n of $C(n-1)$ into $M(m+1, m/2+1)$ ($m = 2^{n/2}$) as shown in Fig. 6. From Theorem B (Conditions 2 and 3) and

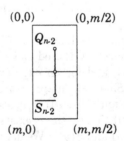

Fig. 6. Embedding U'_n

Lemma 4 (Condition 7), it is not difficult to see that U'_n satisfies the following condition:

Condition 8. $U'_n|U'_n$, U'_n/U'_n, and $\{(m, 0), (0, m/2), (m, m/2)\} \cap \phi^{U'_n}(V(C(n-1))) = \emptyset$.

Thus, we can construct an embedding of $C(n-1)$ into $D(m, m/2)$ with unit congestion by a similar argument of the proof of Lemma 8. Therefore, although the details are omitted here, we have the following lemma:

Lemma 10. *For an odd integer n, $C(n)$ can be embedded into $D(2^{(n+1)/2}, 2^{(n-1)/2})$ with unit congestion.* $\square$

Lemmas 9 and 10 complete the proof of Theorem 1.

References

1. D. Gordon: Efficient embeddings of binary trees in VLSI arrays. IEEE Trans. Computers **C-36** (1987) 1009–1018
2. Y. M. Kim and T.-H. Lai.: The complexity of congestion-1 embedding in a hypercube. J. Algorithms **12** (1991) 246–280
3. S.-K. Lee and H.-A. Choi: Embedding of complete binary trees into meshes with row-column routing. IEEE Trans. Parallel and Distributed Systems. **7** (1996) 493–497
4. P. Zienicke: Embeddings of treelike graphs into 2-dimensional meshes. In: R. H. Möhring (eds.): Graph-Theoretic Concepts in Computer Science (Proceedings of WG'90). Lecture Notes in Computer Science, Vol. 484. Springer-Verlag, Berlin (1991) 182–192

Maximum Stabbing Line in 2D Plane[*]

Francis Y. L. Chin[1], Cao An Wang[2], and Fu Lee Wang[3]

[1] Department of Computer Science and Information Systems,
The University of Hong Kong, Pokfulam Road, Hong Kong
chin@csis.hku.hk
[2] Department of Computer Science,
Memorial University of Newfoundland,
St. John's, NFLD, Canada.
wang@cs.mun.ca
[3] Department of Computer Science and Information Systems,
The University of Hong Kong, Pokfulam Road, Hong Kong
flwang@csis.hku.hk

Abstract. Two line estimator problems using the perpendicular and vertical distance measure are considered in this paper. (1) Given a set of n points in a plane, the *line estimator* problem is to determine a straight line which best fits this set of points. (2) Given a sequence of n points in a plane, the *ordered line estimator* problem is to determine a straight line which can best fit this set of points in order. Depending on the perpendicular/vertical distance measure, these two problems are related to the stabbing line problems of n circles/vertical line segments. With arrangement of curves, both problems can be solved in $O(n^2)$ and $O(n^3)$ time respectively.

1   Introduction

Given n simple geometric objects in a 2D plane, a *transversal* (stabbing line) is a straight line that intersects all n objects. Active research on transversal has been carried out in the past, e.g., Edelsbrunner has developed an algorithm [6] to find a transversal for a set of n simple geometric objects in $O(n \log n)$ time. Edelsbrunner et. al. [7] and Avis and Wenger [2] have independently presented algorithms to find a transversal for n line segments in $O(n \log n)$ time.

The *transversal* problem is related to the *line estimator* problem [9,14,16], which is to find a straight line approximating a set of points $P = \{p_i = (a_i, b_i) \mid 1 \le i \le n\}$ in a 2D plane. The distance $d(p, l)$ between point p and line l can be either *perpendicular distance* or *vertical distance* (Fig. 1).

When the objects in the transversal problem are circles with radius ε and centered at p_i's, the transversal l will approximate P with perpendicular distance error ε. On the other hand, if the objects are vertical line segments, of length 2ε with the mid-point at p_i's, then the transversal l will approximate P

---

[*] This research is supported by RGC Research Grant HKU7024/98E

T. Asano et al. (Eds.): COCOON'99, LNCS 1627, pp. 379–388, 1999.
© Springer-Verlag Berlin Heidelberg 1999

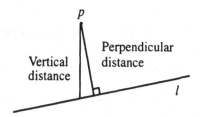

Fig. 1. Perpendicular and Vertical Distance

with the vertical distance error ε. When transversal does not exist, the *line estimator* problem is to find a straight line which intersects the maximum number of circles/vertical line segments [9,14,16].

In some applications, such as polygonal curves [3,10,13], objects are represented by a sequence of points. An *ordered line estimator* (OLE) for an "ordered" set of points, besides best "fits" the given set of points with respect to a given distance error, has to consider the order of stabbing or visitation. Fig. 2 gives a sequence of 8 points representing the polygonal curve (solid line). Both line estimators, l_1 and l_2, can best fit six points within distance error ε. If the order of stabbing is also considered, l_1 can visit four points, e.g., 1236 or 1237 or 1246 or 1247, within error ε and in order, while l_2 can visits six points 1, 2, 4, 5, 7, 8 within distance error ε and in order. In other words, l_2 is the ordered line estimator that can better fit this sequence of points.

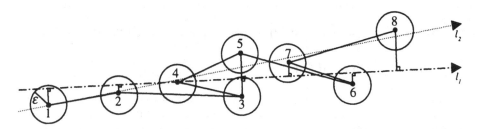

Fig. 2. Approximation of Polygonal Curves

Given a directed line l and a sequence of n points $P = \{p_1, p_2, \ldots p_n\}$, each of which is associated with a circle of radius ε or a vertical line segment of length 2ε, the *visiting order* of P', $P' \subset P$, w.r.t. l can be determined by the "stabbing order" of these circles or vertical line segments by l. We say l *visits* P'' *in order*, when P'' is an increasing subsequence of the visiting order of P' w.r.t. l. For example, in Fig. 2, the visiting order of l_1 is 124376 and l_1 visits 1236 or 1237 or 1246 or 1247 in order. Similarly, the visiting order of l_2 is 124578 and l_2 also visits these points in order.

The *minimum k-error line estimator* (MkLE) which is a variation of the line estimator problem can be defined in one of the following ways:

- to find a straight line that stabs through at least k out of n vertical line segments with minimum length or circles with minimum radius.
- to find a straight line which minimizes the distance to its kth closest point, in other words, ignores the $(n-k)$ outlying points and minimizes the maximum distance of its k closest points.
- to find a strip of region, bounded by two parallel lines with minimum width, that can cover at least k points.

The *LMS (Least Median-of-Square) line estimator* problem [9,16], which ignores 50% of the outlying points, is a special case of the MkLE problem for vertical distance with $k=\frac{n}{2}$ that can be solved in $O(n^2)$ time. Recently, an $O(n\log n)$ time randomized Las Vegas approximation algorithm and another more practical $O(n^2 \log n)$ time algorithm for the LMS problem were presented in [14]. A similar problem for perpendicular distance was presented in [4]. Basically MkLE and the line estimator problem studied in this paper are dual problems that find the optimal value of k or ε by fixing the other. As far as we know, the order line estimator problem has not been studied previously.

Section 2 gives an $O(n^2)$ algorithm to solve the line estimator problem for perpendicular distance, or equivalently, to find a transversal which intersects the maximum number of circles. Section 3 describes an $O(n^3)$ algorithm to solve the ordered line estimator problem for perpendicular distance. These two line estimator problems for vertical distance, i.e., the transversal problem for n vertical line segments, can also be solved with similar time complexities (Section 4). Section 5 concludes.

2   The Line Estimator Problem of Perpendicular Distance

2.1   Stabbing Lines of Circles

A straight line l can be expressed in the Normal Form $x \cos\theta + y \sin\theta - \rho = 0$, and each circle C_i by its radius r_i and center (ρ_i, θ_i). In this paper, we assume $r_i = \varepsilon$ for all $1 \leq i \leq n$. Thus a straight line l in the primal plane can be transformed to a point (ρ, θ) in the dual plane (Fig. 3). A straight line l intersects circle C_i iff $\rho_i \cos(\theta - \theta_i) - r_i \leq \rho \leq \rho_i \cos(\theta - \theta_i) + r_i$. Thus, for each circle centered at (ρ_i, θ_i) and with radius r_i, we shall construct two curves which form a strip, *upper curve* $\rho_i \cos(\theta - \theta_i) + r_i$ and *lower curve* $\rho_i \cos(\theta - \theta_i) - r_i$ in the dual plane. A point within the upper and lower curves represents an intersecting line to the circle (the shaded strip in Fig. 3b). It is easy to see that the shaded strip in the dual plane has constant vertical width of $2r_i$, and all the lines in the primal plane can be represented as (ρ, θ) in the dual plane, where $(0° \leq \theta < 360°)$. Note that the lines on the dual plane are symmetric about the 180° line as they represent lines of opposite direction.

Line $l = (\rho, \theta)$ is a transversal of n circles iff $\rho_i \cos(\theta - \theta_i) - r_i \leq \rho \leq \rho_i \cos(\theta - \theta_i) + r_i$ for all $1 \leq i \leq n$, or equivalently, $\max\{\rho_i \cos(\theta - \theta_i) - r_i \mid 1 \leq i \leq n\} \leq \rho \leq \min\{\rho_i \cos(\theta - \theta_i) + r_i \mid 1 \leq i \leq n\}$.

Atallah and Bajaj [1] proved that, given n circles, a transversal l can be found in $O(n \log n \cdot \alpha(n))$ time, where $\alpha(n)$ is the inverse Ackermann's function. However, if there does not exist a transversal $l = (\rho, \theta)$ which satisfies all the inequalities, the line estimator problem is to find a straight line $l = (\rho, \theta)$, which stabs through as many circles as possible, i.e., to find a solution $l = (\rho, \theta)$ which satisfies the maximum number of the above inequalities.

If we plot all the inequalities, the strips will divide the plane into many zones (*arrangement* of curves), each of which is covered by different number of strips (Fig. 3b). The line estimator problem is to find a zone that is covered by the maximum number of strips. Points in that zone represent those straight lines which stab through the maximum number of circles.

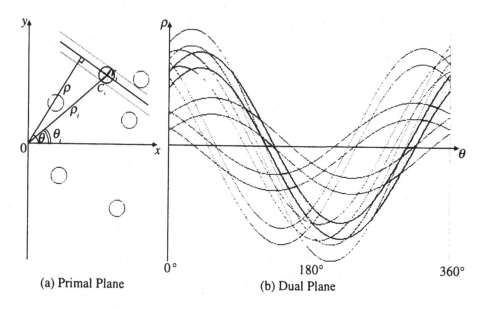

(a) Primal Plane   (b) Dual Plane

Fig. 3. Arrangement of Stabbing Line Family to Circles

2.2 Arrangement Representation

Similar to the approach described in [5,8], an arrangement is represented by an incidence graph and the arrangement of curves can be constructed incrementally, i.e., by adding each curve one by one to an already existing arrangement. Assume l is the curve to be added.

Step 1: To locate the leftmost edge e_0 in the already existing arrangement, which l intersects. The process starts with an arbitrary edge in the vertical line at $0°$ and proceeds edge by edges closer to l until e_0 is reached.

Step 2: (a) Set $i \leftarrow 0$;

       (b) On the other side of the zone associated with e_i (e_i is usually on the left boundary of the zone), locate the other edge e_{i+1} that intersects with l. The incidence graph is updated with all the changes in the arrangement, i.e., the splitting of edges e_i and e_{i+1} into two, the insertion of a new edge which is part of l inside the zone, the insertion of a new vertex (intersection point of e_{i+1} and l) and the update of the zone information.

       (c) Repeat (b) with $i \leftarrow i + 1$ until l intersects with the vertical line at $180°$.

The time complexity of the construction depends very much on the size of the arrangement. Since all circles are of the same size, the strips in the dual plane corresponding to these circles must have the same vertical distance and every two curves can intersect with each other at no more than one position. Thus, similar to the arrangement of straight lines [5,8], the size of the arrangement and the time of its construction must be $O(n^2)$.

2.3   Covering Number Calculation

Each zone in the arrangement is associated with a *covering number* which indicates the number of strips covering that zone, alternatively, the number of circles stabbed through by those lines corresponding to points in that zone. Thus, points in the zone with the maximum covering number represent those lines which stab through the maximum number of circles.

To calculate the covering numbers, we can start at the uppermost zone which is not covered by any strip, so its covering number must be zero. We shall then sweep through the arrangement from top to bottom traversing from one zone to another one by one, and at the same time calculate the covering number of each zone. When we cross over an upper curve, the zone will be covered by one more strip, so its covering number will be increased by one. On the other hand, if we cross over a lower curve, the covering number is reduced by one.

Lemma 1. *The covering numbers of all zones in the arrangement can be calculated in $O(n^2)$ time.*

$\square$

Theorem 2. *The line estimator problem of n points with perpendicular distance error can be solved in $O(n^2)$ time.*

$\square$

3   The Ordered Line Estimator Problem of Perpendicular Distance

The ordered line estimator problem is solved by a similar approach as the line estimator problem. In addition to the covering number, each zone in the arrangement is associated with a visiting order of the circles. Our approach is to find the longest increasing subsequence (LIS) of the visiting order of those stabbing circles in each zone, and then to find the longest LIS among the LIS's of all zones. As there are $O(n^2)$ zones and each LIS problem takes $O(n \log n)$ time [12,15], the ordered line estimator problem can be solved in $O(n^3 \log n)$ time. In this section, we shall show that since the difference of the visiting orders of circles between two neighboring zones is small, based on the information of the neighboring zones, the LIS problem in each zone can be solved in $O(n)$ time and thus the ordered line estimator problem can be reduced to $O(n^3)$ time.

Theorem 3. *The visiting orders of the circles corresponding to two neighboring zones differ by at most one circle, i.e., equivalent to an insertion/deletion of a circle.*

Proof. Let's consider a point p in a zone with covering number c, i.e., the line l_p corresponding to p visits these c circles in a particular order. Since any two points, p_1 and p_2, in the same zone are connected by a line (which may not be a straight line) without leaving the zone, line l_{p_1} can be transformed to line l_{p_2} by a series of translations and rotations. As the circles are disjoint and the stabbing line never leaves any circles or enters any new circle during the transformation, lines l_{p_1} and l_{p_2} should visit the same set of circles in the same order.

Now let's consider two points, p_1 and p_2 in two neighboring zones of different covering numbers (the difference is one). Lines l_{p_1} and l_{p_2} must stab through two sets of circles, which differ by one circle. The transformation of l_{p_1} to l_{p_2} has to leave a circle or enter into a new circle. Since the circles are disjoint, the visiting orders of the other circles w.r.t. l_{p_1} and l_{p_2} should remain the same, and these two visiting orders can be transformed to each other by an insertion or deletion of a circle. □

Let sequence $A = (a_1 a_2 \ldots a_n)$, $n > 1$, be the visiting order of P' w.r.t. a stabbing line. As given in [15], $to(i)$ denotes the maximum length of an increasing subsequence ending with a_i (for $1 \le i \le n$), e.g.

a_i:	8	3	6	5	2	9	1
$to(i)$:	1	1	2	2	1	3	1

Theorem 4. *[15] The length of the LIS of A is $\max\{to(i) \mid 1 \le i \le n\}$.* □

$to(i)$ can be computed as $1 + \max\{to(j) \mid 1 \le j < i$ and $a_j < a_i\}$. Since each $to(i)$, $1 \le i \le n$, can be computed in $O(i)$ time, all $to(i)$, $1 \le i \le n$, can be found in $O(n^2)$ time by brute force. In [15], an efficient algorithm which computes all $to(i)$, $1 \le i \le n$, in $O(n \log n)$ time was presented.

In the following, we will show how to recompute all the $to(i)$, $1 \le i \le n$, after an insertion/deletion of an element into/from the sequence in $O(n)$ time.

Lemma 5. *Let $to'(i)$ or $to''(i)$ be the new $to(i)$, after insertion or deletion of an element a_j, then $to'(i) = to''(i) = to(n)$ for $i < j$ and $to(i) \le to'(i) \le to(n) + 1$ or $to(i) - 1 \le to''(i) \le to(n)$ for $j < i$.*

Proof. (Sketch) Since there is no change of the sequence a_i with $i < j$, $to'(i) = to''(i) = to(i)$. If the increasing subsequence of the new sequence does not include the inserted element a_j, $to'(i) = to(i)$. If the increasing sequence of the new sequence includes a_j, then obviously $to'(i) \le 1 + to(i)$. Similarly if the increasing sequence of the old sequence includes a_j, then $to''(i) \ge to(i) - 1$, otherwise $to''(i) = to(i)$.

□

In order to speed up the calculation of $to(n)$, we define M[k] as the smallest ending element among all increasing sequences of length k, i.e., $\min\{a_i \mid 1 \le i \le n \cup to(i) = k\}$ when $a_1 \ldots a_i$ is considered, and $k \le \max\{to(n)\}$. Initially all M[k] $= \infty$ for $k \ge 1$. Assume a_j is the new element inserted to A, we shall construct all $to'(i)$ and compute M[$to'(i)$] at the same time when scanning each element a_i, for $i = 1, 2, \ldots$ When $i < j$, $to'(i) = to(i)$ (Lemma 2) and M[$to'(i)$] is replaced by a_i. When $i = j$, $to'(j) = 1 + \max\{to(k) \mid k < j$ and $a_k < a_j\}$ and M[$to'(j)$] $= a_i$. When $i > j$, $to'(i)$ and M[$to'(i)$] will be updated accordingly. In particular, by Lemma 2, we have

$$to'(i) \leftarrow \text{if } a_i < M[to(i)] \text{ then}$$
$$to(i) \text{ else } to(i) + 1$$
$$M[to'(i)] \leftarrow a_i$$

Let us consider the previous example and insert $a_4 = 4$ into A, then we have

				j				
a_i:	8	3	6	4	5	2	9	1
$to(i)$:	1	1	2	/	2	1	3	1
$to'(i)$:	1	1	2	2	3	1	4	1

M[1]=8 M[1]=3 M[1]=3 M[1]=3 M[1]=3 M[1]=2 M[1]=2 M[1]=1
        M[2]=6 M[2]=4 M[2]=4 M[2]=4 M[2]=4 M[2]=4
                        M[3]=5 M[3]=5 M[3]=5 M[3]=5
                                          M[4]=9 M[4]=9

Theorem 6. *After insertion of a_j, $to'(i)$, $1 \le i \le n$ and the LIS of the new sequence can be found in O(n) time.*

□

Similarly, for the deletion of a_j, $to''(i)$, $1 \le i \le n$ and the LIS of the new sequence can also be found in O(n) time.

Theorem 7. *The OLE problem can be solved in O(n^3) time.*

Proof. Based on the arrangement of transversals through circles, we can exhaust all zones in the arrangement by visiting the neighboring zones one by one. When a zone is visited, the LIS of the visiting order of the circles for the transversals corresponding to points in that zone can be found in O(n) time (Theorem 6). As there are at most O(n^2) zones, the OLE problem can be solved in O(n^3) time.

□

4 Line Estimator Problem (ordered and unordered) of Vertical Distance

In this section, we shall consider the line estimator problem of vertical distance for a set of points $P = \{p_i = (a_i, b_i) \mid 1 \leq i \leq n\}$ in a 2D plane, which is equivalent to finding a stabbing line to n vertical line segments of length 2ε and centered at $p'_i s$.

Each vertical line segment S_i is expressed by its half-length ε_i and center (ρ_i, θ_i) (Fig. 4). Again, in this paper, we assume $\varepsilon_i = \varepsilon$ for all $1 \leq i \leq n$.

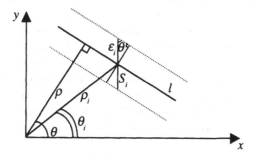

Fig. 4. Stabbing Line to a Line Segment S_i

A straight line l stabs a vertical line segment S_i (Fig. 4) iff

$$\rho_i \cos(\theta - \theta_i) - \mid \varepsilon_i \sin \theta \mid \; \leq \rho_i \leq \rho_i \cos(\theta - \theta_i) + \mid \varepsilon_i \sin \theta \mid$$

Similar to the line estimator problem of perpendicular distance, the line estimator problem of vertical distance can be solved by construction of arrangement of the sinusoidal envelop curves $\rho_i \cos(\theta - \theta_i) \pm \varepsilon_i \sin \theta$, i.e., $\rho_i \cos \theta_i \cos \theta + (\rho_i \sin \theta_i \pm \varepsilon_i) \sin \theta$.

Theorem 8. *The arrangement of $2n$ curves representing the family of stabbing lines of n vertical line segments can be constructed in $O(n^2)$ time, the line estimator and the ordered line estimator problems of vertical distance can be solved in $O(n^2)$ and $O(n^3)$ respectively.*

Proof. Since the envelop curves are also sinusoidal (as the envelop curves for circles), the line estimator problem can be solved in $O(n^2)$ and $O(n^3)$ time by the algorithms similar to those given in Sections 2 and 3.   □

5 Conclusion

When the radius ε of the circles or the length 2ε of line segments approaches zero, i.e., $\varepsilon \rightarrow 0$, the line estimator problem can be used to solve the 3SUM-hard problem [11] of determining whether any 3 points in a given set of n

points are collinear. Thus, it is unlikely that there exists a faster than quadratic time algorithm for the line estimator problem. In this paper, we have proposed an $O(n^3)$ algorithm for the ordered line estimator problem of n points. A more detailed analysis of the algorithm reveals that its time complexity is $O(\sum_{All\ Zones} Covering\ Number)$ which can still be $O(n^3)$. It is obvious that the ordered line estimator problem can be solved without considering the visiting order of the stabbing lines in all zones of the arrangement. The same result can be obtained if only those zones which give "maximal" visiting orders are considered. A sequence is *maximal* if it is not a proper subsequence of another sequence. Unfortunately, there are still examples that the sum of the length of the maximal visiting orders is $O(n^3)$. This case occurs when the set of points are divided into approximately 3 parts: part A contains points with small y-coordinates but a common large positive x-coordinate, part B with small y-coordinates but a common large negative x-coordinate, part C are points with small x-coordinates and a common zero y-coordinate. It is easy to see that each maximal sequence contains one point from A, all points from B and one point from C, and there are at least $\frac{n^2}{9}$ such maximal sequences with length $\frac{n}{3} + 2$. It remains open whether the ordered line estimator problem can be solved in less than $O(n^3)$ time.

References

1. M. Atallah, C. Bajaj: Efficient Algorithms for Common Transversals. Information Processing Letters, Vol. 25, No. 2, (1987) 87-91
2. D. Avis, R. Wenger: Algorithms for Line Transversals in Space. Proc. 3rd ACM Symp. on Computational Geometry, (1987) 300-306
3. W. Chan, F. Chin: Approximation of Polygonal Curves with Minimum Number of Line Segments or Minimum Error. International Journal of Computational Geometry and Applications Vol. 6, No. 1, (1996) 59-77
4. S. Chattopadhyhy, P. Das: The *k*-dense corridor problems. Pattern Recognition Letters Vol. 11, (1996) 463-469
5. H. Edelsbrunner: Algorithms in Combinatorial Geometry. EATCS Monographs on Theoretical Computer Science Springer-Verlag, Berlin, Heidelberg, Vol. 10, (1987) 123-138
6. H. Edelsbrunner: Finding Transversals for Sets of Simple Geometric Figures. Theoretical Computer Science, Vol. 35, (1985) 55-69
7. H. Edelsbrunner, H. A. Maurer, F. P. Preparata, A.L. Rosenberg, E. Weizl, D. Wood: Stabbing Line Segments. BIT, Vol. 22, (1982) 274-281
8. H. Edelsbrunner, J. O'Rourke, R. Seidel: Constructing Arrangements of Lines and Hyperplanes with Application. SIAM Journal of Computer Science. Vol. 15, (1986) 341-363
9. H. Edelsbrunner, D. L. Souvaine: Computing Least Median of Squares Regression Line and Guided Topological Sweep. Journal of the American Statistical Association, Vol. 85, No. 409, (1990) 115-119
10. L. J. Guibas, J. E. Hershberger, J. S. B. Mitchell, J. S. Snoeyink: Approximating Polygons and Subdivisions with Minimum-Link Paths. International Journal of Computational Geometry and Application, Vol. 3, No. 4, (1993) 383-415

11. A. Gajentaan, M. H. Overmars: On a Class of $O(n^2)$ Problems in Computational Geometry. Computational Geometry Theory Application, Vol. 5, (1995) 165–185
12. J. Hunt, T. Szymanski: A Fast Algorithm for Computing Longest Common Subsequences. Commnucation of the ACM, Vol. 20, No. 5, (1977) 350–353
13. H. Imai, M. Iri: Polygonal Approximations of a Curve - Formulations and Algorithms. Computational Morphology, (1988) 71–86
14. D. M. Mount, N. S. Netanyahu, K. Romanik, R. Silverman, A. Y. Wu: A Practical Approximation Algorithm for the LMS Line Estimator. Proceedings of the 8th Annual ACM-SIAM Symposium on Discrete Algorithms (1997) 473–482
15. P. Pritchard: Another Look at the "Longest Ascedding Subsequence" Problem. Acta Informatica, Vol. 16, (1981) 87–91
16. P. J. Rousseeuw: Least median-of-squares regression. Journal of American Statistical Association, Vol. 79, (1984) 871–880

Generalized Shooter Location Problem

Jeet Chaudhuri[1] and Subhas C. Nandy[2]

[1] Wipro Limited, Technology Solns, Bangalore 560034, India
[2] Indian Statistical Institute, Calcutta - 700 035, India[***]

Abstract. In this paper, we propose an efficient algorithm for the *generalized shooter location problem*. A prior work on this problem appeared in [5], where an $O(n^5 \log n)$ algorithm was proposed. The time complexity of our algorithm is $O(n^4 + nK)$, where K is the number of dominating cells inside the floor. The concept of dominating cell will be introduced later in the text. Though the worst case number of K may be $O(n^4)$, it is much less in the actual practice.

1 Introduction

The *transversal* problem is an well-studied algorithmic paradigm in computational geometry where a number of objects are assumed to be distributed arbitrarily on a two- or three-dimensional region. The objective is to check whether there exists a single line that passes through all the objects, and if so, determine the same [2]. A detailed survey on this topic appeared in [4,5].

An interesting variation of *transversal* problem is proposed in [4]. It assumes targets as arbitrary simple polygons positioned on a 2-D floor, and a shooter has to stab all the targets. The shooter can fire or emit rays along straight line in arbitrary direction. A ray stabs all the objects in its linear path of motion from its origin up to infinity. The objective is to choose an appropriate placement of the shooter on the floor such that the number of shots necessary to exhaust (stab) all the objects is minimum. An $O(n^3)$ time algorithm for a restricted version of the problem was proposed in that paper, where the shooter can assume its placement on a specified line and targets are line segments, called sticks. Here, n is the number of sticks placed on the floor. For the generalized version of this problem, an $O(n^5 \log n)$ algorithm is proposed in [5]. They have also proposed an $O(n^5)$ time algorithm using geometric duality which produces a factor-2 approximate solution for the problem.

In this paper, we improve the worst case time complexity of the generalized version of shooter location problem. We consider line segments as targets. The time complexity of our algorithm depends on the distribution of the targets on the floor. It runs in $O(n^4 + nK)$ time, where K is the number of dominating cells inside the floor. The concept of dominating cell will be introduced in Section 3 of the paper. Though K may be $O(n^4)$ in the worst case, it is much less in the

---

[***] Currently at School of Information Science, Japan Advanced Institute of Science and Technology, Ishikawa 923-1292, Japan.

actual practice where the distribution of the targets are random. Surely in the worst case also, our algorithm improves the time complexity of the problem by a factor of $O(\log n)$ in comparison to the existing results [5].

2   Preliminaries

We consider a finite set S of n straight line segments, called sticks, inside a rectangular bounding box, called *floor*. Each stick has its finite length but its orientation is arbitrary. A single shooter can be positioned anywhere inside the floor and can emit multiple rays in arbitrary directions. The *generalized shooter location problem* is as follows:

Locate an appropriate position (point) p inside the floor so that the number of shots required by the shooter to exhaust all the objects from p is minimum.

For the sake of self completeness, first we mention the more primitive problem, called the *fixed shooter problem*. The formulation of our *generalized shooter location problem* will follow subsequently.

2.1   Fixed shooter problem

Consider a single shooter, positioned at a predefined fixed point inside the floor. The shooter is allowed to take multiple shots in arbitrary directions. The objective is to determine a set of rays hitting all the targets at least once, and the cardinality of the set is minimum. In [4], it is shown that the fixed shooter problem can easily be mapped as the location of minimal linear clique cover of a circular arc graph which is constructed as follows:

Let C be a circle that encloses all the sticks and also the point p. For each stick ℓ_i, we join both its end points with p and extend them up to the boundary of C; the resulting arc will be referred to as the projection of ℓ_i on C with respect to p. The projections of all sticks on C with respect to p evidently form a *circular arc family* and is denoted by $\mathcal{F}_p$. In order to impose an ordering on the members of $\mathcal{F}_p$, we consider the orientation of each arc in clockwise direction. An end point of an arc is said to be its *left end point* if the other end point of the same arc is accessible by a clockwise visit along the periphery of C. Surely, the other end point of that arc will be referred to as its *right end point*.

Now let us consider a graph $G(p)$, called *circular arc graph*, whose each vertex corresponds to a member of $\mathcal{F}_p$, and between two vertices there is an edge if the corresponding two arcs overlap.

Definition 1. *A set of nodes $V^* \in V$ in $G(p)$ is said to form a linear clique, if their corresponding arcs in $\mathcal{F}_p$ have a common point along the boundary of C, and further, V^* is maximal if it is not subsumed by any other bigger set of this type.*

In this context it is important to mention that, a clique in a circular arc graph may not always correspond to a linear clique. For notational simplicity, a clique (in graph theoretic context) in a circular arc graph will be referred to as a *graphical clique*. Thus a linear clique is always a graphical clique, but the converse is not always true.

For a given circular arc family, say $\mathcal{F}_p$, the construction of $G(p)$ requires $O(n \log n)$ time, and further, the minimum linear clique cover of the circular arc graph can be obtained in $O(n)$ time [3].

2.2   Some useful definitions

Consider a pair of non-intersecting sticks ℓ_i and ℓ_j. A straight line τ_{ij} is said to be a *separator* of ℓ_i and ℓ_j if ℓ_i and ℓ_j lie in the opposite sides τ_{ij}. It is easy to observe that there exists an infinite number of separators of (ℓ_i, ℓ_j). A separator τ_{ij} is said to be *extremal* if it touches the end points of ℓ_i and ℓ_j. Needless to say, there exists exactly two extremal separators for every pair of non-intersecting sticks (as shown in Figure 1). This is also to notice that τ_{ij} partitions the plane into a pair of half planes, one containing ℓ_i completely, called *region-1*(τ_{ij}), and the other contains ℓ_j completely, called *region-2*(τ_{ij}).

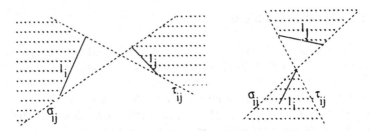

Fig. 1. Demonstration of extremal separators

Observation 1 *Let ℓ_i and ℓ_j be a pair of sticks and ℓ be a line that contains ℓ_i. Then ℓ intersects ℓ_j if and only if the extremal separators of ℓ_i and ℓ_j intersect at an end point of ℓ_i (as shown in Figure 1b).*

Definition 2. *For a pair of non-intersecting sticks ℓ_i and ℓ_j, the component of the active region corresponding to ℓ_i, denoted by AR_{ij}^i, is the intersection of region-1(τ_{ij}), region-1(σ_{ij}) and the half plane induced by ℓ which does not contain ℓ_j. If the line ℓ containing ℓ_i intersects ℓ_j, then AR_{ij}^i is simply the intersection of region-1(τ_{ij}) and region-1(σ_{ij}).*

In an exactly similar manner we can define AR_{ij}^j, the component of active region corresponding to ℓ_j. Needless to say, the component of active regions are necessarily convex. Given a pair of sticks ℓ_i and ℓ_j, the active region(ℓ_i, ℓ_j) is the union of AR_{ij}^i and AR_{ij}^j, and is denoted by AR_{ij}.

Depending on whether the line containing a stick, say ℓ_i, intersects another stick ℓ_j, or not, the nature of the active region changes. In Figure 1, the shaded area demonstrates the active region for a pair of sticks ℓ_i and ℓ_j in two different cases.

Definition 3. *The dead region corresponding to a pair of sticks ℓ_i and ℓ_j, denoted by DR_{ij}, is the collection of all those points in the plane that do not belong to AR_{ij}.*

Result 1 *Given a point $p \in AR_{ij}$, both the sticks ℓ_i and ℓ_j can be shot by a single firing from it. But if $p \in DR_{ij}$, two firings are needed for the same.* □

It is perhaps not impertinent to comment that in case ℓ_i and ℓ_j are two intersecting sticks, then corresponding to any point p on the plane, a single shot from p is possible to hit both the sticks. Due to this fact, we have defined the *active region* of non-intersecting sticks only.

Definition 4. *Let ℓ be a line that is either an extremal separator for the pair of sticks ℓ_i and ℓ_j, or simply a line containing any one of them. Then ℓ is said to be a good line for the pair of sticks ℓ_i and ℓ_j if it is a bounding line for AR_{ij}.*

Let $\mathcal{L}_{ij}$ denote the set of good lines bounding the active region AR_{ij}. From Figure 1a and 1b, it is obvious that $\mathcal{L}_{ij}$ may consist of either four or three lines.

Definition 5. *Consider a member ℓ of $\mathcal{L}_{ij}$. The active segment of ℓ is the portion of it that actually bounds AR_{ij}, while its dead segment is constituted by the remaining portion(s).*

It is easy to observe that each of the vertices of AR_{ij} is either an end point of ℓ_i or that of ℓ_j. Thus a member ℓ of $\mathcal{L}_{ij}$ is dissected into active and dead segments at the end points of ℓ_i and/or ℓ_j. It is also obvious that the number of such segments in ℓ is always equal to three.

3   Construction of Pseudo Arrangement

A *pseudo arrangement* with respect to the given set S of sticks is denoted by $\mathcal{A}'$. The vertices, edges and pseudo-cells of $\mathcal{A}'$ correspond to the maximally connected pieces of dimensions 0, 1 or 2 respectively, which are induced by the bounding edges of the active regions corresponding to each pair of non-intersecting sticks. Note that, a pseudo-cell lies either completely inside or completely outside the active region of any pair of sticks.

Lemma 6. *Consider a pair of sticks ℓ_i and ℓ_j and a point p on the floor. The arcs on the circle C corresponding to ℓ_i and ℓ_j with respect to the point p, overlap along the periphery of C if and only if $p \in AR_{ij}$.* □

Thus, the circular arc graph $G(p)$ with respect to a point p on the plane has an edge joining the nodes, say v_i and v_j, if and only if p lies in the active region of the corresponding pair of sticks ℓ_i and ℓ_j.

From the construction of circular arc family it follows that if f_i and f_j be two arcs on C corresponding to ℓ_i and ℓ_j with reference to point p, and the end points of f_i and f_j are denoted by $(left_i, right_i)$ and $(left_j, right_j)$ respectively, then the following four situations may happen.

$left_i < left_j < right_j < right_i$: f_i properly subsumes f_j.
$left_i < left_j < right_i < right_j$: right end of f_i overlaps with left end of f_j.
$left_j < left_i < right_j < right_i$: left end of f_i overlaps with right end of f_j.
$left_i < right_i < left_j < right_j$: f_i and f_j are non-overlapping.

Theorem 7. *For any pair of points p and q in a pseudo-cell R of $\mathcal{A}'$, minimum number of shots required to hit all the sticks from p is equal to that of q.*

Proof: In view of Lemma 1, it is enough to show that $G(p)$ and $G(q)$ has same number of linear cliques.

As discussed above, a pseudo-cell lies completely within or outside the active region corresponding to each pair of sticks. Then by Lemma 1, the circular graph remains unchanged over every point in R, or in other words, $G(p)$ and $G(q)$ are isomorphic. However, we need to prove the stronger result that the circular arc families corresponding to $G(p)$ and $G(q)$ has the same set of linear cliques. Thus we need to show that a graphical clique in $G(p)$ is observed to be a linear clique if and only if the same set of nodes form a linear clique in $G(q)$.

Let $V^* = \{v_1, v_2, \ldots v_k\}$ be a set of vertices which constitute a linear clique in $G(p)$, but only a graphical clique in $G(q)$. We denote by $\mathcal{F}_p(\mathcal{F}_q)$, the circular arc family with respect to p (q), corresponding to the set of vertices V^*. Let $f_1, f_2, \ldots, f_k$ be the arcs in $\mathcal{F}_p$ clockwise ordered with respect to their left end points along C. The corresponding arcs in $\mathcal{F}_q$ are denoted by $f'_1, f'_2, \ldots, f'_k$. As V^* forms a linear clique with respect to p, $f_1, f_2, \ldots f_k$ intersect at a common point. Again since V^* does not form a linear clique with respect to q, there exists at least a pair i and j such that left end of f_i overlaps with the right end of f_j, but the left end of f'_i does not overlap with the right end of f'_j. However since V^* forms a graphical clique in $G(q)$, the right end of f'_i surely overlaps with the left end of f'_j. Thus there must exists a point r on the joining straight line of p and q such that in the circular arc family $\mathcal{F}_r = \{f''_1, f''_2, \ldots, f''_k\}$ corresponding to the point r, one of the following two situations must have occurred.

Case 1 The right end of f''_i gets overlapped with the left end of f''_j while the left end of f''_i remains overlapped with the right end of f''_j.

Case 2 The left end of f''_i leaves the right end of f''_j and the right end of f''_i is not yet overlapped with the left end of f''_j.

But the construction of circular arc family suggests that Case 1 is infeasible. Again since every point in R must have isomorphic circular arc graph, Case 2 implies that r can not belong to R. Thus the contradiction follows from the fact that R is a connected region and the line joining p and q lies completely in R.□

The above result says that the minimum number of shots required to hit all the sticks from a point on the plane remains invariant over a pseudo-cell of $\mathcal{A}'$. Now we introduce the notion of dominance among the pseudo cells.

Definition 8. *A pseudo-cell R in the pseudo-arrangement $\mathcal{A}'$ is said to be dominated by another pseudo-cell R' in the same pseudo-arrangement if corresponding to each pair of sticks ℓ_i and ℓ_j, either (i) both of R and R' lie in their dead region, or (ii) both of R and R' lie in their active region, or (iii) R lies in their dead region but R' lies completely in their active region.*

Lemma 9. *Let R and R' be a pair of pseudo cells such that R is dominated by R'. Then all the edges of the circular arc graph G corresponding to R are present in the circular arc graph G' corresponding to R'.*   □

Theorem 10. *For a pair of pseudo-cells R and R' having a common point α on the plane, and R dominated by R', the number of shots required to hit all the sticks from a point in R is greater than or equal to that needed for a point in R'.*

Proof: Follows from Lemma 2.   □

Definition 11. *Let R and R' be a pair of pseudo cells in $\mathcal{A}'$ which intersect at a vertex α. Also suppose that R and R' are to the opposite sides of the vertical line through α. Then α is a dominated vertex if R lies to the left of this vertical line and further, R is dominated by R'.*

Thus, by virtue of Theorem 1, we need to compute minimum number of shots required to hit all the given sticks only for a single point in each pseudo cell. Again since each pseudo cell contributes at least one vertex in $\mathcal{A}'$, it follows that the necessary computation at every vertex of $\mathcal{A}'$ will suffice. Further, in a left to right sweep, the computation at the dominated vertices can also be avoided in view of Theorem 2. This leads to our algorithm for finding the optimal point p from which all the sticks can be hit with minimum number of shots.

4   Theme of the Algorithm

In the above discussions, we have considered the pseudo arrangement of the active regions of all pairs of sticks. In actual implementation, we considered the set of all good lines for all pairs of sticks, instead of working with their active segments only.

Let $\mathcal{A}$ be the arrangement of all good lines $\mathcal{L}$. As the number of good lines is $O(n^2)$, the number of vertices, edges and cells of this arrangement is $O(n^4)$. Note that, all the vertices of the pseudo arrangement $\mathcal{A}'$ defined in Section 3 are contained in the set of vertices of the arrangement $\mathcal{A}$. There are however, vertices in $\mathcal{A}$, that are absent in $\mathcal{A}'$. These vertices are called *numb vertices*, and hence no computation is needed for those vertices. We use topological plane sweep [1] for processing the vertices of the arrangement. For simplicity, we assume that the good lines have their slopes finite and distinct, and further, no three of them are concurrent. The degenerate cases may be tackled using the same technique as suggested in [1].

4.1 Data structure

The slope and intercept parameters of all the good lines are stored in an array. The array is sorted with respect to the slope parameters. For each line, (i) its active segments are highlighted, and (ii) the pair of sticks, which define this line as their maximal separator, need to be attached. These two fields may accelerate the algorithm. The role of these two fields will be mentioned in Subsection 4.3.

The topological plane sweep maintains three data structures, namely *upper horizon tree* (UHT), *lower horizon tree* (LHT) and a *stack* as described below [1].

UHT: It is an array whose i-th element is a 2-tuple that stores the following information about a line which is currently the i-th one in the ordering of lines with respect to their slope.
- The left and right end points of the delimited portion of the line that is present in the upper horizon tree. In case, it is leftmost (rightmost), we store -1 (0) as the corresponding entry.

LHT: This array represents the lower horizon tree in an analogous manner.

stack: A set of vertices of $\mathcal{A}$ that are present in both UHT and LHT. The vertices in *stack* will be referred to as *cut vertices*, which are observed up to current position of the sweep line, but not yet processed. During execution, the top element of *stack* indicates the next position where the sweep line stops.

Two more variables $\mathcal{P}$ and Min_Num are also used. $\mathcal{P}$ stores the optimum point so far obtained, Min_Num stores the number of shots required at that point.

4.2 Algorithm

The initialization of UHT, LHT and *stack* are done as described in [1]. Initially, the top element of *stack* corresponds to the left most cut vertex, where from the topological sweep starts. At each cut vertex π, we compute the number of shots required to hit all the sticks, as stated in subsection 2.1, provided π is not a numb vertex. Min_Num and $\mathcal{P}$ are initialized accordingly.

At the end of processing π, the sweep line proceeds towards right; the UHT and RHT are updated as in [1]. This gives birth to new cut vertices which will be available on the top of *stack*. Now, the following observation is very important.

Lemma 12. *The cut vertices along the sweep line are present in stack in increasing order of their y-coordinates from top to bottom.* □

At each stage, a cut vertex is popped from the top of *stack*. The same computation need to be done at this vertex, *Min_Num* and $\mathcal{P}$ are updated if necessary. The UHT and LHT data structures are also updated. This may cause insertion of new cut vertices on the top of *stack*. The same process is repeated until all the vertices of $\mathcal{A}$ are exhausted. At the end of the sweep, $\mathcal{P}$ gives the optimal position of the shooter. Finally, a new circular arc family is constructed with reference to $\mathcal{P}$ and the minimum linear clique cover is computed to obtain the resulting shooting directions.

Incremental Update of circular arc family In order to determine the number of shots necessary to hit all the sticks from a vertex v of the arrangement, we need the configuration of the circular arc family with respect to v. Fortunately for us, this does not mean that we need to construct the circular arc family afresh at every vertex of the arrangement. By Theorem 1, the minimum number of shots required to hit all the objects from a point remains invariant over a cell of the arrangement $\mathcal{A}$. In view of this, we can obtain the circular arc family corresponding to each cell from that of its neighboring cell by doing some local modifications as stated below.

Theorem 13. *Let R and R' be a pair of cells in $\mathcal{A}$ with corresponding circular arc families $\mathcal{F}_R$ and $\mathcal{F}_{R'}$ respectively. Then*

(i)   if R and R' have a common edge, then there is at most one pair of arcs, say f_i and f_j, which intersect in exactly one of the circular arc families.

(ii) if R and R' have a common vertex but not a common edge, then there is at most two pairs of arcs, say f_i and f_j, and f_i^* and f_j^* such that each of these pairs intersect in exactly one of the circular arc families.

Proof: Follows from the concept of maximal separators of a pair of sticks. $\square$

Thus, the relative order of the end points of the members in the circular arc family can easily be updated in constant time when the control moves from one cell to its neighboring one in the arrangement $\mathcal{A}$.

Consider now Figure 2. The current position of the sweep line is indicated by a dotted line. After processing the vertex v, the next vertex to be processed by the sweep line is w. In order to obtain the circular arc family corresponding to w, the circular arc families corresponding to the cells R_1, R_2, R_3, and R_4 are to be obtained in that order. Further, after processing w, the circular arc family corresponding to the cell R_5 is to be obtained from that of R_4. Thus we arrive two cases where the circular arc family is to be modified:

Case 1 The sweep line crosses a vertex of the arrangement.

Case 2 In order to reach the sweep line to the next cut vertex, a cell is to be reached from its neighboring cell, separated by an edge which is observed to be a cut edge.

It follows that, the modification in the circular arc family in Case 1 takes place exactly once for each vertex of $\mathcal{A}$. Lemma 3 and Lemma 4 (stated below) prove

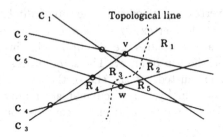

Fig. 2. Order of processing the cut vertices guided by topological line sweep

that the modification of the circular arc family in Case 2 also takes place at most once for each edge of $\mathcal{A}$.

Lemma 14. *The circular arc family $\mathcal{F}$ is modified at each edge of the arrangement $\mathcal{A}$ at most once.*

Proof: Consider that $\mathcal{F}$ is modified along an edge e. Two cases may arise here.

The circular arc family of the region below e is computed from that of the region above e. By Lemma 3, it follows that there is no cut vertex corresponding to any pair of edges above e in the present position of the sweep line. Consequently, we shall have neither the necessity nor the opportunity to cross over from the region below e to the one above it. This shows that corresponding to the edge e considered here, there will be at most one modification of $\mathcal{F}$.

The circular arc family of the region above e is computed from that of the region below e. Note that, this happens only when e ends in a cut vertex. This vertex is immediately processed and the sweep line proceeds to the right of the end point of e. Hence modification of $\mathcal{F}$ along this edge e is also performed at most once. $\qquad\square$

Complexity Analysis At the initial step of this procedure we need to generate all good lines which are $O(n^2)$ in number, for an assembly of n sticks; this step requires $O(n^2)$ time. The initialization of horizon trees, and *stack* with $O(n^2)$ good lines also require $O(n^2)$ time.

Next, consider the processing at a generic vertex encountered by the sweep line in its current position. The modification of the horizon trees at each cut vertex requires time linear to the number of lines, which is also $O(n^2)$ in our case. As far the update of entries of the stack is concerned a local check around the concerned vertex is sufficient. As the total number of cut vertices generated in the entire process is $O(n^4)$, the total time required is $O(n^6)$. But following an amortized analysis proposed in [1], it can be shown that the sweep can be completed in $O(n^4)$ time.

The complexity for computing the minimum number of shots required to hit all arcs in a circular arc family corresponding to each vertex is $O(n)$ under the assumption that the end points of the arcs are sorted along the boundary of C.

This procedure is invoked at $O(n^4)$ vertices. But as we go from one vertex to the other one, the ordering of the end points get changed. But in the previous subsection we have introduced an incremental procedure for updating the circular arc family. From Theorem 3, any single modification in $\mathcal{F}$ involves either one pair or two pairs of edges in the circular arc graph G, which can be performed in constant time. Thus we have the following theorem:

Theorem 15. *The time and space complexities of our proposed algorithm are $O(n^5)$ and $O(n^2)$ respectively.*

Proof: Lemma 4 and the discussions preceding to it implies that the circular arc family need to be modified at most $O(n^4)$ times. Each modification can be done in constant time. So, the result regarding the time complexity follows from the fact that at each of the $O(n^4)$ cut vertices, an $O(n)$ time computation is required to solve the *fixed shooter problem*.

Regarding the space complexity, each of the two horizon trees requires space linear in the number of lines, which is $O(n^2)$ in our case. The stack may require $O(n^2)$ space in the worst case. The circular arc family is incrementally updated and a single copy is to be maintained. This requires $O(n)$ space.   □

4.3   A further refinement

The algorithm, proposed above, can further be refined using Theorem 2. Unlike the method stated above, we invoke the fixed shooter problem only when we encounter a dominating cut vertex.

As soon as we enter to a new cell from its neighboring cell through a cut vertex or a cut edge, we update the circular arc graph using constant time computation as stated in Subsection 4.2. Next, we check whether it is an active region of a new pair of sticks or a dead region of a new pair of sticks. This can be done in constant time by observing (i) the coordinates of the active segments of the concerned good line, and (ii) by considering the pair of sticks attached to that good line. In the former case, we proceed, while in the later case, the predecessor cell is dominating the current cell. So, we select a point inside the dominating cell and find the minimum number of shots required from that point by calling the procedure for fixed shooter problem. This requires $O(N)$ time since the relative order of the end points for the arcs in the circular arc family is already available. As in the earlier case, $\mathcal{P}$ and Min_Num parameters contain the same information, and these are updated at each invoke of fixed shooter algorithm, if required. Thus, during the entire sweep if K such dominating cells are encountered, the fixed shooter algorithm is to be invoked K times, which require $O(nK)$ time. As discussed earlier, the topological line sweep requires $O(n^4)$ time. Thus the time complexity of our proposed Algorithm can further be improved to $O(n^4 + nK)$. We explicitly mention that K may be $O(n^4)$ in the worst case; such an example is very easy to construct. But in actual practice, where targets are arbitrarily distributed, K is observed to be much less.

5   Conclusion

In this paper, we have improved the time complexity for the *generalized shooter location problem*, introduced in [5]. If the targets are arbitrary polygons instead of sticks, the extremal separators of a pair of target polygons will be the common tangents of their convex hulls which are mutually crossing. Then our algorithm can easily be tailored so that it may work on the arrangement of those extremal separators to find the optimal placement of the shooter. The most important thing to mention is that, the worst case time complexity of our proposed algorithm is $O(n^5)$, which is reasonably high for practical implementation. So, it is worth paying attention to design an efficient heuristic which is expected to produce considerably good results. Surely, improving the worst case time complexity of this problem is a challenging open problem.

References

1. H. Edelsbrunner and L. J. Guibas *Topological sweeping an arrangement*, Journal of Computer and System Sciences, vol. 38, pp. 165-194, 1989
2. H. Edelsbrunner, M. H. Overmars and D. Wood, *Graphics in flatland: a case study*, in F. P. Preparata ed., Advances in Computing Research vol. 1, JAI Press, London, 1983 pp. 35-59.
3. W. L. Hsu and K. H. Tsai, *Linear time algorithms on circular-arc graphs*, Information Processing Letters, vol. 40, 1991, pp. 123-129.
4. S. C. Nandy, K. Mukhopadhyaya and B. B. Bhattacharya, *Shooter location problem*, Proc. 8th Canad. Conf. on Computational Geometry, pp. 93-98, August 1996.
5. Cao An Wang and Binhai Zhu, *Shooter Location Problems Revisited*, Proc. 9th Canad. Conf. on Computational Geometry, pp. 223-228, 1997.

A Competitive Online Algorithm for the Paging Problem with "Shelf" Memory.*

Sung-Pil Hong

School of Business, Chung-Ang University, An-sung-gun, Kyung-gi-do, 456-756, Korea, tel: +82-334-70-3214, fax: +82-334-676-8460, sphong@cau.ac.kr.

Abstract. We consider an extension of the two-level paging problem. Besides a fast memory (the cache) and a slow memory, we postulate a third memory, called the "shelf", near the fast memory so that it is more cost-efficient to store or retrieve a page from the shelf than to retrieve a page from the slow memory. Unlike the standard two-level paging problem, the extension is not a special case of the K-server problem as it is not embedded in a space with a symmetric metric. Our goal is to establish an upper bound on the competitive ratio of this "three-level" memory paging problem. We show that unless per page storage costs more than per page retrieval from the shelf, a simple extension of the well-known LRU algorithm has competitive ratio of $2k + l + 1$, where k and l are, respectively, the capacities of the cache and the shelf. If, in addition, $k \leq l$, then the same algorithm is $k + l + 2$-competitive.

1   Introduction

In this short paper, the traditional two-level memory paging problem is extended with an additional memory level, called the *shelf*, to the two-level store consisting of a fast memory (the *cache*) and a slow memory. For a meaningful extension, it is assumed that per page interaction (storage or retrieval) with the shelf costs less than per page retrieval from the slow memory. This extended problem may have many potential applications. For instance, besides the small desktop space and the large-size main book shelves remote from the desk, the author owns an additional small bookshelf near the desk. Similar situations may arise in the information storage and retrieval systems with multiple hierarchy. In this paper, the problem is defined in the context of the paging problem so that a cost incurs only in three cases: when a page is retrieved from the slow memory, when a page is stored in the shelf, or when a page is retrieved from the shelf. An example may be a network computer extended with a small memory for its own page storage other than the large slow memory shared with other computers in the network.

This paper is organized as follows. Section 2 describes the proposed three-level memory paging problem in detail. In Section 3 we extend an online algorithm called *Least-Recently-Used* (LRU) for the two-level paging problem to an online algorithm for the three-level paging problem. In Section 4 we establish

---

* The research is partially supported by Chung-Ang University Research Grant 96-106.

T. Asano et al. (Eds.): COCOON'99, LNCS 1627, pp. 400–408, 1999.
© Springer-Verlag Berlin Heidelberg 1999

a competitive ratio of the extended algorithm. Finally, Section 5 provides some concluding remarks and discusses some directions of future research.

2   The three-level paging problem

The three-level paging problem consists of a cache that can hold k pages, a shelf that can store l pages, and a slow memory. The input sequence are requests to pages. If the page requested is in the cache (a *hit*) then we need not do anything and no cost is incurred. If the page is in the shelf (*OK*), we can bring it from the shelf to the cache at the cost of α. Finally, if the page is in the slow memory (a *fault*), it takes the cost of γ to bring it to the cache. Whenever a new page needs to be brought to the cache, we must decide which page currently in the cache to be replaced by the incoming page. We can simply remove the replaced page from the cache or store it to the shelf at the cost of β for a future request to it. In the latter case, we also need to decide which page in the shelf to be removed if there is no room for the incoming page. (See Figure 1.) We assume that the shelf is more cost-efficient than the slow memory. Namely, the cost per page interaction (storage or retrieval) with the shelf is no greater than the cost per page retrieval from the slow memory:

$$\alpha , \beta \leq \gamma. \tag{1}$$

Notice that the three-level paging problem is not a special case of the K-*server problem* [1]. To see this, consider two distinct pages, say, a in the cache and b in the slow memory. Now suppose b, as requested, is retrieved from the slow memory and a is evicted from the cache and stored in the shelf. So it costs $\gamma + \alpha$ to get the new configuration. To recover the old configuration it costs β. Therefore, if we consider the space, in which each page request corresponds to a single point, the distance defined by the movements of the two kinds of servers (the cache and shelf locations) is not symmetric. Thus the upper-bound obtained in [2], which depends on the symmetry of the metric, does not apply to our case. More generally, our problem, however, can be regarded as a *metrical task system* [3] with an asymmetric distance.

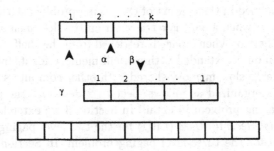

Fig. 1. Paging problem with shelf memory.

There are several previous works with some similarity to the three-level paging problem done by Aggarwal et al [4], and Chrobak and Noga [5]. The *multilevel caching problem* [4], or equivalently *relaxed list update problem* [5], considers a caching system consisting of m caches of increasing size and per page retrieval cost, where the level-1 cache is smallest and fastest, and the main memory is viewed as the largest and slowest cache at level-m. An essential difference is, however, that in their model, assuming that m caches are fully associative, once a page accessed, it can be swapped at no cost with a page currently in any higher level cache. But in the three-level paging problem proposed in this paper, this is not the case since additional storage cost β is incurred when two distinct pages are swapped between the cache and the shelf. Thus in this sense, the three-level paging problem may be viewed as a generalization of the multilevel caching problem with $m = 3$.

3 A competitive online algorithm

Consider the following online algorithm for the three-level paging problem.

On the request to a page which is not in the cache, select the page, p currently in the cache whose most recent access is earliest. If there is room for p, store p in the shelf (Initially, the cache and shelf are empty, as we may assume). Otherwise, consider the page q whose most recent access is earliest among the pages currently in the shelf. If the access to p is earlier than that to q, then we simply evict p from the cache to the slow memory. Otherwise, replace q with p in the shelf.

In the absence of the shelf, the algorithm described above reduces to the LRU algorithm for the traditional two-level problem. It is well-known that the LRU algorithm is k-competitive for the two-level paging problem: the LRU algorithm can service any input sequence at the cost less than k times the cost incurred by an optimal offline algorithm [7]. Now we will show that the (extended) LRU algorithm is also competitive for the three-level paging problem with an additional assumption on α and β:

Theorem 31 *If per page storage does not cost more than per page retrieval from the shelf, namely, if $\alpha \geq \beta$ then the LRU algorithm is $2k + l + 1$-competitive for the three-level paging problem.*

4 The proof of competitiveness

As in [7], the proof of Theorem 31 relies on the powerful technique of potential function. Denote our algorithm and an optimal offline algorithm by LRU and OPT, respectively. Let $\text{cost}_{LRU}(\sigma)$ and $\text{cost}_{OPT}(\sigma)$ be the costs incurred by LRU and OPT, respectively, on a given request sequence σ. To establish a competitive ratio c, namely, to show

$$\text{cost}_{LRU}(\sigma) \leq c \cdot \text{cost}_{OPT}(\sigma) + \text{a constant,} \tag{2}$$

for any σ, it suffices to find a function Φ mapping the current configurations of LRU and OPT to the real numbers with following properties [6] [7].

1. Φ is nonegative, and 0 initially.
2. For any single request in the sequence, let the additional costs of LRU and OPT be $\Delta\mathrm{cost}_{LRU}$ and $\Delta\mathrm{cost}_{OPT}$, respectively. Also let $\Delta\Phi$ be the change in Φ due to the processing of the page request. Then

$$\Delta\mathrm{cost}_{LRU} + \Delta\Phi \leq c \cdot \Delta\mathrm{cost}_{OPT}. \tag{3}$$

To do so, it is convenient to define an order relation between the pages in the LRU memories: for a pair of pages p and q, if the most recent access to p is later than to q, we will say that p *is ahead of* q and write $p \prec q$. Let p_1, p_2, $\cdots p_k$ be the pages in the LRU cache sorted so that $p_i \prec p_j$ if and only if $i < j$. Then we will say that p_i is the ith place in the LRU cache. Define $\Omega_1(p_i) = i - 1$. For any page p not in the cache, define $\Omega_1(p) = k$. Similarly, let $q_1, q_2, \cdots q_{k+l}$ be the pages currently in the LRU cache and shelf sorted so that $q_i \prec q_j$ if and only if $i < j$. Define $\Omega_2(q_j) = j - 1$. If p is not in either the cache or the shelf, let $\Omega_2(q) = k + l$. If the most recent access to q is jth earliest among the pages in the shelf, we will say q *is in the jth place in the LRU shelf*. In the LRU configurations, it is easy to see by induction that any page in the cache is ahead of any page in the shelf and this order relation is maintained through the application of algorithm. It implies that q is in the jth place in the LRU shelf if and only if $\Omega_2(q_j) = k + j - 1$.

Now we are ready to define our potential function: let S and T respectively be the set of the pages in the cache and shelf of OPT, then our potential function defined as follows:

$$\Phi = (\alpha + \beta) \sum_{p \in S} \Omega_1(p) + (\gamma - \alpha) \sum_{p \in S \cup T} \Omega_2(p) \tag{4}$$

Using this potential function, we show the condition in Eq. (3) holds for some value of c in the all the nine possible cases summarized in Table 1. To do so, it suffices to show that Eq. (3) is satisfied for the minimum and maximum possible values of $\Delta\mathrm{cost}_{OPT}$ and $\Delta\mathrm{cost}_{LRU}$, respectively. The values are easily computed from the problem and LRU descriptions. They are also provided in the third column of the same table. The corresponding sufficient conditions for the nine cases are summarized in the last column of the same table.

Now we compute an upperbound on $\Delta\Phi$ for each of the nine cases to find a value of c for which all the nine conditions are satisfied. Then the value should be a competitive ratio of LRU

Case 1: Let p be the requested page. As OPT hits, $p \in S$ of OPT and there is no change in S or T due to processing p. Since LRU also hits, p is in the cache of LRU. Let p be in the jth place in the LRU cache. Then $\Omega_1(p) = \Omega_2(p) = j - 1$. After processing p, both $\Omega_1(p)$ and $\Omega_2(p)$ reduce to 0 and accordingly both $\sum_{p \in S} \Omega_1(p)$ and $\sum_{p \in S \cup T} \Omega_2(p)$ will decrease by $j - 1$.

On the other hand, each of the $j - 1$ pages which were ahead of p in the LRU cache has its Ω_1 and Ω_2 values increased by 1 after processing. Thus there are

Cases	OPT	LRU	Min Δcost_{OPT}	Max Δcost_{LRU}	Condition for Eq. (3)
1	hit	hit	0	0	$\Delta\Phi \leq 0$
2	hit	OK	0	$\alpha + \beta$	$\Delta\Phi \leq -(\alpha + \beta)$
3	hit	fault	0	$\gamma + \beta$	$\Delta\Phi \leq -(\gamma + \beta)$
4	OK	hit	α	0	$\Delta\Phi \leq c\alpha$
5	OK	OK	α	$\alpha + \beta$	$\Delta\Phi \leq c\alpha - (\alpha + \beta)$
6	OK	fault	α	$\gamma + \beta$	$\Delta\Phi \leq c\alpha - (\gamma + \beta)$
7	fault	hit	γ	0	$\Delta\Phi \leq c\gamma$
8	fault	OK	γ	$\alpha + \beta$	$\Delta\Phi \leq c\gamma - (\alpha + \beta)$
9	fault	fault	γ	$\gamma + \beta$	$\Delta\Phi \leq c\gamma - (\gamma + \beta)$

Table 1. Possible cases of OPT and LRU .

at most $j - 1$ pages in S or $S \cup T$ of OPT whose Ω values increases by 1. This results in an increase in both $\sum_{p \in S} \Omega_1(p)$ and $\sum_{p \in S \cup T} \Omega_2(p)$ by at most $j - 1$ Thus $\Delta\Phi \leq 0$ in sum.

Case 2: Again, as OPT hits, neither S nor T will change. On the other hand, LRU brings the requested page p from the shelf to cache. Then $\Omega_1(p)$ decreases from k to 0. Due to this the k pages currently in the cache have their Ω_1 values increased by 1. This means that there can be at most $k - 1$ pages (other than p) of S whose Ω_1 values increased by 1. Hence the total change in $\sum_{p \in S} \Omega_1(p)$ is ≤ -1.

Let $\Omega_2(p) = k + j - 1$ before processing p (*i.e.* p is in the jth place in the shelf). Then it decreases to 0 after processing. By a similar argument in Case 1, there are at most $k + j - 1$ pages in $S \cup T$ whose Ω_2 values increase by 1. Hence the total change in $\sum_{p \in S \cup T} \Omega_2(p)$ is ≤ 0. Thus in total $\Delta\Phi \leq -(\alpha + \beta)$.

Case 3: As in the above cases, there is no change in OPT's configuration, S and T. In the LRU configuration, p is currently in the slow memory and hence $\Omega_1(p) = k$ and $\Omega_2(p) = k + l$. As LRU brings p from the slow memory to the cache, both decrease to 0. By the same argument as in Case 2, there are at most $k - 1$ pages in S whose Ω_1 values increase by 1. Hence the total change in $\sum_{p \in S} \Omega_1(p)$ is ≤ -1. Similarly total change in $\sum_{p \in S \cup T} \Omega_2(p)$ is also ≤ -1. In sum, $\Delta\Phi \leq -(\alpha + \beta) -(\gamma - \alpha) = -(\gamma + \beta)$.

Case 4: In this case, the requested page, p is brought from T to S by OPT and accordingly some page q is evicted from S to T or to the slow memory. First notice that the eviction of q from S never introduce a new member (other than p) to S or $S \cup T$. Therefore, as we are concerned about an upper on the increase in $\Omega_1(p)$ and $\Omega_2(p)$, we may assume that there is no change (other than p) in either S or T after processing p in Case 4,5, and 6.

Suppose the requested page p is in the jth place in the LRU cache so that $\Omega_1(p) = j-1$. Although p moves from T to S in OPT's configuration, it does not increase the total Ω_1 values as $\Omega_1(p)$ decreases to 0. But it makes each Ω_1 value of the other $j - 1$ pages which are currently ahead of p in the cache increase by 1. Thus after processing at most $j - 1$ pages in S have their Ω_1 values increased by 1. Thus the total change in Ω_1 is $\leq j - 1$.

As $\Omega_2(p)$ decreases from $j-1$ to 0, $\sum_{p \in S \cup T} \Omega_2(p)$ decreases by $j-1$ By the same argument as above, at most $j-1$ pages in $S \cup T$ have their Ω_2 values increased by 1. Hence, in sum, the total increase in $\sum_{p \in S \cup T} \Omega_2(p)$ is nonpositive. Hence $\Delta\Phi \le (j-1)(\alpha+\beta) \le 2(j-1)\alpha$ with $j \le k$, where the last inequality follows from the assumption of the theorem: $\alpha \ge \beta$.

Case 5: Suppose the requested page p is in the jth place in the LRU shelf so that $\Omega_1(p) = k$ and $\Omega_2(p) = k+j-1$. Again OPT's moving p from T to S does not contribute to Ω_1 values as $\Omega_1(p)$ is 0 after processing. But it makes the Ω_1 values of other $k-1$ pages increase by 1 and then at most $k-1$ pages in S have their Ω_1 values increased by 1. Thus the total change in Ω_1 is $\le k-1$.

On the other hand, as $\Omega_2(p)$ decreases to 0, each of at most $k+j-1$ pages currently ahead of p in $S \cup T$ has its Ω_2 value increased by 1 after processing. Then the total change in Ω_2 is ≤ 0.

Hence, in sum, $\Delta\Phi \le (k-1)(\alpha+\beta) = k(\alpha+\beta) - (\alpha+\beta) \le 2k\alpha - (\alpha+\beta)$.

Case 6: Before processing, $\Omega_1(p) = k$ and $\Omega_2(p) = k+l$. Both decrease to 0 after processing. As p moves to the top of the LRU cache, each of the k pages currently in the cache has its Ω_1 value increased by 1. Thus there can be at most k pages in S whose Ω_1 values increased by 1. Hence the total change in Ω_1 is $\le k$.

On the other hand, as p moves to the top of the cache, $\Omega_2(p)$ reduces to 0. This results in a unit increase in the Ω_2 value of each of the $k+l$ pages currently in the cache or shelf. Since p is in T before processing, at most $k+l-1$ pages in $S \cup T$ may have their Ω_1 values increased by 1. Therefore, the total increase in Ω_2 is ≤ -1.

Thus $\Delta\Phi \le k(\alpha+\beta) - (\gamma-\alpha) = (k+1)(\alpha+\beta) - (\gamma+\beta) \le 2(k+1)\alpha - (\gamma+\beta)$.

Case 7: As OPT faults, p is brought from the slow memory to S in OPT's configuration. Accordingly some pages may be evicted from a higher to a lower level memory. Again notice that such eviction only may reduce the members (other than p) of S or $S \cup T$. Hence in the following tree cases, we may assume that there is no change (other than p itself) in S or T after processing p.

Suppose the requested page p is in the jth place in the LRU cache so that $\Omega_1(p) = \Omega_2(p) = j-1$. Although p moves from the slow memory to S in OPT's configuration, it does not increase the total Ω_1 or Ω_2 value since $\Omega_1(p)$ and $\Omega_2(p)$ decreases to 0. But it makes the Ω_1 and Ω_2 values of the other $j-1$ pages currently ahead of p in the LRU cache increase by 1. Thus after processing at most $j-1$ pages in S or $S \cup T$ have their Ω values increased by 1. Hence $\Delta\Phi \le (j-1)(\alpha+\beta) + (j-1)(\gamma-\alpha) \le (j-1)(\beta+\gamma) \le 2(j-1)\gamma$ with $j \le k$. The last inequality follows from Eq. (1).

Case 8: Suppose the requested page p is in the jth place in the LRU shelf so that $\Omega_1(p) = k$ and $\Omega_2(p) = k+j-1$. Again moving p from T to S in OPT's configuration does not increase $\sum_{p \in S} \Omega_1(p)$ or $\sum_{p \in S \cup T} \Omega_2(p)$.

But it makes each Ω_1 value of the k pages in the LRU cache increase by 1. Thus after processing at most k pages in S have their Ω_1 values increased by 1.

Similarly, each Ω_2 value of $k+j-1$ pages currently ahead of p in the LRU cache or shelf increase by 1. Thus at most $k+j-1$ pages in $S \cup T$ may have their Ω_2 values increased by 1.

In sum,

$$
\begin{aligned}
\Delta\Phi \leq \quad & k(\alpha+\beta)+(k+j-1)(\gamma-\alpha) & (5) \\
= & (k+j-1)\gamma+(k+1)\beta-(j-2)\alpha-(\alpha+\beta) \\
\leq & (k+j-1)\gamma+(k+1)\alpha-(j-2)\alpha-(\alpha+\beta) \\
= & (k+j-1)\gamma+(k-j+1)\alpha+2\alpha-(\alpha+\beta).
\end{aligned}
$$

When $j \leq k$, the last expression of Eq. (5) is $\leq (k+j-1)\gamma+(k-j+1)\gamma+2\gamma-(\alpha+\beta) = (2k+2)\gamma-(\alpha+\beta)$. When $j \geq k+1$, the same expression is $\leq (k+j-1)\gamma+2\gamma-(\alpha+\beta) = (k+j+1)\gamma-(\alpha+\beta)$. Thus in any case, we have $\Delta\Phi \leq (2k+j+1)\gamma-(\alpha+\beta)$.

Case 9: Since LRU faults $\Omega_1(p) = k$ and $\Omega_2(p) = k+l$. As both $\Omega_1(p)$ and $\Omega_2(p)$ decrease to 0, p does not increase $\sum_{p \in S} \Omega_1(p)$ or $\sum_{p \in S \cup T} \Omega_2(p)$.

Each Ω_1 value of the k pages in the LRU cache increases by 1. Thus at most k pages in S have their Ω_1 values increased by 1.

Similarly, each Ω_2 value of the $k+l$ pages in the LRU cache and shelf increase by 1. Thus at most $k+l$ pages in $S \cup T$ may have their Ω_2 values increased by 1.

Hence

$$
\begin{aligned}
\Delta\Phi \leq \quad & k(\alpha+\beta)+(k+l)(\gamma-\alpha) & (6) \\
= \quad & l(\gamma-\alpha)+k(\gamma+\beta) \\
= & (k+l+1)\gamma+(k+1)\beta-l\alpha-(\gamma+\beta).
\end{aligned}
$$

Since $\alpha \geq \beta$, $\Delta\Phi \leq (2k+l+1)\gamma-(\gamma+\beta)$.

In Table 2, the upperbounds on $\Delta\Phi$ obtained for all the cases are summarized and compared with the corresponding sufficient conditions for a constant c to be a competitive ratio of LRU.

Cases	Obtained upperbound	Required condition
1	$\Delta\Phi \leq 0$	$\Delta\Phi \leq 0$
2	$\Delta\Phi \leq -(\alpha+\beta)$	$\Delta\Phi \leq -(\alpha+\beta)$
3	$\Delta\Phi \leq -(\gamma+\beta)$	$\Delta\Phi \leq -(\gamma+\beta)$
4	$\Delta\Phi \leq 2(j-1)\alpha$ with $j \leq k$	$\Delta\Phi \leq c\alpha$
5	$\Delta\Phi \leq 2k\alpha-(\alpha+\beta)$	$\Delta\Phi \leq c\alpha-(\alpha+\beta)$
6	$\Delta\Phi \leq 2(k+1)\alpha-(\gamma+\beta)$	$\Delta\Phi \leq c\alpha-(\gamma+\beta)$
7	$\Delta\Phi \leq 2(j-1)\gamma$ with $j \leq k$	$\Delta\Phi \leq c\gamma$
8	$\Delta\Phi \leq (2k+j+1)\gamma-(\alpha+\beta)$ with $j \leq l$	$\Delta\Phi \leq c\gamma-(\alpha+\beta)$
9	$\Delta\Phi \leq (2k+l+1)\gamma-(\gamma+\beta)$	$\Delta\Phi \leq c\gamma-(\gamma+\beta)$

Table 2. Upperbounds and corresponding sufficient conditions for c to be a competitive ratio.

From the table, we can see that the sufficient condition for Eq. (3) is satisfied in each of all the 9 cases when $c = 2k + l + 1$. Therefore, LRU is $2k + l + 1$-competitive.

Now we impose an additional but natural assumption that $k \leq l$. Then from Table 2, it is easily seen that the sufficient condition is satisfied in each of the first 7cases if $c = k + l + 2$. So to show that the algorithm is $k + l + 2$-competitive when $k \leq l$, it suffices to show that $\Delta\Phi \leq (k + l + 2)\gamma - (\alpha + \beta)$ in Case 8, and $\Delta\Phi \leq (k + l + 2)\gamma - (\gamma + \beta)$ in Case 9.

In Case 8, from the second equation of Eq. (5), we have $\Delta\Phi \leq (k + j - 1)\gamma + k\beta - (j - 1)\beta$, which is $\leq (k + j - 1)\gamma \leq (k + l + 2)\gamma - (\alpha + \beta)$ if $j - 1 \geq k$. If $j - 1 < k$, then $\leq (k + j - 1)\gamma + (k - j + 1)\gamma = 2k\gamma \leq (k + l)\gamma - (\alpha + \beta) + 2\gamma = (k + l + 2)\gamma - (\alpha + \beta)$.

In Case 9, the last equation of Eq. (6) is equal to $(k + l + 1)\gamma + \beta + k\beta - l\alpha - (\gamma + \beta)$. Since $k \leq l$, this is $\leq (k + l + 1)\gamma + \beta - (\gamma + \beta) \leq (k + l + 2)\gamma - (\gamma + \beta)$.

Corollary 41 *If, in addition, $k \leq l$, then LRU algorithm is $k + l + 2$-competitive.*

5   Conclusions and future research

We have proposed a three-level memory paging problem which has many potential applications. We also have shown that a straightforward extension of the well-known LRU online algorithm for two-level paging problem is also competitive for the problem when per page storage does not cost more than per page retrieval from the additional level of memory.

It is not known, however, whether LRU-type algorithm is still competitive in the general case. Also an optimal offline algorithm or a tight lower or upper bound on the competitive ratio are still unknown.

As discussed, the three-level paging problem is a generalization of three-level caching problem in the sense that the cost of swapping pages between caches of different levels is nonzero. In this context, it may be interesting to extend the proposed model to a multilevel problem. This problem seems to have more practicality especially when the caches are not fully associative.

Acknowledgments

We gratefully acknowledge helpful comments from anonymous referees; especially pointing out the previous works on the multilevel caching problems.

References

1. M. Manasse, L. A. McGeoch, and D. Sleater, "Competitive algorithms for server problems," In *Journal of Algorithms*, 11:208-230, 1990.
2. E. Koutsoupias and C. Papadimitriou, "On the k-server conjecture," In *Proc. 25th Symposium on Theory of Computing*, 507-511, 1994.

3. A. Borodin, N. Linial, and M. Saks, "An optimal online algorithm for metrical task systems," In *Proc. 19th Annual ACM Symposium on Theory of Computing*, 373-382, 1987.

4. A. Aggarwal, B. Alpern, A. K. Chandra, and M. Snir, "A model for hierarchical memory," In *Proc. 19th Symposium on Theory of Computing*, 305-313, 1987.

5. M. Chrobak and J. Noga, "Competitive algorithms for multilevel caching and relaxed list update," In *Proc. 9th Annual ACM-SIAM Symposium on Discrete Algorithms*, 87-96, 1998.

6. S. Irani and A. R. Karlin, "Online computation," in *Approximation Algorithms for NP-hard problems* edited by D. S. Hochbaum, PWS, Boston, 1997 (521-564).

7. D. Sleater and R. Tarjan, "Amortized efficiency of list update and paging rules," *Communications of the ACM*, 28(2):202-208, 1985.

Using Generalized Forecasts for Online Currency Conversion

Kazuo Iwama and Kouki Yonezawa

School of Informatics
Kyoto University
Kyoto 606-8501, Japan
{iwama,yonezawa}@kuis.kyoto-u.ac.jp

Abstract. El-Yaniv et al. presented an optimal on-line algorithm for the unidirectional currency conversion problem based on the threat-based strategy. Later, al-Binali pointed out that this algorithm, while certainly optimal in terms of the competitive ratio, may be too conservative from "gamblers' view points." He proposed the trading strategy using the forecast that the rate will increase to at least some level M_1. If the forecast is true, the algorithm achieves a better competitive ratio than the ratio r_0^* of the threat-based strategy. Otherwise, the algorithm suffers from a loss (a worse ratio than r_0^*) but it can be controlled within a certain factor.

In this paper, we generalize this notion of forecasts: (i) Not only the forecast that the rate will increase to some level, called an *above-forecast*, but the forecast that the rate will never increase to some level, a *below-forecast*, is also allowed. (ii) Forecasts are allowed twice or more. Several different algorithms can be designed using this extension, e.g., an algorithm making two rounds of forecasts where one can regain r_0^* by the second forecast even if the first forecast is known to be false. Furthermore, we discuss the bidirectional conversion (i.e., both conversions from dollar to yen and yen to dollar are allowed) which helps even for a monotonically-changing rate, if below-forecasts are involved.

1 Introduction

As far as their performance is measured by the (worst-case) competitive ratio, it is not surprising that on-line algorithms tend to be conservative or tend to be risk-avoiding. A typical example is the *threat-based* algorithm, by El-Yaniv et al., for the unidirectional currency conversion problem (UCCP) [EFKT92]. Although it achieves the optimal competitive ratio, denoted by r_0^* throughout this paper, it does not reflect common practice of traders. For example, investors usually make some assumptions about the future [CCEKL95] and also are willing to take "risk" while trading [RS70,RS71,SSW95,Mor96], but it is hard to evaluate those actions within the framework of the conventional competitive ratio.

In his recent paper [alB97], al-Binali proposed competitive analysis which allows a trader to make a *forecast*. If the forecast is true, then the trader can get a better ratio than $r_0^\star$; otherwise the trader suffers from a worse ratio than $r_0^\star$. The important point is, however, that the risk can be controlled within the factor of t, say, $t = 1.01$. Namely the worse ratio is guaranteed not to be worse than $tr_0^\star$. In [alB97], al-Binali mainly discusses UCCP and gives the following algorithm: (i) The algorithm takes, as a part of parameters, a forecast given by a rate M_1 and the risk-tolerance factor t (> 1). (The forecast means that the rate will increase to at least M_1.) (ii) If the forecast is true then the algorithm achieves the competitive ratio $r_1^\star$ ($< r_0^\star$) that is determined by t and is optimal under the following condition: (iii) If the forecast is not true, the ratio of the algorithm is not worse than $tr_0^\star$. Namely, one can take a risk which is controlled within a certain factor and can get an optimal reward from the risk.

In this paper, we generalize this risk-taking strategy for UCCP in two respects: (1) A forecast in [alB97] is limited to the type that the rate will increase to some level. We also allow the opposite type, i.e., the forecast that the rate will *never* increase to some level. Forecasts of the former type are called *above-forecasts* and the latter *below-forecasts*. (2) In the course of trading, forecasts can be made twice or more. By these extensions, it is now possible to take several different types of flexible strategies, for example, the one such that the player makes an above-forecast first and then, if it is successful, he/she makes a second forecast, say, a below-forecast, to aim at an even better competitive ratio. Another example is the double-below-forecast strategy where even if the first below-forecast fails, i.e., if the rate has increased to the level of the first forecast, one can regain $r_0^\star$ if the second below-forecast is successful. To assist designing these variety types of algorithms, we provide a scheme which enables a systematic generation of on-line algorithms for UCCP including several forecasts. For this purpose we need a new necessary and sufficient condition for being able to design optimal on-line conversion algorithms, not for the entire trading period but for its partial period.

Another new finding in this paper is the usefulness of the bidirectional conversion. It should be noted that under the condition of unidirectional conversion, we can assume that the exchange rate is monotonically increasing (and finally drops suddenly) without loss of generality. In other words, one can see easily that the unidirectional conversion is enough for the threat-based strategy. Also, it is not hard to prove that the unidirectional conversion is enough even if above-forecasts are introduced. However, the situation does differ for below-forecasts: It is shown that the bidirectional conversion enables us to design an algorithm whose competitive ratio for successful forecasts is much better than we can do using only unidirectional conversion.

2   Threat-Based Strategy

In the *unidirectional currency conversion problem* (*UCCP*), an on-line player converts dollars to yen over some period of time and must convert all the remaining dollars at the end of a trading period. It is prohibited to convert already purchased yen back to dollars, i.e., the trade is unidirectional. Under this set-

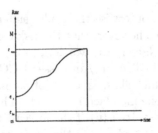

Fig. 1.

ting, we can assume, without loss of generality [EFKT92], that the exchange rate starts from e_0, increases monotonically up to some rate e_{max}, drops suddenly to e_{fin} and stay there forever (see Fig. 1). M and m denote the upper and lower bounds, respectively, of the exchange rate. The player initially holds one dollar and wishes to get as much yen as possible at the end of the game.

The performance of on-line algorithms is measured by *the competitive ratio*: Suppose that the player carries out trading using some algorithm, A, which determines the trades. $D(x)$ and $Y(x)$ denote such amounts of dollars and yen, respectively, when the rate is x. Thus the player obtains $Y(e_{max}) + e_{fin} \cdot D(e_{max})$ yen at the end of the game. The off-line player can get e_{max} yen by changing the original one dollar at the highest rate and hence the competitive ratio of the algorithm A is the worst-case value of $e_{max}/(Y(e_{max}) + e_{fin} \cdot D(e_{max}))$.

The threat-based strategy [EFKT92] requires the on-line player to make the following action: To attain a given competitive ratio r_0, the player should simply defend himself/herself against the threat that the exchange rate may drop to its minimum value m at the next moment. More concretely, (i) the player does not trade at all when $x \leq r_0 m$. (ii) The player carries out the trading so that his/her $D(x)$ and $Y(x)$ will always satisfy that $mD(x) + Y(x) = x/r_0$ when $x > r_0 m$. As a result, $D(x)$ and $Y(x)$ can be given as

$$D(x) = 1 - \frac{1}{r_0} \ln \frac{x - m}{r_0 m - m}, \tag{1}$$

$$Y(x) = \frac{1}{r_0} \left(m \ln \frac{x - m}{r_0 m - m} + x - r_0 m \right). \tag{2}$$

It is not hard to see that the optimal r_0, denoted by $r_0^\star$, must satisfy that $D(M) = 0$, which implies that

$$r_0^\star = \ln \frac{M - m}{r_0^\star m - m}. \tag{3}$$

This is the case for $e_0 \leq r_0 m$. If $e_0 > r_0 m$, formulas (1)-(3) change a little as follows:

$$D(x) = \frac{e_0 - e_0/r_0}{e_0 - m} - \frac{1}{r_0} \ln \frac{x - m}{e_0 - m}, \tag{4}$$

$$Y(x) = \frac{e_0(e_0 - r_0 m)}{r_0(e_0 - m)} + \frac{1}{r_0}\left(m\ln\frac{x - m}{e_0 - m} + x - e_0\right), \qquad (5)$$

$$r_0^\star = 1 + \frac{e_0 - m}{e_0}\ln\frac{M - m}{e_0 - m}. \qquad (6)$$

The value of $r_0^\star$ in (6) is now a function of e_0 and is different from $r_0^\star$ in (3). Unless otherwise stated, however, $r_0^\star$ always means the value of (3) in this paper.

The model so far described is called the *continuous model*, where the player can trade at any moment. The other model is the *discrete model*, where the player can trade only at discrete time steps. In this paper, we restrict ourselves to the continuous model but similar results also hold for the discrete model.

3   Designing Multi-Phase Conversion

3.1   An Useful Property

Suppose that the player has the above-forecast M_1, i.e., the forecast that the rate will increase up to at least M_1 ($m < M_1 < M$). Then, during the period when the rate x is between m and M_1, the player exchanges less dollars to yen compared to the threat-based algorithm. This allows the player to hold more dollars when the rate reaches M_1 (i.e., if the forecast is successful). After that, the player can exchange more dollars, which gives him/her a better competitive ratio than that of threat-based. Thus, trading based on forecasts consists of two or more phases of different trading strategies which should be combined at the moments when the rate has reached the value of the forecast. In this section, we first give a useful property when designing each single phase. Also we shall see how to use that property to design a whole multi-phase conversion algorithm using a simple example, i.e., a conversion algorithm which involves a single above-forecast. More sophisticated algorithms will be designed in the next section.

An *interval* is denoted by a pair $[N_1, N_2]$, $m \leq N_1 < N_2 \leq M$, of exchange rates. Suppose that we fix some conversion algorithm, A, that works for this interval. Then A determines $D(N_1)$ and $Y(N_1)$, the amounts of dollars and yen at the beginning of the interval, $D(N_2)$ and $Y(N_2)$, the amounts of them at the end of the interval (if the rate reaches there), and the competitive ratio r. Conversely speaking, it would be possible to design A from such values as $D(N_1)$, $Y(N_1)$, $D(N_2)$, $Y(N_2)$, and r. Actually those five parameters are not independent, so that we only need some of them. For example, the threat-based algorithm is a unique algorithm, A, which satisfies (i) $D(rm) = 1$, $Y(rm) = 0$ and $D(M) = 0$, and (ii) A is optimal in the sense that A achieves the same competitive ratio, r, whenever the rate drops to m between rm and M. The remaining values, $Y(M)$ and r, are determined automatically. Here is a little more general statement for an optimal algorithm to be designed:

Theorem 1. Suppose that an algorithm, A, satisfies that $D(N_1) = d_1$, $Y(N_1) = y_1$ and a competitive ratio r is given. Then an "optimal" A, by which the player can achieve the competitive ratio r and he/she can continue trading as far as possible, is uniquely determined.

Proof. Observing the argument of the threat-based algorithm in [EFKT92], it is not hard to see that an optimal conversion algorithm can be designed by the following basic policy: (i) The competitive ratio during the trading should not be greater than r, whenever the rate drops to the minimum rate m. (ii) If the current amounts of dollars d and yen y satisfy that $md + y \geq x/r$ for the current rate x and the target ratio r, then one should not trade now. (By this one can keep more dollars to exchange at higher rate.) Just recall the threat-based algorithm: Policy (ii) applies while $x \leq r_0 m$, and after that the trade proceeds based on policy (i).

Now we apply these policies to the present situation. (1) If $N_1 \leq rm$ then policy (ii) applies. One should not trade until the rate reaches rm, because if $N_1 \leq rm$, one should have 1 dollar and no yen. After the rate reaches rm, the trade can be governed by the following formulas which are exactly the same as (1) and (2) except that r_0 is replaced by r.

$$D(x) = 1 - \frac{1}{r} \ln \frac{x - m}{rm - m}, \tag{7}$$

$$Y(x) = \frac{1}{r} \left(m \ln \frac{x - m}{rm - m} + x - rm \right). \tag{8}$$

Note that trade can continue until the moment x such that $D(x) = 0$.

(2) If $N_1 > rm$, then we have furthermore two cases: *Case* (2-1) $md_1 + y_1 \leq N_1/r$. In this case, we have to exchange some amount of dollars to yen at the beginning, since otherwise we cannot achieve the ratio r if the rate drops now. The amount of dollars that we have to trade at this moment is determined so that the amounts of the dollars d_1' and yen y_1' after the exchange will satisfy that $md_1' + y_1' = N_1/r$. And it must be satisfied that $m(d_1 - d_1') = y_1' - y_1$. From these formulas, it turns out that $d_1' = \frac{1}{N_1 - m} \left(N_1 d_1 + y_1 - \frac{N_1}{r} \right)$, $y_1' = \frac{1}{N_1 - m} \left(\frac{N_1^2}{r} - m(N_1 d_1 + y_1) \right)$. After that, the trade proceeds by

$$D(x) = d_1' - \frac{1}{r} \ln \frac{x - m}{N_1 - m}, \tag{9}$$

$$Y(x) = y_1' + \frac{1}{r} \left(m \ln \frac{x - m}{N_1 - m} + x - N_1 \right). \tag{10}$$

Case (2-2) $md_1 + y_1 > N_1/r$. In this case the policy (ii) applies. We have to wait until the rate goes up to $\tilde{N}_1$ ($> N_1$) such that $md_1 + y_1 = \tilde{N}_1/r$, because otherwise we cannot gain a certain ratio during $[N_1, N_2]$. After that we should follow the following formulas:

$$D(x) = d_1 - \frac{1}{r} \ln \frac{x - m}{\tilde{N}_1 - m}, \tag{11}$$

$$Y(x) = y_1 + \frac{1}{r} \left(m \ln \frac{x - m}{\tilde{N}_1 - m} + x - \tilde{N}_1 \right). \tag{12}$$

3.2   Single Above-Forecast

Let us design an optimal algorithm, A, using a single above-forecast M_1: Namely, if the rate drops before the rate reaches M_1 then A may achieve tr_0^* ($t > 1$). (Note that $M_1 \geq tr_0^* m$ must be satisfied for such a trade to be possible.) Under this condition, we wish to design A so that we can achieve the optimal competitive ratio after the rate gets to M_1. We assume that the starting rate e_0 is less than rm. (The other case is similar and omitted. This is the same also in the rest of the paper.)

(1) While the rate is between $[e_0, M_1]$, we design the optimal algorithm of the competitive ratio tr_0^* by the case (1) of Theorem 1. Let $D_1(x)$ and $Y_1(x)$ be the amount of dollars and yen the player should hold. Then

$$D_1(x) = 1 - \frac{1}{tr_0^*} \ln \frac{x - m}{tr_0^* m - m}, \tag{13}$$

$$Y_1(x) = \frac{1}{tr_0^*} \left(m \ln \frac{x - m}{tr_0^* m - m} + x - tr_0^* m \right). \tag{14}$$

Now we can calculate $D_1(M_1)$ and $Y_1(M_1)$.

(2) For the interval $[M_1, M]$, we once assume a tentative competitive ratio r_1 (which must be better than r_0^*) and design the optimal algorithm from $D_1(M_1)$, $Y_1(M_1)$ and r_1, by using Case (2-1) of Theorem 1. Let $D_2(x)$ and $Y_2(x)$ be the amounts of dollars and yen in this phase. Then

$$D_2(x) = d - \frac{1}{r_1} \ln \frac{x - m}{M_1 - m}, \tag{15}$$

$$Y_2(x) = y + \frac{1}{r_1} \left(m \ln \frac{x - m}{M_1 - m} + x - M_1 \right). \tag{16}$$

As mentioned in the proof of Theorem 1, a certain amounts of dollars are sold at this moment to get d dollars and y yen, which must satisfy

$$1 - \frac{1}{tr_0^*} \ln \frac{M_1 - m}{tr_0^* m - m} - \frac{M_1}{M_1 - m} \left(\frac{1}{r_1} - \frac{1}{tr_0^*} \right) - \frac{1}{r_1} \ln \frac{M - m}{M_1 - m} = 0. \tag{17}$$

By solving these equations, we can obtain d and y as $d = D_1(M_1) - \frac{M_1}{M_1 - m} \left(\frac{1}{r_1} - \frac{1}{tr_0^*} \right)$ and $y = Y_1(M_1) + \frac{M_1^2}{M_1 - m} \left(\frac{1}{r_1} - \frac{1}{tr_0^*} \right)$. Then by substituting them into (15) and (16), one can get $D_2(x)$ and $Y_2(x)$ without d or y. Now the optimal value for r_1 should satisfy that $D_2(M) = 0$, or

$$1 - \frac{1}{tr_0^*} \ln \frac{M_1 - m}{tr_0^* m - m} - \frac{M_1}{M_1 - m} \left(\frac{1}{r_1} - \frac{1}{tr_0^*} \right) - \frac{1}{r_1} \ln \frac{M - m}{M_1 - m} = 0. \tag{18}$$

By solving this equation, one can get

$$r_1^* = \frac{M_1 - m}{(M_1 - m)\left(1 - \frac{1}{tr_0} \ln \frac{M_1 - m}{tr_0 m - m}\right) + \frac{M_1}{tr_0}} \left(\frac{M_1}{M_1 - m} + \ln \frac{M - m}{M_1 - m} \right). \tag{19}$$

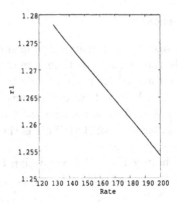

Fig. 2.

When $m = 100, M = 200$ and $t = 1.01$, the relation between r_1^* and M_1 is expressed as in Fig. 2. (Note that $r_0^* \approx 1.28$ in this setting.) Thus, the value of r_1^* decreases (is getting better) almost in linear as M_1 grows. In other words, if the player takes more risk (selects higher M_1) then he/she can get linearly more benefit if the forecast is true. It should be noted that in [alB97], al-Binali did the similar argument for the discrete model.

4   Several Multi-phase Algorithms

4.1   Double Above-Forecasts

The player uses two above-forecasts M_1 and M_2 $(m < M_1 < M_2 < M)$ aiming at a furthermore better competitive ratio after the first above-forecast succeeds. If the first forecast fails, then the competitive ratio is tr_0^*. If the second forecast fails, it is r_1. Finally, the ratio is r_2 if the second forecast is true. The design procedure is exactly the same as before, i.e., one can design the optimal algorithm from the first phase $[m, M_1]$, then $[M_1, M_2]$ and finally $[M_2, M]$ sequentially. This gives us r_2 as a function of r_1. If we assume that $r_1 = tr_1^1$, where r_1^1 is the optimal value of r_1 in the case of a single above-forecast discussed in Section 3.2, then the optimal value of r_2 (i.e., by which the value of dollars becomes zero at M) can be given as follows:

$$r_2^* = \frac{\frac{M_2}{M_2 - m} + \ln \frac{M - m}{M_2 - m}}{1 - \frac{1}{tr_0^*} \ln \frac{M_1 - m}{tr_0^* m - m} - \frac{M_1}{M_1 - m}\left(\frac{1}{r_1} - \frac{1}{tr_0^*}\right) - \frac{1}{r_1} \ln \frac{M_2 - m}{M_1 - m} + \frac{M_2}{r_1(M_2 - m)}}. \quad (20)$$

For example, when $M_1 = 150$ and $M_2 = 175$, the values of $r_1 = tr_1^1$ and r_2^* are about 1.288 and 1.266, respectively. Recall that $r_1^1 \approx 1.271$ if the player uses only a single above-forecast at $M_1 = 150$.

4.2   Single Below-Forecast

The player can get a better ratio r_1 ($< r_0^*$) if the rate drops before M_1. Otherwise the ratio may be as worse as tr_0^*.

(1) For interval $[m, M_1]$, we apply case (1) of Theorem 1, by using r_1 as the competitive ratio in this interval. For $r_1 m \le x \le M_1$, $D_1(x)$ and $Y_1(x)$ are given as

$$D_1(x) = 1 - \frac{1}{r_1} \ln \frac{x - m}{r_1 m - m}, \tag{21}$$

$$Y_1(x) = \frac{1}{r_1} \left(m \ln \frac{x - m}{r_1 m - m} + x - r_1 m \right) \tag{22}$$

and we can get $D_1(M_1)$ and $Y_1(M_1)$.

(2) Now we design the second phase using the values of $D_1(M_1)$, $Y_1(M_1)$ and tr_0^*. Since $r_1 < r_0^* < tr_0^*$, it follows that $m D_1(M_1) + Y_1(M_1) > M_1/tr_0^*$, which means we should apply Case (2-2) of Theorem 1. Let $\widetilde{M_1}$ be the rate such that $m D_1(M_1) + Y_1(M_1) = \widetilde{M_1}/tr_0^*$. Then $D_2(x)$ and $Y_2(x)$ are given as follows:

$$D_2(x) = D_1(M_1) - \frac{1}{tr_0^*} \ln \frac{x - m}{\widetilde{M_1} - m}, \tag{23}$$

$$Y_2(x) = Y_1(M_1) + \frac{1}{tr_0^*} \left(m \ln \frac{x - m}{\widetilde{M_1} - m} + x - \widetilde{M_1} \right). \tag{24}$$

Note that the value of M_1 must be selected such that $M_1 \le \frac{r_1}{tr_0^*} M$, since $\widetilde{M_1}$ should be at most M.

The optimal value r_1^* for r_1 is obtained by setting $D_2(M) = 0$, or as the solution of equation

$$1 - \frac{1}{r_1} \ln \frac{M_1 - m}{r_1 m - m} - \frac{1}{tr_0^*} \ln \frac{M - m}{\widetilde{M_1} - m} = 0, \tag{25}$$

where $\widetilde{M_1}$ takes the value given above.

Let us set $t = 1.01$ again and recall that $r_0^* \approx 1.28$. Then r_1^* is a function of M_1 as shown in Fig. 3. The value of r_1^* increases (becomes worse) linearly as M_1 grows, namely, the merit when the forecast succeeds becomes smaller as M_1 grows. This is reasonable since as M_1 grows the forecast "is more likely" to be successful. However, this growth of r_1^* is saturated after $M_1 = 140$. Therefore, it might not be a good idea to select $M_1 = 140$, since the risk-reward is very similar to the selection that, say, $M_1 = 170$. $M_1 = 170$ is much easier for the success of the forecast.

4.3   Double Below-Forecasts

In the case of above-forecasts, when the player knows that his/her forecast fails, the game is over. Conversely, in the case of below-forecasts, the game still continues after the forecast is known to be false, but of course the guaranteed competitive ratio goes down under r_0^*. Therefore, it is a natural idea for the player

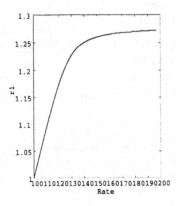

Fig. 3.

to try to regain r_0^* after he/she knows that the below-forecast M_1 has failed, by making another below-forecast M_2.

(1) We assume that the first phase $[m, M_1]$ is exactly the same as the first phase of the single below-forecast. $D_1(M_1)$ and $Y_1(M_1)$ are obtained as before.

(2) The competitive ratio for second phase $[M_1, M_2]$ is r_0^*. Use Case (2-2) of Theorem 1.

$$D_2(x) = D_1(M_1) - \frac{1}{r_0^*} \ln \frac{x - m}{\widetilde{M_1} - m}, \tag{26}$$

where

$$\widetilde{M_1} = \frac{r_0^*}{r_1^*} M_1. \tag{27}$$

(3) The third phase guarantees the competitive ratio r_2. It then follows that

$$D_3(x) = D_2(M_2) - \frac{1}{r_2} \ln \frac{x - m}{\widetilde{M_2} - m}, \tag{28}$$

where

$$\widetilde{M_2} = \frac{r_2}{r_0} M_2. \tag{29}$$

The optimal r_2, r_2^*, is obtained when $D_3(M) = 0$, namely as the solution of the following equation

$$1 - \frac{1}{r_1^*} \ln \frac{M_1 - m}{r_1^* m - m} - \frac{1}{r_0^*} \ln \frac{M_2 - m}{\widetilde{M_1} - m} - \frac{1}{r_2} \ln \frac{M - m}{\widetilde{M_2} - m} = 0. \tag{30}$$

This implies that

$$r_2^* = \frac{\ln \frac{M - m}{M_2 - m}}{1 - \frac{1}{r_1^*} \ln \frac{M_1 - m}{r_1^* m - m} - \frac{1}{r_0^*} \ln \frac{M_2 - m}{M_1 - m}}. \tag{31}$$

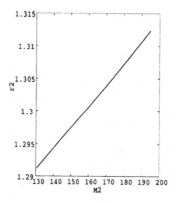

Fig. 4.

When $M_1 = 120$, the relation between r_2^* and M_2 is illustrated in Fig. 4. For example, when $M_2 = 160$, The competitive ratio for each phase is 1.17 for the first phase (see Fig. 3), 1.28 ($=r_0^*$) for the second phase and 1.30 for the third phase. The difference between the second and the third phases is small since t is close to 1. (Actually, the competitive ratio for the single below-forecast is 1.18 and 1.29 for the first and second phases, respectively. Thus the double below-forecast is not so interesting in this setting.) If we set $t = 1.05$, then the above values change to 1.13, 1.28 and 1.38 for the three phases. In the case of a single below-forecast, the values are 1.13, 1.34 and 1.34 (the second and the third phases are the same). Thus the difference becomes much clearer.

5   Allowing Bidirectional Conversion

5.1   Basic Idea

If the rate changes up and down (not monotonically), then allowing bidirectional conversion (i.e., both conversions from dollar to yen and yen to dollar are allowed) makes the nature of the problem completely different. However, if the rate change is monotone, we cannot get any additional merit by allowing bidirectional conversion [EFKT92]. It also turns out that bidirectional conversion does not help for above-forecasts either (proof is omitted).

In this section, we show that bidirectional conversion does help for below-forecasts. Recall Case (2-2) of Theorem 1, where the player has to "wait" at the beginning of the phase. This happens since the values of dollars and yen have the relation that $md + y < N_1/r$. Intuitively, this means that the value d of dollars is not enough to start the normal trading in the phase. Now one can see that if we are allowed to trade both way, then the player can buy a certain amount of dollars back and create the relation $md' + y' = N_1/r$ immediately. Now the trading can start without any waiting.

The problem is whether this gives us any merit or not. The answer is yes. To see this, let us take a look at an easier case: In the threat-based algorithm it is also the case that the player has to wait if the rate is lower than $r_0^* m$ ($r_0^* \approx 1.28$). Suppose that $m = 100$ and the starting rate is 110. Then the player has to wait until the rate reaches 128. However, one can calculate the hypothetical value d (> 1) of dollars and y (< 0) of yen at $x = 110$ by means of substituting $x = 110$ to $D(x)$ and $Y(x)$ of (1) and (2). The actual values are $d \approx 2.02$ and $y \approx -116$. Now one can see that the player needs only 0.97 dollars at $x = 110$ if the bidirectional conversion (and negative values in this special case) is allowed. (Namely, the player can exchange 0.97 dollars to 2.02 dollars and -116 yen immediately. It should be noted that this kind of trade is actually impossible since we do not allow negative value.) Recall that the player has to hold one dollar to achieve the competitive ratio r_0^* previously.

Thus, if bidirectional conversion is allowed then the player does not have to wait, which usually requires us to hold a smaller amount of dollars. This means the player can spend a more amount of dollars in the previous phase, which should result in a better competitive ratio in that phase.

Theorem 2. Suppose that an algorithm, A, satisfies that $D(N_1) = d_1$, $Y(N_1) = y_1$. Then an optimal A is uniquely determined under the assumption that the player is allowed to trade bidirectionally.

Proof. Very similar to the proof of Theorem 1. Only one difference is in Case (2-2). The player can exchange some amount of yen back to dollars so that resulting d'' and y'' will satisfy that $md'' + y'' = N_1/r$. (Such a trade is always possible.) After that $D(x)$ and $Y(x)$ are given as follows:

$$D(x) = d'' - \frac{1}{r} \ln \frac{x - m}{N_1 - m}, \tag{32}$$

$$Y(x) = y'' + \frac{1}{r} \left(m \ln \frac{x - m}{N_1 - m} + x - N_1 \right). \tag{33}$$

∎

5.2 Single Below-Forecast

As before, let r_1 be the ratio when the forecast succeeds and tr_0^* be the ratio when it fails.

(1) For the first interval $[m, M_1]$, our design procedure is exactly the same as (1) of section 4.2.

(2) For the second interval $[M_1, M]$, we apply Theorem 2. At rate M_1, we buy some dollars back so as to have d dollars and y yen that should satisfy

$$md + y = \frac{M_1}{tr_0^*}, \tag{34}$$

$$d - D_1(M_1) = \frac{1}{M_1}(Y_1(M_1) - y). \tag{35}$$

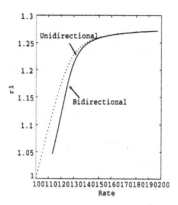

Fig. 5.

Now the trade in the second phase is expressed by

$$D_2(x) = d - \frac{1}{tr_0^\star} \ln \frac{x - m}{M_1 - m}, \tag{36}$$

$$Y_2(x) = y + \frac{1}{tr_0^\star} \left(m \ln \frac{x - m}{M_1 - m} + x - M_1 \right). \tag{37}$$

Note that $D_1(M_1) \geq 0$ must be met, which restricts the selection of M_1. The optimal value $r_1^\star$ for r_1 is obtained by setting $D_2(M) = 0$, which is the solution of the next equation:

$$1 - \frac{1}{r_1} \ln \frac{M_1 - m}{r_1 m - m} + \frac{M_1}{M_1 - m} \left(\frac{1}{r_1} - \frac{1}{tr_0^\star} \right) - \frac{1}{tr_0^\star} \ln \frac{M - m}{M_1 - m} = 0. \tag{38}$$

When $t = 1.01$, the value of $r_1^\star$ is illustrated as the solid line in Fig. 5. It should be noted that it is much better than the value of $r_1^\star$ obtained in Section 4.2 (see the dotted line in Fig. 5).

Multiple below-forecasts can be designed similarly.

6   Concluding Remarks

In this paper, it is assumed that a forecast is just successful or not. In practice, however, people usually think a specific forecast will come true with a certain *probability* p. Hence, the merit one can enjoy when the forecast succeeds is to be given as a function of p. Although the probability distribution is hard to guess, this is obviously one possible extension of this work. Another possible extension is to introduce forecasts in the real bidirectional-conversion game where the rate can change up and down [DS97,EFKT92].

References

alB97. Sabah al-Binali. "The Competitive Analysis of Risk Taking with Applications to Online Trading", *Proc. 38th IEEE Symposium on Foundations of Computer Science (FOCS)*, pp.336-344, 1997.

CCEKL95. A. Chou, J. Cooperstock, R. El-Yaniv, M. Klugerman, and F. T. Leighton. "The statistical adversary allows optimal money-making trading strategies", *Proc. 6th ACM-SIAM Symposium on Discrete Algorithms (SODA)*, pp.467-476, 1995.

DS97. E. Dannoura and K. Sakurai. "On-line versus off-line in money-making strategies with brokerage", In *Eighth Annual International Symposium on Algorithms and Computation (ISAAC)*, 1997.

EFKT92. R. El-Yaniv, A. Fiat, R. Karp, and G. Turpin. "Competitive Analysis of Financial Games", *Proc. 33th IEEE Symposium on Foundations of Computer Science (FOCS)*, pp.327-333, 1992.

Mor96. J. P. Morgan. "RiskMetrics ^{TM} -- Technical Document", fourth edition, 1996.

RS70. M. Rothschild and J. E. Stiglitz. "Increasing risk I: A definition", *Journal of Economic Theory*, 2:225-243, 1970.

RS71. M. Rothschild and J. E. Stiglitz. "Increasing risk II: Its economic consequences", *Journal of Economic Theory*, 3:66-84, 1971.

SSW95. C. W. Smithson and C. W. Smith Jr. , and D. S. Wilford. "Managing Financial Risk: A Guide to Delivative Products, Financial Engineering and Value Maximization", Irwin, 1995.

On S-Regular Prefix-Rewriting Systems and Automatic Structures

Friedrich Otto

Fachbereich Mathematik/Informatik, Universität Kassel, D-34109 Kassel, Germany
otto@theory.informatik.uni-kassel.de

Abstract. Underlying the notion of an automatic structure is that of a synchronously regular (s-regular for short) set of pairs of strings. Accordingly we consider s-regular prefix-rewriting systems showing that even for fairly restricted systems of this form confluence is undecidable in general. Then a close correspondence is established between the existence of an automatic structure that yields a prefix-closed set of unique representatives for a finitely generated monoid and the existence of an s-regular canonical prefix-rewriting system presenting that monoid.

1 Introduction

A fundamental issue in computational algebra is the quest for classes of finite descriptions for infinite algebraic structures such that these descriptions allow algorithmic solutions for various decision problems. In the case of monoids and groups, infinite structures can be presented through finite presentations of the form $(\Sigma; S)$, where Σ is a finite set of *generators* and S is a finite set of *defining relations*. However, for the class of all such presentations essentially all interesting decision problems are known to be undecidable in general. Therefore various subclasses of presentations have been considered in the literature. Prominent among them are the following two:

1. finite presentations $(\Sigma; S)$ that admit a finite (or simple infinite) convergent string-rewriting system, and
2. finite presentations that admit an automatic structure.

A *convergent string-rewriting system* R for a presentation $(\Sigma; S)$ generates the same Thue congruence as the given set of defining relations S, but it has the additional property that the reduction relation $\rightarrow_R$ induced by R is noetherian and confluent (see Section 2 for the definitions of these concepts). As a consequence the set of irreducible strings $\mathrm{IRR}(R)$ is a *cross-section*, that is, a complete set of unique representatives, for the monoid M_S presented by $(\Sigma; S)$. If R is sufficiently simple, then the representative corresponding to a string $w \in \Sigma^*$ can effectively be obtained from w. For example, this is the case if the system R is finite or *synchronously regular*, s-regular for short. Hence, based on an s-regular string-rewriting system the *word problem* as well as many other decision problems can be solved effectively [1].

T. Asano et al. (Eds.): COCOON'99, LNCS 1627, pp. 422–431, 1999.
© Springer-Verlag Berlin Heidelberg 1999

An *automatic structure* for a presentation $(\Sigma; S)$ consists of a regular set C of (not necessarily unique) representatives for the monoid M_S presented by $(\Sigma; S)$ and finite-state acceptors A_a, $a \in \Sigma \cup \{\lambda\}$, such that A_a accepts those suitably encoded pairs (u, v) of elements $u, v \in C$ that satisfy the equality $u \cdot a =_{M_S} v$. Thus, these finite-state acceptors 'realize' the right-multiplication in the monoid M_S. It follows that the word problem for M_S can be solved in quadratic time if an automatic structure is given (see, e.g., [3]).

However, these two concepts do not relate well. On the one hand, there exists a finitely presented monoid, in fact a group, that is presented through a finite convergent string-rewriting system, but that does not admit an automatic structure, because its Dehn function grows too fast [4]. On the other hand, there exists a finitely presented monoid that is even *bi-automatic*, but that cannot be presented through any finite convergent string-rewriting system, because it does not meet the homotopical finiteness condition FDT [9]. By the way, for groups it is still an open question whether or not each automatic group admits a presentation through some finite convergent string-rewriting system.

From an automatic structure $(C, A_a(a \in \Sigma \cup \{\lambda\}))$ for a presentation $(\Sigma; S)$ another automatic structure $(D, B_a(a \in \Sigma))$ for the same presentation can be obtained such that the set of representatives D is actually a cross-section for the monoid M_S. Accordingly the latter structure is called an automatic structure *with uniqueness* [3]. Using an automatic structure with uniqueness, the word problem is solved by reducing a given string to its unique representative by processing the given string strictly from left to right: repeatedly the shortest prefix that is not itself a representative is replaced by its representative using the corresponding multiplier automaton. Hence, intuitively automatic structures correspond to *prefix-rewriting systems*.

In this paper the investigation of this correspondence is initiated. First the notion of a prefix-rewriting system and the prefix-reduction relation induced by it are restated in short. In fact, we will consider three classes of prefix-rewriting systems: the finite systems, the f-regular systems, and the s-regular systems (see Section 2 for the definitions).

If each letter of the underlying alphabet is interpreted as a unary function symbol (and a single constant is added), then a prefix-rewriting system corresponds to a *ground term-rewriting system* over this signature. Hence, it follows from known results that a prefix-rewriting system is *canonical* if and only if it is *interreduced*. Further, it is easily seen that for each prefix-rewriting system P there exists an equivalent system P_0 that is *right-reduced*, that is, a system P_0 for which each right-hand side is irreducible. If the system P is finite or f-regular, then P_0 can effectively be constructed from P. For s-regular prefix-rewriting systems, however, the situation is more involved. By replacing the right-hand side of each rule of an s-regular prefix-rewriting system by one of its irreducible descendants we will in general obtain a left-regular prefix-rewriting system that is no longer s-regular.

In Section 3 we consider the notion of *regularity preservation*. A prefix-rewriting system P on an alphabet Σ is called regularity preserving if, for each regular subset $L \subseteq \Sigma^*$, the set $\Delta_P^*(L)$ of all descendants of strings in L modulo the prefix-rewriting relation induced by P is itself regular. As it turns out each interreduced prefix-rewriting system that has only finitely many different right-hand sides preserves regularity. On the other hand, even s-regular prefix-rewriting systems that are length-reducing and interreduced do in general not preserve regularity. In fact, each recursively enumerable language $L_0 \subseteq \Sigma^*$ can be obtained as an intersection of the form $\Delta_P^*(L) \cap \Sigma^*$, where P is an s-regular prefix-rewriting system on a finite alphabet strictly containing Σ such that P is length-reducing and interreduced, and L is a regular language.

For s-regular prefix-rewriting systems in general the confluence property is easily seen to be undecidable. However, it remains the question of whether confluence is decidable for each s-regular prefix-rewriting system that is right-reduced. In Section 4 we will answer this question in the negative by showing that it is undecidable in general whether an s-regular prefix-rewriting system that is length-reducing, right-reduced and unambiguous is confluent. Here a prefix-rewriting system is called *unambiguous* if no two different rules have identical left-hand sides.

Finally in Section 5 we come to the intended correspondence between automatic structures and prefix-rewriting systems. We show that a finitely generated monoid-presentation $(\Sigma; S)$ has an automatic structure such that the set of representatives is a prefix-closed cross-section for the monoid M_S if and only if there exists an s-regular canonical prefix-rewriting system P on Σ such that P is equivalent to S. It follows that the class of finitely presented monoids that admit a finite convergent string-rewriting system and the class of finitely presented monoids that admit an s-regular canonical prefix-rewriting system are incomparable under set inclusion.

2   Prefix-rewriting systems

Here we introduce the basic notions and notation concerning prefix-rewriting systems, and we establish some properties of these systems.

Let Σ be a finite alphabet. Then by Σ^* we denote the set of strings over Σ including the empty string λ. As usual the concatenation of two strings u and v is denoted by uv, $|u|$ denotes the length of the string u, and u^n is used as an abbreviation for the string $uu \cdots u$ (n-times).

A *prefix-rewriting system* P is a subset of $\Sigma^* \times \Sigma^*$, the elements of which are called (prefix-rewrite) rules. It induces a *single-step prefix-reduction relation* $\Rightarrow_P$ on Σ^* that is defined as follows:

$$u \Rightarrow_P v \text{ iff } \exists (\ell, r) \in P \; \exists x \in \Sigma^* : u = \ell x \text{ and } v = rx.$$

The *prefix-reduction relation* $\Rightarrow_P^*$ induced by P is the reflexive, transitive closure of $\Rightarrow_P$, and $\mathrm{dom}(P)$ and $\mathrm{range}(P)$ denote the sets $\mathrm{dom}(P) := \{\ell \in \Sigma^* \mid \exists r \in \Sigma^* : (\ell, r) \in P\}$ and $\mathrm{range}(P) := \{r \in \Sigma^* \mid \exists \ell \in \Sigma^* : (\ell, r) \in P\}$.

A string u is *reducible* modulo P, if $u \Rightarrow_P v$ holds for some $v \in \Sigma^*$; otherwise, u is *irreducible* modulo P. By $\mathrm{IRR}(P)$ we denote the set of all irreducible strings.

Obviously, $\text{IRR}(P) = \Sigma^* \setminus \text{dom}(P) \cdot \Sigma^*$. Hence, if P is finite or *left-regular*, that is, $\text{dom}(P)$ is a regular language, then the set $\text{IRR}(P)$ is a regular language.

The prefix-rewriting system P is called *noetherian* if there is no infinite $\Rightarrow_P$-reduction sequence, *confluent* if the relation $\Rightarrow_P$ is confluent, *interreduced* if $\text{range}(P) \subseteq \text{IRR}(P)$ and $\ell \in \text{IRR}(P \setminus \{(\ell, r)\})$ for each rule $(\ell, r) \in P$, and *canonical* if it is noetherian, confluent, and interreduced.

It is easily seen that $\Rightarrow_P$ is a partial function on Σ^* if P is interreduced. Further, in this case each reduction sequence $u_0 \Rightarrow_P u_1 \Rightarrow_P \cdots \Rightarrow_P u_m$ satisfies $m \leq |u_0|$. Hence, we obtain the following result, which is a special case of a result for ground term-rewriting systems [11].

Proposition 1. *A prefix-rewriting system is canonical if and only if it is interreduced.*

In this paper we are concerned with certain infinite prefix-rewriting systems. Following the convention for string-rewriting systems [6] a prefix-rewriting system P on Σ is called *f-regular* if it has the form $P = \bigcup_{i=1}^{k} L_i \times \{r_i\}$, where $L_1, \ldots, L_k \subseteq \Sigma^*$ are regular languages and $r_1, \ldots, r_k \in \Sigma^*$ are strings, and it is called *synchronously regular*, *s-regular* for short, if its encoding $\nu(P) \subseteq \Sigma_{\#}^*$ is a regular language. Here $\#$ is an additional symbol, $\Sigma_{\#}$ is the finite alphabet $\Sigma_{\#} := ((\Sigma \cup \{\#\}) \times (\Sigma \cup \{\#\})) \setminus \{(\#, \#)\}$, which is called the *padded extension* of Σ, and the encoding $\nu : \Sigma^* \times \Sigma^* \to \Sigma_{\#}^*$ is defined as follows: if $u := a_1 a_2 \cdots a_n$ and $v := b_1 b_2 \cdots b_m$, where $a_1, \ldots, a_n, b_1, \ldots, b_m \in \Sigma$, then

$$\nu(u,v) := \begin{cases} (a_1, b_1)(a_2, b_2) \cdots (a_m, b_m)(a_{m+1}, \#) \cdots (a_n, \#), & \text{if } m < n, \\ (a_1, b_1)(a_2, b_2) \cdots (a_m, b_m). & \text{if } m = n, \\ (a_1, b_1)(a_2, b_2) \cdots (a_n, b_n)(\#, b_{n+1}) \cdots (\#, b_m), & \text{if } m > n. \end{cases}$$

Obviously, for each f- or s-regular prefix-rewriting system P, $\text{dom}(P)$ and $\text{range}(P)$ are regular languages. In particular, if finite-state acceptors (fsa's) are given for the languages $L_1, \ldots, L_k$, respectively the language $\nu(P)$, then fsa's for $\text{dom}(P)$ and $\text{range}(P)$ can be constructed effectively. Thus, it is easily decidable whether or not an f- or s-regular prefix-rewriting system P is *right-reduced*, that is, whether $\text{range}(P) \subseteq \text{IRR}(P)$ holds.

If a noetherian string-rewriting system is not right-reduced, then an equivalent noetherian system that is right-reduced can be obtained by replacing the right-hand side of each rule by one of its irreducible descendants [1]. The same idea also works for prefix-rewriting systems. Thus, if P is a prefix-rewriting system on Σ that is noetherian but not right-reduced, and for each string $r \in \text{range}(P)$, $r_0 \in \text{IRR}(P)$ denotes an irreducible descendant of r, then the prefix-rewriting system $P_0 := \{(\ell, r_0) \mid (\ell, r) \in P\}$ is *equivalent* to P, that is, $\Leftrightarrow_P^* = \Leftrightarrow_{P_0}^*$, P_0 is right-reduced, and $\text{dom}(P) = \text{dom}(P_0)$, which means that also $\text{IRR}(P) = \text{IRR}(P_0)$. In addition, if P is confluent, then so is P_0. Here $\Leftrightarrow_P^*$ denotes the reflexive, symmetric, and transitive closure of the single-step prefix-reduction relation $\Rightarrow_P$, which is a *left-congruence* on Σ^*, that is, $u \Leftrightarrow_P^* v$ implies that $uw \Leftrightarrow_P^* vw$ for all $u, v, w \in \Sigma^*$.

If the system P is finite or f-regular, then the system P_0 is finite or f-regular as well, and P_0 can effectively be obtained from P. For s-regular prefix-rewriting systems the situation is quite different, however, as the following example shows.

Example 1. Let $\Sigma := \{a, b, c\}$, and let R denote the string-rewriting system $R := \{ac \to \lambda, bbc \to cb\}$ on Σ. Then R is finite, length-reducing, and confluent. From R we obtain the prefix-rewriting system P_1 as follows:

$$P_1 := \{xac \to x \mid x \in \text{IRR}(R)\} \cup \{xbbc \to xcb \mid x \in \text{IRR}(R)\}.$$

Obviously, P_1 is length-reducing, s-regular, and confluent satisfying $\Leftrightarrow^*_{P_1} = \leftrightarrow^*_R$. Here $\leftrightarrow^*_R$ denotes the Thue congruence generated by the string-rewriting system R. However, P_1 is not right-reduced. By applying the above process to P_1 we obtain the following canonical prefix-rewriting system P_2, which is not s-regular:

$$P_2 := \{xac \to x \mid x \in \text{IRR}(R)\}$$
$$\cup \{xab^{2n}c \to xb^n \mid x \in \text{IRR}(R), n \geq 1\}$$
$$\cup \{xb^{2n}c \to xcb^n \mid x \in \text{IRR}(R), n \geq 1, x \text{ does not end in } b^2 \text{ or in } a\}.$$

$\square$

3 Preserving regularity

Let P be a prefix-rewriting system on Σ, and let $L \subseteq \Sigma^*$. Then by $\Delta^*_P(L)$ we denote the language $\Delta^*_P(L) := \{v \in \Sigma^* \mid \exists u \in L : u \Rightarrow^*_P v\}$ of all *descendants* of L modulo P.

A prefix-rewriting system P on Σ is said to *preserve Σ-regularity* if $\Delta^*_P(L)$ is a regular language for each regular language $L \subseteq \Sigma^*$, and P is called *regularity preserving*, if P preserves Σ-regularity for each finite alphabet Σ containing all those letters that have occurrences in the rules of P. In contrast to the situation for term-rewriting systems in general [5] it has been observed by the author that a string-rewriting system is regularity preserving if and only if it preserves Σ_0-regularity, where Σ_0 is the alphabet consisting of those letters that have occurrences in the rules of the string-rewriting system considered [10]. Since the same techniques apply, this result also holds for prefix-rewriting systems.

Since a prefix-rewriting system corresponds to a ground term-rewriting system, it follows that every finite prefix-rewriting system preserves regularity [2]. Which infinite left-regular systems do share this nice property? To derive a first positive result we need the following simple observation.

Lemma 1. *Let P be a prefix-rewriting system that is interreduced, and let $u, v \in \Sigma^*$ such that $v \in \Delta^*_P(u)$. Then there exist a rule $(\ell, r) \in P$ and strings $x, w \in \Sigma^*$ such that $u = wx$, $v = rx$, and $w \Rightarrow^*_P \ell \Rightarrow_P r$, or $u = v$.*

For each $r \in \text{range}(P)$, let $L_r := \{w \in \Sigma^* \mid \exists (\ell, r) \in P : w \Rightarrow^*_P \ell\}$. If P is interreduced and finite or f-regular, then a left-linear grammar can easily be constructed from P and $r \in \text{range}(P)$ that generates the language L_r. For a language $L \subseteq \Sigma^*$, membership in the language $\Delta_P^*(L)$ can be characterized as follows by Lemma 3:

$$v \in \Delta_P^*(L) \text{ iff } v \in L \text{ or } \exists r \in \text{range}(P) \exists x \in \Sigma^* \exists w \in L_r : wx \in L \text{ and } v = rx.$$

Hence, $\Delta_P^*(L) = L \cup \bigcup_{r \in \text{range}(P)} r \cdot (L_r \smallsetminus_\ell L)$, where $L_r \smallsetminus_\ell L := \{x \in \Sigma^* \mid \exists w \in L_r : wx \in L\}$ denotes the left-quotient of L by L_r. Thus, we obtain the following result.

Proposition 2. *Let P be an interreduced prefix-rewriting system. If* range(P) *is finite, then P is regularity preserving. If in addition P is f-regular, then an fsa for $\Delta_P^*(L)$ can be constructed effectively from an fsa for L.*

However, a corresponding result does not hold for those interreduced prefix-rewriting systems that are s-regular. In fact, the sets of the form $\Delta_P^*(L)$, where P is an s-regular prefix-rewriting system that is length-reducing and interreduced and L is a regular language, are sufficient to describe all recursively enumerable languages.

To see this let $L_0 \subseteq \Sigma^*$ be a recursively enumerable language, and let $T = (Q, \Gamma, q_0, q_a, \delta)$ be a deterministic single-tape Turing machine that accepts this language. Here Q is the finite set of states, $\Gamma \supsetneq \Sigma$ is the finite tape alphabet including the blank symbol $b \notin \Sigma$, $q_0 \in Q$ is the initial state, $q_a \in Q$ is the unique halting state, and $\delta : (Q \smallsetminus \{q_a\}) \times \Gamma \to (Q \smallsetminus \{q_0\}) \times (\Gamma \cup \{\ell, r\})$ is the transition function. As usual a *configuration* of T will be described as uqv, where $uv \in \Gamma^*$ is the actual tape inscription and $q \in Q$ is the actual state. Thus, for all $w \in \Sigma^*$, $w \in L_0$ iff $\exists u, v \in \Gamma^* : q_0w \vdash_T^* uq_av$, where $\vdash_T^*$ denotes the reflexive, transitive closure of the single-step computation relation $\vdash_T$ of T.

Since T is deterministic, each non-halting configuration uqv has a unique immediate successor configuration $u'q'v'$. In general, however, there may be several configurations that have the same immediate successor configuration. However, it is possible to turn T into a deterministic single-tape machine for which the single-step computation relation $\vdash_T$ is an injective (partial) function on the set of configurations [7]. Thus, we can assume that already T has this property, that is, each configuration of T has at most one immediate predecessor configuration.

From T we constuct a prefix-rewriting system $P(T)$ as follows. Let $\Omega := Q \cup \Gamma \cup \{\mathfrak{c}, \$, d, \#\}$, where we assume without loss of generality that $Q \cap \Gamma = \emptyset$, and that $\mathfrak{c}, \$, d,$ and $\#$ are four additional letters. On Ω we define the prefix-rewriting system $P(T)$:

$$P(T) := \{\mathfrak{c}u_1q_1v_1\$d^3 \to \mathfrak{c}u_2q_2v_2\$ \mid u_2q_2v_2 \vdash_T u_1q_1v_1\} \cup$$
$$\{\mathfrak{c}q_0w\$\# \to w \mid w \in \Sigma^*\}.$$

Then $P(T)$ is length-reducing and right-reduced, and it is easily seen that $P(T)$ is s-regular. The first group of rules of $P(T)$ just simulates the single-step computation relation of T in reverse order, and the second group of rules transforms encodings of initial configurations into the corresponding input string. Because of the above assumption on T $P(T)$ is interreduced.

Let $FINAL := \mathbf{c} \cdot \Gamma^* \cdot q_a \cdot \Gamma^* \cdot \$$, and let $L := FINAL \cdot (d^3)^* \cdot \#$. Then L is a regular language, and $\Delta_P^*(L) \cap (\Omega \smallsetminus \{\#\})^* = L_0$, since, for each $\mathbf{c} u q_a v \$ \in FINAL$, if the final configuration $u q_a v$ is reached from an initial configuration $q_0 w$ in m steps, then $\mathbf{c} u q_a v \$ \cdot (d^3)^m \cdot \# \Rightarrow_P^m \cdot \mathbf{c} q_0 w \$ \cdot \# \Rightarrow_P w \in L_0$, while all other descendants of strings from L do still contain the suffix $\#$. Thus, we have the following result.

Proposition 3. *For each recursively enumerable language $L_0 \subseteq \Sigma^*$, there exist an alphabet $\Omega \supsetneq \Sigma$, an s-regular prefix-rewriting system P on Ω that is length-reducing and interreduced, and a regular language $L \subseteq \Omega^*$ such that $L_0 = \Delta_P^*(L) \cap \Sigma^*$.*

4   Confluence of s-regular prefix-rewriting systems

Confluence of f-regular, length-reducing string-rewriting systems is undecidable in general [8]. If R is such a string-rewriting system on Σ, then $P := \{(x\ell, xr) \mid x \in IRR(R), \text{ and } (\ell, r) \in R\}$ is an s-regular prefix-rewriting system on Σ that is length-reducing such that the binary relation $\Leftrightarrow_P^*$ coincides with the Thue congruence $\leftrightarrow_R^*$ generated by R. In particular, it follows that P is confluent if and only if R is. Thus, confluence is undecidable in general for s-regular prefix-rewriting systems that are length-reducing. However, in general P will not be right-reduced, and as Example 2 indicates, when replacing the right-hand side of each rule of P by one of its irreducible descendants, the property of s-regularity will usually be lost. Hence, it remains the question of whether confluence is decidable for s-regular prefix-rewriting systems that are length-reducing and right-reduced. In this section we will answer this question in the negative.

Theorem 1. *It is undecidable in general whether an s-regular prefix-rewriting system that is length-reducing, right-reduced, and unambiguous is confluent.*

To prove this result we proceed as follows. With a Turing machine T we associate an s-regular prefix-rewriting system P_T that is length-reducing and interreduced, and that simulates the computations of T step by step. Then, given an input string w for the Turing machine T, we extend the system P_T to an s-regular prefix-rewriting system $P_T(w)$ that is length-reducing, right-reduced, and unambiguous. As it will turn out $P_T(w)$ is confluent if and only if T accepts on input w.

Following the outline described above let $T = (Q, \Gamma, q_0, q_a, \delta)$ be a deterministic single-tape Turing machine, where the transition function δ is a function from $(Q \smallsetminus \{q_a\}) \times \Gamma_b \to (Q \smallsetminus \{q_0\}) \times (\Gamma \cup \{\ell, r\})$, where $\Gamma_b = \Gamma \cup \{b\}$ and b denotes the blank symbol. By $L(T)$ we denote the language accepted by T, that is, $L(T) = \{w \in \Gamma^* \mid \exists u, v \in \Gamma_b^* : q_0 w \vdash_T^* u q_a v\}$, and we assume that this language is non-recursive.

Let $\mathbf{c}, \$, d$, and $\#$ be four additional symbols, and let $\Sigma := \Gamma_b \cup Q \cup \{\mathbf{c}, \$, d, \#\}$. As before a configuration $u q v$ $(u, v \in \Gamma_b^*, q \in Q)$ of the Turing machine T will be encoded as $\mathbf{c} u q v \$$. The prefix-rewriting system P_T on Σ consists of the following two groups of rules:

(1.) Rules to simulate the stepwise behavior of the Turing machine T:

$$\text{\textcent} uqv\$d^3 \to \text{\textcent} u_1 q_1 v_1 \$, \text{ if } uqv \vdash_T u_1 q_1 v_1 \ (u, v \in \Gamma_b^*, q \in Q \smallsetminus \{q_a\}).$$

(2.) Rules to delete configurations: $\text{\textcent} uqv\$\# \to \text{\textcent}\$ \ (u, v \in \Sigma_b^*, q \in Q).$

It is easily seen that P_T is s-regular, length-reducing, and interreduced. Now, for $w \in \Gamma^*$, let $P_T(w)$ denote the following prefix-rewriting system:

$$P_T(w) := P_T \cup \{\text{\textcent} q_0 w\$d^{3m}\# \to \text{\textcent}\$ \mid m \geq 1\}.$$

From w the system $P_T(w)$ is easily obtained. Obviously, $P_T(w)$ is s-regular, length-reducing, right-reduced, and unambiguous. Since the left-hand side of each rule of the form $(\text{\textcent} q_0 w\$d^{3m}\# \to \text{\textcent}\$) \in P_T(w) \smallsetminus P_T$ contains the proper prefix $\text{\textcent} q_0 w\$d^3 \in \text{dom}(P_T)$, it follows that the system $P_T(w)$ is confluent if and only if $\text{\textcent} q_0 w\$d^{3m}\# \Rightarrow_{P_T}^* \text{\textcent}\$$ holds for all $m \geq 1$.

Lemma 2. $\text{\textcent} q_0 w\$d^{3m}\# \Rightarrow_{P_T}^* \text{\textcent}\$$ *holds for all* $m \geq 1$ *if and only if* $w \notin L(T)$.

Proof. If $w \in L(T)$, then there is an accepting computation of the form $q_0 w \vdash_T u_1 q_1 v_1 \vdash_T \cdots \vdash_T u_m q_m v_m = u_m q_a v_m$ for some $m \geq 1$. Hence, $\text{\textcent} q_0 w\$d^{3m+3}\# \Rightarrow_{P_T} \text{\textcent} u_1 q_1 v_1 \$d^{3m}\# \Rightarrow_{P_T} \cdots \Rightarrow_{P_T} \text{\textcent} u_m q_a v_m \$d^3\# \in \text{IRR}(P_T)$. Since P_T is canonical, we see that $\text{\textcent} q_0 w\$d^{3m+3}\# \Rightarrow_{P_T}^* \text{\textcent}\$$ does not hold.

Conversely, if $w \notin L(T)$ and $m \geq 1$, then starting from the initial configuration $q_0 w$, T reaches some non-halting configuration $u_m q_m v_m$ after m steps. Hence, $\text{\textcent} q_0 w\$d^{3m}\# \Rightarrow_{P_T}^* \text{\textcent} u_m q_m v_m \$\# \Rightarrow_{P_T} \text{\textcent}\$.$ □

It follows that $P_T(w)$ is confluent if and only if $w \notin L(T)$. Since $L(T)$ is non-recursive, and since the prefix-rewriting system $P_T(w)$ is easily obtained from w, this proves the above undecidability result.

5   Automatic structures

After introducing the notion of monoid-presentation in short we will consider automatic structures for monoid-presentations. Under certain restrictions we will be able to establish a correspondence between the existence of an automatic structure and the existence of an s-regular interreduced prefix-rewriting system for the monoid-presentation under consideration.

Let Σ be a finite alphabet, and let S be a *string-rewriting system* on Σ. Then $\leftrightarrow_S^*$ denotes the *Thue congruence* generated by S. For $w \in \Sigma^*$, $[w]_S = \{u \in \Sigma^* \mid u \leftrightarrow_S^* w\}$ is the *congruence class* of w, and by M_S we denote the factor monoid $\Sigma^*/\leftrightarrow_S^*$. If M is a monoid that is isomorphic to M_S, then the ordered pair $(\Sigma; S)$ is called a *monoid-presentation* of M with *generators* Σ and *defining relations* S. A monoid is called *finitely generated* if it has a presentation with a finite set of generators.

An *automatic structure* for a finitely generated monoid-presentation $(\Sigma; S)$ consists of a finite-state acceptor (fsa) W over Σ, a fsa $M_=$ over $\Sigma_\#$, and fsa's $M_a \ (a \in \Sigma)$ over $\Sigma_\#$ satisfying the following conditions:

(0.) $L(W) \subseteq \Sigma^*$ is a complete set of (not necessarily unique) representatives for the monoid M_S, that is, $L(W) \cap [u]_S \neq \emptyset$ holds for each $u \in \Sigma^*$,

(1.) $L(M_=) = \{\nu(u,v) \mid u,v \in L(W) \text{ and } u \leftrightarrow^*_S v\}$, and

(2.) for all $a \in \Sigma$, $L(M_a) = \{\nu(u,v) \mid u,v \in L(W) \text{ and } u \cdot a \leftrightarrow^*_S v\}$.

Actually, one may require that the set $L(W)$ is a cross-section for M_S, in which case we say that we have an *automatic structure with uniqueness* [3]. In this situation the fsa $M_=$ is trivial, and hence, it will not be mentioned explicitly.

A set $C \subseteq \Sigma^*$ is called *prefix-closed* if the prefix of any element of C also belongs to C. Recall that the set $\text{IRR}(P)$ of strings that are irreducible with respect to a prefix-rewriting system P is prefix-closed. Finally, we say that the prefix-rewriting system P on Σ is *equivalent* to a string-rewriting system S on the same alphabet Σ, if the relations $\Leftrightarrow^*_P$ and $\leftrightarrow^*_S$ coincide.

Lemma 3. *Let $(W, A_a\ (a \in \Sigma))$ be an automatic structure with uniqueness for $(\Sigma; S)$ such that the set $L(W)$ is in addition prefix-closed. Then there exists an s-regular canonical prefix-rewriting system P on Σ that is equivalent to S such that $\text{IRR}(P) = L(W)$.*

Proof. Define $P := \{(ua,v) \mid u,v \in L(W), a \in \Sigma, ua \leftrightarrow^*_S v, \text{ but } ua \neq v\}$.

If $u,v \in C := L(W)$ and $a \in \Sigma$ such that $ua \leftrightarrow^*_S v$, but $ua \neq v$, then $ua \notin C$, since C is a regular cross-section for $(\Sigma; S)$. Since C is prefix-closed, we see that P is interreduced, and that $\text{IRR}(P) = C$. Obviously, $\Leftrightarrow^*_P \subseteq \leftrightarrow^*_S$.

Since P is noetherian, $w \Rightarrow^*_P w_0 \in \text{IRR}(P)$ for some w_0. But from the definition of P we see that $w \leftrightarrow^*_S w_0$ and that $w_0 \in C$. Since C is a cross-section for $(\Sigma; S)$, $w_0 \in C$ is the unique representative of $[w]_S$. Thus P is a canonical prefix-rewriting system on Σ that is equivalent to S such that $\text{IRR}(P) = C = L(W)$ holds.

Finally, P is s-regular, since $\nu(P) = \{\nu(ua,v) \mid u,v \in C, a \in \Sigma, ua \leftrightarrow^*_S v, ua \neq v\} = \bigcup_{a \in \Sigma} \{\nu(ua,v) \mid \nu(u,v) \in L(A_a), \text{ but } v \neq ua\}$. $\square$

On the other hand we have the following implication.

Lemma 4. *If there exists an s-regular canonical prefix-rewriting system P on Σ that is equivalent to the string-rewriting system S, then $C := \text{IRR}(P)$ is part of an automatic structure with uniqueness for $(\Sigma; S)$.*

Proof. By hypothesis C is a regular cross-section for $(\Sigma; S)$. Thus, it remains to verify the existence of the multiplier automata $A_a\ (a \in \Sigma)$.

Let $a \in \Sigma$. Then, for $u,v \in C$, $\nu(u,v) \in L(A_a)$ iff $ua \leftrightarrow^*_S v$ iff $ua \Rightarrow^*_P v$ iff $ua = v$ or $ua \Rightarrow^*_P v$ (Recall that P is canonical !) iff $ua = v$ or $(ua,v) \in P$. Hence, $L_a := L(A_a) = \{\nu(u,ua) \mid ua \in C\} \cup \{\nu(u,v) \mid (ua,v) \in P\}$. Since C is regular, the first of these two sets is regular, and since P is s-regular, the other set is regular as well. Thus, the multiplier automata $A_a\ (a \in A)$ exist as required, and $C = \text{IRR}(P)$ is indeed part of an automatic structure with uniqueness for $(\Sigma; S)$. Observe that additionally $C = \text{IRR}(P)$ is prefix-closed. $\square$

Together these two lemmata yield the following characterization.

Theorem 2. *Let $(\Sigma; S)$ be a finitely generated monoid-presentation. Then the following two statements are equivalent:*

(a) *There exists an automatic structure $(W, A_a(a \in \Sigma))$ with uniqueness for $(\Sigma; S)$ such that the set $L(W)$ is prefix-closed.*

(b) *There exists an s-regular canonical prefix-rewriting system P on Σ that is equivalent to S.*

Is there a corresponding characterization for automatic structures (with uniqueness), where $L(W)$ is **not** prefix-closed? Observe that it is still an open question whether or not each automatic structure can be turned into one for which the set of representatives has the uniqueness property and is prefix-closed [3].

References

1. Book, R.V., Otto, F.: *String-Rewriting Systems*. Springer-Verlag, New York (1993)
2. Brainerd, W.J.: Tree generating regular systems. *Information and Control* 14 (1969) 217–231
3. Epstein, D.B.A., Cannon, J.W., Holt, D.F., Levy, S.V.F., Paterson, M.S., Thurston, W.P.: *Word Processing In Groups*. Jones and Bartlett Publishers (1992)
4. Gersten, S.M.: Dehn functions and l1-norms of finite presentations. In: Baumslag, G., Miller III, C.F. (eds.): *Algorithms and Classification in Combinatorial Group Theory*. Math. Sciences Research Institute Publ. 23. Springer-Verlag, New York (1992) 195–224
5. Gyenizse, P., Vágvölgyi, S.: Linear generalized semi-monadic rewrite systems effectively preserve recognizability. *Theoretical Computer Science* 194 (1998) 87–122
6. Otto, F., Katsura, M., Kobayashi, Y.: Infinite convergent string-rewriting systems and cross-sections for monoids. *Journal Symbolic Computation* 26 (1998) 621–648
7. Ó'Dúnlaing, C.: *Finite and infinite regular Thue systems*. Ph.D. dissertation, Department of Math., Univ. of California at Santa Barbara (1981)
8. Ó'Dúnlaing, C.: Infinite regular Thue systems. *Theoretical Computer Science* 25 (1983) 171–192
9. Otto, F., Sattler-Klein, A., Madlener, K.: Automatic monoids versus monoids with finite convergent presentations. In: Nipkow, T. (ed.): *Rewriting Techniques and Applications, Proceedings RTA'98*. Lecture Notes in Computer Science, Vol. 1379. Springer-Verlag, Berlin (1998) 32–46
10. Otto, F.: Some undecidability results concerning the property of preserving regularity. *Theoretical Computer Science* 207 (1998) 43–72
11. Snyder, W.: Efficient ground completion: an $O(n \log n)$ algorithm for generating reduced sets of ground rewrite rules equivalent to a set of ground equations E. In: Deshowitz, N. (ed.): *Rewriting Techniques and Applications, Proceedings RTA'89*. Lecture Notes in Computer Science, Vol. 355. Springer-Verlag, Berlin (1989) 419–433

Tractable and Intractable
Second-Order Matching Problems

Kouichi Hirata, Keizo Yamada, and Masateru Harao *

Department of Artificial Intelligence,
Kyushu Institute of Technology,
Kawazu 680-4, Iizuka 820-8502, Japan
{hirata,yamada,harao}@dumbo.ai.kyutech.ac.jp

Abstract. The *second-order matching problem* is the problem of determining, for a finite set $\{\langle t_i, s_i \rangle \mid i \in I\}$ of pairs of a second-order term t_i and a first-order closed term s_i, called a *matching expression*, whether or not there exists a substitution σ such that $t_i\sigma = s_i$ for each $i \in I$. It is well-known that the second-order matching problem is NP-complete. In this paper, we introduce the following restrictions of a matching expression: *k-ary*, *k-fv*, *predicate*, *ground*, and *function-free*. Then, we show that the second-order matching problem is NP-complete for a unary predicate, a unary ground, a ternary function-free predicate, a binary function-free ground, and an 1-fv predicate matching expressions, while it is solvable in polynomial time for a binary function-free predicate, a unary function-free, a *k-fv* function-free ($k \geq 0$), and a ground predicate matching expressions.

1 Introduction

The *unification problem* is the problem of determining whether or not any two terms possess a common instance. The *matching problem*, on the other hand, is the problem of determining whether or not a term is an instance of another term. Both the unification and the matching play an important role in many research areas, including theorem proving, term rewriting systems, logic and functional programming, database query language, program synthesis, and so on.

The *second-order unification problem* is formulated as the problem of determining, for a finite set $\{\langle t_i, s_i \rangle \mid i \in I\}$ of pairs of second-order terms t_i and s_i, called a *unification expression*, whether or not there exists a substitution σ such that $t_i\sigma = s_i\sigma$ for each $i \in I$. The *second-order matching problem* is the special unification problem that t_i is a second-order term and s_i is a first-order closed term. A unification expression for the second-order matching problem is called a *matching expression*. The second-order matching has been applied to program synthesis and transformation, schema-guided proof, analogical reasoning, and machine learning [4,5,9,13,14].

* This work is partially supported by the Japanese Ministry of Education, Grant-in-Aid for Exploratory Research 10878055 and Kayamori Foundation of Informational Science Advancement.

T. Asano et al. (Eds.): COCOON'99, LNCS 1627, pp. 432–441, 1999.
© Springer-Verlag Berlin Heidelberg 1999

It is well-known that the second-order unification problem is undecidable [12], and various researchers have separated decidable from undecidable unification problems by introducing several restrictions of a unification expression [1,6,7,8,11,12,15,16,17]. It is also well-known that the second-order matching problem is NP-complete [2]. Huet and Lang [14] have designed a complete and nonredundant second-order matching algorithm. However, there exist few researches to analyze deeply the complexity of the matching problem. It is one of the reason that the interest of the researchers [3,5,9,13,14] is rather the matching algorithm than the matching problem itself. In this paper, by introducing the several restrictions of a matching expression, we give a sharp characterization between tractable and intractable second-order matching problems.

A matching expression is called k-*ary* if any function variable in it is at most k-ary, and k-*fv* if it includes at most k distinct function variables. Furthermore, a matching expression is called *predicate* if any argument's term of function variables in it includes no function variables, *ground* if it includes no individual variables, and *function-free* if it includes no function constants.

In this paper, we show that the second-order matching problem is NP-complete for a unary predicate, a unary ground, a ternary function-free predicate, and a binary function-free ground matching expressions, while it is solvable in polynomial time for a binary function-free predicate and a unary function-free matching expressions. We also show that it is NP-complete for an 1-fv predicate matching expression, while it is solvable in polynomial time for a k-fv function-free matching expression for $k \geq 0$. Furthermore, we show that it is solvable in polynomial time for a ground predicate matching expression.

2   Preliminaries

Instead of considering arbitrary second-order languages, we shall restrict our attention to languages containing just simple terms (i.e., terms without variable-binding operators like the λ operator). Throughout this paper, we deal with the term languages introduced by Goldfarb [12] and Farmer [8].

Let a *term language* L be a quadruple (IC_L, IV_L, FC_L, FV_L), where IC_L is a set of *individual constants* (denoted by $a, b, c, \cdots$); IV_L is a set of *individual variables* (denoted by $x, y, z, \cdots$); FC_L is a set of *function constants* (denoted by $f, g, h, \cdots$); FV_L is a set of *function variables* (denoted by $F, G, H, \cdots$). Each element of $FC_L \cup FV_L$ has a fixed arity ≥ 1, and IC_L, IV_L, FC_L and FV_L are mutually disjoint. We call an element of $IV_L \cup FV_L$ a *variable* simply. Let BV_L be an infinite collection $\{w_i\}_{i \geq 1}$ of symbols not in L called *bound variables*.

The *L-terms* and *L^*-terms* are defined inductively by:

1. Each $d \in IC_L \cup IV_L$ (*resp.* $IC_L \cup IV_L \cup BV_L$) is an L-term (*resp.* L^*-terms).
2. If $d \in FC_L \cup FV_L$ has arity $n \geq 1$ and $t_1, \cdots, t_n$ are L-terms (*resp.* L^*-terms), then $d(t_1, \cdots, t_n)$ is an L-term (*resp.* L^*-terms).

The *rank* of an L^*-term t is the largest n such that w_n occurs in t.

For L^*-terms $t, t_1, \cdots, t_n$, we write $t[t_1, \cdots, t_n]$ for the L^*-term obtained by replacing each occurrence of w_i in t with t_i for all i $(1 \leq i \leq n)$ simultaneously. The *head* of t, denoted by $hd(t)$, is the outermost symbol occurring in t. An L^*-term is *closed* if it contains no variables.

A *substitution* (in L) is a function σ with a finite domain $dom(\sigma) \subseteq IV_L \cup FV_L$ which maps individual variables to L-terms and n-ary function variables with $n \geq 1$ to L^*-terms of rank $\leq n$. A substitution σ is denoted by $\{s_1/v_1, \cdots, s_m/v_m\}$, where $dom(\sigma) = \{v_1, \cdots, v_m\}$. Each element s_i/v_i of σ is called a *binding* of σ. The result $t\sigma$ of applying σ to an L^*-term t is defined inductively by:

1. If $t \in IC_L \cup IV_L \cup BV_L$ but $t \notin dom(\sigma)$, then $t\sigma = t$.
2. If $t = x \in IV_L$ and $x \in dom(\sigma)$, then $t\sigma = x\sigma$.
3. If $t = d(t_1, \cdots, t_n)$ $(d \in FC_L \cup FV_L)$ but $d \notin dom(\sigma)$, then $t\sigma = d(t_1\sigma, \cdots, t_n\sigma)$.
4. If $t = F(t_1, \cdots, t_n)$ and $F \in dom(\sigma)$, then $t\sigma = (F\sigma)[t_1\sigma, \cdots, t_n\sigma]$.

A *matching expression* E (in L) is a finite set $\{\langle t_i, s_i \rangle \mid i \in I\}$, where t_i is an L-term and s_i is a closed L-term for each $i \in I$. For a substitution σ, $E\sigma$ denotes the matching expression $\{\langle t_i\sigma, s_i \rangle \mid i \in I\}$. The *size* of E, denoted by $|E|$, is the number of symbols of L occurring in E. Furthermore, for a function variable F, E_F denotes a matching expression $\{\langle t, s \rangle \in E \mid hd(t) = F\}$.

A matching expression E is called *matchable* if there exists a substitution σ such that $t_i\sigma = s_i$ for each $i \in I$. Such a σ is called a *matcher* of E.

The *transformation rules* [14] are defined as follows:

1. **simplification**:
   $\{\langle f(t_1, \cdots, t_n), f(s_1, \cdots, s_n) \rangle\} \cup E \Rightarrow \{\langle t_1, s_1 \rangle, \cdots, \langle t_n, s_n \rangle\} \cup E$ $(n \geq 0)$,
2. **projection** (on F): $E \Rightarrow E\{w_i/F\}$,
   if $\langle F(t_1, \cdots, t_n), s \rangle \in E$ $(n \geq 1, 1 \leq i \leq n)$,
3. **imitation** (on F): $E \Rightarrow E\{f(H_1(w_1, \cdots, w_n), \cdots, H_m(w_1, \cdots, w_n))/F\}$,
   if $\langle F(t_1, \cdots, t_n), f(s_1, \cdots, s_m) \rangle \in E$ $(n, m \geq 0)$.

Theorem 1 ((Huet&Lang [14])). *E is matchable iff $E \Rightarrow^* \emptyset$.*

The *second-order matching problem* is defined as follows:

SECOND-ORDER MATCHING (MATCHING)
INSTANCE: A matching expression E.
QUESTION: Is E matchable?

Theorem 2 ((Baxter [2])). MATCHING *is NP-complete*.

In this paper, we reduce the following problem MONOTONE 1-IN-3 3SAT [10] to the restricted problems of MATCHING:

MONOTONE 1-IN-3 3SAT [10]
INSTANCE: A set X of variables and a collection C of monotone 3-clauses (clauses consisting of three positive literals) over X.
QUESTION: Is there a truth assignment to X that makes exactly one literal of each clause in C true?

Hereafter, we refer $X = \{x_1, \cdots, x_n\}$ and $C = \{c_1, \cdots, c_m\}$ to an instance of MONOTONE 1-IN-3 3SAT, where $c_j \in C$ consists of the variables x_1^j, x_2^j and x_3^j.

3   The Restricted Second-Order Matching Problems

Let E be a matching expression. E is k-ary if any function variable in E is at most k-ary. E is k-fv if E includes at most k distinct function variables. E is predicate if any argument's term of function variables in E includes no function variables. E is ground if E includes no individual variables. E is function-free if E includes no function constants.

We introduce the following restricted problems of MATCHING:

> kARY (resp. kFV, PRED, GROUND, FFREE) MATCHING
> INSTANCE: A k-ary (resp. k-fv, predicate, ground, function-free) matching expression E.
> QUESTION: Is E matchable?

Theorem 3. UNARYPREDMATCHING is NP-complete.

Proof. For each clause $c \in C$, let z_1, z_2 and z_3 be the variables in c. Then, let E_c be the following unary predicate matching expression:

$$E_c = \left\{ \begin{array}{l} \langle F(f(z_3, f(z_2, f(z_1, y)))), \ f(0, f(0, f(1, f(0, f(0,0))))) \rangle, \\ \langle F(y), \ f(0, f(0,0)) \rangle \end{array} \right\}.$$

Suppose that c is satisfiable and let (a_1, a_2, a_3) be a truth assignment to (z_1, z_2, z_3) satisfying c, where there exists exactly one index i $(1 \le i \le 3)$ such that $a_i = 1$ and $a_l = 0$ $(l \ne i)$. We can construct the matcher σ of E_c as follows:

1. If $(a_1, a_2, a_3) = (1, 0, 0)$, then $\sigma = \{w_1/F, 1/z_1, 0/z_2, 0/z_3, f(0, f(0,0))/y\}$;
2. If $(a_1, a_2, a_3) = (0, 1, 0)$, then $\sigma = \{f(0, w_1)/F, 0/z_1, 1/z_2, 0/z_3, f(0,0)/y\}$;
3. If $(a_1, a_2, a_3) = (0, 0, 1)$, then $\sigma = \{f(0, f(0, w_1))/F, 0/z_1, 0/z_2, 1/z_3, 0/y\}$.

Conversely, suppose that E_c is matchable and let σ be a matcher of E_c. Then, σ includes the binding t/F, where t is w_1, $f(0, w_1)$, or $f(0, f(0, w_1))$.

Suppose that $w_1/F \in \sigma$. Since $E_c\{w_1/F\}$ is of the form

$$\{\langle f(z_3, f(z_2, f(z_1, y))), f(0, f(0, f(1, f(0, f(0,0))))) \rangle, \langle y, f(0, f(0,0)) \rangle\},$$

and by a simplification, σ includes the bindings $1/z_1$, $0/z_2$ and $0/z_3$.

Suppose that $f(0, w_1)/F \in \sigma$. Since $E_c\{f(0, w_1)/F\}$ is of the form

$$\{\langle f(0, f(z_3, f(z_2, f(z_1, y)))), f(0, f(0, f(1, f(0, f(0,0))))) \rangle, \langle y, f(0,0) \rangle\},$$

and by a simplification, σ includes the bindings $0/z_1$, $1/z_2$ and $0/z_3$.

Suppose that $f(0, f(0, w_1))/F \in \sigma$. Since $E_c\{f(0, f(0, w_1))/F\}$ is of the form

$$\{\langle f(0, f(0, f(z_3, f(z_2, f(z_1, y))))), f(0, f(0, f(1, f(0, f(0,0))))) \rangle, \langle y, 0 \rangle\},$$

and by a simplification, σ includes the bindings $0/z_1$, $0/z_2$ and $1/z_3$.

Then, we can construct the truth assignment (a_1, a_2, a_3) to (z_1, z_2, z_3) satisfying c such that $a_i = 1$ if $1/z_i \in \sigma$; $a_i = 0$ if $0/z_i \in \sigma$ $(1 \le i \le 3)$. Hence, (a_1, a_2, a_3) satisfies c, where exactly one of a_1, a_2 and a_3 is 1 and others are 0.

For C, let E be the unary predicate matching expression $\bigcup_{j=1}^{m}(E_c, \{F_j(w_1)/F, y_j/y\})$. Then, C is satisfiable by a truth assignment that makes exactly one literal in each clause in C true iff E is matchable.   □

Theorem 4. UNARYGROUNDMATCHING *is NP-complete.*

Proof. For each clause $c \in C$, let z_1, z_2 and z_3 be the variables in c. Then, let E_c be the following unary ground matching expression:

$$E_c = \{\langle F_{z_1}(F_{z_2}(F_{z_3}(0))), f(0)\rangle, \langle F_{z_1}(F_{z_2}(F_{z_3}(1))), f(1)\rangle\}.$$

For a truth assignment (a_1, a_2, a_3) to (z_1, z_2, z_3) and a matcher σ of E_c, there exists exactly one index i $(1 \leq i \leq 3)$ such that $a_i = 1$ iff $f(w_1)/F_{z_i} \in \sigma$, and $a_l = 0$ iff $w_1/F_{z_i} \in \sigma$ $(l \neq i)$. Then, c is satisfiable by a truth assignment that makes exactly one literal true iff E_c is matchable.

For C, let E be the unary ground matching expression $\bigcup_{j=1}^{m} E_{c_j}$. Then, C is satisfiable by a truth assignment that makes exactly one literal in each clause in C true iff E is matchable. □

Theorem 5. TERNARYFFREEPREDMATCHING *is NP-complete.*

Proof. For each clause $c \in C$, let z_1, z_2 and z_3 be the variables in c. Then, let E_c be the following ternary predicate function-free matching expression:

$$E_c = \{\langle F(z_1, z_2, z_3), 1\rangle, \langle F(z_2, z_3, z_1), 0\rangle, \langle F(z_3, z_1, z_2), 0\rangle\}.$$

For a truth assignment (a_1, a_2, a_3) to (z_1, z_2, z_3) and a matcher σ of E_c, there exists exactly one index i $(1 \leq i \leq 3)$ such that $a_i = 1$ iff $1/z_i \in \sigma$, and $a_l = 0$ iff $0/z_l \in \sigma$ $(l \neq i)$. Then, c is satisfiable by a truth assignment that makes exactly one literal true iff E_c is matchable.

For C, let E be the ternary function-free predicate matching expression $\bigcup_{j=1}^{m} (E_{c_j}, \{F_j(w_1, w_2, w_3)/F\})$. Then, C is satisfiable by a truth assignment that makes exactly one literal in each clause in C true iff E is matchable. □

Theorem 6. BINARYFFREEPREDMATCHING *is solvable in polynomial time.*

Proof. We reduce BINARYFFREEPREDMATCHING to 2SAT [10]. Let E be a binary function-free predicate matching expression. Without loss of generality, we can suppose that E_F includes pairs $\langle t_1, s_1\rangle, \langle t_2, s_2\rangle$ such that $s_1 \neq s_2$.

Let IC_E, IV_E, and FV_E be the sets of all individual constants, individual variables, and function variables in E, respectively. Suppose that E_F is of the form $\{\langle F(t_1^i, t_2^i), s_i\rangle \mid i \in I\}$. Note that each s_i is in IC_L. For each pair $\langle F(t_1^i, t_2^i), s_i\rangle \in E_F$, construct the following formula T_j^i $(j = 1, 2)$:

1. If $t_j^i \in IC_E$ and $t_j^i = s_i$, then $T_j^i = \mathbf{true}$;
2. If $t_j^i \in IC_E$ and $t_j^i \neq s_i$, then $T_j^i = \mathbf{false}$;
3. If $t_j^i = v \in IV_E$, then $T_j^i = x_{vs_i} \wedge (\bigwedge_{c \in IC_E - \{s_i\}} \overline{x_{vc}})$.

For E_F, the DNF formula $(\bigwedge_{i \in I} T_1^i) \vee (\bigwedge_{i \in I} T_2^i)$ is denoted by T_{E_F}. The 2CNF formula equivalent to T_{E_F} is denoted by C_{E_F}. For E, $\bigwedge_{F \in FV_E} C_{E_F}$ is denoted by C_E. The number of clauses in C_E is at most $(\#IC_E \times \#E)^2 \times \#FV_E \leq |E|^5$.

Suppose that C_E is satisfiable and let a be a truth assignment to variables $\{x_{vc} \mid v \in \mathrm{IV}_E, c \in \mathrm{IC}_E\}$ satisfying C_E. By the definition of C_E, a satisfies C_{E_F} for any $F \in \mathrm{FV}_E$, so it satisfies T_{E_F}. Then, it also satisfies $\bigwedge_{i \in I} T_1^i$, $\bigwedge_{i \in I} T_2^i$, or both. If a satisfies $\bigwedge_{i \in I} T_j^i$ $(j = 1, 2)$, then we add the bindings w_j/F and c/v to σ for each positive literal $x_{vc} \in \bigwedge_{i \in I} T_j^i$. By the construction of σ and by the definition of $\bigwedge_{i \in I} T_j^i$, σ is a matcher of E_F for any $F \in \mathrm{FV}_E$. Hence, σ is a matcher of E.

Conversely, suppose that E is matchable and let σ be a matcher of E. For $v \in \mathrm{IV}_E$ and $c \in \mathrm{IC}_E$, let a truth assignment a_{vc} to the variable x_{vc} be 1 if $c/v \in \sigma$; 0 otherwise. By the supposition, σ includes the binding either w_1/F or w_2/F for any function variable $F \in \mathrm{FV}_E$. Suppose that $w_j/F \in \sigma$ $(j = 1, 2)$. Since $E_F\{w_j/F\}$ is of the form $\{\langle t_j^i, s_i \rangle \mid i \in I\}$, it holds that $t_j^i \sigma = s_i$. If $t_j^i \in \mathrm{IC}_E$, then it holds that $T_j^i = \mathbf{true}$, since $t_j^i \sigma = t_j^i = s_i$. Then, T_j^i is always satisfiable. If $t_j^i = v_i \in \mathrm{IV}_E$, then it holds that $s_i/v_i \in \sigma$, since $t_j^i \sigma = s_i$. Since T_j^i is of the form $x_{v,s} \wedge (\bigwedge_{c \in \mathrm{IC} - \{s_i\}} \overline{x_{v,c}})$, the truth assignment $\{a_{v,c} \mid c \in \mathrm{IC}_E\}$ satisfies T_j^i. By the definition of T_{E_F}, the truth assignment $\{a_{v,c} \mid c \in \mathrm{IC}_E, i \in I\}$ satisfies T_{E_F}, so it satisfies C_{E_F}. Then, by collecting the truth assignment $\{a_{v,c} \mid c \in \mathrm{IC}_E, i \in I\}$ for every function variable $F \in \mathrm{FV}_E$, C_E is satisfiable. $\square$

Theorem 7. BINARYFFREEGROUNDMATCHING *is NP-complete.*

Proof. For each clause $c \in C$, let z_1, z_2 and z_3 be the variables in c. Then, let E_c be the following binary function-free matching expression:

$$E_c = \left\{ \begin{array}{l} \langle F_{z_1}(G_{z_1}(H_{z_2}(0), H_{z_3}(0)), H_{z_1}(0)), 1 \rangle, \\ \langle F_{z_1}(H_{z_1}(1), H_{z_2}(1)), 0 \rangle, \quad \langle G_{z_1}(0,0), 0 \rangle, \\ \langle F_{z_2}(G_{z_2}(H_{z_3}(0), H_{z_1}(0)), H_{z_2}(0)), 1 \rangle, \\ \langle F_{z_2}(H_{z_2}(1), H_{z_3}(1)), 0 \rangle, \quad \langle G_{z_2}(0,0), 0 \rangle, \\ \langle F_{z_3}(G_{z_3}(H_{z_1}(0), H_{z_2}(0)), H_{z_3}(0)), 1 \rangle, \\ \langle F_{z_3}(H_{z_3}(1), H_{z_1}(1)), 0 \rangle, \quad \langle G_{z_3}(0,0), 0 \rangle \end{array} \right\}.$$

For a truth assignment (a_1, a_2, a_3) to (z_1, z_2, z_3) and a matcher σ of E_c, there exists exactly one index i $(1 \le i \le 3)$ such that $a_i = 1$ iff $1/H_{z_i} \in \sigma$, and $a_l = 0$ iff $0/H_{z_i} \in \sigma$ $(l \ne i)$. Then, c is satisfiable by a truth assignment that makes exactly one literal true iff E_c is matchable.

For C, let E be the binary function-free matching expression $\bigcup_{j=1}^{m} E_{c_j}$. Then, C is satisfiable by a truth assignment that makes exactly one literal in each clause in C true iff E is matchable. $\square$

Theorem 8. UNARYFFREEMATCHING *is solvable in polynomial time.*

Proof. Let E be a unary function-free matching expression. For the transformation rules, we adopt the constraint that a projection on F is applied to E if there exist pairs $\langle t_1, s_1 \rangle, \langle t_2, s_2 \rangle \in E_F$ such that $s_1 \ne s_2$. Since E is unary, the transformation rules can be applied deterministically to E. $\square$

Theorem 9. 1FVPREDMATCHING *is NP-complete.*

Proof. Let q_1, q_2 and q_3 be terms $g(1,0,0)$, $g(0,1,0)$ and $g(0,0,1)$, respectively. Then, let E be the following 1-fv matching expression:

$$E = \left\{ \begin{array}{l} \langle F(g(x_1^1, x_2^1, x_3^1), y_1, z_1, \cdots g(x_1^m, x_2^m, x_3^m), y_m, z_m), \\ \quad f(q_1, q_2, q_3, \cdots, q_1, q_2, q_3) \rangle, \\ \langle F(d_1, d_1, d_1, \cdots, d_m, d_m, d_m), \ f(d_1, d_1, d_1, \cdots, d_m, d_m, d_m) \rangle \end{array} \right\}.$$

Suppose that C is satisfiable and let $(a_1, \cdots, a_n)$ be a truth assignment to X satisfying C. From $(a_1, \cdots, a_n)$, we obtain the m 3-tuples (a_1^j, a_2^j, a_3^j) $(1 \le j \le m)$ assigned to the variables x_1^j, x_2^j, x_3^j in c_j. For each j $(1 \le j \le m)$, there exists exactly one index i_j $(1 \le i_j \le 3)$ such that $a_{i_j}^j = 1$ and $a_l^j = 0$ $(l \ne i_j)$. Then, E is matchable by the following substitution σ:

$$\sigma = \{ f(w_{\mu(1,i_1,1)}, w_{\mu(1,i_1,2)}, w_{\mu(1,i_1,3)}, \cdots, w_{\mu(m,i_m,1)}, w_{\mu(m,i_m,2)}, w_{\mu(m,i_m,3)})/F \}$$
$$\cup \{ a_1^j/x_1^j, a_2^j/x_2^j, a_3^j/x_3^j, q_{\rho(i_j,2)}/y_j, q_{\rho(i_j,3)}/z_j \mid 1 \le j \le m \},$$

where $\rho(l,n) = ((l + n - 2) \bmod 3) + 1$ and $\mu(j,l,n) = 3(j-1) + ((3 - l + n) \bmod 3) + 1$ for $1 \le l, n \le 3$ and $1 \le j \le m$.

Conversely, suppose that E is matchable. By Theorem 1, E is matchable iff so is the following matching expression E':

$$E' = \left\{ \begin{array}{l} \langle H_1^j(g(x_1^1, x_2^1, x_3^1), y_1, z_1, \cdots, g(x_1^m, x_2^m, x_3^m), y_m, z_m), q_1 \rangle, \\ \langle H_1^j(d_1, d_1, d_1, \cdots d_m, d_m, d_m), d_j \rangle, \\ \langle H_2^j(g(x_1^1, x_2^1, x_3^1), y_1, z_1, \cdots, g(x_1^m, x_2^m, x_3^m), y_m, z_m), q_2 \rangle, \\ \langle H_2^j(d_1, d_1, d_1, \cdots d_m, d_m, d_m), d_j \rangle, \\ \langle H_3^j(g(x_1^1, x_2^1, x_3^1), y_1, z_1, \cdots, g(x_1^m, x_2^m, x_3^m), y_m, z_m), q_3 \rangle, \\ \langle H_3^j(d_1, d_1, d_1, \cdots d_m, d_m, d_m), d_j \rangle \end{array} \middle| 1 \le j \le m \right\}.$$

Let σ be a matcher of E'. Then, σ includes the bindings w_{r_1}/H_1^j, w_{r_2}/H_2^j, and w_{r_3}/H_3^j such that $3(j-1) + 1 \le r_1, r_2, r_3 \le 3j$ for each j $(1 \le j \le m)$. Let t_i^j $(1 \le i \le 3, 1 \le j \le m)$ be the following L-term:

$$H_i^j(g(x_1^1, x_2^1, x_3^1), y_1, z_1, \cdots, g(x_1^m, x_2^m, x_3^m), y_m, z_m),$$

By the definition of t_i^j, for each j, there exists exactly one index i_j $(1 \le i_j \le 3)$ such that $t_{i_j}^j \sigma = g(x_1^j, x_2^j, x_3^j)\sigma = q_{i_j}$. Then, we obtain the truth assignment (a_1^j, a_2^j, a_3^j) to (x_1^j, x_2^j, x_3^j) as $q_{i_j} = g(a_1^j, a_2^j, a_3^j)$. By collecting all of (a_1^j, a_2^j, a_3^j) from q_{i_j} $(1 \le j \le m)$, we obtain the truth assignment $(a_1, \cdots, a_n)$ to X satisfying C such that $(a_1, \cdots, a_n)$ makes exactly one literal in each clause in C true. $\square$

Theorem 10. kFVFFREEMATCHING *is solvable in polynomial time for $k \ge 0$.*

Proof. Let E be a k-fv function-free matching expression with k function variables $F_1, \cdots, F_k$ and n be the maximum arity of F_i $(1 \le i \le k)$. We adopt the same constraint of Theorem 8. Since E is function-free, once applying an imitation or a projection to E decreases at least one function variable in E. Furthermore, a projection is applied to E at most n times for every function variable. Then, we can determine whether E is matchable by checking at most n^k first-order matching expressions. $\square$

Theorem 11. GROUNDPREDMATCHING *is solvable in polynomial time.*

Proof. Let E be a ground predicate matching expression. Consider the following two projections, instead of a projection:

1. **projection 1** (on F): $E \Rightarrow E\{w_i/F\}$,
   if E_F is of the form $\{\langle F(t_1^1, \cdots, t_n^1), t_i^1\rangle, \cdots, \langle F(t_1^m, \cdots, t_n^m), t_i^m\rangle\}$,
2. **projection 2** (on F): $E \Rightarrow$ `fail`,
   if a projection 1 on F cannot be applied to E and there exists pairs
   $\langle F(t_1, \cdots, t_n), s_1\rangle, \langle F(u_1, \cdots, u_n), s_2\rangle \in E_F$ but $\mathrm{hd}(s_1) \neq \mathrm{hd}(s_2)$.

An imitation on F is applied to E if the above projections 1 and 2 on F cannot be applied. Then, the transformation rules is applied deterministically to E.

Since E is ground and predicate, E is transformed to `fail` by a projection 2 iff $E \not\Rightarrow^* \emptyset$ by only an imitation and a simplification. Furthermore, by an imitation on F and a simplification, the right-hand term of pairs in E_F is decomposed into the subterms. By Theorem 1, the statement holds. $\square$

4   The Comparison between Second-Order Matching and Unification Problems

In this section, we compare the tractability/intractability of the restricted second-order matching problems with the decidability/undecidability of the restricted second-order unification problems.

1. Amiot [1] (and implicitly Farmer [8]) has shown that the unification problem is undecidable for an unary predicate unification expression with at least one binary function constant. On the other hand, by Theorem 3, the matching problem is NP-complete for a unary predicate matching expression with at least one binary function constant.

2. A matching expression is *monadic* if any function constant in it is unary, and *nonmonadic* if it is not monadic. Farmer has shown that the unification problem is decidable for a monadic unification expression [7], but undecidable for a nonmonadic unary one with at least one binary function constant [8]. On the other hand, by Theorem 4, the matching problem is NP-complete for a monadic matching expression. Also by Theorem 3, it is NP-complete for a nonmonadic unary one with at least one binary function constant.

3. Goldfarb [12] has shown that the unification problem is undecidable for a ternary ground unification expression. On the other hand, by Theorem 4, the matching problem is NP-complete for a unary ground matching expression. Note that Amiot's and Farmer's results [1,8] do not imply that the unification problem is undecidable for a unary ground unification expression, because the existence of individual variables is essential in their proofs.

4. As pointed by Goldfarb [12], the unification problem is decidable for a function-free unification expression. On the other hand, by Theorem 5 or 7, the matching problem is NP-complete for a function-free matching expression.

However, if a function-free matching expression is binary predicate, unary, or k-fv ($k \geq 0$), then it is solvable in polynomial time by Theorem 6, 8 or 10.

5. Ganzinger et al. [11] have shown that the unification problem is undecidable for an 1-fv unification expression, where the function variable occurs at most twice. On the other hand, by Theorem 9, the matching problem is NP-complete for an 1-fv matching expression, where the function variable occurs at most twice. Also Levy and Veanes [17] have shown that the unification problem is undecidable for an 1-fv ground and an 1-fv unary unification expressions. Whether the corresponding matching problems are NP-complete is still open.

6. A matching expression is k-*linear* if any function variable occurs at most k times. Dowek [6] has shown that the unification problem is decidable for an 1-linear unification expression, and the matching problem is solvable in linear time for an 1-linear matching expression. Furthermore, Levy [16] has shown that the unification problem is undecidable for a 2-linear unification expression. On the other hand, Theorem 3 or 9 claims that the matching problem is NP-complete for a 2-linear matching expression.

5   Conclusion

We summarize the results obtained by this paper in the following table.

expression	matching	ref.	unification	ref.
UNARYPRED	NP-complete	Theorem 3	undecidable	[1,8]
TERNARYGROUND	NP-complete	Theorem 4	undecidable	[12]
UNARYGROUND	NP-complete	Theorem 4		
TERNARYFFREEPRED	NP-complete	Theorem 5	decidable	[12]
BINARYFFREEPRED	polynomial	Theorem 6	decidable	[12]
BINARYFFREEGROUND	NP-complete	Theorem 7	decidable	[12]
UNARYFFREE	polynomial	Theorem 8	decidable	[12]
1FVPRED	NP-complete	Theorem 9	undecidable	[11]
kFVFFREE ($k \geq 0$)	polynomial	Theorem 10	decidable	[12]
GROUNDPRED	polynomial	Theorem 11		
MONADIC	NP-complete	Theorem 4	decidable	[7]
NONMONADIC	NP-complete	Theorem 3	undecidable	[8]
1LINEAR	linear	[6]	decidable	[6]
2LINEAR	NP-complete	Theorem 3,9	undecidable	[16]

In this paper, we have dealt with the second-order matching problem for a matching expression consisting of L-terms, not L^*-terms. The matching expression consisting of L^*-terms follows the matching problem of *second-order patterns* [18,19], which is related to the problem GROUNDPREDMATCHING.

Curien et al. [3] have designed a complete second-order matching algorithm which works more efficient than the one of Huet and Lang [14] in most cases. When it is necessary to obtain the complete set of matchers for a given matching expression, we know no more efficient algorithm than their algorithms although it is not a polynomial-time algorithm. It is a future work to give the trade-off between completeness and efficiency of the second-order matching adequate for each research field.

References

1. Amiot, G.: *The undecidability of the second order predicate unification problem*, Archive for Mathematical Logic **30**, 193–199, 1990.
2. Baxter, L. D.: *The complexity of unification*, Doctoral Thesis, Department of Computer Science, University of Waterloo, 1977.
3. Curien, R., Qian, Z. and Shi, H.: *Efficient second-order matching*, Proc. 7th International Conference on Rewriting Techniques and Applications, LNCS **1103**, 317–331, 1996.
4. Défourneaux, G., Bourely, C. and Peltier, N.: *Semantic generalizations for proving and disproving conjectures by analogy*, Journal of Automated Reasoning **20**, 27–45, 1998.
5. Donat, M. R. and Wallen, L. A.: *Learning and applying generalized solutions using higher order resolution*, Proc. 9th International Conference on Automated Deduction, LNCS **310**, 41–60, 1988.
6. Dowek, G.: *A unification algorithm for second-order linear terms*, manuscript, 1993, also available at `http://coq.inria.fr/~dowek/`.
7. Farmer, W. M.: *A unification algorithm for second-order monadic terms*, Annals of Pure and Applied Logic **39**, 131–174, 1988.
8. Farmer, W. M.: *Simple second-order languages for which unification is undecidable*, Theoretical Computer Science **87**, 25–41, 1991.
9. Flener, P.: *Logic program synthesis from incomplete information*, Kluwer Academic Press, 1995.
10. Garey, M. R. and Johnson, D. S.: *Computers and intractability: A guide to the theory of NP-completeness*, W. H. Freeman and Company, 1979.
11. Ganzinger, H., Jacquemard, F. and Veanes, M.: *Rigid reachability*, Proc. 4th Asian Computing Science Conference, LNCS **1538**, 1998, also available at `http://www.mpi-sb.mpg.de/~hg/`.
12. Goldfarb, W. D.: *The undecidability of the second-order unification problem*, Theoretical Computer Science **13**, 225–230, 1981.
13. Harao, M.: *Proof discovery in LK system by analogy*, Proc. 3rd Asian Computing Science Conference, LNCS **1345**, 197–211, 1997.
14. Huet, G. P. and Lang, B.: *Proving and applying program transformations expressed with second-order patterns*, Acta Informatica **11**, 31–55, 1978.
15. Levy, J.: *Linear second-order unification*, Proc. 7th International Conference on Rewriting Techniques and Applications, LNCS **1103**, 332–346, 1996.
16. Levy, J.: *Decidable and undecidable second-order unification problem*, Proc. 9th International Conference on Rewriting Techniques and Applications, LNCS **1379**, 47–60, 1998.
17. Levy, L. and Veanes, M.: *On unification problems in restricted second-order languages*, Proc. Annual Conference of the European Association for Computer Science Logic (CSL'98), 1998, also available at `http://www.iiia.csic.es/~levy/`.
18. Miller, D.: *A logic programming language with lambda-abstraction, function variables, and simple unification*, Journal of Logic and Computation **1**, 497–536, 1991.
19. Prehofer, C.: *Decidable higher-order unification problems*, Proc. 12th International Conference on Automated Deduction (CADE 12), LNAI **814**, 635–649, 1994.

Efficient Fixed-Size Systolic Arrays for the Modular Multiplication*

Sung-Woo Lee[1], Hyun-Sung Kim[1], Jung-Joon Kim[2],
Tae-Geun Kim[2], and Kee-Young Yoo[1]

[1]Department of Computer Engineering
Kyungpook National University
Taegu, Korea, 702-701
{swlee, hskim}@purple.kyungpook.ac.kr
yook@bh.kyungpook.ac.kr
[2]Wireless Comm. Research Lab., Korea Telecom

Abstract. In this paper, we present an efficient fixed-size systolic array for Montgomery's modular multiplication. The array is designed by the LPGS (Locally Parallel Globally Sequential) partition method [14] and can perform efficiently modular multiplication for the input data with arbitrary bits. Also, we address a computation pipelining technique, which improves the throughput and minimizes the buffer size used. With the analysis of VHDL simulation, we discuss a gap between a theoretical optimal number of partition and an empirical one.

1   Introduction

With the increasing popularity of electronic communications, security is becoming a more and more important issue. In 1978, Rivest, Shamir, and Adleman proposed the RSA public key cryptosystem which is relatively easy to implement and has been satisfactorily secure so far [2]. But it requires the fast modular exponentiation with a large modulus that has 512 or more bits in encryption and decryption.

In the last few years, a number of algorithms for the modular multiplication have been proposed and some of them have been realized. One of the most attractive modular multiplication algorithms was proposed by Peter L. Montgomery[3]. By using this algorithm, software and hardware implementations for the fast modular multiplication have been researched [4-9,12,13]. Walter [9] developed a systolic array with $(n+1)(n+2)$ PEs using the Montgomery's algorithm. Heo[12] proposed a systolic array with $n+1$ PEs by transforming Walter's algorithm. The structure is well suited to VLSI implementation. Iwamura[7] and Chen[13] also designed a systolic array with $n+1$ PEs, respectively. Since n is 512 or more, these full-size systolic arrays require a very large space so that is difficult to implement the array in a single chip.

In this paper, we present a efficient fixed-size systolic arrays for Montgomery's modular multiplication. The array is designed by the LPGS(Locally Parallel Globally

*This work is supported by the Korea Telecom, KOREA under a grant NO. 98-13.

T. Asano et al. (Eds.): COCOON'99, LNCS 1627, pp. 442–451, 1999

Sequential) partition method[14] and perform efficiently modular multiplication for input data with arbitrary bits. Also, we address a computation pipelining technique, which improve the throughput and minimize the buffer size. With the analysis of VHDL simulation, we discuss a gap between a theoretical optimal number of partitions and a empirical that.

2   Modular multiplication algorithm

The Montgomery's modular multiplication algorithm (MMM) was represented. Let A, B, and N be multi-digit numbers with radix r, but let n_0 be a single digit number which satisfies the condition $n_0 * N[0] \mod r = -1$, where $N[0]$ is the least significant digit of N. The number of digits of A, B, and N is n, $n+1$, and n, respectively. $A[i]$ means the i th digit of A. The Montgomery's algorithm MMM generates $T=ABR^{-1} \mod N$. R^{-1} is the multiplicative inverse of $R(=r^{n+1})$ modulo N and satisfies $GCD(N,R)=1$. The number of digits of T is $n+1$.

> MMM (A, B, N, r)
> $n_0 = -N[0]^{-1} \mod r$
> $T=0$
> *for* $i=0$ n-1 {
> $M = ((T[0] + A[i]B[0])\, n_0) \mod r$
> $T = (T+A[i]B + MN)\, div\, r$
> }
> *if* $(T \geq N)$ *then* return T-N
>   *else* return T

The Montgomery's algorithm is potentially faster and simpler than an ordinary modular computation $AB \mod N$. However, the conversion from an ordinary residue to an N residue, computation of n_0, and the conversion back to an ordinary residue are time consuming tasks. Thus, it is not a good idea to use Montgomery's algorithm when only a single modular multiplication will be performed. It is more suitable for cases requiring several modular multiplication with respect to the same modulus, as when one needs to compute modular exponentiations. In the modular exponentiation, the output T of MMM can be directly used repeatedly as the next input B. But T can not be used as the input A. We determine that the number of digits of A is $n+1$. This does not affect the result value T of MMM. So the output of MMM can be directly used as the next input for the squaring and multiplication. In the algorithm MMM, it is necessary to convert $A[i]B$ and MN into digit-level computations. Thus the MMM is converted into the following recurrence algorithm MMM1 [9].

> Initail value :
> $n[0,j] = N[j]$ $0 \leq j \leq n-1$, $n[0,n] = n[0,n+1] = 0$
> $b[0,j] = B[i]$ $0 \leq j \leq n$, $b[0,n+1] = 0$
> $a[i,0] = A[i]$ $0 \leq i \leq n$, $a[n+1,0] = 0$
> $T[0,j] = 0$ $0 \leq j \leq n+1$, $T[i,n+1] = 0$ $0 \leq i \leq n+1$

MMM1 $(A, B, N, r, n0)$

for $i = 0$ *to* $n+1$ *do*
  for $j = 0$ *to* $n+1$ *do* {
    if $(j=0)$ {
      $m[i, j+1] = ((T[i, j] + a[i, j]b[i,0])\ n0[i,0])\ \mathrm{mod}\ r$
      $c_1[i, j+1] = ((T[i, j] + a[i, j]b[i, j]) + m[i, j+1]n[i, j])\ \mathrm{div}\ r\ \mathrm{div}\ r$
      $c_0[i, j+1] = ((T[i, j] + a[i, j]b[i, j]) + m[i, j+1]n[i, j])\ \mathrm{div}\ r\ \mathrm{mod}\ r$
      $n0[i+1,0] = n0[i,0]$

    } *else* {
      $m[i, j+1] = m[i, j]$
      $c_0[i, j+1] = ((T[i, j] + a[i, j]b[i, j]) + m[i, j]n[i, j] + c_0[i, j] + c_1[i, j]r)\ \mathrm{div}\ r\ \mathrm{mod}\ r$
      $c_1[i, j+1] = ((T[i, j] + a[i, j]b[i, j]) + m[i, j]n[i, j] + c_0[i, j] + c_1[i, j]r)\ \mathrm{div}\ r\ \mathrm{div}\ r$
      $T[i+1, j-1] = (T[i, j] + a[i, j]b[i, j] + m[i, j]n[i, j] + c_0[i, j])\ \mathrm{mod}\ r$

    }
    $a[i, j+1] = a[i, j]$ $b[i+1, j] = b[i, j]$ $n[i+1, j] = n[i, j]$

}

In MMM1, each computation occurs at each index point $[i, j]^T$, $(0 \le i, j \le n+1)$ the so-called a computation point. The final result value $T[n+2, 0] \dots T[n+2, n]$ is $ABr^{n \cdot 2}$ *mod N*, and carry is two digits, i.e. $c_0[i, j+1]$, $c_1[i, j+1]$. The input n_0 is precomputed from $-N[0]^{-1}\ mod\ r$ and always is not 0. If r is binary, n_0 is always 1.

The recurrence algorithm can be represented by the data dependence graph (DG) as shown in Fig.1. In the DG, each node means computation point and each edge means data flow.

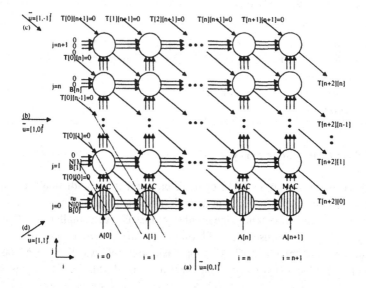

Fig. 1. Dependence Graph of MMM1. The vector $\bar{u}$ is valid projection vectors

3   Fixed size systolic array design

3.1   Full size systolic array

A linear systolic array can be designed according to the systematic procedure [10,11]. From the DG systolic arrays can be designed by a space-time linear transformation. Heo [12] had proved that the systolic array derived by a space vector $S=[1, 0]$ and a schedule vector $\Pi=[2,1]$ is optimal. Let $c=[i, j]^T$ be a computation point, i.e. a node in the DG. The product Sc gives the coordinate of the PE that the node c is mapped into. And the product Πc gives the time step at which it's computation is performed.

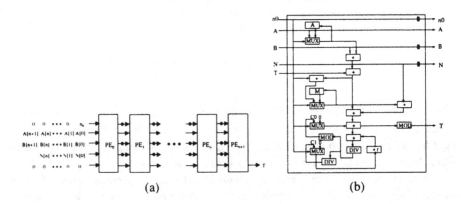

(a)   (b)

Fig. 2. Full-size Systolic Array (a) Array structure with input data. (b) PE structure

The full-size systolic array derived by the space vector S and the schedule vector Π is shown in Fig. 2. Note the input A is not preloaded, but supplied to the first PE. If the input A is preloaded on each PEs, the number of I/O pins depends on the problem size, causing a implementation to be hard. The systolic array has $n+2$ PEs and the computation time step is $3n+4$.

3.2   A fixed-size systolic array

As the RSA cryptosystem uses 512 or more bit numbers, it is difficult to implement such a large full-size systolic array in a single RSA hardware chip. Thus, it is important to design systolic array who's the number of PEs is much smaller than that of full-size array. For this, we use the well-known *locally parallel, globally sequential* (LPGS) partitioning scheme [14]. The approach is as follows: First of all, assign an index point $c=[i, j]^T$ to a ($\lfloor Sc/ p \rfloor$)th band, and a (Sc mod p)th PE where p is the number of PEs. Next, arrange the input data layout of the fixed-size systolic array by partitioning total input data and arranging them for each band computation. As shown in Fig. 3, the DG can be partitioned into the fixed-size systolic array.

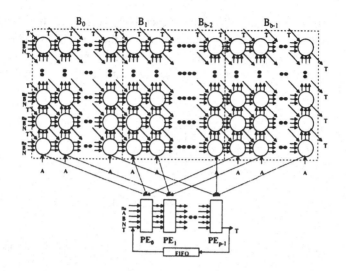

Fig. 3. LPGS partition method. The rectangles imply a band which is referred as B_i.

The following facts are observed from the computation behavior of the derived fixed-size systolic array. (1)Each PE computes all nodes consisting of a corresponding column of each band B_i in the DG. (2)The p digits of A must be inputted at each band B_i. (3) In each band computation, all digits of B and N should be inputted because all digits of B and N are needed for each band computation. (4) Control signals consisting of n_0 and some 0s depending on the time step at which each band B_i is computed, are needed.

From the above facts whenever each band B_i is computed, values are inputted as follows: if p is the number of PEs of fixed-size array, the p digits from $p i$th to $(p(i+1)-1)$th digit of A are inputted at every time step $(2p+n)i+1$. Because all digits of B and N are needed for all band computations, all digits of B and N are inputted. The $n+p+1$ digits of T are need for every band computation. $n+1$ digits among them are inputted from the FIFO buffer but other p digits of T are inputted from the outside at time step $(2p+n)i+n+2$. The $2p+n$ values of control signals are inputted. The first value of control signals is n_0 and the others are all 0s.

Let $I = (P_{ctl}, P_A, P_B, P_N, P_T)$ denote the layout of the input values for the full-size systolic array. P_{ctl}, P_A, P_B, P_N, and P_T are data sequences of control signals, A, B, N, and T, respectively. Each sequence is as follows: note $|P_A|$ denotes the length of sequence P_A. $P_{ctl} = \{ n_0, 0, 0, \dots, 0 \}$, $P_A = \{A[0], A[1], \dots, A[n], 0\}$, $P_B = \{B[0], B[1], \dots, B[n], 0\}$, $P_N = \{N[0], N[1], \dots, N[n-1], 0, 0\}$, $P_T = \{T[0][0], \dots, T[0][n], T[1][n], \dots, T[n][n]\} = \{ 0, 0, \dots, 0, 0 \}$. $|P_A| = |P_B| = |P_N| = n+2$. $|P_{ctl}| = |P_T| = 2n+3$.

Let $I^k = (P^k_{ctl}, P^k_A, P^k_B, P^k_N, P^k_T)$ denote the layout of input values for the fixed-size systolic array with k-th band. P^k_{ctl}, P^k_A, P^k_B, P^k_N, and P^k_T are data sequences of control signals, A, B, N, and T for k th bands, respectively. The sequences P^k_B and P^k_N are the same as P_B, P_N for all bands, respectively. But the data sequences P^k_{ctl}, P^k_A, and P^k_T are as follows: $P^k_{ctl} = \{ n_0, 0, \dots, 0 \}$, $|P^k_{ctl}| = n+1+p$. Sequence $P^k_A = \{A[kp+0],$

$A[kp+1], ..., A[kp+(p-1)]\}$ and $|P^k_A| = p$. Sequence $P^k_T = \{0, 0, ..., 0, 0\}$ and $| P^k_T | = n+1+p$. The sequences P_T and P^k_T are generated by the first PE instead of being in-putted. The Fig. 4 shows the fixed-systolic array structure and the input layout for the first band computation. The structures of PE are the same as PE in the Fig. 2(b) except for the last PE.

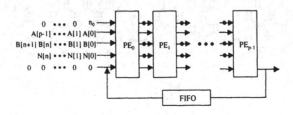

Fig. 4. Derived fixed-size systolic array with input layout for the first band computation

3.2.1 Non-pipelined fixed-size systolic array

After mapping algorithm into a fixed-size systolic array, we should arrange input data. According to a formal method of input data arrangement [11], after each band B_i is completely computed, the computation of the next band B_{i+1} begins in the fixed-size systolic arrays. The intermediate results of band B_i are stored into the FIFO buffer and will be used for the computation of the next band. We analyze the size of buffer and the computation time step.

Let q be the number of bands in the DG. For convenience' sake, assume that $(n+2)$ is a multiple of q.

Lemma 1. Let p be the number of PEs of the non- pipelined fixed-size systolic array. The size of the FIFO buffer, $FIFO_{non}$ of the derived non-pipelined fixed-size systolic array is as follows:

$$FIFO_{non} = n. \tag{1}$$

Proof: Suppose that the first value of T is computed on PE_{p-1} for band B_i on time step t. If the output value T is input into PE_0 for the computation of its next band B_{i+1} at time step $t+1$, the buffer does not needed. However, the computation of band B_{i+1} can be started after the computation of band B_i is finished. So, the buffer size depends on only the difference between the time step of the first output from band B_i and the time step which the first input for the next band B_{i+1} is inputted into. The first output from band B_i can be computed at time step $(2p+n)i+2p$ and the first input for band B_{i+1} can be processed at time step $(2p+n)(i+1)+1$ (note: i is the index of all bands, $0 \le i < q$). The difference between two time steps is $n+1$. Therefore, the size of the FIFO buffer be-comes n. ∎

The computation time steps of the non-pipelined fixed-size systolic array is summa-rized in the following theorem.

Theorem 1. The total computation time step, t_{non} of the derived non-pipelined fixed-size systolic array is as follows:

$$t_{non} = 3n+4+n\times((n+2)/p-1). \tag{2}$$

Proof: The total computation time of the derived non-pipelined fixed-size systolic array increases more than the full-size systolic array, t_{full} due to delay of the FIFO buffer. So the computation of the band B_{i+1} has to start after the previous band B_i is computed. This delayed time is repeated as the number of bands-1, i.e. $q-1$. Note that $q=(n+2)/p$. The overhead for partitions, t_{delay} becomes $n\times(q-1)$, i.e. $n\times((n+2)/p-1)$. Therefore, the total computation time step of the derived non-pipelined fixed-size systolic array results in $t_{full} + t_{delay}$. So, it is equal to $3n+4+n\times((n+2)/p-1)$. ∎

3.2.2 Pipelined fixed-size systolic array

The performance of the non-pipelined array is not efficient, because many idle PEs exist between each band computation. From observation about the computation behavior of the non-pipelined array, the following fact can be obtained: all data flows (links) are the same direction [1,0]. Thus, the first PE (PE_0) can start the computation of the next band immediately after computing of the last node in the last column of the previous band on the DG.

Though the non-pipelined array supplies the input data of a band to the first PE after the last PE performs computation of the last node of the previous band, the pipelined array supplies them after the first PE performs computation of the last node of the previous band. Consequently, we can optimize a total computation time and a PE utilization and minimize a FIFO size.

Lemma 2. The computation time step of value T, which is needed for the computation of the first node in each band is equal to the computation time step of the last node in the previous band.

Proof: Suppose that the time step of the last computation on the PE_0 is t_{last} and the time step of the second computation on the PE_{p-1} is t_{first}. For example, as shown in Fig. 5(a), t_{last} is the time step of node A and t_{first} is the time step of node B. If the first time step on the PE_0 is t_0, t_{first} is t_0+n+1 and t_{last} is $t_0+2(p-1)+1$. In the pipelined fixed-size systolic array, t_{last} is equal to t_{first} because $p = (n+2)/2$. ∎

Lemma 3. The FIFO buffer size, $FIFO_{pipe}$ for the pipelined fixed-size systolic array is as follows:

$$FIFO_{pipe} = (n+2)-2p \tag{3}$$

Proof: Suppose that the first value of T is computed on PE_{p-1} within B_i at time step t, the buffer does not used when value T is inputted at PE_0 for the computation of the next band B_{i+1} at time step t+1. On the other band, every value T which will be used for the next time step should be stored in the buffer if PE_0 is working on a node in B_i at time step t+1. So the needed size of the buffer is expected to be the difference between the computation time step of the last node on PE_0 in band B_j, t_{last} and the computation time step of second node on PE_{p-1}, t_{first}. If the computation time step of a band is t_0, t_{first} becomes t_0+n+1 and t_{last} becomes $t_0+2(p-1)+1$. Therefore, the size of the buffer is $(n+2)-2p$. ■

As shown in fig. 5(a), if the first output value T of band B_0 can be used in the computation of band B_1 on PE_0 as soon as the computation of band B_0 on PE_{p-1} is finished, the buffer size is to be 0. Note that if a pipelined array can process the next band without the buffer, the same performance as the full-size systolic array can be obtained just using the only half number of PEs.

If the DG is partitioned into more than 2 bands, the buffer is needed as described in the equation (3). The fig.5(b) shows the case of partition with 3 bands. In fig5(b), the computation nodes on diagonal line can be computed at the same time and the values T computed at the black colored nodes have to be stored in the buffer. Therefore the buffer size is 4.

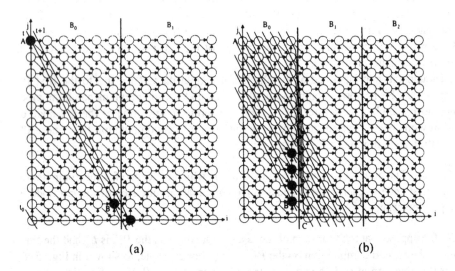

(a)   (b)

Fig. 5. Partioning of DG.(a) Partitioned DG with 2band, (b) Partitioned DG with 3band

Theorem 2. Let p be the number of PEs of the pipelined fixed-size systolic array. The total computation time step, t_{pipe} for the array is as follows:

$$t_{pipe} = 3n+4+(n+2-2p)((n+2)/p-1) \qquad (4)$$

Proof: The proof of this theorem is similar to that of **Theorem 1**. In the pipelined fixed-size systolic array, the buffer size is $(n+2)-2p$. The delayed time is repeated as the number of bands−1. So the delay t_{delay} is $(n+2-2p)((n+2)/p-1)$. Therefore, the time step of the derived pipelined fixed-size systolic array results in $t_{full}+t_{delay}$. So the total computation time step is equal to $3n+4+(n+2-2p)((n+2)/p-1)$. ■

The difference of computation time steps between the non-pipelined and pipelined fixed-size systolic array is $(2p-2)((n+2)/p-1)$ and the difference of the buffer size is $2p-2$. It shows that the pipelined fixed-size systolic array has more efficient performance than the non-pipelined fixed-size systolic array. Moreover when DG is partitioned into 2 bands, the buffer is no longer needed.

3.3   Examples of a pipelined fixed-size systolic array

Let n be 10. The $(n+2)\times(n+2)$ DG of the Montgomery's algorithm is partitioned into 2 bands, which are mapped into a fixed-size systolic array with 6 PEs. In the nonpipelined array, the size of the FIFO buffer is 10, each band computation time step is 22 and the total computation time step is 44. In the pipelined fixed-size systolic array, the total computation time step is 34 which is equal to the full-size systolic array's computation time step and the buffer is no longer needed.

4   VHDL simulation

We simulate the pipelined fixed-size systolic array using ALTERA MAX+PLUS II which is a programmable logic device simulation tools. The FLEX10KA families of ALTERA's FPGA device is used. Fig.6 shows the result of simulation when base is 16 and the bit number of input data A, B, and N is 126. In the VHDL simulation of the pipelined fixed-size systolic array, the array is partitioned into 2, 4, and 8 bands. As shown fig.6, there are some differences between the theoretical analysis and experimental results. The reason of it is that the larger a chip area is, the longer a minimum clock cycle of the tool. If a execution time is only considered, a 2 bands partitioning is best. But if the chip area and execution time are all considered, a 4 bands partitioning is the best.

Table 1. Simulation result

Pes	FIFO size	# of logic cell	Clock cycle (ns)	Execution time(us)
128 (full size)	0	31101	214	81.748
64 (2 bands)	0	15309	151	57.833
32 (4 bands)	64	7809	104.4	59.925
16 (8 bands)	96	3928	90.2	95.071

5   Conclusion

We presented a pipelined fixed-size systolic arrays for Montgomery's modular multiplication which has input data with arbitrary bits. Also we show that our pipelining technique improves the throughput and minimize the buffer size. It's total computation time step is $3n+4+(n+2-2p)((n+2)/p-1)$, and a buffer size is $(n+2)-2p$. Moreover, when the DG is partitioned into 2 bands, the pipelined fixed-size systolic array has the same time steps as the full-size systolic array and the number of PEs are reduced by half. As the analysis of VHDL simulation, for 504 bits input data, when partitioning the algorithm into 4bands, the proposed pipelined fixed-size systolic array requires only 1/4 chip area of full-size array's area and its computation time is approximately best.

References

1. W. Diffie and M. Hellman, "New Directions in Cryptography," *IEEE Trans. on Info. Theory*, vol. IT-22(6) (1976) 644-654
2. R.L. Rivest, A. Shamir, and L. Adleman, "A Method for Obtaining Digital Signatures and Public-key Cryptosystems," *Commu. ACM.* vol. 21 (1978) 120-126
3. P.L. Montgomery, "Modular Multiplication Without Trial Division," *Mathemat. of Computat.* vol. 44 (1985) 519-521
4. Antoon Bosselaers, Rene Govaerts, and Joos Vandewalle, "Comparison of three modular reduction functions," *Advances in Cryptology-CRYPTO '93* (1993) 175-186
5. S.R. Dusse and B.S. Kaliski Jr., "A Cryptographic Library for the Motorola DSP56000," in *Advances in Cryptology EUROCRYPT '90*, Springer-Verlag (1991) 230-244
6. S.E. Eldridge and C.D. Walter, "Hardware Implementation of Montgomery's Modular Multiplication Algorithm," *IEEE Trans. Comput.* vol. 42 (1993) 693-699
7. K. Iwamura, T. Matsumoto, and H. Imai, "Systolic Arrays for Modular Exponentiation using Montgomery Method," *Proc. Euro CRYPT '92* (1992) 477-481
8. J. Sauerbrey, "A Modular Exponentiation unit based on Systolic Arrays," *Abst. AUSCRYPT '92* (1992) 1219-1224
9. C.D. Walter, "Systolic Modular Multiplication," *IEEE Trans. on Computers*, vol. 42 (1993) 376-378
10. S.Y. Kung, *VLSI Array Processors*, Prentice-Hall (1987)
11. K.Y. Yoo, "*A Systolic Array Design Methodology for Sequential Loop Algorithms*," Ph.D. thesis, Rensselaer Polytechnic Institute (1992)
12. Y.J. Heo, K.J. Lee and K.Y. Yoo, "An optimal linear systolic array for computing montgomery modular multiplication", *PDCS '98* (1998)
13. Po-Song Chen, Shih-Arn Hwang and Cheng-Wen Wu, "A systolic RSA public key cryptosystem", *IEEE International Symposium on Circuits and Systems*, vol. 4 (1996)
14. D.I. Moldovan and J.A.B. Fortes, " Partitioning and Mapping Algorithms into Fixed Size Systolic Arrays", IEEE transaction on Computers, vol. C-35, no.1, January (1986) 1-12

Improving Parallel Computation with Fast Integer Sorting

Ka Wong Chong[1], Yijie Han[2], Yoshihide Igarashi[3], and Tak Wah Lam[4]

[1] Department of Computer Science, The University of Hong Kong,
Hong Kong, China. kwchong@cs.hku.hk
[2] Electronic Data Systems, Inc., 750 Tower Drive, CPS, Mail Stop 7121,
Troy, MI 48098, USA.
yhan01@cps.plnin.gmeds.com, http://welcome.to/yijiehan
[3] Department of Computer Science, Gunma University, Kiryu, 376-8515 Japan.
igarashi@comp.cs.gunma-u.ac.jp
[4] Department of Computer Science, The University of Hong Kong,
Hong Kong, China. twlam@cs.hku.hk

Abstract. This paper presents results which improve the efficiency of parallel algorithms for computing minimum spanning trees. These results are obtained by mainly applying fast integer sorting. For an input graph with n vertices and m edges our EREW PRAM minimum spanning tree algorithm runs in $O(\log n)$ time with $O((m + n)\sqrt{\log n})$ operations. Our CRCW PRAM minimum spanning tree algorithm runs in $O(\log n)$ time with $O((m + n) \log \log n)$ operations. These complexities relate to the complexities of parallel integer sorting. We also show that for dense graphs we can achieve $O(\log n)$ time with $O(n^2)$ operations on the EREW PRAM.

Keywords: Parallel algorithms, graph algorithms, minimum spanning tree, integer sorting, PRAM.

1 Introduction

Let $G = (V, E)$ be the input graph, where V is the vertex set and E is the edge set. Also let $|V| = n$ and $|E| = m$ and each edge is assigned a real-valued weight. In this paper we consider the problems of computing the minimum spanning trees for G on the CRCW and EREW PRAM models. Many researchers have designed parallel algorithms for computing minimum spanning trees [3,4,5,6,9,13,14,15,17]. Cole, Klein and Tarjan have obtained a randomized algorithm with $O(\log n)$ time and linear operations (time processor product). The best previous deterministic CRCW minimum spanning tree algorithm runs in $O(\log n)$ time with $O((m+n) \log n)$ operations[3,17]. For a period of time it is known that minimum spanning trees can be computed on the CREW and EREW PRAM in $O(\log^2 n)$ time[4,13]. Johnson and Metaxas were the first to show a deterministic EREW PRAM algorithm with time complexity $O(\log^{3/2} n)$ and $m + n$ processors (therefore $O((m + n) \log^{3/2} n)$ operations) for computing minimum spanning trees[14]. Johnson and Metaxas' result for minimum spanning

T. Asano et al. (Eds.): COCOON'99, LNCS 1627, pp. 452–461, 1999.
© Springer-Verlag Berlin Heidelberg 1999

trees[14] was improved by Anderson, Beame and Sinha (private communication) and Chong[5] to time complexity $O(\log n \log^{(2)} n)$ with $O((m + n) \log n \log^{(2)} n)$ operations $(\log^{(1)} n = \log n, \log^{(c)} n = \log \log^{(c-1)} n)$. Recently Chong, Han and Lam [6] presented an EREW minimum spanning tree algorithm with $O(\log n)$ time and $O((m + n) \log n)$ operations.

In this paper we further improve the efficiency of previous deterministic parallel algorithms for computing minimum spanning trees. We achieve our results by mainly applying fast parallel integer sorting[2,11,12]. In addition, we also make use of known results for parallel selection[8] and parallel approximate compaction[10]. We present an EREW algorithm for computing minimum spanning trees with time complexity $O(\log n)$ using $O((m + n)\sqrt{\log n})$ operations. We also show that our algorithm runs on the CRCW PRAM with time complexity $O(\log n)$ and operation complexity $O((m + n) \log \log n)$. We relate the operation complexity for minimum spanning tree to the operation complexity of integer sorting[2,11,12]. The operation complexities of our minimum spanning tree algorithms can be improved further if more efficient parallel integer sorting algorithms are available. For dense graphs our algorithm is optimal and runs in $O(\log n)$ time with $O(n^2)$ operations on the EREW PRAM.

2   Preliminaries

Given an undirected graph $G = (V, E)$, we assume that vertices in V are labeled with integers $\{1, 2, ..., n\}$. Every vertex v is associated with an adjacency list $L(v)$. Each edge (u, v) is represented twice, once in the adjacency list of u (denoted $\langle u, v \rangle$) and once in the adjacency list of v (denoted $\langle v, u \rangle$). We call $\langle v, u \rangle$ the dual of $\langle u, v \rangle$. These two representations are linked to each other. The weight of (u, v) is denoted $w(u, v)$. As computation proceeds, components will be formed which will be represented by supervertices. Internal and multiple edges will also be generated. An edge is an internal edge if it is incident to only one supervertex. Edge e_1 and edge e_2 are multiple edges if they are incident to the same two supervertices.

Our algorithm has $\log n$ stages. For the moment we assume that each stage takes constant time. Let B_i be the set of minimum spanning tree edges computed at the end of stage i. B_i induces a set of trees $F_i = \{T_1, T_2, ..., T_k\}$. We maintain the property that each T_j, $1 \leq j \leq k$, contains at least 2^i vertices. At stage $i+1$, the minimum external edge for each tree in $F'_i \subseteq F_i$ is computed and used to hook trees in F_i together to form larger trees. F'_i is a subset of F_i guaranteed to contain all trees in F_i of size $< 2^{i+1}$. Thus at the end of stage $i + 1$ each tree in F_{i+1} contains at least 2^{i+1} vertices. A tree in F_i of size $\geq 2^s$, $s > i$, is said grown for stage s.

The main problem in obtaining efficient algorithms is that internal and multiple edges will be generated as supervertices are formed. Therefore we have to perform compression repeatedly to remove internal and multiple edges. Compression is done by first sorting the edges so that multiple edges are consecutive

in the adjacency list, then removing internal and multiple edges and then packing the remaining edges to consecutive positions in the adjacency list. In order
to shorten the time for compression we take a partial list containing the (say
$2^s - 1$) minimum edges from the adjacency list. We form a graph where each
vertex has only a partial adjacency list containing $2^s - 1$ minimum edges from its
original adjacency list (in the case where the adjacency list of a vertex has less
than $2^s - 1$ edges then all the edges in the adjacency list will be in the partial
adjacency list). We call this partial list the active list of the vertex, and edges
adjacent to the vertex not in the active list are in the inactive list. Edges in
the active list are active edges while edges in the inactive list are inactive edges.
Note that it is possible that $\langle u, v \rangle$ is an active edge of u while $\langle v, u \rangle$ is an inactive
edge of v. In this case (u, v) is a half edge. We call $\langle u, v \rangle$ a half edge and $\langle v, u \rangle$
a dual half edge. When both $\langle u, v \rangle$ and $\langle v, u \rangle$ are active edges (u, v) is called a
full edge. When both $\langle u, v \rangle$ and $\langle v, u \rangle$ are inactive edges (u, v) is called a lost
edge. For each vertex v we define threshold $h(v)$. If all edges incident to v are
in active list then $h(v) = \infty$. Otherwise $h(v) = w(e_0)$, where e_0 is the external
edge of minimum weight in v's inactive list. An active list is clean if it does not
contain internal edges (full or half) and full multiple edges. It is possible that a
clean active list contains a full external edge e and several half edges which are
multiple of e.

Suppose that clean active list for each vertex (we use vertex instead of supervertex here because we will use supervertex for later stages) is constructed at
the end of stage i. Denote this set of active lists by A. Suppose that each active
list in A contains $2^s - 1$ edges (or all edges if the number of incident edges is no
more than $2^s - 1$). The essence of the following lemma is given in [14].

Lemma 1. *At any stage* $i + j$, $1 \leq j < s$, *for each supervertex* v *either the
minimum external edge of* v *is in a list in* A *or* v *contains at least* 2^s *vertices.*

Proof. Omitted.

The clean active list containing at least $2^s - 1$ edges (or all incident edges)
is called a $(2^s - 1)$-active-list.

Because of Lemma 1 we will maintain the following property for our algorithm:

Property 1: In stage $i + j$, $1 \leq j < s$, only edges in A are used for hooking.

In fact we can strengthen Property 1. For any $j < s$ let T be a tree at stage
$i + j$. Let $u \in T$ be the vertex with the smallest threshold among vertices in T.
Let S be the set of active edges incident to T with weight $< h(u)$. Then either
the minimum external edge is in S or T contains at least 2^s vertices. Thus we
maintain

Property 1′ : In stage $i + j$, $1 \leq j < s$, only edges in S are used for hooking.

When a tree T uses a half edge e to hook to another tree T', e is the minimum
external edge of T while $w(e)$ is no less than $h(u) = \min\{h(w)|w \in T'\}$. Because
of Property 1′ we maintain the following property for our algorithm.

Property 2: If a tree T uses a half edge to hook to another tree, then T and
all vertices and edges incident to T will be labeled as inactive.

If for a particular j for each tree T computed at stage $i + j$ we build a new clean active list L_T which contains the minimum $2^{s-j} - 1$ edges e incident to T with $w(e) < \min\{h(u)|u \in T\}$. It is obvious that edges in L_T are in A. We also update the threshold for each new active list constructed. Let us use A_1 to denote the set of all (edges in) L_T's constructed for supervertices. By Lemma 1 at any stage $i + j + k$ with $j + k < s$ for each supervertex v either the minimum external edge is in A_1 or v contains at least 2^s vertices. Property 1 is refined to read: In stage $i + j + k$, $j + k < s$, only edges in A_1 are used for hooking.

Let us consider the situation that edges in lists in A are compressed at the beginning of stage $i + j + 1$, $j \le s$. If no more than 2^s vertices are hooked together all by full edges to form a supervertex, then the active lists of these vertices can be merged into one list L and the compression of the edges on L takes $O(s)$ time because there are no more than 2^{2s} edges to be sorted. In other words, if we expend $O(s)$ time and cannot finish compression then the tree we computed contains more than 2^s vertices. In this case the tree is grown for stage $i + s$ and we mark all edges incident on this tree inactive.

It is possible that half edges are used for hooking. Because of Property 1 it is not possible that dual half edges or lost edges are used for hooking.

We need the notion of root component. When every vertex uses the minimum external edge for hooking. We obtain a set of tree-loops. The loop in each tree-loop contains two edges which are dual of each other. By breaking the loop and choosing an arbitrary vertex in the loop to become the root we obtain a rooted tree. Initially each vertex is its own root component. When minimum external edges are used to hook vertices together to form a rooted tree the fully connected component is a maximal set of vertices hooked together by full edges. The root component of the tree is the fully connected component containing the root of the tree. A nonroot component is a fully connected component not containing the root of the tree. We also say that the active edges incident to vertices in a fully connected component are in that component.

For a set S of vertices we let $h(S) = \min\{h(u)|u \in S\}$.

Lemma 2. *Let T' be the root component of T. Then $h(T') = h(T)$. Also the minimum external edges of T with weight $< h(T)$ are all in the root component.*

Proof. Omitted.

By Lemma 2 our algorithm uses only minimum external edges in the root component for hooking.

When half edges as well as full edges are used for hooking, the tree T we obtained can be decomposed into a set S of fully connected components. By apply compression the vertices in each nonroot component will find out that either they are in a fully connected component containing at least 2^s vertices when compression cannot finish in $O(s)$ time, or they hook to an inactive vertex, or they hook to another fully connected component by a half edge. In either of these cases all vertices and edges in the nonroot component will be labeled as inactive. The root component in S does not hook to another fully connected component in S although other nonroot component in S may hook to the root component. If we cannot finish compression in $O(s)$ time for active edges in the root component then all edges incident to the root component will be marked as inactive. In this case the tree is grown and contains at least 2^s vertices. If we finish compression in $O(s)$ time then all full internal and full multiple edges of the root component will be removed. In this case we build the active list for the supervertex v representing T as follows. First compress edges in the root component C. We then compute $h(C)$ and then all active edges of the root component whose weight no less than $h(C)$ will be moved to inactive list. Among the remaining edges in the root component we then pick the minimum $2^{s-j} - 1$ edges (note that all full internal and full multiple edges have already been removed) and form the active list of v. We also update the threshold of the new active list. This process is called the active list construction.

Lemma 3. *The active list of v constructed does not contain half internal edges.*

Proof. Omitted.

Thus the active list constructed does not contain full or half internal edges and does not contain full multiple edges. That is, the active list is clean.

We denote the set of active lists constructed by A_1. By Lemma 1 at stage $i + j + k$ either the minimum external edge of a supervertex v is in A_1 or v contains at least 2^s vertices.

Refering to Lemmas 1 and 3 we say that A (A_1) can be used to grow supervertices to contain 2^s vertices.

If in the active list construction we keep only the minimum $2^{s-j-t} - 1$ non-internal and non-full multiple edges in each active list constructed. Then these new active lists can be used to grow supervertices to contain 2^{s-t} vertices.

3   An EREW Algorithm with $O(\log n)$ Time and $O(n \log \log n)$ Processors

We arrange the edge compression into $(1/2) \log \log n$ levels. Each level is a process. When we say process i we mean level i and vice versa. Each level is divided into intervals. An interval at level i contains $\log n/4^i$ consecutive stages. Therefore level 1 has 4 intervals, level 2 has 16 intervals and so on. A run of process i has 5 intervals of level i (except for the last run). A run of process i starts at the point where an interval of level $i - 1$ begins. Therefore a run of process i always starts at stage $(j - 1) \log n/4^{i-1} + 1$, $j = 1, 2, 3....$ We call the run of process i which starts at stage $(j - 1) \log n/4^{i-1} + 1$ the j-th run of process i. We note that the last run has only 4 intervals. Note that because each run has 5 intervals, the last interval of the j-th run at level i overlaps with the first interval of the $(j + 1)$-th run at level i. During this overlapped interval two runs of the process execute in parallel. (We may think of each process contains two subprocesses which can run in parallel). The first interval of a run (except for the first run) of process i is called the intialization interval of the run. The remaining intervals of a run are called the effective intervals of the run. The first interval of the first run is also an effective interval. Note that effective intervals for all runs at a level do not overlap. These effective intervals cover the overall $\log n$ stages.

Our intention is to let a run at level i get $(2^{12 \log n/4^i})$-active-lists from the active lists at level $i-1$ at the beginning of the run and then repeatedly compress the active lists for 5 times in 5 intervals. There is only one run at level 1. That run gets the input graph at the beginning of the run. At the beginning of each interval the run updates the hooking among active lists with the minimum spanning tree edges found in the stages covered by the previous interval (i.e. the run accomplishes the hooking of supervertices by the minimum spanning tree edges found). Because an interval at level i has $\log n/4^i$ stages the $(2^{12 \log n/4^i})$-active-lists can be compressed in an interval if no more than $2^{6 \log n/4^i}$ vertices are hooked together by full edges into the root component of a supervertex. This is to say that if we cannot finish compressing within the interval then the supervertex has already contained enough vertices (the supervertex is already grown for the run) and we need not do compressing for the remaining intervals of the run. In this case the compressing is done in lower numbered levels. Because a run at level i compresses the edges within an interval, the remaining edges after compressing contains newly generated internal and multiple edges within the interval. If the interval starts at stage t and finishes at stage $t + \log n/4^i - 1$, then the active lists constructed is clean at stage t but it is not clean at stage $t + \log n/4^i$. We say that the run has a lag of $\log n/4^i$ stages. At the beginning of the first interval of a run at level i, the run constructs a set H_1 of $(2^{12 \log n/4^i})$-active-lists from the set H_2 of active lists at the end of an interval I at level $i-1$. Because each of these active list at level i is obtained from level $i-1$, it has a lag of $\log n/4^{i-1}$ stages. That is, vertices in H_1 have already hooked together to form supervertices such that each supervertex contains $2^{\log n/4^{i-1}}$ vertices. Since H_2 is clean at the beginning of interval I, H_1 is also clean at the beginning of interval I. If interval I starts from stage t then active lists in H_1 can be used to grow supervertices to contain at least $2^{t+12 \log n/4^{i-1}}$ vertices. That is, edges in H_1 can be used until stage $t + 12 \log n/4^{i-1}$. Because H_1 is obtained at the end of interval I which contains $\log n/4^{i-1}$ stages, the active lists in H_1 can be used for another $8 \log n/4^i$ stages. We construct new $(2^{8 \log n/4^i})$-active-lists from the compressed edges by the end of the first interval for the run at level i. This set of active lists is the input to the second interval. At the end of the first interval, the run has a lag of only $\log n/4^i$ stages. By the same reason we construct $(2^{(8-j+1) \log n/4^i})$-active-lists by the end of the j-th interval (which is the input to the $(j+1)$-th interval), $j = 2,3,4$. At the end of the 5-th interval the remaining edges are discarded. Note that the first interval of the next run at the same level overlaps with the 5-th interval of the current run. Note that at the beginning of the first(initialization) interval of a run at level i the active lists are obtained from level $i-1$, and it therefore has a lag of $\log n/4^{i-1}$ stages. These active lists are not passed to level $i+1$ because of their lag. At the beginning of each effective interval the active lists are passed to level $i+1$ (by picking some minimum edges to make smaller active lists for level $i+1$). A run at level i does not pass its active lists to level $i+1$ at the beginning of its initialization interval.

At that point the previous run at level i is at the beginning of its 5-th interval and it passes the active lists to level $i + 1$. Because a run at level i has a lag of $\log n/4^i$ stages, a run at level $(1/2)\log\log n$ has a lag of only 1 stage. That is, we can obtain the minimum external edge for a supervertex from level $(1/2)\log\log n$ in constant time. Because each stage grows the supervertex to contain at least two vertices, after $\log n$ stages the minimum spanning tree for the input graph is computed.

Because there are $O(\log\log n)$ processes, the processors used is $(m+n)\log\log n$. The algorithm can be run on the EREW PRAM. The only place where we need avoid concurrent read is to obtain hooking edges from the minimum spanning trees found so far for each level. Because level i consults the minimum spanning trees (obtaining the hooking edges) $c4^i$ times, c is a constant. Thus the number of consults forms a geometric series. The bottom level consults the minimum spanning tree $c\log n$ times. Therefore if we allow constant time for one consult the total time for all consults will be $O(\log n)$ although some stages will now take more than constant time (when there are more than constant number of concults in the stage).

Lemma 4. *Property 1 is maintained in our algorithm. That is, edges used for hooking in an interval are in the active lists of the interval.*

Proof. Omitted.

Because the way we structured our algorithm Property 1$'$ and Property 2 are also maintained at each level in our algorithm.

Theorem 1. *There is an EREW minimum spanning tree algorithm with time complexity $O(\log n)$ using $(m+n)\log\log n$ processors (or $O((m+n)\log n\log\log n)$ operations).*

4   Reducing the Operation (Processor) Complexity

We can assume that each adjacency list contains at most m/n edges in the input graph. We can always achieve this by converting the input graph.

We use several phases to compute the minimum spanning trees. Because each adjacency list has only m/n edges, in the first phase we need only build levels $(1/2)\log\log n$ down to level i_1, where $2^{12\log n/4^{i_1}} = m/n$ or $i_1 = (1/2)\log\dfrac{12\log n}{\log(m/n)}$

If we were to build lower numbered levels, they would only duplicate the compressing process at level i_1. Because we build only to level i_1 and because the number of edges in the active lists at each level forms a geometric series, the number of processors used is $O(m + n)$. Also because we build only to level i_1, in the first phase we need to execute only 4 intervals of level i_1. However, after the first phase there will be only $n/2^{4\log n/4^{i_1}} = n/(m/n)^{1/3}$ supervertices and each supervertex has $(m/n)^{4/3}$ edges. We then execute the second phase. We rebuild each level, This time we build levels $(1/2)\log\log(n/(m/n)^{1/3})$ down to i_2, where $i_2 = (1/2) \log \dfrac{12 \log(n/(m/n)^{1/3})}{\log(m/n)^{4/3}}$. Again because the number of edges at each level forms a geometric series, we need only $O(m + n)$ processors. We then execute 4 intervals at level i_2 in this phase. We then execute the third phase, and so on. In each next phase we build up to lower numbered levels. The number of processors used is always $O(m + n)$ in every phase. After we executed $\log n$ stages, the minimum spanning trees are computed. It is not difficult to see that for the sparsest graphs where $m = O(n)$ we need at most $O(\log \log n)$ phases. Therefore we have:

Theorem 2. *There is an EREW minimum spanning tree algorithm with time complexity $O(\log n)$ using $O((m + n) \log n)$ operations.*

For dense graphs with $m = \Omega(n^2)$ we can obtain an optimal algorithm by using a selection algorithm. By adapting Cole's parallel selection algorithm[8] we can show that the selection of any minimum i items from n items can be done in $O(\log n)$ time with $O(n)$ operations on the EREW PRAM provided that $i \leq n/\log n$. The details of this adaptation is omitted here.

We select $\log n$ minimum edges from the adjacency list of each vertex. We build levels $(1/2) \log \log n$ down to i_3, where $i_3 = (1/2)(\log \log n - \log \log \log n)$, using selected edges. We then execute 1 interval at level i_3. After that there will be $n/\log n$ supervertices left. We then renumber vertices such that vertices being hooked into one supervertex are numbered consecutively. Edges are renumbered according to the numbers given to the vertices. This renumbering enables us to move internal edges and multiple edges incident to a supervertex into consecutive memory locations and therefore they can be removed. Note that renumbering can be done by prefix computation and it takes only $O(\log n)$ time and $O(n^2)$ operations. We therefore avoid the sorting operation. After internal and multiple edges are removed there are $n/\log n$ supervertices and $O((n/\log n)^2)$ edges remaining. Now we can use Theorem 2 to compute the minimum spanning trees in additional $O(\log n)$ time and $O((n/\log n)^2 \log n) = O(n^2/\log n)$ operations.

Theorem 3. *There is an EREW minimum spanning tree algorithm with $O(\log n)$ time using $O(n^2)$ operations.*

We now explain how to reduce further the number of processors in Theorem 2. There are two places where we used $O((m + n) \log n)$ operations. The first is merging the active lists of several vertices into one active list of the supervertex after hooking. The second is the sorting used to sort internal and multiple edges together and used to select minimum edges in the active list.

One way to merge the active lists is to use pointer jumping as this is used in [5] [14]. Pointer jumping will need $O((m+n) \log n)$ operations. We do not do pointer jumping directly. We store each active list in an array such that edges in an active list are stored in consecutive memory locations. In merging active lists, we first construct a tree T of vertices (instead of the active lists) representing the hooking from the hooking of the active lists. If each adjacecy list contains s edges and t vertices are hooked into a supervertex, the tree T we build has only those t vertices instead of ts edges. T can be obtained by a prefix computation on each active list. For the s-active-lists at a level we need to execute only $O(\log s)$ steps (because each interval at that level has only $O(\log s)$ stages) for pointer jumping along the Euler tour of T[16]. The number of operations involved is $O(ts + t \log s) = O(ts)$.

Sorting is used to sort edges such that internal edges and multiple edges incident to two supervertices are moved into consecutive locations in memory for removal. Sorting is also used to find the minimum (say s) edges so that the active list can be constructed. Here we use integer sorting algorithms to sort internal and multiple edges. This is possible because each edge is represented by (a, b), where a and b are integers in the range $\{1, 2, ..., n\}$ representing vertices. On the CRCW PRAM n integers in the range $\{0, 1, 2, ..., m - 1\}$ can be sorted in $O(\log n)$ time with $O(n \log \log n)$ operations[2,12]. For a very sparse input graph, we use $O(\log \log n)$ phases. If the b-th phase builds down to level i, then there are $O(m)$ edges at level i at the beginning of the phase (the number of edges at higher numbered levels are geometrically decreasing and therefore is dominated by the edges at level i). To sort all these edges it takes $O(\log n/4^i)$ time (in one interval) and $O((m + n) \log \log n)$ operations. For $O(\log \log n)$ phases the time is $O(\log n)$ and the number of operations is $O((m + n)(\log \log n)^2)$. We use the following method to further reduce the number of operations. In the sorting each integer incurs $O(\log \log n)$ operations. Before sorting we use concurrent write to determine for each edge e whether e is the unique edge incident to two supervertices s_1 and s_2. If e is the unique edge then e does not participate in sorting and therefore e incurs only a constant number of operations. If e is not the unique edge (there are other edges incident to s_1 and s_2), then e participates in the sorting and therefore e incurs $O(\log \log n)$ operations. Because multiple edges are removed (except for the minimum edge among the mutiple) after sorting, each removed edge incurs $O(\log \log n)$ operations. We can associate the $O(\log \log n)$ operation incurred by the minimum edge among the multiple with a multiple edge which is removed after sorting and therefore the minimum edge now incurs only constant number of operations. Thus if e is not removed it incurs a constant number of operations in one phase. For $O(\log \log n)$ phases e incurs $O(\log \log n)$ operations. Each edge can be removed only once, therefore the total number of operations becomes $O((m + n) \log \log n)$. Note that processors have to be reallocated because there are two sets of edges, a set S_1 of unique edges and a set S_2 of multiple edges and we have to reallocate processors associated with S_1 to work on S_2. We cannot use standard prefix computation to reallocate processors because such reallocatoin need $O(\log n)$ time while at level i we can expend only $O(\log n/4^i)$ time. However, processor reallocation can be done in $O(\log \log n)$ time and $O(m + n)$ operations[10] on the CRCW PRAM. Thus for level $i < (1/2)(\log \log n - \log \log \log n)$ we can use integer sorting to remove interval and multiple edges because each interval at such level has at least $\Omega(\log \log n)$ stages. For level $i \geq (1/2)(\log \log n - \log \log \log n)$ we use comparison sorting[1,7] which takes $O(m \log \log n)$ operations because each interval at such level has no more than $O(\log \log n)$ stages.

On the EREW PARM n integers in the range $\{0, 1, 2, ..., m - 1\}$ can be sorted in $O(\log n)$ time with $O(n\sqrt{\log n})$ operations[11]. Therefore the operation complexity for removing internal and multiple edges becomes $O((m + n)\sqrt{\log n})$ while the time complexity is kept at $O(\log n)$.

We can also show that the sorting used for selecting minimum edges to construct active lists can be done in $O(m + n)$ operations by sorting on a fraction of the edges in the adjacency lists. The details are omitted here.

The ideas explain above gives us the following theorem.

Theorem 4. *There is an EREW minimum spanning tree algorithm with time complexity $O(\log n)$ with $O((m + n)\sqrt{\log n})$ operations. Also there is an CRCW minimum spanning tree algorithm with time complexity $O(\log n)$ with $O((m + n)\log\log n)$ operations.*

References

1. M. Ajtia, J. Komlós, E. Szemerédi. Sorting in $c \log n$ parallel steps. *Combinatorica*, 3, pp. 1-19(1983).
2. A. Andersson, T. Hagerup, S. Nilsson, R. Raman. Sorting in linear time? *Proc. 1995 Symposium on Theory of Computing*, 427-436(1995).
3. B. Awerbuch and Y. Shiloach. New connectivity and MSF algorithms for shuffle-exchange network and PRAM. *IEEE Trans. on Computers*, C-36, 1258-1263, 1987.
4. F. Y. Chin. J. Lam and I-N. Chen. Efficient parallel algorithms for some graph problems. *Comm. ACM*, 25(1982), pp. 659-665.
5. K. W. Chong. Finding minimum spanning trees on the EREW PRAM. *Proc. 1996 International Computer Symposium (ICS'96)*, Taiwan, 1996, pp. 7-14.
6. K. W. Chong, Y. Han, T. W. Lam. On the parallel time complexity of undirected connectivity and minimum spanning trees. *Proc. 1999 Tenth Annual ACM-SIAM Symposium on Discrete Algorithms (SODA'99)*, Baltimore, Maryland, 225-234(1999).
7. R. Cole. Parallel merge sort. *SIAM J. Comput.*, 17(1988), pp. 770-785.
8. R. Cole. An optimally efficient selection algorithm. *Information Processing Letters*, **26**, 295-299(1987/88).
9. R. Cole, P. N. Klein, R. E. Tarjan. Finding minimum spanning forests in logarithmic time and linear work using random sampling, *SPAA'96*, pp. 243-250.
10. T. Goldberg and U. Zwick. Optimal deterministic approxiamate parallel prefix sums and their applications. *Proc. 3rd Israel Symposium on Theory and Computing Systems*, 220-228(1995).
11. Y. Han, X. Shen. Parallel integer sorting is more efficient than parallel comparison sorting on exclusive write PRAMs. *Proc. 1999 Tenth ACM-SIAM Symposium on Discrete Algorithms (SODA'99)*, Baltimore, Maryland, 419-428(1999).
12. Y. Han, X. Shen. Conservative algorithms for parallel and sequential integer sorting. *Proc. 1995 International Computing and Combinatorics Conference (COCOON'95)*, Lecture Notes in Computer Science **959**, 324-333(August, 1995).
13. D. S. Hirshberg, A. K. Chandra and D. V. Sarwate. Computing connected components on parallel computers. *Comm. ACM*, 22(1979), pp. 461-464.
14. D. B. Johnson and P. Metaxas. A parallel algorithm for computing minimum spanning trees. *J. of Algorithms*, 19, 383-401(1995).
15. C. K. Poon, V. Ramachandran. A randomized linear work EREW PRAM algorithm to find a minimum spanning forest. *Proc. 8th Annual International Symposium on Algorithms and Computation*, 1997, pp. 212-222.
16. R.E. Tarjan and U. Vishkin. An efficient parallel biconnectivity algorithm. *SIAM J. Comput.*, 14, pp. 862-874(1985).
17. C.D. Zaroliagis. Simple and work-efficient parallel algorithms for the minimum spanning tree problem. *Parallel Processing Letters*, Vol. 7, No. 1, 25-37(1997).

A Combinatorial Approach to Performance Analysis of a Shared-Memory Multiprocessor *

Sajal K. Das[1], Bhabani P. Sinha[2], and Rajarshi Chaudhuri[2]

[1] Department of Computer Science, University of North Texas, TX 76203, USA
[2] Advanced Computing Unit, Indian Statistical Institute, Calcutta 700 035, INDIA

Abstract. A unified approach is proposed for performance analysis of an $N \times M$ shared-memory multiprocessor system, consisting of N processors and M memory modules, each of which may be 'hot' and/or 'favorite'. Processors are allowed to have non-uniform memory access patterns and unsatisfied requests are queued up in the buffers of the corresponding modules. The system performance is measured in terms of the effective *bandwidth*, which is the average number of busy memory modules in one cycle. Our analytical approach, based on the combinatorial arguments as well as queuing models, estimates the bandwidth with good accuracy for arbitrary values of N and M. These estimations tally very well with the simulation results. Since the presence of a hot module almost always leads to an accumulation of memory requests in its direction and thus deteriorates the system performance, one may expect that the hotness should be spread over as many (say, K) modules as possible. However, simulation results showed an upper bound on K beyond which the bandwidth either drops or saturates. From the approximate queuing model, we derive those saturation values in terms of N, M and p_h (the probability of accessing a hot module by a processor).

1 Introduction

An $N \times M$ shared-memory multiprocessor architecture consists of a set of N processors and a set of M memory modules. The underlying processor-memory communication scheme plays a vital role in characterizing the performance of such a system. Memory contentions occur when more than one processor generate requests for a common module in the same memory cycle. By some arbitration policy (usually random), one of the conflicting requests is granted, while others are either rejected (*non-buffered* system), or queued up for future processing (*buffered* system). Contentions degrade the acceptance rate of memory requests and hence the overall system performance. One measure of the performance is the effective memory *bandwidth* (BW), i.e., the number of busy modules in a memory cycle. Other metrics include mean waiting time for a memory request and expected queue length. In the past two decades, the performance

* This work is partially supported by Texas Advanced Research Program grant TARP-97-003594-013. E-mail: das@cs.unt.edu, bhabani@isical.ac.in

of processor-memory interconnections has been studied widely with the help of various analytical models, some of which are exact [1,18,19,20] while others are probabilistic [2,4,8,14]. These models mostly use Markov chains [1,4,18] or queuing networks [2,4,11,14].

The memory access pattern of the processors can be uniform or non-uniform. A *uniform reference model* is one in which all the processors have the same probability of accessing any module [1,9,10,14,19]. However, many practical applications exhibit *non-uniform* memory accesses [2,6,20] due to, for example, the *locality of reference* in the computation requirements, or the need for accessing *shared variables and codes.*

The non-uniform memory access patterns can again be intra-cycle or inter-cycle [17]. There are four types of *intra-cycle* non-uniformity.

- **Type I** [2]: each processor may have *favorite* memory module(s) which it accesses more often than the other modules.
- **Type II** [4,15]: all the processors may access a subset of modules more often than the others. These are called the *hot spots*, due to Pfister and Norton [15]. Physically, this captures the effect of sharing global variables or primitives like barrier synchronization.
- **Type III** [4,18]: the memory module accessed by a processor in the current cycle is its favorite memory in the next cycle and accessed with a probability greater than $\frac{1}{M}$. This captures the *locality* of references.
- **Type IV** [13]: this *spatial* non-uniformity implies the events that a particular memory module has at least one request, are not independent of each other.

On the other hand, the *inter-cycle* non-uniformity occurs when the event that a processor's requests in the current cycle, is dependent on whether its request was satisfied in the previous cycle.

Classifying each module as one of the types – H (Hot), F (Favorite), NH (Non-Hot), NF (Non-Favorite), a shared-memory machine can have any one of the four combinations of memory modules: $\{(H, NH) \times (F, NF)\} = \{(NH, NF), (H, NF), (NH, F), (H, F)\}$. Multiprocessors of the type (NH, NF) have been analyzed in [1,19]. The concept of hot spots was introduced in [15] in the context of analyzing contentions in the multistage interconnection networks in order to avoid *tree saturation,* a phenomenon occurring in systems of the type (H, NF). The analysis of (NH, F) systems has been proposed in [2,18].

1.1   Our Contributions

In this paper, we take a unified approach to solve the problem of memory contentions for a buffered, multiprocessor system in which all the four combinations of memory modules are possible. We assume that the processor-memory interconnection mechanism does not create any bottleneck, i.e., a request has successfully passed through it. Starting from a combinatorial view point, we arrive at closed-form solutions for memory bandwidth for a general $N \times M$ multiprocessor. To evaluate the average delay for all the memory requests made by each processor, with or without of hot spots (but no favorite memory), we use an approximate queuing model to derive another closed-form expression for bandwidth.

Intuitively, the presence of multiple hot spots should lead to an improvement in the overall system performance in the (H, NF) environment. Our analytical results show that the bandwidth increases with the increase in the number, K, of hot spots up to a certain value (say, K_{sat}), beyond which the bandwidth either drops or *saturates*. For $N \geq M$ (which is more common in practice), the saturation is given by $K_{sat} = M \left\lceil \frac{M + 2p_h N}{2(M+N)} \right\rceil$, where p_h is the probability with which a processor accesses the hot modules. These results closely tally the simulation results of [4]. In the (NH, F) and (H, F) environments, we observe that as the probability for accessing the favorite memory increases, the system bandwidth tends to increase which also agrees with the observations in [2].

2   The Model and Notations

- The system architecture consists of N identical processors and M globally accessible identical memory modules.
- Each module has a buffer for queuing up unsatisfied memory requests.
- Each processor generates memory requests randomly and independently. If multiple processors make simultaneous requests to the same module, only one of them is serviced and the remaining requests are queued up.
- There is no contention in the interconnection mechanism.
- The system is synchronous in the sense that the processors make memory requests at the beginning of the cycle, with probability p_m.
- The *cycle time* of all the memory modules is β (a constant).
- All the K hot modules in the system are identical and the memory reference is uniform among them.
- Let p_h and p_f be the *hot* and *favorite* memory request rates of a processor.
- Let λ_h (resp. λ_{nh}) be the total arrival rate of requests to each hot (resp. non-hot) memory module queue. Also, let μ be the total service rate of requests at memory queues.
- The reference pattern among the 'non-hot, non-favorite' memory modules are also uniform.

3   A Combinatorial Analysis of Bandwidth

For simplicity, let us assume that $p_m = 1$ such that the number of requests generated at the beginning of a memory cycle is N. This does not loose the generality, since N will be replaced with $N' = p_m N$ where $0 \leq p_m \leq 1$. Then the bandwidth (BW) is defined by

$$BW = \sum_{k=1}^{M} k * Pr(k) \tag{1}$$

where k is the number of *busy* modules in a memory cycle and $Pr(k)$ is the probability that they are busy. First we analyze the (NH, NF) class.

3.1   (NH, NF) class

We can choose k out of M memory modules in $\binom{M}{k}$ ways. The number of ways to distribute N requests into k modules such that each receives at least one request, is the same as the number of ways of distributing N different objects into k different cells with no cell empty. This is given by $k!S(N,k)$, where S is the *Stirling number of the second kind* [5,16]. In the (NH, NF) class, each processor has the same probability of $\frac{1}{M}$ for requesting a module. So,

$$Pr(k) = \binom{M}{k} k! S(N,k) \frac{1}{M^N} \tag{2}$$

3.2   (H, NF) class

Let h be the number of *busy hot* modules in a particular memory cycle and $k - h$ be the number of busy non-hot modules. Let i out of N requests be directed to the hot modules while the remaining $N - i$ requests to the non-hot modules. There are $\binom{N}{i}$ ways to choose these i requests for the hot modules and $h!\binom{K}{h}S(i,h)$ ways to distribute i requests into the total of K hot modules among which only h of them are busy. Similarly, the remaining $N - i$ requests can be distributed in $(k - h)! \binom{M-K}{k-h} S(N-i, k-h)$ ways to the $M - K$ non-hot modules among which $k - h$ modules are busy. For each processor, the probability for requesting a hot module is $\frac{p_h}{K}$ and that for a non-hot module is $\frac{1-p_h}{M-K}$. Thus,

$$Pr(k) = \sum_{h=0}^{\min\{k,K\}} \sum_{i=0}^{N-k+h} \binom{N}{i}\binom{K}{h} h! \times S(i,h) \binom{M-K}{k-h}(k-h)!$$

$$\times S(N-i, k-h) \left(\frac{p_h}{K}\right)^i \left(\frac{1-p_h}{M-K}\right)^{N-i} \tag{3}$$

3.3   (NH, F) class

In this case, each processor accesses its favorite memory with a probability p_f, and a non-favorite memory with a probability $p_{nf} = \frac{1-p_f}{M-1}$. We also assume that any particular module is favorite to a single processor only. Thus, for $N \geq M$, only M processors have favorite memory modules while the remaining $N - M$ processors have a uniform access pattern. Let $p(N, k, f)$ denote the probability that out of the N requests, f are favorite memory requests and k is the total number of busy modules. Then $p(N, k, f)$ can be determined from the *Principle of Inclusion and Exclusion* [16]. If we fix the position of f requests then the total number, Γ_f, of all such selections, without any restrictions on the placement of the remaining $N - f$ requests, satisfies the following relation:

$$k!S(N,k) * \Gamma_f = \binom{k}{f} \sum_{j=0}^{\min\{N-k,f\}} (k - f + j)! \, S(N - f, k - f + j).$$

Therefore, $p(N, k, f) = \sum_{l=f}^{k}(-1)^{l-f}\binom{l}{l-f} * \Gamma_l$. Since f can vary from 0 to k, the probability that k modules are busy in a memory cycle is given by

$$Pr(k) = \binom{M}{k} k! \, S(N, k) \sum_{f=0}^{k} p(N, k, f)(p_f)^f (p_{nf})^{M-f} \frac{1}{M^{N-M}} \qquad (4)$$

Equation (4) implies $BW = 1$ if $p_f = 1$. This is justified because in that case, all the M modules are busy due to the M distinct favorite requests.

3.4   (H, F) class

In this case, only $\mathcal{F} = M - K$ processors have favorite memory modules and the remaining $N - \mathcal{F}$ processors have no favorite memory. Let i processors generate requests for the hot modules. Also, let j out of these i requests come from the group $\mathcal{F}$ of processors and the remaining requests from the other processors. For the group with favorite memory, the probability for accessing the favorite modules is $(1 - p_h)p_f$ and that for accessing any 'non-hot, non-favorite' module is $\frac{(1-p_h)(1-p_f)}{(\mathcal{F}-1)}$. For the processors with no favorite modules, the probability for accessing any 'non-hot' module is $\frac{1-p_h}{\mathcal{F}}$. Hence,

$$Pr(k) = \sum_{h=0}^{K} \sum_{i=0}^{N-k+h} \sum_{j=\max\{i-N+\mathcal{F},0\}}^{\min\{\mathcal{F},i\}} \binom{\mathcal{F}}{j}\binom{N-\mathcal{F}}{i-j}\binom{K}{h} h! S(i,h) \binom{\mathcal{F}}{k-h}(k-h)!$$

$$S(N-i, k-h) \sum_{f=0}^{\min\{k-h,\mathcal{F}-j\}} p(N-i, k-h, f)(\alpha_1)^i(\alpha_2)^f(\alpha_3)^{\mathcal{F}-j-f}(\alpha_4)^{N-\mathcal{F}-i+j}$$

$$(5)$$

where $\alpha_1 = \frac{p_h}{K}$; $\alpha_2 = (1 - p_h)p_f$; $\alpha_3 = \frac{(1-p_h)(1-p_f)}{(\mathcal{F}-1)}$; $\alpha_4 = \frac{(1-p_h)}{\mathcal{F}}$.

4   An Approximate Queuing Model

This section analyzes the (NH, NF) and (H, NF) classes. The aggregate queuing model used in our analysis is described in Figure 1(a), assuming all the arrival processes are Poisson. In this case, the expression for the bandwidth is given by $BW = N\lambda_m$ where N is the number of processors and λ_m is the total arrival rate of requests to the aggregate memory queue. Additionally, we use a flow-equivalent aggregation [3] to find the average delay of the memories as a function of the request rate. For simplicity, once again we assume $p_m = 1$.

4.1   (NH, NF) class

The decomposed queuing model of memory is shown in Figure 1(b), which is drawn from the viewpoint of a single processor. All memory requests are uniformly distributed over all modules. The load-dependent *average memory delay*

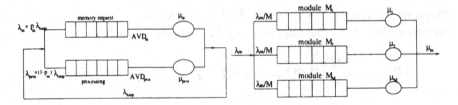

Fig. 1. (a) Aggregate and (b) Decomposed queuing model

as seen by each processor can be written as $AVD_m = \sum_{j=0}^{\infty} Pr(j) * delay(j)$ where $Pr(j)$ is the probability of j requests being in the queue at any time, and $delay(j)$ is the associated delay until completion of service for newly-arriving requests. Recalling that $\rho_i = \frac{\lambda_i}{\mu_i}$ is the *utilization factor* for each module M_i, we obtain $Pr(j) = (1 - \rho_i)\rho_i^j$ [7] and $delay(j) = (j + 1)\beta$, where β is the memory cycle time. This yields $AVD_m = \sum_{j=0}^{\infty} (1 - \rho_i)\rho_i^j (j + 1)\beta$. Since $\mu_i = \frac{1}{\beta}$, for $1 \le i \le M$, after simplification $AVD_m = \frac{\beta}{1 - \rho_i} = \frac{1}{\mu_i - \lambda_i}$.

Because of the symmetry among the modules and the fact that memory references are uniformly directed to each module, the arrival rate of requests to a module is given by $\lambda_i = \frac{N\lambda_m}{M}$. Thus,

$$AVD_m = \frac{M\beta}{M - N\lambda_m\beta}. \tag{6}$$

Referring to Figure 1, the total average loop delay can be approximated as

$$AVD_{loop} = p_m \times AVD_m + (1 - p_m) \times AVD_{proc}$$

and the loop rate is given by $\lambda_{loop} = \frac{1}{AVD_{loop}}$. Since $p_m = 1$,

$$\lambda_m = \frac{1}{AVD_m} \tag{7}$$

Solving Equations (6) and (7), we obtain $\lambda_m = \frac{M}{(M+N)\beta}$. A closed-form expression for the bandwidth is derived as

$$BW = \frac{MN}{(M + N)\beta} \tag{8}$$

4.2   (H, NF) class

The decomposed queuing model of K hot memory modules is shown in Figure 2. The model for the $M - K$ non-hot modules is exactly the same except the subscript 'h' is replaced with 'nh'. Again, this is from the viewpoint of a single processor. The modified request rate of a processor to the hot and non-hot modules are $p_h\lambda_m$ and $(1 - p_h)\lambda_m$, respectively. All requests directed to hot (or non-hot) modules are uniformly distributed over them.

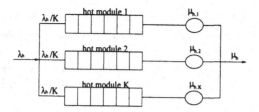

Fig. 2. Decomposed model of hot modules

The load-dependent average delay for the hot and non-hot modules (as seen by each processor) are obtained as

$$AVD_h = \frac{K\beta}{K - Np_h\beta\lambda_m} \quad \text{and} \quad AVD_{nh} = \frac{(M-K)\beta}{(M-K) - N(1-p_h)\beta\lambda_m} \quad (9)$$

The average delay for all memory requests made by each processor is then

$$AVD_m = p_h \times AVD_h + (1 - p_h) \times AVD_{nh} \quad (10)$$

Equations (7), (9), and (10) lead to the following quadratic equation in λ_m, which is the total arrival rate of requests to the aggregate memory queues.

$$a_2\lambda_m^2 - a_1\lambda_m + a_0 = 0 \quad (11)$$

where $a_2 = N(M+N)p_h(1-p_h)\beta^2$; $a_1 = (M-K)(N+K)p_h\beta + K(M+N-K)(1-p_h)\beta$; and $a_0 = K(M-K)$.

Solving Equation (11) we obtain $\lambda_m = \frac{a_1 - \sqrt{a_1^2 - 4a_2a_0}}{2a_2}$. The other solution of this equation is ignored because it does not fall within the interval $[0, 1]$. Hence a closed-form expression for the memory bandwidth is

$$BW = N \left[\frac{a_1 - \sqrt{a_1^2 - 4a_2a_0}}{2a_2} \right]. \quad (12)$$

Optimizing Equation (11) with respect to K, we derive

$$K_{sat} = M \left[\frac{M + 2p_hN}{2(M+N)} \right] \quad \text{for} \quad N \geq M \quad (13)$$

which is the saturation value beyond which the rate of increase of bandwidth with increase in K is not significant. As we shall see, this value K_{sat} is in agreement with the simulation results presented in [4,17].

5   Comparison of Analytical and Simulation Results

Simulation results reported in [2] and [4,17] are used to validate our analytic models. Table 1 computes the bandwidth values from the combinatorial as well as queuing models for an $N \times N$ multiprocessor system in the (NH, NF) environment. It also shows the simulation results from [2]. The bandwidth obtained from the combinatorial model (Equation 2) closely matches the simulation results, whereas, the queuing model (Equation 8) under-estimates it. The difference in performance between these two models is demonstrated in Figure 3.

Table 1. Memory bandwidth for an $N \times N$ system (NH, NF) with $\beta = 1$

N	$BW(Eqn.\ 2)$	$BW(Eqn.\ 8)$	$BW(Simulation)$ [2]
2	1.5	1.0	1.5
4	2.73	2.0	2.61
8	5.25	4.0	4.91
16	10.3	8.0	9.72

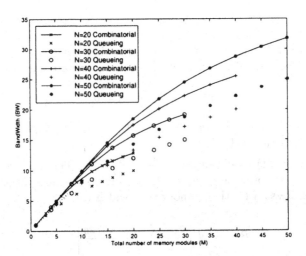

Fig. 3. Variation of bandwidth with M for class (NH, NF)

Table 2 shows the bandwidth computed by the queuing model (Equation 12) for an (H, NF) environment with varying K (the number of hot modules), and compares these results with the simulation results presented in [4,17], when $N = 50$, $M = 10$, and $p_h = 0.75$. We observe that the bandwidths obtained from our queuing analysis agrees very well with the simulation results. The plots from combinatorial and queuing models are depicted in Figure 4.

In Table 3 the saturation values of K are computed from Equation (13) for an (H, NF) environment, for $N = 50$, $p_h = 0.8$ and varying M. This table also shows the simulation results obtained in [4,17]. Our analytical expression for K_{sat} strongly agrees with the simulation results. In fact, for $p_h = 0.5$ we obtain $K_{sat} = \frac{M}{2}$, which is also highly expected.

Table 2. Bandwidth for an (H, NF) system with $\beta = 1$ $N = 50$, $M = 10$, $p_h = 0.75$

K	BW(Eqn. 12)	BW(Simulation) [4]
1	1.31	1.47
2	2.56	2.81
3	3.77	4.18
4	4.92	4.96
5	6.03	6.51
6	7.09	7.11

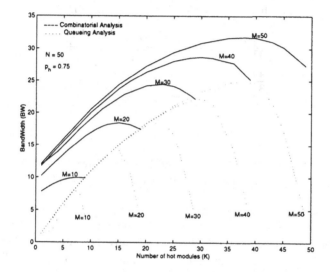

Fig. 4. Variation of bandwidth with K for class (H, NF), for fixed N and varying M

For an $N \times N$ multiprocessor of type (NH, F), Table 4 shows the bandwidth values obtained from our combinatorial analysis as well as simulation studies [2]. The bandwidth obtained from the analysis closely matches with the simulation results. Furthermore, with the increase in p_f, the overall bandwidth of the system increases which is also in accordance with [2].

Let us plot the of bandwidth, as derived in Equation (5), for an (H, F) environment. We observe from Figure 5(a) that for large $p_h \geq 0.75$, the bandwidth slowly increases with p_f; the reverse effect is noted for $0.5 \leq p_h \leq 0.7$. For $p_h < 0.5$, the bandwidth decreases rapidly with p_f. If p_f remains fixed and p_h varies, the bandwidth decreases as shown in Figure 5(b).

6 Conclusions

We have analyzed the performance of an (N, M) shared-memory multiprocessor system under various types of non-uniform memory access patterns. Analytical

Table 3. Saturation values of K in an (H, NF) Environment for $N = 50$, $p_h = 0.8$

M	K_{sat} (Eqn. 13)	K_{sat} (Simulation) [4]
10	7	6
20	14	14
30	21	24
40	27	32

Table 4. Memory bandwidth for an $N \times N$ system (NH, F) with $p_f = 0.8$

N	BW(Eqn. 4)	BW (Simulation) [2]
2	1.68	1.61
4	3.34	3.13
8	6.69	6.25
16	13.38	12.49

expressions for bandwidth and average memory delay are derived. The simulation results in [4,17] and [2] are found to tally well with our analysis.

The presence of multiple hot spots (say, K in number) in the system leads to an improvement in the overall system performance in the (H, NF) environment. However, for $N \geq M$, the rate of increase in bandwidth is not significant beyond a particular value of K, and this saturation value is given by $K_{sat} = \frac{M(M + 2p_h N)}{2(M+N)}$, where p_h is the probability with which a processor accesses the hot modules. In other words, in a system with hot spots, it is better to distribute the shared program and data into no more than K_{sat} memory modules.

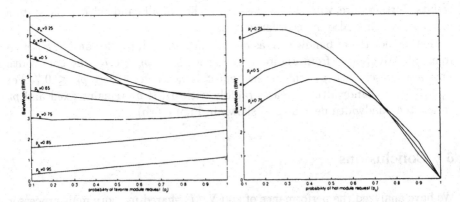

Fig. 5. Variation of bandwidth (a) with p_f and (b) with p_h, for class (H, F)

References

1. D. P. Bhandarkar, "Analysis of memory interference in multiprocessors," *IEEE Trans. Comput.*, vol c-24, pp. 897-908, Sept. 1975.
2. L. N. Bhuyan, "An analysis of processor-memory interconnection networks", *IEEE Trans. Comput.*, vol. C-34, pp. 279-283, Mar. 1985.
3. M. K. Chandy and C. Sauer, "Approximate methods for analyzing queuing network models of computer systems", *ACM Comput. Surv.*, vol. 10, pp. 281-317, 1978.
4. S. K. Das and S. K. Sen, "Analysis of Memory Interference in Buffered Multiprocessor Systems in the Presence of Hot Spots and Favorite Memories," *Proceedings of IEEE International Parallel Processing Symposium*, Hawaii, pp. 281-285, 1996.
5. R. L. Graham, D. E. Knuth, and O. Patashnik, *Concrete Mathematics*, Addison-Wesley, 2nd Ed., 1994.
6. H. S. Kim and A. L. Garcia, "Performance of buffered banyan networks under non-uniform traffic patterns," *IEEE Trans. Commun.*, vol 38, pp. 648-656, 1990.
7. L. Kleinrock, Queuing Systems. New York: John Wiley, 1975
8. M. Kumar and J. R. Jump, "Performance of unbuffered shuffle-exchange networks," *IEEE Trans. Comput.*, vol c-35, pp. 573-578, June 1986.
9. T. Lang., M. Valero, I. Alegre, "Bandwidth of crossbar and multiple-bus connections for multiprocessors", *IEEE Trans. Comput.*, vol. C-31, pp. 1227-1234, 1982.
10. Y. C. Liu and C. J. Jou, "Effective memory bandwidth and processor blocking probability in multiple-bus systems", *IEEE Trans. Comput.*, vol. C-36, pp. 761-764, June 1987.
11. M. A. Marsan and M. Gerla, "Markov models for multiple bus multiprocessor systems," *IEEE Trans. Comput.*, vol c-31, pp. 239-248, Mar. 1982.
12. T. N. Mudge and H. B. Al-Sadoun, "A semi-Markov model for the performance of multiple bus systems," *IEEE Trans. Comput.*, vol c-34, pp. 934-942, Oct. 1987.
13. T. N. Mudge, J. P. Hayes, G. D. Buzzard, and D. C. Winsor, "Analysis of multiple-bus interconnection networks", *J. Parallel and Distributed Computing*, vol. 3, pp. 328-343, 1986.
14. J. H. Patel, "Performance of processor-memory interconnections for multiprocessors," *IEEE Trans. Comput.*, vol c-30, pp. 771-780, Oct. 1981.
15. G. F. Pfister and V. A. Norton, "Hot Spot contention and combining in multistage interconnection networks", *IEEE Trans. Comput.*, vol. C-34, pp.943-948, Oct. 1985.
16. J. Riordan, *An Introduction to Combinatorial Analysis*, John Wiley, NY, 1958.
17. S. K. Sen, "Analysis of memory interference in buffered multiprocessor systems in presence of hot spots and favorite memories", *Master's Thesis*, Dept. of Comp. Sc., University of North Texas, Denton, TX, Aug. 1995.
18. A. S. Sethi and N. Deo, "Interference in multiprocessor systems with localized memory access probabilities", *IEEE Trans. Comput.*, vol. C-28, pp. 17-163, 1979.
19. C. E. Skinner and J. R. Asher, "Effects of storage contention on system performance", *IBM Syst. Journal*, vol. 8, pp. 319-333, 1969.
20. D. W. L. Yen, J. H. Patel and E. S. Davidson, "Memory interference in synchronous multiprocessor system", *IEEE Trans. Comput.*, vol. C-31, pp. 1116-1121, 1982.

A Fast Approximation Algorithm for TSP with Neighborhoods and Red-Blue Separation

Joachim Gudmundsson and Christos Levcopoulos

Department of Computer Science
Lund University, Box 118, S-221 00 Lund, Sweden

Abstract. In TSP with neighborhoods (TSPN) we are given a collection X of k polygonal regions, called *neighborhoods*, with totally n vertices, and we seek the shortest tour that visits each neighborhood. The Euclidean TSP is a special case of the TSPN problem, so TSPN is also NP-hard. In this paper we present a simple and fast algorithm that, given a start point, computes a TSPN tour of length $O(\log k)$ times the optimum in time $O(n+k \log k)$. When no start point is given we show how to compute a "good" start point in time $O(n^2 \log n)$, hence we obtain a logarithmic approximation algorithm that runs in time $O(n^2 \log n)$. We also present an algorithm which performs at least one of the following two tasks (which of these tasks is performed depends on the given input): (1) It outputs in time $O(n \log n)$ a TSPN tour of length $O(\log k)$ times the optimum. (2) It outputs a TSPN tour of length less than $(1+\epsilon)$ times the optimum in cubic time, where ϵ is an arbitrary real constant given as an optional parameter.

The results above are significant improvements, since the best previously known logarithmic approximation algorithm runs in $\Omega(n^5)$ time in the worst case.

1 Introduction

A salesman wants to meet potential buyers. Each buyer specifies a region, his *neighborhood*, within which he is willing to meet. The salesman wants to find a tour of shortest length that visits all of the buyers neighborhoods and finally returns to his initial departure point. The TSP with neighborhoods (TSPN) asks for the shortest tour that visits each of the neighborhoods. The problem generalizes the Euclidean Traveling Salesman Problem in which the areas specified by the buyers are single points, and consequently it is NP-hard [6,12] (for approximability results concerning TSP see [3]).

One can think of the TSPN as an instance of the "One-of-a-Set TSP" also known as the "Multiple Choice TSP", and the "Covering Salesman Problem". This problem is widely studied for its importance in several applications, particularly in communication network design and VLSI routing [9,14]. Arkin and Hassin [1] gave an $O(1)$-approximation algorithm for the special case in which the neighborhoods all have diameter segments that are parallel to a common direction, and the ratio between the longest and the shortest diameter is bounded

T. Asano et al. (Eds.): COCOON'99, LNCS 1627, pp. 473–482, 1999.

by a constant. Recently, Mata and Mitchell [11] provided a general framework that gives an $O(\log k)$-approximation algorithm for the general case, when no start point is given, with polynomial time complexity $\Omega(n^5)$ in the worst case. In this paper we give several results: First we show a simple and practical algorithm that produces a TSPN tour with logarithmic approximation factor in the case when a start point is given. If no start point is given we show how a good start point can be computed in $O(n^2 \log n)$. Hence, by combining these two results we obtain a logarithmic approximation algorithm for the general case with running time $O(n^2 \log n)$. Our main result is an algorithm that, given an arbitrary real constant ϵ as an optional parameter, performs at least one of the following two tasks (depending on the instance): (1) It outputs in time $O(n \log n)$ a TSPN tour of length $O(\log k)$ times the optimum. (2) It outputs a TSPN tour of length less than $(1+\epsilon)$ times the optimum in cubic time.

The first part of our method builds upon the idea in [11], in that our logarithmic approximation algorithm produces a guillotine subdivision. However, we produce a quite different guillotine partition (partly inspired from [7,10]) and show that it has some nice "sparseness" properties, which guarantee the $O(\log k)$ approximation bound. The method described can also be applied to other problems as suggested by [11]. We also consider the Red-Blue Separation Problem (RBSP). The RBSP asks for the minimum-length simple polygon that separates a set of red points from a set of blue points, a problem shown to be NP-hard [2,5]. Mata and Mitchell showed that the general framework presented in [11] gives an $O(\log m)$-approximation algorithm for this problem in time $O(n^5)$, where $m < n$ is the minimum number of sides of a minimum-perimeter rectilinear polygonal separator for the points. By using the methods suggested in this paper we obtain an $O(\log m)$ approximation algorithm that runs in time $O(n \log n)$.

This paper is organized as follows. In Section 3 the approximation algorithm is presented for the TSPN case where a start point is given. First we compute a bounding square that includes or touches all neighborhoods. Next, a guillotine subdivision algorithm operating within this square is presented that runs in time $O(n+k \log k)$. We prove that the subdivision is of length $O(\log k)$ times the length of an optimal tour. In Section 4 we show how to compute start points that are included in TSPN-tours of length within a constant times the optimal in time $O(n^2 \log n)$. In Section 5, we describe an algorithm that, depending on the given input, decides which method to apply to obtain a TSPN tour. Finally, in Section 6, we show how the methods presented in Section 3-4 can be used to obtain approximation algorithms for some separation problems. The algorithms presented in Section 3-4 and 6 are very practical and easy to implement. The TSPN algorithm has been implemented and runs efficiently even for quite large instances [8]. The proofs omitted in this extended abstract can be found in [8].

2   Definitions and Preliminaries

Throughout this paper we will denote by X the collection of k possibly overlapping simple polygons, called *neighborhoods*, with totally n vertices in the plane.

A polygonal subdivision D of a polygon P is said to be "guillotine" if it is a binary planar partition of P, i.e. either there exist no edges of D interior to P, or there exists a straight edge of D such that this edge is a chord of P dividing P into two regions, P_1 and P_2, such that D restricted to P_i, $1 \leq i \leq 2$, is a guillotine subdivision of P_i. If all the faces of the subdivision are rectangles, we obtain a guillotine rectangular subdivision. From now on we will denote by GRS the guillotine rectangular subdivision obtained from the procedure described in Section 3.1. We will denote by $|p|$ the total length of the segments of p, where p may be a polygon, a tour or a subdivision.

3   A Fast Approximation Algorithm for TSPN

This section is devoted to the proof of the following theorem.

Theorem 1. *Let X be a collection of k possibly overlapping simple polygons, having a total of n vertices. Let SP be any start point. One can compute in time $O(n+k \log k)$ a TSPN tour starting at SP, whose length is $O(\log k)$ times the length of a minimum-length tour starting at SP.*

The general idea is simple. Compute a bounding square that includes or touches all neighborhoods. Next a binary partition is performed within the square such that every neighborhood is intersected either by the binary partition or the bounding square. By traversing the bounding square and the binary partition we obtain a tour. We show how to do this in some more details.

First we detect each neighborhood whose closed region contains SP. This is done in linear time by using standard techniques (see, e.g. [13]). These neighborhoods are visited without departing from SP, and we may exclude them from further consideration. The next step is to compute the minimal isothetic square, called *bounding square*, or $\mathcal{B}$, with center at SP such that each neighborhood is at least partially contained in or touched by $\mathcal{B}$. This bounding square can be constructed in linear time by computing, for each neighborhood $x_i \in X$, $1 \leq i \leq k$, the minimum isothetic square centered at SP and reaching x_i, and then taking the maximum over all these squares, Fig. 1a.

If all the polygons intersect the bounding square then we output the tour obtained by walking from SP to the bounding square, then following the perimeter of the bounding square and returning to SP. This tour is at most five times longer than an optimal TSPN-tour. (In the optimal tour, the salesman has to reach the boundary of $\mathcal{B}$ and then return.) Otherwise, one or more neighborhoods lie entirely inside $\mathcal{B}$. In this case we construct a rectangular binary planar partition, which we call GRS, of the bounding square $\mathcal{B}$, such that every polygon lying entirely within $\mathcal{B}$ is intersected or touched by the GRS. In [11], Mata and Mitchell described how a guillotine subdivision and its bounding box can be traversed such that the obtained tour visits all neighborhoods in X and has length at most twice the length of the subdivision plus $|\mathcal{B}|$.

To prove Theorem 1 it remains to define the GRS, show how it can be computed in time $O(n+k \log k)$ (section 3.1), and prove that its length is $O(\log k)$ longer than the length of a shortest tour (section 3.2).

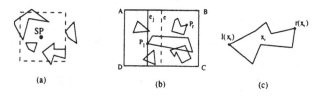

Fig. 1. (a) Constructing a bounding square. (b) A guillotine cut e_j. (c) $l(x_i)$ and $r(x_i)$ of a polygon x_i.

3.1   The guillotine rectangular subdivision

In this subsection we first describe how the subdivision algorithm works. Then, we show how this algorithm can be implemented to run in time $O(n+k \log k)$.

The partition is done as follows. Let T_j be the rectangle in the GRS induced by all segments drawn by the procedure up to but not including the segment e_j, such that e_j is drawn in T_j during the next step. Let A, B, C and D be the corners of T_j in clockwise order, such that AB is the top side of T_j. The segment e_j is drawn parallel to a shortest side of T_j, and it is vertical if T_j is a square. Assume for simplicity that $|AD| \leq |AB|$, i.e. e_j will be a vertical segment, Fig. 1a.

Let e be the vertical chord splitting T_j into two rectangles of equal size. Let $N = \{x_1, \ldots, x_{k'}\}$ be the set of k' polygonal neighborhoods entirely inside T_j, and let $l(x_i)$ respectively $r(x_i)$, $1 \leq i \leq k'$, be the leftmost, respectively rightmost, point in the polygonal neighborhood x_i, Fig. 1b. Two cases may occur:

1. If the region to the left of e is empty then the procedure draws e_j such that it intersects the leftmost point $r(x_i)$, $1 \leq i \leq k'$. The corresponding procedure is performed symmetrically if the region in T_j to the right of e is empty.

2. If none of the regions are empty then let N_l, respectively N_r, be the polygons in N with non-empty intersection with the closed region of T_j to the left, respectively to the right, of e. Let P_l be the point in $\{\forall x \in N_l \quad l(x)\}$ with shortest distance to e (ties are broken arbitrarily). Symmetrically, P_r is the point in $\{\forall x \in N_r \quad r(x)\}$ with shortest distance to e. If the horizontal distance between P_l and e is shorter than the horizontal distance between P_r and e, then the procedure draws the vertical chord e_j such that it passes through P_l. Otherwise, it draws the vertical chord e_j such that it passes through P_r.

This is done recursively within the two resulting rectangles until all the polygonal neighborhoods are intersected. The final result is called the GRS.

Lemma 1. *The guillotine algorithm presented above can be implemented to run in time $O(n+k \log k)$.*

In [8] we give a detailed description of the algorithm and prove Lemma 1.

3.2   Proving a logarithmic bound

Consider a minimum-length tour, L, with start point SP, and let B_L be its minimal isothetic bounding box. Let B be the bounding square of X computed

Fig. 2.

by the algorithm, as described above, and let L' be a shortest tour in $\mathcal{B}$. Since the length of the segments of a guillotine rectangular subdivision can be $\Omega(\sqrt{n})$ times the length of the perimeter of the bounding box, we first have to prove that there exists a tour in $\mathcal{B}$ of length $O(|L|)$. Then we show that our subdivision is within a logarithmic factor longer than a shortest tour in $\mathcal{B}$.

Proposition 1. *If $\mathcal{B}$ is a bounding square with perimeter of length within a constant factor longer than the perimeter of B_L, and $\mathcal{B}$ includes a point in an optimal tour, then there exists a TSPN-tour of length $O(|L|)$ within $\mathcal{B}$.*

Recall that the perimeter of the bounding square produced by the algorithm is within a constant factor (4) longer than the perimeter of a minimal bounding box. Note that a minimum-length tour is a simple (not self-intersecting) polygon, having at most k vertices. This result also holds for L'. It is easy to see that a shortest rectilinear TSPN-tour L_R, including the start point SP, is a simple rectilinear polygon and has length at most $\sqrt{2}|L|$. Furthermore, in [4,10,11], it was shown that there exists a guillotine rectangular subdivision of the rectilinear tour L_R and its minimal bounding box B_{L_R} such that the length of the subdivision is $O(|L_R| \log k)$, Fig. 2. Hence, we obtain the following fact:

Fact 1. *If $OPT(X, B_L)$ is a rectangular subdivision of minimum length of X and B_L, then it holds that $|OPT(X, B_L)|=O(|L| \log k)$.*

According to the above results we have that Fact 1 also holds if we replace B_L and L with $\mathcal{B}$ and L'. This means that an optimal rectangular subdivision of the polygonal neighborhoods X within the bounding square $\mathcal{B}$ is within a logarithmic factor longer than an optimal TSPN tour. Hence, to prove Theorem 1 it remains to show that the GRS presented in the previous section produces a subdivision which is at most within a constant factor longer than an optimal subdivision.

Lemma 2. *Let L' be a minimum-length tour within a bounding square $\mathcal{B}$, let L be an optimal tour and let B_L be its minimal isothetic bounding box. If the length of the perimeter of $\mathcal{B}$ is within a constant factor longer than the perimeter of B_L, then it holds that the guillotine rectangular subdivision of X within $\mathcal{B}$ is of length $O(|L'| \log k)$.*

4   Finding a Good Start Point

In the previous section we showed a fast algorithm that produces a TSPN tour when a start point is given. To use this algorithm in the general case we will in this section show how to compute a nearly optimal start point, i.e. a start point that is included in a TSPN-tour of length $O(|L|)$, where L is a minimum-length tour of X. We describe a method which, given an arbitrary straight-line segment, computes in time $O(n \log n)$ a starting point on this segment which is a nearly optimal point under the condition that the optimal tour has to pass through this segment. By observing that there is always an optimal tour with non-empty intersection with the boundary of at least one neighborhood, we obtain as a corollary an $O(n^2 \log n)$-time method for finding a globally nearly-optimal start point. Combining this result with the algorithm described in the previous section we obtain a logarithmic approximation algorithm that runs in time $O(n^2 \log n)$.

4.1   The length function

The idea is that we will for each of the n edges of the neighborhoods find the minimum L_∞-distance to all polygonal neighborhoods. This is done by constructing a length function $L(g, f)$ for each pair of edges $g, f \in E$ (not belonging to the same neighborhood), where E is the set of edges in X. This function describes the shortest distance in L_∞-metric from each point on g to f. Let $g, f \in E$ be two edges in E. We want to calculate $L(g, f)$, that is for each point in g we want to know the shortest distance to f in L_∞-metric. Rotate g and f such that g is horizontal. The endpoints of g are denoted G_1 and G_2. The length function is a piecewise linear function containing at most three different pieces. We will calculate the value of $L(g, f)$ in at most four points. Compute the values for G_1 and G_2. Next compute the shortest distance from f's endpoints to g. Denote the points on f with shortest distance to G_1 respectively G_2, by p_1 respectively p_2. $L(g, f)$ is obtained by drawing straight line segments between these four points, from left to right. Note that p_1 and/or p_2 may coincide with the endpoints of g. This function may now be used to compute the shortest distance between an edge $f \in E$ and a neighborhood $x \in X$, where $f \notin x$. First compute the length function for each edge in x and the edge f. Since we only are interested in the shortest distance between every point on f and the neighborhood x, we calculate the lower envelope of these $|x|$ functions, denoted $LEnv(f, x)$. Calculating the lower envelope of the $|x|$ functions can be done in time $O(|x| \log |x|)$ according to Sharir in [15]. If f intersects the closed region described by x then we have to adjust the lower envelope. This can easily be done in linear time.

This new piecewise linear function describes the shortest distance between every point on f and the polygon x, and can be computed in time $O(|x| \log |x|)$.

4.2   Using the length function to obtain a good start point

We extend the above computation such that we obtain a function that for each point on f describes the shortest distance to all the neighborhoods in X.

The algorithm is just an extension of the above described algorithm. Let f be an edge as above and let $\{x_1, \ldots, x_k\}$ be the neighborhoods in X. For each neighborhood $x_i \in X$ and $f \notin x_i$ compute the lower envelope, $LEnv(f, x_i)$, as described above. Combine the $k-1$ lower envelopes for all the neighborhoods in X to an upper envelope, $UEnv(f, X)$. This new function describes the shortest distance between each point in f and all the neighborhoods in X. In the case when we know that an optimal TSPN-tour intersects f we can select the point on f with minimum $UEnv(f, X)$-value as start point (SP). Note that the function also gives us the size of an optimal bounding square with center at SP.

Calculating the envelope can be done in time $O(n \log n)$ [15]. The idea is to partition the collection of functions into two subcollections. Then calculate recursively the envelopes of each subcollection separately. Finally merge the two partitions into a single refined partition. The total time complexity for this algorithm is $O(n \log n)$.

If the $UEnv$-function is computed for every edge in E we can in linear time find a good start point for the general case, by just selecting the best of the n suggested start points. We obtain the following results.

Proposition 2. *The computed bounding square centered at SP includes a tour of length $O(|L|)$ and the length of the perimeter of the square is at most four times the length of a minimum-length TSPN-tour.*

Hence the computed bounding square fulfills the requirement needed for Lemma 2 to hold and we obtain the following corollary.

Corollary 1. *Let X be a collection of k possibly overlapping simple polygons, having a total of n vertices. One can compute in time $O(n^2 \log n)$ a start point that is included in a tour of length within a constant factor times the optimal.*

5   TSPN When No Start Point Is Given

We have already described an approximation algorithm for the general case. By finding a start point and then applying the GRS we obtain an approximation algorithm that runs in time $O(n^2 \log n)$ and produces a tour that is within a logarithmic factor longer than an optimal tour. In this section we describe an algorithm that, depending on the given input, decides which method to apply to obtain a TSPN tour. The decision depends on the value of ϵ, an optional parameter given as input, and the ratio between the size of the smallest neighborhood in X and the length of a minimum-length tour. Recall that the minimal bounding box, B_L, is the smallest axis-aligned rectangle that contains L, where L is a minimum-length tour of X. The algorithm works as follows.

First a minimal axis-aligned bounding box for every neighborhood in X is constructed. Let $x \in X$ be the neighborhood, such that its bounding box, B_x, has shortest long side. Let c denote the center of B_x, and let l be the length of B_x's longest side, Fig. 3a. Next, a minimal isothetic bounding square B with center at c is constructed such that each neighborhood is at least partially contained in or touched by B, Fig. 3b. This is done in linear time as described in Section 3. Two main cases may occur.

1. If $|B| > 4l$ then it holds that there exists a neighborhood $x' \in X$ such that x' lies outside B_x. Hence, an optimal tour must intersect B_x, that means that we

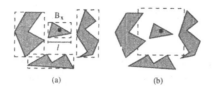

Fig. 3.

can find a good start point, SP, that lies on one of the four sides of B_x. This can be done in time $O(n \log n)$ as explained in Section 4.2. By applying the subdivision algorithm defined in Section 3.1 with SP as start point we obtain a tour of length $O(\log k)$ times the optimal in time $O(n \log n)$.

2. If $|B| < 4l$ we first make a refined search in time $O(n)$ for much smaller bounding boxes in the vicinity of x. For this reason, we imagine a square-grid centered at c, with side length $2l$, and where each square has side length equal to $l/10$. For each one of the $O(1)$ corners of the small squares, we find the minimal isothetic bounding square, as if this corner was a given starting point. Let D be the smallest of these bounding squares and, let l' be the side length of D. Again, two cases may occur:

(a) If $l' > l/3$, then output D as the TSPN tour. We can show that it is within a constant factor from the minimum, and we have found it in linear time.

(b) If $l' \leq l/3$ then we run the approximation scheme described in Section 5.1. We obtain a TSPN tour of length $1 + \epsilon$ in cubic time.

5.1 An approximation scheme

In this section we show how to obtain a polynomial-time approximation scheme, in the case when $l' < l/3$ (defined in the previous paragraph).

Observation 1. *For every TSPN tour of length less than $2l$, it holds that the convex hull of that tour also is a TSPN tour.*

Now, we find (at most) a quadratic number of approximately minimal bounding squares (i.e. with perimeter at most a constant factor larger than the minimal tour), possibly overlapping, such that an optimal tour has to be totally inside at least one of them. This is done as follows. Find all *essential points* in X. These points are the vertices of the polygons of X and the intersections between the polygons. Since there are n edges there can be $O(n^2)$ such intersections. Hence the number of essential points is $O(n^2)$. Then the algorithm selects every minimal bounding square with center at an essential point and with perimeter at most a constant factor longer than a minimal tour (a constant factor longer than the smallest found bounding square). Finally, for reasons explained in the proof of Observation 2, expand each selected minimal bounding square such that the sides of a square is a constant C, say 10, times the length of the original bounding square. There are $O(n^2)$ such bounding squares and each square takes $O(n)$ time to construct. Thus, the time complexity of this step is $O(n^3)$.

Observation 2. *An optimal tour is included in at least one of the $O(n^2)$ bounding squares produced above.*

For each selected bounding square b, let $l(b)$ be the length of the minimum tour overlapping with b and denote by $h(b)$ the length of the sides of b. For each b, we find a tour of length less than $(1+\epsilon)\cdot l(b)$ in linear time, and in this way in cubic total time we find a tour of length $(1+\epsilon)$ times the optimal.

A simple, although impractical, linear-time method to find a tour of length $(1+\epsilon)\cdot l(b)$, is as follows: We partition b into $f(\epsilon) \times f(\epsilon)$ equal-sized squares, where $f(\epsilon)$ is a sufficiently large constant only depending on ϵ. Let S be the set of all points which are corners of these small squares. Now, for each subset S' of S, we construct the convex hull of S'. Next, among those of the constructed convex hulls which are TSPN tours, we select the shortest one. (Of course, in practice we don't really have to consider all subsets of S.) It is easily seen that by choosing $f(\epsilon)$ sufficiently large, say, $f(\epsilon) = \frac{4h}{\epsilon}$, we will find a tour of length $(1+\epsilon)\cdot l(b)$.

6   Other Applications

The methods described in this paper apply also to other problems. Here we show how to use our method for the Red-Blue Separation Problem (RBSP). The objective is to find a minimum-length simple polygon that separates a set of red points, R, from a set of blue points B. This problem is seen to be NP-hard by using a reduction from the Euclidean TSP [2,5]. In this section we give the idea of an $O(\log m)$ approximation algorithm for RBSP, where $m<n$ is the minimum number of sides of a minimum-perimeter rectilinear polygonal separator for the points. The total number of points is $n=|R|+|B|$. We will solve two problems and pick the shorter length: find a minimum-length polygon that encloses red, while excluding blue; and find a minimum-length polygon that encloses blue, while excluding red.

First construct the minimum bounding box for the red points respectively the blue points. Let P be a minimum-length separating simple polygon, of length $|P|$. Without loss of generality, assume that P surrounds the blue points and excludes the red points. Let $\mathcal{B}$ be the bounding box for B. Note that P lies within $\mathcal{B}$ and that $\mathcal{B}$ is the bounding box of P. If no red points lie within $\mathcal{B}$ then we are finished. The length of the perimeter of this polygon is at most a factor $\sqrt{2}$ longer than $|P|$. Otherwise, one or more red points lie within $\mathcal{B}$. In this case construct a guillotine rectangular subdivision of $\mathcal{B}$, such that every rectangle within the subdivision only contains points of one color. Next we run the GRS-algorithm described in Section 3.1, with some adjustments [8].

In [11], Mata and Mitchell described how a guillotine subdivision and its bounding box can be traversed such that the resulting tour encloses all blue points, excludes all red points, and has length at most twice the length of the subdivision plus $|\mathcal{B}|$. Hence, we obtain the following corollary.

Corollary 2. *Given a set of n points in the plane, where each point is either red or blue, a red-blue separating simple polygon of length $O(\log m)$ times the mini-*

mum can be computed in time $O(n \log n)$, where $m < n$ is the minimum number of sides of a minimum-perimeter rectilinear polygonal separator for the points.

7 Conclusion and Future Work

We have presented approximation algorithms for the TSPN-problem and the red-blue separation problem. For both these results we improved the running-time considerably by computing a simple and fast guillotine rectangular subdivision.

There are several open questions concerning these problems. It is natural to ask whether a constant approximation ratio is achievable for these problems. Further, is there a logarithmic approximation algorithm for TSPN in the case when the neighborhoods are not simple polygons, for example, sets in the plane?

References

1. E. M. Arkin and R. Hassin. Approximation algorithms for the geometric covering salesman problem. *Discrete Applied Math*, 55:197–218, 1994.
2. E. M. Arkin, S. Khuller, and J. S. B. Mitchell. Geometric knapsack problems. *Algoritmica*, 10:399–427, 1993.
3. S. Arora. Polynomial time approximation schemes for Euclidean TSP and other geometric problems. In *FOCS'96*, pages 2–11, 1996.
4. D. Z. Du, L. Q. Pan, and M. T. Shing. Minimum edge length guillotine rectangular partition. Technical Report 02418-86, Math. Sci. Res. Inst., University of California, Berkeley, CA, 1986.
5. P. Eades and D. Rappaport. The complexity of computing minimum separating polygons. *Pattern Recognition Letters*, 14:715–718, 1993.
6. M. R. Garey, R. L. Graham, and D. S. Johnson. Some NP-complete geometric problems. In *STOC'76*, pages 10–21, 1976.
7. T. Gonzalez and S. Q. Zheng. Bound for partitioning rectilinear polygons. In *SoCG'85*, pages 281–287, 1985.
8. J. Gudmundsson and C. Levcopoulos. A fast approximation algorithm for TSP with neighborhoods and red-blue separation. Technical Report LU-CS-TR:97-196, Dept. of Computer Science, Lund University, Lund,Sweden, 1997.
9. J. Hershberger and S. Suri. A pedestrian approach to ray shooting: Shoot a ray, take a walk. *Journal of Algorithms*, 18:403–431, 1995.
10. C. Levcopoulos. Fast heuristics for minimum length rectangular partitions of polygons. In *SoCG'86*, pages 100–108, 1986.
11. C. S. Mata and J. S. B. Mitchell. Approximation algorithms for geometric tour and network design problems. In *SoCG'95*, pages 360–369, 1995.
12. C. H. Papadimitriou. Euclidean TSP is NP-complete. *TCS*, 4:237–244, 1977.
13. F.P. Preparata and M.I. Shamos. *Computational Geometry: An Introduction*. Springer-Verlag, New York, NY, 1985.
14. G. Reich and P. Widmayer. Beyond steiner's problem: A vlsi oriented generalization. In *Proceedings of the 15th International Workshop Graph-Theoretical Concepts in Computer Science*, volume 411 of *LNCS*, pages 196–210, 1989.
15. M. Sharir. Davenport-schinzel sequences and their geometric applications. In R. A. Earnshaw, editor, *Theoretical Foundations of Computer Graphics and CAD*, volume F40 of *NATO ASI*, pages 253–278. Springer-Verlag, Berlin, Germany, 1988.

The Greedier the Better: An Efficient Algorithm for Approximating Maximum Independent Set

H.Y. Lau and H.F. Ting

Department of Computer Science and Information Systems,
The University of Hong Kong, Hong Kong
{hylau,hfting}@csis.hku.hk

The classical greedy heuristic for approximating maximum independent set is simple and efficient. It achieves a performance ratio bound of $(\Delta + 2)/3$, where Δ is the degree of the input graph. All known algorithms for the problem with better ratio bounds are complicated and run slowly for moderately large Δ. In this paper, we describe a natural extension of the greedy heuristic. It is as simple and as efficient as the classical greedy heuristic. By a careful analysis on the structure of the intermediate graphs manipulated by our heuristic, we prove that the ratio bound is improved to $(\Delta + 3)/3.25$.

1   Introduction

An *independent set* of a graph is a subset of nodes in which no two nodes are adjacent. The problem of finding maximum independent sets is NP-complete. Thus, we focus on designing efficient algorithms that return large, but not necessary maximum, independent sets. The effectiveness of such an algorithm is measured by its *performance ratio bound* (or simply ratio bound), which is defined to be the worst-case ratio between the size of a maximum independent set and the size of the independent set returned by the algorithm. When there is no restriction on the input graph, the best polynomial-time algorithm known for approximating maximum independent set is given by Boppana and Halldórsson [3]. The ratio bound of this algorithm is $\theta(n/\log^2 n)$, where n is the number of nodes of the input graph. Recently, Håstad [7] showed that unless $NP = coR$, there does not exist any polynomial-time algorithm approximating maximum independent set with ratio bound smaller than $n^{1-\epsilon}$ for any $\epsilon > 0$.

Given this hardness result, it is natural to impose some restrictions on the input graphs. In particular, we focus on those graphs whose nodes are all of degree no greater than some constant Δ. We call the problem of approximating maximum independent sets for such graphs the MIS-Δ problem. This problem is $MAXSNP$-complete when $\Delta \geq 3$ [1]. Recently, Feige showed that there is a constant $\epsilon > 0$ such that unless $NP = P$, there is no polynomial-time algorithm for MIS-Δ with ratio bound $O(\Delta^\epsilon)$ (see [6]).

The classical greedy heuristic is one of the most popular and practical algorithm for MIS-Δ [4,5]. It works as follows.

T. Asano et al. (Eds.): COCOON'99, LNCS 1627, pp. 483–492, 1999.
© Springer-Verlag Berlin Heidelberg 1999

Select and add to the solution a node of minimum degree. Then, delete the node and all its neighbors. Repeat this process until there is no node left.

In [4], Halldórsson and Radhakrishnan proved that the ratio bound of the greedy heuristic is at most $(\Delta + 2)/3$; and they also gave a worst-case example showing that the bound is tight. In this paper, we show that a greedier version of the heuristic reduces the ratio bound to $(\Delta + 3)/3.25$. Our heuristic is almost the same as the classical one, except that in each iteration, it select a pair of nodes, instead of just a single node, with the minimum number of neighbors. The proof on the ratio bound is non-trivial. It involves a careful analysis on the structure of the subgraphs removed in each iteration. The heuristic can be implemented with running time $O(\Delta^2 n)$. Hence, it is as efficient as the classical greedy heuristic, but with a better ratio bound.

We note that there are already algorithms with better ratio bound. However, these algorithms are much more complicated and run slowly for moderately large Δ. The algorithm with asymptotically the best known ratio bound, namely, $\theta(\Delta / \log \log \Delta)$ is given by Halldórsson and Radhakrishnan [6]. However, the constant hidden in the θ is huge and the algorithm is not practical. Halldórsson and Radhakrishnan [4] also gave an algorithm for $MIS\text{-}\Delta$ with ratio bound $(\Delta + 3)/4$. It runs in $O(\Delta^\Delta n^2)$ time. Berman and Fürer [2] gave an algorithm with better ratio bound, namely $(\Delta + 3)/5 + \epsilon$; but its time complexity is larger than n^{50}. In [6], Halldórsson and Radhakrishnan obtained an algorithm with ratio bound smaller than $\frac{\Delta}{6}(1 + o(1))$. The time complexity of this algorithm is $O(\Delta^k n^2)$, where the value of k determines how small the $o(1)$ will be (for a reasonably good ratio bound, k should be greater than 6). All these algorithms are complicated and they run very slowly for $\Delta \geq 6$.

In the following section, we present our greedier heuristic. In Section 3, we describe a general framework for deriving ratio bound for the greedy heuristic, and based on it, prove that the performance ratio of our greedier heuristic is $(\Delta + 3)/3.25$. Section 4 contains the main result of this paper; we prove an important inequality, which is used in the general framework for deriving the ratio bound. In Section 5, we explain how our heuristic can be implemented with running time $O(\Delta^2 n)$ time.

2   The greedier heuristic

Given any independent set I, define the *degree* of I to be the number of nodes that are adjacent to some node in I. Define $\delta(I)$, the *degree density* of I, to be the degree of I divided by its size. Note that the degree of a node u equals the degree density of the independent set $\{u\}$. Thus, the classical greedy heuristic can be rephrased as follows:

Iteratively select, among all independent sets of size one, the set with the minimum degree density. Remove this set, as well as all of its neighbors. Repeat this process until there is no node left.

The notion of degree density of an independent set immediately suggests a way to generalize the classical greedy heuristic. Fixed some constant k. The Greedy$_k$ heuristic is almost the same as the classical one, except that at each iteration, it selects, among all independent sets of size smaller than or equal to k, the one with the minimum degree density. Thus, the classical greedy heuristic is just Greedy$_1$. It is not obvious that Greedy$_k$ will have ratio bound better than that of Greedy$_l$ for any $l < k$; the set chosen to be added into the solution set can be very different. In this paper, we prove that the ratio bound of Greedy$_2$ is $(\Delta + 3)/3.25$. Since it is shown in [4] that the ratio bound of Greedy$_1$ cannot be smaller than $(\Delta + 2)/3 - \epsilon$ for any ϵ, the ratio bound of Greedy$_2$ is strictly better than Greedy$_1$ for $\Delta > 2$. In Figure 1, we give a formal description of Greedy$_2$. A

```
input: G = (V, E);
output: An Independent I.
begin
    I ← ∅;
    while V ≠ ∅ do
        select H ∈ {X | X is an independent set of G, |X| ≤ 2} with
            minimum degree density;
        I ← I ∪ H;
        S ← H ∪ {u | u ∈ V and is adjacent to some node in H};
        D ← {(u, v) | (u, v) ∈ E and either u or v are in S};
        V ← V - S;
        E ← E - D;
        G ← (V, E);
    end;
    return I;
end.
```

Fig. 1. The new heuristic

natural question is whether Greedy$_k$, $k > 2$ has even better ratio bound. In [8], we prove that the ratio bound of Greedy$_k$ is no greater than $\frac{\Delta + 2k}{3 + (1 - 1/k)^k}$. Thus, the ratio bound of the greedy heuristic does improve if we examine larger set of nodes at each iteration. However, the improvement becomes marginal when $k > 4$.

3 The ratio bound of Greedy$_2$

We adopt the framework suggested in [4] for deriving ratio bound for Greedy$_2$. Assume that the heuristic has totally iterated t times on an input graph of n nodes with maximum degree Δ. Let α be a maximum independent set of the input graph. For $1 \leq i \leq t$, let S_i and D_i be the set of nodes and edges, respectively, removed at the ith iteration. Let $A_i = \alpha \cap S_i$. The following lemma is implied in [4].

Lemma 1. $(n - |\alpha|)\Delta \geq \sum_{1 \leq i \leq t} (|D_i| + |E(S_i - A_i, S_i - A_i)|)$, *where* $E(S_i - A_i, S_i - A_i)$ *denotes the set of edges that have both endpoints in* $S_i - A_i$.

Proof. Consider the input graph. We count the total degree of nodes outside α, which entails counting edges incident on α once and those do not have any end points in α twice. Since all S_i are disjoint, we have $(n - |\alpha|)\Delta \geq |E| + \sum_{1 \leq i \leq t} |E(S_i - A_i, S_i - A_i)|$. Note that $|E| = \sum_{1 \leq i \leq t} |D_i|$. The lemma follows.

If we are using the classical greedy heuristic, we can prove that, for any $1 \leq i \leq t$, $|D_i| + |E(S_i - A_i, S_i - A_i)| \geq |S_i|^2 - |A_i|(|S_i| - |A_i|) - |S_i|$. When substituting these bounds into Lemma 1, we can derive the $(\Delta+2)/3$ ratio bound (see [4]). Our main contribution is to prove that using Greedy$_2$, we can derive a better lower bound on $|D_i| + |E(S_i - A_i, S_i - A_i)|$. More precisely, in Section 4, we prove that, for any $1 \leq i \leq t$,

$$|D_i| + |E(S_i - A_i, S_i - A_i)| \geq |S_i|^2/m_i - \frac{3}{4}|A_i|(|S_i| - |A_i|)/m_i - \frac{3}{2}|S_i| \quad (1)$$

where m_i is the size of the independent set chosen at the ith iteration (thus, m_i can either be one or two). Note that $\sum_{1 \leq i \leq t} m_i$ equals the size of the independent set returned by our heuristic and the ratio bound of Greedy$_2$ is $|\alpha|/\sum_{1 \leq i \leq t} m_i$. Below, we detail the arithmetic for proving the bound.

Applying Lemma 1 and Inequality 1, we have

$$(n - |\alpha|)\Delta \geq \sum_{i=1}^{t} \left(\frac{|S_i|^2}{m_i} - \frac{3|A_i|(|S_i| - |A_i|)}{4m_i} - \frac{3}{2}|S_i| \right)$$

$$= \sum_{i=1}^{t} \frac{(|S_i| - 3/8|A_i|)^2}{m_i} + \frac{39}{64} \sum_{i=1}^{t} \frac{|A_i|^2}{m_i} - \frac{3}{2} \sum_{i=1}^{t} |S_i|.$$

Apply Cauchy's inequality (i.e., $\sum a_i^2 \geq (\sum a_i b_i)^2 / \sum b_i^2$), and rearrange the terms, we have

$$(n - |\alpha|)\Delta \geq \frac{(\sum_{i=1}^{t}(|S_i| - 3/8|A_i|))^2}{\sum_{i=1}^{t} m_i} + \frac{39}{64} \frac{(\sum_{i=1}^{t} |A_i|)^2}{\sum_{i=1}^{t} m_i} - \frac{3}{2} \sum_{i=1}^{t} |S_i|$$

$$= \frac{(\sum_{i=1}^{t} |S_i|)^2 - \frac{3}{4}(\sum_{i=1}^{t} |A_i|)(\sum_{i=1}^{t} |S_i| - \sum_{i=1}^{t} |A_i|)}{\sum_{i=1}^{t} m_i} - \frac{3}{2} \sum_{i=1}^{t} |S_i|.$$

Since $\sum_{1 \leq i \leq t} |S_i| = n$ and $\sum_{1 \leq i \leq t} |A_i| = |\alpha|$, we have

$$(n - |\alpha|)\Delta + \frac{3}{2}n \geq \frac{n^2 - \frac{3}{4}|\alpha|(n - |\alpha|)}{\sum_{i=1}^{t} m_i}.$$

Apply the inequality to $|\alpha|/\sum_{1 \leq i \leq t} m_i$ and simplify, we have the following theorem.

Theorem 1. *Let* τ *be* $|\alpha|/n$. *The ratio bound of Greedy$_2$ is* $\frac{((1-\tau)\Delta + \frac{3}{2})\tau}{1 - \frac{3}{4}\tau + \frac{3}{4}\tau^2}$.

Note that τ is between zero and one. It is easy to verify that $\frac{(1-\tau)\tau}{1-\frac{3}{4}\tau+\frac{3}{4}\tau^2} \leq 4/13$ and $\frac{3/2\tau}{1-\frac{3}{4}\tau+\frac{3}{4}\tau^2} \leq 3/2$ when $\tau \in [0,1]$. From Theorem 1, we have the ratio bound is no greater than $\frac{\Delta+4.875}{3.25}$. If we apply the Nemhauser-Trotter preprocessing [9] before executing the heuristic, we can improve the ratio bound a little. Based on the preprocessing, we can assume without loss of generality that the size of the maximum independent set is at most half of the size of the graph. In other words, we can assume $\tau \leq 1/2$. Then, we have the following theorem.

Theorem 2. *The performance ratio bound of* $Greed_2$ *is no greater than* $\frac{\Delta+3}{3.25}$.

4   Proof of Inequality 1

In this section, we will prove Inequality 1. First, we need to prove a general lemma on graphs, which is crucial to our analysis.

4.1   A general lemma on graphs

Consider any graph G. We say that any two nodes are *non-adjacent* if there is no edge between them. For any two non-adjacent nodes u, v of G, let $C(u,v)$ denote the set of nodes that are adjacent to both u and v. For any two disjoint node sets X, Y, let $E(X,Y)$ denote the set of edges (u,v) such that $u \in X$ and $v \in Y$.

Lemma 2. *Let* X *and* Y *be any two disjoint sets of nodes of* G. *Assume that* X *is an independent set of size greater than one. We have the following two inequalities:*

(I) $|E(X,Y)| - \frac{\sum_{u,v \in X} |C(u,v)|}{|X|-1} \leq \frac{|X||Y|}{2}$;

(II) $|E(X,Y)| - \frac{2\sum_{u,v \in X} |C(u,v)|}{|X|-1} \leq \frac{|X|^2|Y|}{4(|X|-1)}$.

Proof. For any node $u \in X$, let $m(u)$ be the number of nodes in Y that are adjacent to u. Note that $\sum_{u,v \in X}(m(u) + m(v)) = (|X|-1)\sum_{u \in X} m(u)$ and $\sum_{u \in X} m(u) = |E(X,Y)|$. Thus,

$$|E(X,Y)| = \frac{\sum_{u,v \in X}(m(u) + m(v))}{|X|-1}.$$

To prove Inequality (I), observe that for any two nodes $u, v \in X$, $m(u) + m(v) - |C(u,v)| \leq |Y|$. Thus,

$$|E(X,Y)| - \frac{\sum_{u,v \in X} |C(u,v)|}{|X|-1} = \frac{\sum_{u,v \in X}(m(u) + m(v) - |C(u,v)|)}{|X|-1} \leq \frac{|X||Y|}{2}.$$

To prove Inequality (II), we define for any pair of nodes $u, v \in X$, $A(u,v)$ to be the set of nodes in Y that are adjacent to exactly one of u and v. Observe that $m(u) + m(v) - 2|\mathcal{C}(u,v)| \leq |A(u,v)|$. Thus,

$$|E(X,Y)| - \frac{2\sum_{u,v \in X} |\mathcal{C}(u,v)|}{|X| - 1} = \frac{\sum_{u,v \in X}(m(u) + m(v) - 2|\mathcal{C}(u,v)|)}{|X| - 1}$$

$$\leq \frac{\sum_{u,v \in X} |A(u,v)|}{|X| - 1}.$$

Below, we show that $\sum_{u,v \in X} |A(u,v)| \leq |X|^2|Y|/4$, and Inequality (II) follows.

For any node $w \in Y$, let $p(w)$ be the total number of pairs u, v in X such that $A(u,v)$ contains w. Observe that for each $w \in Y$, w is counted $p(w)$ times in $\sum_{u,v \in X} |A(u,v)|$ and thus $\sum_{u,v \in X} |A(u,v)| = \sum_{w \in Y} p(w)$. Fix any $w \in Y$. Let c be the number of nodes in X that are adjacent to w. Note that for any $u, v \in X$, $A(u,v)$ contains w if and only if exactly one of u, v is adjacent to w, and there are exactly $c(|X| - c)$ such u, v pairs. Thus, $p(w) = c(|X| - c)$. Since $c(|X| - c) \leq |X|^2/4$ for any $0 \leq c \leq |X|$, $\sum_{u,v \in X} |A(u,v)| = \sum_{w \in Y} p(w) \leq |X|^2|Y|/4$.

4.2   Bound on $|D_i| + |E(S_i - A_i, S_i - A_i)|$

We are now ready to prove Inequality 1, which asserts that for $1 \leq i \leq t$, $|D_i| + |E(S_i - A_i, S_i - A_i)| \geq |S_i|^2/m_i - 3/4|A_i|(|S_i| - |A_i|)/m_i - 3/2|S_i|$. Since inequalites are proved in the same way for every $1 \leq i \leq t$, we abstract our problem as follows.

Let $G = (V, E)$ be an undirected graph. Let S be any subset of nodes and $A \subseteq S$ be an independent set. Let D be the set of edges having at least one endpoint in S. Find a lower bound on $|D| + |E(S - A, S - A)|$.

For any node u of G, let $d(u)$ denote the degree of u. Let $\delta = \min\{\delta(I) \mid I$ is an independent set of size smaller than or equal to 2$\}$. The following properties are important to our analysis.

- For any $u \in V$, $d(u) \geq \delta$;
- For any non-adjacent pair of nodes u, v, $d(u) + d(v) - |\mathcal{C}(u,v)| \geq 2\delta$ (because the degree density of the independent set $\{u, v\}$ is $(d(u) + d(v) - |\mathcal{C}(u,v)|)/2$).

Because A is an independent set, we have $\sum_{u \in S} d(u) = |D| + |E(S - A, S - A)| + |E(A, S - A)|$, or equivalently, $|D| + |E(S - A, S - A)| = \sum_{u \in S} d(u) - |E(A, S - A)|$. We find it more convenient to derive bounds on $K = \sum_{u \in S} d(u) - |E(A, S - A)|$. Below, we derive two lower bounds on K, one for the general case and the other for the case when $|S| \geq 2\delta$.

Lemma 3. $K \geq |S|\delta - \frac{1}{2}|A|(|S| - |A|) - \frac{1}{2}|S|$.

Proof. The lemma is obviously true when $|A| \leq 1$. Assume that $|A| > 1$. Since for any $u, v \in A$, u, v are non-adjacent, we have $d(u) + d(v) - |\mathcal{C}(u,v)| \geq 2\delta$ and

$$\sum_{u \in A} d(u) = \frac{\sum_{u,v \in A}(d(u) + d(v))}{|A| - 1}$$

$$\geq \frac{\sum_{u,v \in A} 2\delta + |\mathcal{C}(u,v)|}{|A| - 1} = |A|\delta + \frac{\sum_{u,v \in A} |\mathcal{C}(u,v)|}{|A| - 1}.$$

Thus,

$$K = \sum_{u \in S-A} d(u) + \sum_{u \in A} d(u) - |E(A, S - A)|$$

$$\geq (|S| - |A|)\delta + |A|\delta + \frac{\sum_{u,v \in A} |C(u,v)|}{|A| - 1} - |E(A, S - A)|.$$

Applying Inequality (I) of Lemma 2 with $X = A$ and $Y = S - A$, we prove the lemma.

Lemma 4. *If* $|S| \geq 2\delta$, *then* $K \geq |S|\delta - \frac{3}{8}|A|(|S| - |A|) - \frac{1}{2}|S|$.

Proof. Let M be a maximum matching of the graph with node set $(S - A)$ and edge set $(S - A) \times (S - A) - E$. Intuitively, M is the maximum set of pairs of non-adjacent nodes of G that are in $S - A$. Let $B = \bigcup_{(u,v) \in M} \{u, v\}$, and $C = S - A - B$. Note that $S = A \cup B \cup C$. Furthermore, C induces a clique of G; or else we can enlarge M further.

Since for any $(u, v) \in M$, u, v are non-adjacent in G, we have $d(u) + d(v) - |C(u,v)| \geq 2\delta$ and $\sum_{u \in B} d(u) = \sum_{(u,v) \in M} (d(u) + d(v)) \geq |B|\delta + \sum_{(u,v) \in M} |C(u,v)|$. Together with the fact that $\sum_{u \in A} d(u) \geq |A|\delta + \frac{\sum_{u,v \in A} |C(u,v)|}{|A|-1}$ (see the proof of Lemma 3), we have

$$K = \sum_{u \in A} d(u) + \sum_{u \in B} d(u) + \sum_{u \in C} d(u) - |E(A, S - A)|$$

$$\geq |A|\delta + \frac{\sum_{u,v \in A} |C(u,v)|}{|A| - 1} + |B|\delta + \sum_{(u,v) \in M} |C(u,v)| + \sum_{u \in C} d(u) - |E(A, S - A)|$$

$$= |S|\delta - |C|\delta + \sum_{u \in C} d(u) + \frac{\sum_{u,v \in A} |C(u,v)|}{|A| - 1} + \sum_{(u,v) \in M} |C(u,v)| - |E(A, S - A)|.$$

Note that $|E(A, S - A)| = |E(A, B)| + |E(A, C)|$. Let $L_1 = |E(A, S - A)| - \frac{2\sum_{u,v \in A} |C(u,v)|}{(|A|-1)}$, and $L_2 = |E(A, B)| - 2\sum_{(u,v) \in M} |C(u,v)|$. It is easy to verify that

$$\frac{\sum_{u,v \in A} |C(u,v)|}{|A| - 1} + \sum_{(u,v) \in M} |C(u,v)| - |E(A, S - A)| = -\frac{L_1}{2} - \frac{L_2}{2} - \frac{|E(A,C)|}{2},$$

and thus

$$K \geq |S|\delta - |C|\delta + \sum_{u \in C} d(u) - \frac{L_1}{2} - \frac{L_2}{2} - \frac{|E(A,C)|}{2}.$$

From Inequality (II) of Lemma 2, we have

$$\frac{L_1}{2} \leq \frac{|A|^2(|S| - |A|)}{8(|A| - 1)}.$$

Applying a similar argument used in the proof of Inequality (I) of Lemma 2, we have

$$\frac{L_2}{2} \leq \frac{|A||B|}{4} = \frac{|A|(|S| - |A| - |C|)}{4}.$$

Hence,

$$K \geq |S|\delta - |C|\delta + \sum_{u \in C} d(u) - \frac{|A|^2(|S| - |A|)}{8(|A| - 1)} - \frac{|A|(|S| - |A|)}{4}$$
$$+ \frac{|A||C|}{4} - \frac{|E(A, C)|}{2}$$
$$\geq |S|\delta - \frac{3}{8}|A|(|S| - |A|) - \frac{|S|}{4}$$
$$+ \frac{|A||C|}{4} - \frac{|E(A, C)|}{2} + \sum_{u \in C} d(u) - |C|\delta.$$

Below, we show that $L = \frac{1}{4}|A||C| - \frac{1}{2}|E(A, C)| + \sum_{u \in C} d(u) - |C|\delta \geq -\frac{1}{4}|S|$, and hence prove the lemma.

Since $\sum_{u \in C} d(u) - |C|\delta \geq 0$, we have $L \geq -\frac{1}{4}|S|$ if $|E(A, C)| \leq \frac{1}{2}|A||C| + \frac{1}{2}|S|$. So without loss of generality, assume that

$$|E(A, C)| > \frac{1}{2}|A||C| + \frac{1}{2}|S|.$$

Note that $\sum_{u \in C} d(u) \geq |E(A, C)| + |E(B, C)| + 2|E(C, C)|$. Since C is a clique, we have

$$2E(C, C) = |C|(|C| - 1).$$

We claim that

$$E(B, C) \geq \frac{1}{2}|B|(|C| - 1).$$

Recall that B is the set of nodes in the maximum matching M of the graph $(S - A, (S - A) \times (S - A) - E)$. For any $(u, v) \in M$, at least $|C| - 1$ nodes in C must be connected to either u or v in G. Otherwise, there are at least two nodes $x, y \in C$ where u, x and v, y are non-adjacent. Then, $M \cup \{(u, x)\} \cup \{(v, y)\} - \{(u, v)\}$ is a larger matching, contradicting the fact that M is maximum. It follows that $|E(B, C)| \geq |M|(|C| - 1) = \frac{1}{2}|B|(|C| - 1)$. Therefore,

$$L = \frac{1}{4}|A||C| - \frac{1}{2}|E(A, C)| + |E(A, C)| + |E(B, C)| + 2|E(C, C)| - |C|\delta$$
$$\geq \frac{1}{2}|A||C| + \frac{1}{4}|S| + \frac{1}{2}|B|(|C| - 1) + |C|(|C| - 1|) - |C|\delta$$
$$\geq |C|\left(\frac{1}{2}(|A| + |B| + |C|) - \delta\right) - \frac{1}{4}|S|$$
$$\geq -\frac{1}{4}|S|.$$

The last inequality is from the assumption of the lemma that $|S| \geq 2\delta$.

We are ready to prove Inequality 1. For any $1 \leq i \leq t$, let m_i, δ_i be the size of the independent set selected and its degree density, respectively, at the ith iteration.

Lemma 5. *For any* $1 \leq i \leq t$, $|D_i| + |E(S_i - A_i, S_i - A_i)| \geq |S_i|^2/m_i - \frac{3}{4}|A_i|(|S_i| - |A_i|)/m_i - \frac{3}{2}|S_i|$.

Proof. Consider a particular i. We have two cases to consider.

Case 1: $m_i = 1$. Note that $|S_i|$, the number of nodes removed at the ith iteration, equals $\delta_i + 1$. Applying Lemma 3, we have $|D_i| + |E(S_i - A_i, S_i - A_i)| \geq |S_i|\delta_i - \frac{1}{2}|A_i|(|S_i| - |A_i|) - \frac{1}{2}|S_i| = |S_i|^2/m_i - \frac{1}{2}|A_i|(|S_i| - |A_i|)/m_i - \frac{3}{2}|S_i|$.

Case 2: $m_i = 2$. We have $|S_i| = 2(\delta_i + 1)$. Since $S_i \geq 2\delta_i$, we can apply Lemma 4 and we have $|D_i| + |E(S_i - A_i, S_i - A_i)| \geq |S_i|\delta_i - \frac{3}{8}|A_i|(|S_i| - |A_i|) - \frac{1}{2}|S_i| = |S_i|^2/m_i - \frac{3}{4}|A_i|(|S_i| - |A_i|)/m_i - \frac{3}{2}|S_i|$.

The lemma follows.

5   Time complexity

We briefly explain how we can implement Greedy$_2$ in $O(\Delta^2 n)$ time. Consider any graph G of n node with maximum degree Δ. Let u and v be any two nodes of G. Note that if the distance between u and v is larger than 2, than $\delta(\{u\}) + \delta(\{v\}) = 2\delta(\{u, v\})$. (Recall that $\delta(I)$ is the degree density of the independent set I.) Thus, $\min(\delta(\{u\}), \delta(\{v\})) \leq \delta(\{u, v\})$. This implies that to find the set X, $|X| \leq 2$ with the minimum degree density, it suffices to examine $T = \{\{u\} \mid u \in V\} \cup \{\{u, v\} \mid u, v \in V$ and the distance between u, v is exactly 2$\}$. Since for every node u, there are at most Δ^2 nodes with distance two from u, we have $|T| = O(\Delta^2 n)$. By using bucket-sort, we can sort T in $O(\Delta^2 n)$ time. Based on this observation, we can implement Greedy$_2$ in $O(\Delta^2 n)$ time. We will give the details in the full paper.

References

1. S. Arora, C. Lund, R. Motwani, M. Sudan, and M. Szegedy. Proof verification and hardness of approximation problem. In *Proceedings of the 33th IEEE Annual Symposium on Foundations of Computer Science*, 1992.
2. P. Berman and M. Fürer. Approximating maximum independent set in bounded degree graphs. In *Proceedings of the 5th Annual ACM-SIAM Symposium on Discrete Algorithms*, pages 365–371, 1994.
3. R. Boppana and M. Halldorsson. Approximating maximum independent sets by excluding subgraphsr. *BIT*, pages 180–196, 1992.
4. M. Halldórsson and J. Radhakrishnan. Greed is good: Approximating independent sets in sparse and bounded-degree graph. In *Proceedings of the 26th Annual ACM Symposium on Theory of Computing*, pages 439–448, 1994.
5. M. Halldórsson and J. Radhakrishnan. Greedy approximations of independent sets in low degree graphs. In *Proceedings of the 6th International Symposium on Algorithms and Computation*, 1995.

6. M. Halldórsson and J. Radhakrishnan. Improved approximations of independent sets in bounded-degree graphs via subgraph removal. *Nordic Journal of Computing*, pages 275–292, 1995.

7. J. Håstad. Clique is hard to approximate within $n^{1-\epsilon}$. In *Proceedings of the 37th Annual Symposium on Foundations of Computer Science*, pages 627–636, 1996.

8. H.Y. Lau. *The power of greediness–a general methodology for designing approximation algorithms*. MPhil. thesis, The University of Hong Kong, Hong Kong, 1998.

9. G.L. Nemhauser and W.T. Trotter. Vertex packing: Structural properties and algorithms. *Mathematics Programming*, pages 232–248, 1975.

Author Index

Springer
and the
environment

At Springer we firmly believe that an international science publisher has a special obligation to the environment, and our corporate policies consistently reflect this conviction.
We also expect our business partners – paper mills, printers, packaging manufacturers, etc. – to commit themselves to using materials and production processes that do not harm the environment. The paper in this book is made from low- or no-chlorine pulp and is acid free, in conformance with international standards for paper permanency.

 Springer

Lecture Notes in Computer Science

For information about Vols. 1–1548
please contact your bookseller or Springer-Verlag